EPC and 4G Packet Networks

EPC and 4G Packet Networks

Driving the Mobile Broadband Revolution

Second edition

Magnus Olsson

Shabnam Sultana

Stefan Rommer

Lars Frid

Catherine Mulligan

AMSTERDAM • BOSTON • HEIDELBERG • LONDON • NEW YORK • OXFORD
PARIS • SAN DIEGO • SAN FRANCISCO • SINGAPORE • SYDNEY • TOKYO
Academic Press is an imprint of Elsevier

Academic Press is an imprint of Elsevier
The Boulevard, Langford Lane, Kidlington, Oxford, OX5 1GB
225 Wyman Street, Waltham, MA 02451, USA

First published 2009
Second edition 2013

British Library Cataloguing-in-Publication Data
A catalogue record for this book is available from the British Library

Library of Congress Cataloging-in-Publication Data
A catalog record for this book is available from the Library of Congress

ISBN: 978-0-12-394595-2

For information on all Academic Press publications
visit our website at store.elsevier.com

Printed and bound in the United Kingdom

12 13 14 15   10 9 8 7 6 5 4 3 2 1

Working together to grow
libraries in developing countries

www.elsevier.com | www.bookaid.org | www.sabre.org

ELSEVIER   BOOK AID
            International   Sabre Foundation

Contents

Foreword by Dr. Kalyani Bogineni

There are billions of mobile devices operating on various types of 2G and 3G wireless networks. Projections are for several billion more devices in the next few years on newer technologies with expectations of simultaneous services with high throughput and low-latency requirements. There will be multiple wireless devices for each user and there will be wireless devices embedded in machines supporting automation of many functions. In short, the users will be "anywhere, any time, on any device". This has heralded an era of communication and information exchange that is testing the limits of many existing telecommunications and data technologies. Hence the need for implementing concepts born out of disruptive thinking combined with pragmatic application of innovations.

From a service provider point of view, this is a time for building on the foundation for many of the features needed in future generation networks in order to meet the above expectations. For example, the networks need to enable signaling and low-latency media paths across segments of different technologies to support real-time applications like voice and gaming. Fundamentals like mobility and roaming, the pillars of global cellular technologies, need services management based on availability of resources, offered via policy-peering mechanisms between the home and visited networks. Simplicity and ease of using devices and services are enabled by unified authentication and subscription validation mechanisms across various access networks and application platforms. Coexistence and cooperation are needed between end-user-driven intelligent devices and intelligent network elements.

The 3GPP has specified a core network based on the Internet Protocol (IP) that provides numerous operational benefits in addition to meeting the above-mentioned expectations. The specification:

- Allows evolution of any deployed wireless or wired access technology network towards a common architecture with benefits of seamless mobility between various generations of access networks and global roaming capabilities on different technologies.

- Enables network designs based on high availability, reliability, scalability, and manageability paradigms, as well as efficient bandwidth usage on access, backhaul, and core networks.
- Supports delivery of combinations of advanced telephony and Internet services that can be hosted by any access network or application provider.
- Provides user security functions like privacy and confidentiality while protecting the network through functions like mutual authentication and firewalls.
- Minimizes the number of services databases and the number of services controllers, which reduces the number of provisioning points in the network.
- Provides an efficient charging architecture that reduces the number of network elements sending billing records and minimizes the number of billing records formats.

In addition to the above, enhancements are needed to support the following:

- Emergency voice and non-voice services as well as priority services
- Fallback to a circuit-switched network when Voice on IMS is not available
- Multicast and broadcast services on LTE
- Network coverage extensions through femtocells and relay nodes
- Capabilities to transfer traffic between access technologies and to offload core network traffic at different deployment points based on user and operator policies
- Mechanisms to optimize signaling flows and for overload/congestion management.

The scope of the 3GPP specifications is ambitious but essential. The authors have done an excellent job in writing this book and updating it to keep up with the continued specification development in 3GPP. Their familiarity with the requirements, concepts, and solution alternatives, as well as the standardization work, allows them to present the material in a way that provides easy communication between Architecture and Standards groups and Planning/Operational groups within service provider organizations.

Dr. Kalyani Bogineni

Principal Architect, Verizon

Foreword by Dr. Ulf Nilsson

It has now been a few years since the first commercial launches of SAE/LTE networks in the world and since the first publication of this book. A lot of both expected and unexpected things have happened in the mobile industry since then. When 4G was brought to the market by TeliaSonera, we did it because we could easily see that the traffic demands in the coming years would not be possible to handle using 2G and 3G technologies at a reasonable cost. So did the SAE/LTE networks deliver on the expectations we put on it? The answer is undoubtedly yes. Our users are now browsing the Internet wirelessly at speeds many times higher than previously with noticeably lower delays, even outperforming many of the available fixed accesses on the market. As an operator the benefits of the EPC lies in its flexibility, scalability and optimization possibilities that we really need in order to handle the still growing market of mobile broadband – and we do this with lower operational cost than ever before. But this was what we expected and we were confident that SAE/LTE would handle it. What came as an unexpected surprise and that we did not really think would come to the mobile world as quickly as it did, was the change in our users' behavior when smartphones and tablets burst onto the market.

Mobile phones, even in 2009, used to be devices you made voice calls on, sent a few SMS and MMS to friends on and, perhaps if you really needed some information from the Internet quickly, used to browse the web on. Then along came the new generation of sleek smartphones and the first cool tablets. Just like when people once realized that accessing the Internet wirelessly on a laptop instead of having to be chained to a computer at home produced the mobile broadband revolution, users now want to use their smartphones and tablets to do this and we have a smartphone revolution on our hands. The applications are always on, pictures and videos from high-resolution cameras get uploaded, and downloaded, on the fly to and from the cloud the instant they become available. Users are watching movies and TV shows streamed off the Internet wherever they happen to be and whenever they have time to spare. Any ideas of keeping smartphones and tablets on legacy 2G/3G networks while mobile broadband users would benefit from the SAE/LTE networks can be scrapped – we need all the magic of the SAE/LTE networks to support this behavior at, dull as it sounds, a reasonable cost.

The mobile network, through the use of all the connected devices around us is thus becoming such an important part of the life of our users that they expect nothing but the best from their mobile operator – and it is therefore our responsibility to support this. The SAE/LTE network will be an absolutely crucial tool for handling the continued mobile broadband revolution and the just beginning smartphone revolution. It is without a doubt the technology on which the mobile operators, if not the entire mobile world, will base its future. Nothing is therefore more important than understanding both the basics of the SAE/LTE network as well as its more advanced features; they will all be needed in the future.

The excellent first edition of this book quickly became the reference work on the basics of SAE/LTE networks and many people have since its publication benefited greatly from the knowledge of the authors and their skill to pass this knowledge and their experience on to the readers. It is therefore reassuring that there is now an updated and expanded second edition of the book. The additions describe a number of areas that are becoming increasingly important as SAE/LTE starts to dominate the mobile networks around the world. Both beginners in the field as well as seasoned mobile industry professionals will take great joy in being guided though the intricacies of the SAE/LTE network by the authors, who are some of the foremost experts on the topic.

Dr. Ulf Nilsson, PhD

Network Research
Telia Sonera Mobility Services

Preface

The outcome of the 3GPP SAE (System Architecture Evolution) technical study and specification work is a set of standards that specifies the evolution of the packet core network for GSM/GPRS and WCDMA/HSPA to an all-IP architecture, and enables a feature-rich "common packet core" for radio accesses developed within 3GPP and also by other standardization forums. This common packet core is referred to as EPC (Evolved Packet Core) and the full system is known as EPS (Evolved Packet System), which includes support for 3GPP radio access technologies (LTE, GSM, and WCDMA/HSPA) as well as support for non-3GPP access technologies. Unlike its predecessor, EPC provides support for multiple access technologies and provides for mobility between them, allowing end-users to move between, for example, LTE, WLAN, and other 3GPP and non-3GPP accesses. The architecture, in comparison to the one used for the 2G/3G packet core, is also optimized for efficient payload handling, a so-called "flatter" architecture. In addition to these benefits, EPC provides updates to all of the already established parts of the 2G/3G packet core network, for example security and connectivity management. In short, the SAE work has prepared the core network for the mobile broadband revolution, through the specification of EPC.

The 3GPP specifications for EPC can be perceived as complex, as they span several thousands of pages. This makes it difficult for any individual not involved in the development of the standard to examine these specifications in detail. This book provides a concise and comprehensive description of the different aspects of the 3GPP EPC specifications for several different groups of readers with interests in the mobile communications industry.

Our goal was to ensure that reading this book will improve the overall understanding of the network architecture and protocols included in the EPC system. It is, however, significantly more than just annotated 3GPP specifications. It provides a detailed analysis of the network architecture, nodes, and protocols involved in EPC. Since the the first edition of the book was published, we have seen network deployments of EPS in Western Europe, North America, and Japan. We have also seen tremendous momentum towards industry commitment to deployment of

VoLTE, enabling speech and video services via IMS over LTE and EPC. In addition, operators are also emphasizing the need for voice and service continuity between IMS and circuit-switched networks. This version of the book provides a detailed description of the standards work as well as a more focused technical analysis of EPS based on the market direction taken by operators through their deployment choices. In addition, we explain in detail what VoLTE and Voice Call and Service Continuity is all about and how the standardization efforts are ensuring that the EPS, IMS, and CS networks are modified to support these features.

However, voice and IMS are not the only areas of expansion the standards have seen over the years since the first edition of the book was written. We have seen EPS reach a certain maturity and we have seen the addition of a number of new functions. These include support of emergency and priority services via EPS, enhancing support of selected traffic offload, support for Home (e)Node B functions (also sometimes known as Femto), and Local IP Offload, which impact 3GPP-defined radio access technologies directly.

At the same time, the industry is the subject of keen interest with regard to improving the interworking and coordination with WLAN access within EPS. In this respect, we have seen strong cooperation with Broadband Forum to evolve and interwork 3GPP networks with Fixed Broadband technologies such as WLAN. In line with the deployment trends, where the GTP-based EPS has so far taken a lead in the market, the support for GTP in other non-3GPP access technology areas has also received attention in the standards and this version of the book contains the technical solutions regarding support of the GTP protocol on various non-3GPP interfaces.

We have also made an effort to expand on certain key protocol areas such as Diameter, since it is one of the key protocols used in various 3GPP network reference points.

This book provides a thorough grounding for anyone wishing to learn about how operators and other actors in the industry may implement and have implemented/deployed the EPS, and also the different migration paths that have been taken or may be taken in the currently deployed networks. It also provides an overview of the additional services that are being developed, as well as services that are being utilized in LTE and EPC.

Readers who are already familiar with EPC, LTE, or IMS will hopefully also benefit substantially from this book as it identifies how these concepts fit together in order to deliver the promise of mobile broadband. For example, readers familiar with IMS will gain a new depth of insight into how voice services will fit together with the new network architecture and protocols. Appendix A covers the different specifications that are relevant for SAE. It should be noted, however, that this book is not only for readers interested in 3GPP specifications; it also covers the implementation scenarios for 3GPP2 and also interconnection with non-3GPP accesses such as WLAN, or fixed access. Readers interested in only one access technology, or indeed interested in only one protocol, will also gain a good depth of understanding of how their area of interest fits in with the overall network architecture.

We have divided this book into five different parts.

Part I: Introduction – Background and Vision of EPC

This part puts SAE and EPC in the correct context with regard to other technologies that affect the evolution of telecommunications networks. It provides the reasoning within the industry for evolving the core network and the role of different players in the standards bodies.

Chapter 1

This chapter provides the "outside view" of telecommunications networks as they stand today and where EPC sits in relation to this, covering the following points:

- Why evolve the core network?
- Technologies connected to EPC.
- Standards bodies involved in SAE work.
- Terminologies used in the book.

Part II: Overview of EPS

This section provides technical descriptions of EPS, including functional descriptions of the different components of EPC. It also covers different migration and deployment scenarios, and illustrates how the concepts and standards described in other chapters are connected together to create voice and data services in an operator's network.

Chapter 2

Chapter 2 provides a high-level introduction to the main concepts of the EPS system designed to give a basic understanding of SAE/LTE services:

- A brief description of the EPS services.
- Simplified network diagrams to give the reader an initial understanding of the EPS network and where EPC is placed in the overall network.
- Introductory information on the fundamental choices in LTE.
- Terminal perspective.
- Short LTE overview and its relation to EPC.

Chapter 3

This chapter provides descriptions of how EPC may be deployed based on the situation of the market where it is being deployed, as well as its relation to LTE deployment:

- Brief description of the overall NW when deploying EPC/LTE in different operator configurations.

Chapter 4

This chapter provides a description of the data services that will be used on an EPC network, aiming to bring the whole EPS and its concepts together, analyzing it from several different potential evolutionary paths pertaining to services:

- Description of the predicted target services:
  - Data services and applications
  - Messaging services.
- Machine type communication.

Chapter 5

This chapter provides a description of the voice services that will be used on an EPC network, aiming to bring the whole EPS and its concepts together, analyzing it from several different potential evolutionary paths pertaining to services:

- Voice services using IMS technology.
- Single-radio voice call continuity.
- Circuit-switched fallback.
- IMS Emergency Calls and Priority Services.

Part III: Key Concepts and Services

Chapter 6

This chapter provides a description of key concepts within EPS. Owing to the nature of EPC compared with previous core network architectures, the chapter provides a clear description of these new concepts and compares them with previous core networks. This aims to provide readers with a clear point of reference for the key concepts after the evolution of the core network.

Chapter 7

This chapter provides details on security, including user authentication/authorization as well as network security mechanism for both 3GPP and non-3GPP accesses connecting to EPS.

Chapter 8

This chapter provides an in-depth view of quality of service and policies to control and manage services and to differentiate charging. The chapter also includes a high-level overview of 3GPP charging models and mechanisms.

Chapter 9

This chapter provides an in-depth view of the usage of DNS as well as 3GPP-developed mechanisms as tools for the efficient operation of the EPS network by selecting the "right entity" for the right user in an operator's network.

Chapter 10

This chapter provides an in-depth view of subscriber data management functionality in EPS, including a description of the EPS entities handling subscriber data.

Chapter 11

This chapter provides an in-depth view of voice services in EPS, including a description of emergency and priority services.

Chapter 12

This chapter provides a description of the broadcast services supported in EPS, including descriptions of the network architecture and network entities involved in providing broadcast services to end-users.

Chapter 13

This chapter provides an overview of positioning services available in EPS, including descriptions of the architecture, protocols, and supported position methods.

Chapter 14

This chapter provides a description of offload functions defined for EPS, including functions to offload the core network as well as functions to offload the 3GPP radio access.

Part IV: The Nuts and Bolts of EPC

Chapters 15–17

These chapters together illustrate in detail how the EPS system is built end to end by using the network entities, the interfaces connecting them together, and protocols that provide the "meat" for the "backbone" of the system carrying the information between these entities, and then some high-level procedures illustrating key scenarios such as attaching to the EPC, detaching from the EPC, and handover of various kinds between 3GPP and non-3GPP access technologies as well as between 3GPP access technologies.

Part V: Conclusion and Future of EPS

Chapter 18

This chapter includes observations and conclusions regarding the EPC and some discussion on what may lie ahead for its future evolution.

Acknowledgements

A work of this nature is not possible without others' support.

The authors would like to gratefully acknowledge the contribution of many of our colleagues at Ericsson, in particular Per Beming, Paco Cortes, Erik Dahlman, Jesús De Gregorio, Göran Hall, David Hammarwall, Maurizio Iovieno, Ralf Keller, Torsten Lohmar, Reiner Ludwig, Anders Lundström, Lars Lövsen, Peter Malm, György Miklós, Daniel Molander, Karl Norrman, Mats Näslund, Zu Qiang, Anki Sander, Louis Segura, Iana Siomina, Mike Slssingar, John Stenfelt, Patrik Teppo, and Stephen Terrill.

We would also like to thank our families. Writing this book would not have been possible without their generosity and support throughout the process.

List of Abbreviations

0-9
2G	2nd Generation
3G	3rd Generation
3GPP	Third Generation Partnership Project
3GPP2	Third Generation Partnership Project 2

A
AAA	Authentication, Authorization and Accounting
ABMF	Account Balance Management Function
AECID	Adaptive Enhanced Cell Identity
AF	Application Function
A-GNSS	Assisted Global Navigation Satellite System
A-GPS	Assisted GPS
AH	Authentication Header
AKA	Authentication and Key Agreement
AMBR	Aggregate Maximum Bit Rate
AN	Access Network
ANDSF	Access Network Discovery and Selection Function
AP	Application Protocol
API	Application Programming Interface
APN	Access Point Name
APN-NI	APN Network Identifier
APN-OI	APN Operator Identifier
ARIB	Association of Radio Industries and Businesses (Japan)
ARP	Allocation and Retention Priority
ARQ	Automatic Repeat ReQuest
AS	Application Server
ASME	Access Security Management Entity
ATIS	Alliance for Telecommunications Industry Solutions

ATM	Asynchronous Transfer Mode
AuC	Authentication Centre
AUTN	Authentication Token
AV	Authentication Vector
AVP	Attribute Value Pair

B

BA	Binding Acknowledgement
BBERF	Bearer Binding and Event Reporting Function
BBF	Bearer Binding Function; Broadband Forum
BGCF	Breakout Gateway Control Function
BRA	Binding Revocation Acknowledgement
BRI	Binding Revocation Indication
BS	Base Station
BSC	Base Station Controller
BS	ID Base Station Identity
BSS	Base Station Subsystem
BSSID	Basic Service Set Identifier
BU	Binding Update

C

CAMEL	Customized Application for Mobile network Enhanced Logic
CAP	CAMEL Application Part
CBC	Cell Broadcast Centre
CCSA	China Communications Standards Association
CDF	Charging Data Function
CDMA	Code Division Multiple Access
CDR	Charging Data Records
CGF	Charging Gateway Function
CGI	Cell Global Identity
CHAP	Challenge Handshake Authentication Protocol
CK	Cipher Key
CN	Core Network; Correspondent Node
CoA	Care-of Address
CS	Circuit-Switched
CSCF	Call Session Control Function
CSFB	Circuit Switched Fall Back
CTF	Charging Trigger Function

D

DAD	Duplicate Address Detection
DCCA	Diameter Credit Control Application
DHCP	Dynamic Host Configuration Protocol
DL	DownLink
DM	Device Management
DNS	Domain Name System
DPI	Deep Packet Inspection
DRA	Diameter Routing Agent
DRX	Discontinuous Reception
DSCP	DiffServ Code Point
DSL	Digital Subscriber Line
DSMIPv6	Dual Stack Mobile IPv6
DTF	Domain Transfer Function
DTX	Discontinuous Transmission

E

EAP	Extensible Authentication Protocol
E-CID	Enhanced Cell ID
ECM	EPS Connection Management
EDGE	Enhanced Data rates for GSM Evolution
eHRPD	Evolved High Rate Packet Data
EIR	Equipment Identity Register
eMBMS	Evolved Multicast Broadcast Multimedia Service
EMM	EPS Mobility Management
eNB	E-UTRAN NodeB
EPC	Evolved Packet Core
ePDG	Evolved Packet Data Gateway
EPS	Evolved Packet System
E-RAB	E-UTRAN Radio Access Bearer
ESM	EPS Session Management
E-SMLC	Enhanced Serving Mobile Location Center
ESP	Encapsulated Security Payload
ETSI	European Telecommunications Standards Institute
ETWS	Earthquake and Tsunami Warning System
E-UTRAN	Evolved Universal Terrestrial Radio Access Network
EV-DO	Evolution-Data Only

enodeB- evolved Node B, evolution of Node B in UTRA of UTMS

F

FA	Foreign Agent
FDD	Frequency Division Duplex
FEC	Forward Error Correction
FMC	Fixed Mobile Convergence
FQDN	Fully Qualified Domain Names

G

GAD	Geographical Area Description
GBR	Guaranteed Bit Rate
GERAN	GSM EDGE Radio Access Network
GGSN	Gateway GPRS Support Node
GMLC	Gateway Mobile Location Center
GPRS	General Packet Radio Service
GRE	Generic Routing Encapsulation
GRX	GPRS Roaming eXchange
GSM	Global System for Mobile communications
GSMA	GSM Association
GSN	GPRS Support Node
GTP	GPRS Tunnelling Protocol
GTP-C	GPRS Tunnelling Protocol for Control Plane
GTP-U	GPRS Tunnelling Protocol for User Plane
GUMMEI	Globally Unique MME Identifier
GUTI	Globally Unique Temporary Identifier
GW	Gateway

H

HA	Home Agent
HLR	Home Location Register
HO	Handover
HoA	Home Address
HOM	Higher Order Modulation
H-PCRF	Home PCRF
HPLMN	Home Public Land Mobile Network
HRPD	High Rate Packet Data
HSDPA	High Speed Downlink Packet Access
HSGW	HRPD Serving Gateway
HSPA	High Speed Packet Access
HSS	Home Subscriber Server
HSUPA	High Speed Uplink Packet Access

I

IAB	Internet Architecture Board
IANA	Internet Assigned Numbers Authority
ICMP	Internet Control Message Protocol
ICS	IMS Centralised Services
I-CSCF	Interrogating-CSCF
ICV	Integrity Check Value
IEEE	Institute of Electrical and Electronics Engineers
IESG	Internet Engineering Steering Group
IETF	Internet Engineering Task Force
IK	Integrity key
IKEv1	Internet Key Exchange version 1
IKEv2	Internet Key Exchange version 2
IMEI	International Mobile Equipment Identity
IMS	IP Multimedia Subsystem
IMSI	International Mobile Subscriber Identity
IMT-2000	International Mobile Telecommunications 2000
IMT-Advanced	International Mobile Telecommunications-Advanced
IP	Internet Protocol
IP-CAN	IP Connectivity Access Network
IPMS	IP Mobility Mode Selection
IPSec	IP Security
IPX	IP Packet eXchange
I-RAT	Inter Radio Access Technology
ISAKMP	Internet Security Association and Key Management Protocol
ISDN	Integrated Services Digital Network
ISIM	IP Multimedia Services Identity Module
ISP	Internet Service Provider
ISR	Idle mode Signalling Reduction
ITU	International Telecommunication Union
ITU-R	ITU Radiocommunication Sector
ITU-T	ITU Telecommunication Sector
IWF	Interworking Function
I-WLAN	Interworking Wireless LAN

L

LA	Location Area
LAC	Location Area Code
LAN	Local Area Network
LBO	Local Breakout

LBS	Location-based service
LCS	Location services
LEA	Law Enforcement Agencies
LI	Lawful Intercept
LMA	Local Mobility Anchor
LPPa	LPP Annex
LTE	Long-Term Evolution

M

M2M	Machine-to-Machine
MAG	Mobile Access Gateway
MAP	Mobile Application Part
MBMS	Multimedia Broadcast Multicast Service
MCC	Mobile Country Code
MGCF	Media Gateway Control Function
MGW	Media Gateway
MH	Mobility Header
MID	Mobile Internet Device
MIMO	Multiple Input, Multiple Output
MIP	Mobile IP
MIPv4	Mobile IPv4
MIPv6	Mobile IPv6
MM	Mobility Management
MME	Mobility Management Entity
MMEC	MME Code
MMEGI	MME Group Identifier
MMEI	MME Identifier
MMS	Multimedia Messaging Service
MMTel	MultiMedia Telephony
MN	Mobile Node
MNC	Mobile Network Code
MO	Managed Object
MOBIKE	IKEv2 Mobility and Multi-homing Protocol
MPLS	Multi-Protocol Label Switching
MRFC	Media Resource Function Controller
MRFP	Media Resource Function Processor
MS	Mobile Station
MSC	Mobile Switching Centre
MSC-S	MSC Server
MSIN	Mobile Subscriber Identification Number
MSISDN	Mobile Subscriber ISDN Number

N

NAI	Network Access Identifier
NAP-ID	Network Access Provider Identity
NAPTR	Name Authority Pointer
NAS	Non-Access Stratum; Network Access Server
NAT	Network Address Translation
NB	NodeB
NDS	Network Domain Security
NID	Network Identification
NRI	Network Resource Identifier
NW	Network

O

OCF	Online Charging Function
OCS	Online Charging System
OFDM	Orthogonal Frequency Division Multiplexing
OFCS	Offline Charging System
OMA	Open Mobile Alliance
OTDOA	Observed Time Difference of Arrival

P

PBA	Proxy Binding Acknowledgement
PBU	Proxy Binding Update
PC	Personal Computer
PCC	Policy and Charging Control
PCEF	Policy and Charging Enforcement Function
PCO	Protocol Configuration Options
PCRF	Policy and Charging Rules Function
P-CSCF	Proxy-CSCF
PDCP	Packet Data Convergence Protocol
PDN	Packet Data Network
PDN	GW Packet Data Network Gateway
PDP	Packet Data Protocol
PDU	Protocol Data Unit
P-GW	PDN GW
PIN	Personal Identification Number
PKI	Public Key Infrastructure
PLMN	Public Land Mobile Network
PMIP	Proxy Mobile IP
PMM	Packet Mobility Management

PON	Passive Optical Networks
PPP	Point-to-Point Protocol
PS	Packet-Switched

Q

QAM	Quadrature Amplitude Modulation
QCI	QoS Class Identifier
QoS	Quality of Service

R

RA	Router Advertisement; Routing Area
RAB	Radio Access Bearer
RAC	Routing Area Code
RADIUS	Remote Authentication Dial In User Service
RAI	Routing Area Identity
RAN	Radio Access Network
RANAP	Radio Access Network Application Protocol
RAT	Radio Access Technology
RAU	Routing Area Update
Rel-8	Release 8
Rel-9	Release 9
Rel-99	Release 99
rSRVCC	Return SRVCC
RF	Rating Function
RFC	Request For Comments
RLC	Radio Link Control
RNC	Radio Network Controller
RO	Route Optimization
RRC	Radio Resource Control
RRM	Radio Resource Management
RTSP	Real Time Streaming Protocol

S

SA	Security Association
SAE	System Architecture Evolution
SBC	Session Border Controller
SCC	AS Service Centralization and Continuity Application Server
S-CSCF	Serving-CSCF
SCTP	Stream Control Transmission Protocol

SDF	Service Data Flow
SDP	Session Description Protocol
SDM	Subscriber Data Management
SEG	Security Gateway
SGSN	Serving GPRS Support Node
S-GW	Serving GW
SID	System Identification
SIM	GSM Subscriber Identity Module
SIP	Session Initiation Protocol
SLA	Service Level Agreement
SMS	Short Message Service
SMS-C	Short Message Service Centre
SN	Serving Network
SN ID	Serving Network Identity
S-NAPTR	Straightforward-NAPTR
SPI	Security Parameters Index
SPR	Subscription Profile Repository
SQN	Sequence Number
SRNS	Serving Radio Network Subsystem
SRVCC	Single-Radio Voice Call Continuity
SS7	Signalling System No 7
SSID	Service Set Identifier
SUPL	Secure User Plane Location

T

TA	Tracking Area
T-ADS	Terminating-Access Domain Selection
TAC	Tracking Area Code
TAI	Tracking Area Identity
TAP	Transferred Account Procedure
TAS	Telephony Application Server
TAU	Tracking Area Update
TCP	Transmission Control Protocol
TDD	Time Division Duplex
TDMA	Time-Division Multiple Access
TEID	Tunnel End Point Identifier
TFT	Traffic Flow Template
TISPAN	Telecommunications and Internet converged Services and Protocols for Advanced Networking

TLS	Transport Layer Security
TMSI	Temporary Mobile Subscriber Identity
TOS	Type of Service
TR	Technical Report
TS	Technical Specification
TSG	Technical Specification Group
TTA	Telecommunication Technology Association of Korea
TTC	Telecommunication Technology Committee (Japan)

U

UDC	User Data Convergence
UDM	User Data Management
UDP	User Datagram Protocol
UE	User Equipment
UICC	Universal Integrated Circuit Card
UL	UpLink
UMTS	Universal Mobile Telecommunications System
USB	Universal Serial Bus
USIM	Universal Subscriber Identity Module
UTDOA	Uplink TDOA
UTRAN	Universal Terrestrial Radio Access Network (3G)

V

VCC	Voice Call Continuity
VoIP	Voice over IP
VoLTE	Voice over LTE
V-PCRF	Visited PCRF
VPLMN	Visited Public Land Mobile Network
VPN	Virtual Private Network
vSRVCC	Video SRVCC

W

WCDMA	Wideband Code Division Multiple Access
WG	Working Group
WiMAX	Worldwide Interoperability for Microwave Access
WLAN	Wireless Local Area Network

X

XRES	eXpected RESult

PART I

Introduction – Background and Vision of EPC

CHAPTER 1

Mobile Broadband and the Core Network Evolution

The telecommunications industry is undoubtedly in a period of radical change with the advent of mobile broadband radio access and the rapid convergence of Internet and mobile services. Some of these changes have been enabled by a fundamental shift in the underlying technologies; mobile networks are now increasingly based on a pure Internet Protocol (IP) network architecture. Since the first edition of this book was published in 2009, a multitude of connected devices from eBook readers to smartphones and even Machine-to-Machine (M2M) technologies have all started to benefit from mobile broadband. The sea change over the last few years is only the beginning of a wave of new services that will fundamentally change our economy, our society, and even our environment. The evolution towards mobile broadband is one of the core underlying parts of this revolution and is the focus of this book.

The phenomenal success of GSM (Global System for Mobile Communications) was built on the foundation of circuit switching, providing voice services over cellular networks. Services, meanwhile, were built by developers specializing in telecommunication applications. During the early 1990s, usage of the Internet also took off, which in later years led to a demand for "mobile Internet", Internet services that can be accessed from an end-user's mobile device. The first mobile Internet services had limitations due to the processing capacity of terminals and also a very limited bandwidth on the radio interface. This has now changed as the evolution of radio access networks (RANs) provide high data rates delivered by High-Speed Packet Access (HSPA) and Long-Term Evolution (LTE) radio access technologies. The speed of this change is set to increase dramatically as a number of other developments emerge in addition to the new high-speed radio accesses: rapid advances in the processing capacity of semiconductors for mobile terminals and also in the software that developers can use to create services. IP and packet-switched technologies are soon expected to be the basis for data and voice services on both the Internet and mobile communications networks.

The core network is the part that links these worlds together, combining the power of high-speed radio access technologies with the power of the innovative application

EPC and 4G Packet Networks.
DOI: http://dx.doi.org/10.1016/B978-0-12-394595-2.00001-3
Copyright © 2013, 2009 Elsevier Ltd. All rights reserved.

development enabled by the Internet. The evolution of the core network, or Evolved Packet Core (EPC), is a fundamental cornerstone of the mobile broadband revolution; without it, neither the RANs nor mobile Internet services would realize their full potential. The new core network was developed with high-bandwidth services in mind from the outset, combining the best of IP infrastructure and mobility. It is designed to truly enable mobile broadband services and applications and to ensure a smooth experience for both operators and end-users as it also connects multiple radio access technologies.

This chapter introduces the reasoning behind the evolution of the core network and a brief introduction to the technologies related to EPC. We also briefly touch on how EPS is beginning to change the industrial structure of the mobile industry.

System Architecture Evolution (SAE) was the name of the Third Generation Partnership Project (3GPP) standardization work item that was responsible for the evolution of the packet core network, more commonly referred to as EPC. This work item was closely related to the LTE work item, covering the evolution of the radio network. The Evolved Packet System (EPS) covers the radio access, the core network, and the terminals that comprise the overall mobile system. EPC also provides support for other high-speed access network technologies that are not based on 3GPP standards, for example WiFi, or fixed access. This book is all about Evolved Packet Core and Evolved Packet System – the evolution of the core network in order to support the mobile broadband vision and an evolution to IP-based core networks for all services.

The broad aims of the SAE work item were to evolve the packet core networks defined by 3GPP in order to create a simplified all-IP architecture, providing support for multiple radio accesses, including mobility between the different radio standards. So, what drove the requirement for evolving the core network and why did it need to be a globally agreed standard? This is where we start our discussion.

1.1 A Global Standard

There are many discussions today regarding the evolution of standards for the communications industries, in particular when it comes to convergence between IT and telecommunications services. A question that pops up occasionally is why is a global standard needed at all? Why does the cellular industry follow a rigorous standards process, rather than, say, the de-facto standardization process that the computer industry often uses? There is a lot of interest in the standardization process for work items like LTE and SAE, so there is obviously a commercial reason for this, or very few companies would see value in participating in the work.

The necessity for a global standard is driven by many factors, but there are two main points. First of all, the creation of a standard is important for interoperability in a truly global, multi-vendor operating environment. Operators wish to ensure that they are able to purchase network equipment from several vendors, ensuring competition. For this to be possible nodes and mobile devices from different vendors must work with one another; this is achieved by specifying a set of "interface descriptions", through which the different nodes on a network are able to communicate with one another. A global standard therefore ensures that an operator can select whichever network equipment vendors they like and that end-users are able to select whichever handset they like; a handset from vendor A is able to connect to a base station from vendor B and vice versa. This ensures competition, which in itself attracts operators and drives deployments by ensuring a sound financial case through avoiding dependencies on specific vendors.

Secondly, the creation of a global standard reduces fragmentation in the market for all the actors involved in delivering network services to end-users: operators, chip manufacturers, equipment vendors, etc. A global standard ensures that there will be a certain market for the products that, for example, an equipment vendor develops. The larger the volume of production for a product, the greater the volume there is to spread the cost of design and production across the operators that will use the products. Essentially, with increased volumes a vendor should be able to produce each node at a cheaper per-unit cost. Vendors can then achieve profitability at lower price levels, which ultimately leads to a more cost-effective solution for both operators and end-users. Global standards are therefore a foundation stone of the ability to provide inexpensive, reliable communications networks, and the aims behind the development of EPC were no different.

There are several different standards bodies that have been directly involved in the standardization processes for the SAE work. These standards bodies include the 3GPP, the lead organization initiating the work, the Third Generation Partnership Project 2 (3GPP2), the Internet Engineering Task Force (IETF), Open Mobile Alliance (OMA), Broadband Forum (BBF), and also the WiFi Alliance. 3GPP "owns" the EPS specifications and refers to IETF and occasionally OMA specifications where necessary, while 3GPP2 complements these EPS specifications with their own documents that cover the impact on 3GPP2-based systems.

Readers who are not familiar with the standardization process are referred to Appendix 1, where we provide a brief description of the different bodies involved and the processes that are followed during the development of these specifications. We also provide a very brief history of the development of the SAE specifications.

1.2 Origins of the Evolved Packet Core

Over the years, many different radio standards have been created worldwide, the most commonly recognized ones being GSM, CDMA, and WCDMA/HSPA.

The GSM/WCDMA/HSPA and CDMA radio access technologies were defined in different standards bodies and also had different core networks associated with each one, as we describe below.

EPS is composed of the EPC, End-User Equipment (more commonly known as the UE), Access Networks (including 3GPP access such as GSM, WCDMA/HSPA and LTE as well as CDMA, etc.). The combination of these enables access to an operator's services and also to the IMS, which provides voice and multimedia services.

In order to understand why evolution was needed for 3GPP's existing packet core, we therefore also need to consider where and how the various existing core network technologies fit together in the currently deployed systems. The following sections present a discussion around why the evolution was necessary. While the number of acronyms may appear daunting in this section for anyone new to 3GPP standards, the rest of the book explains the technology in great detail. The following sections highlight only some of the main technical reasons for the evolution.

1.2.1 3GPP Radio Access Technologies

GSM was originally developed within the European Telecommunication Standards Institute (ETSI), which covered both the RAN and the core network supplying circuit-switched telephony. The main components of the core network for GSM were the Mobile Switching Centre (MSC) and the Home Location Register (HLR). The interface between the GSM BSC (Base Station Controller) and the MSC was referred to as the "A" interface. It is common practice for interfaces in 3GPP to be given a letter as a name; in later releases of the standards there are often two letters, for example "Gb" interface. Using letters is an easy shorthand method of referring to a particular functional connection between two nodes.

Over time, the need to support IP traffic was identified within the mobile industry and the General Packet Radio Services (GPRS) system was created as an add-on to the existing GSM system. With the development of GPRS, the concept of a packet-switched core network was needed within the specifications. The existing GSM radio network evolved, while two new logical network entities or nodes were introduced into the core network – SGSN (Serving GPRS Support Node) and GGSN (Gateway GPRS Support Node).

GPRS was developed during the period of time when PPP, X.25, and Frame Relay were state-of-the-art technologies (mid to late 1990s) for packet data transmission on data communications networks. This naturally had some influence on the standardization of certain interfaces, for example the Gb interface, which connects the BSC in the GSM radio network with the GPRS packet core.

During the move from GSM EDGE Radio Access Network (GERAN) to WCDMA/ UMTS Terrestrial Radio Access Network (UTRAN), an industry initiative was launched to handle the standardization of radio and core network technologies in a global forum, rather than ETSI, which was solely for European standards. This initiative became known as the 3GPP and took the lead for the standardization of the core network for UTRAN/WCDMA, in addition to UTRAN radio access itself. 3GPP later also took the lead in the creation of the Common IMS specifications. IMS is short for IP Multimedia Subsystem, and targets network support for IP-based multimedia services. We discuss the IMS further in Chapter 11.

The core network for UTRAN reused much of the core network from GERAN, with a few updates. The main difference was the addition of the interface between the UTRAN Radio Network Controller (RNC), the MSC and the SGSN, the Iu-CS and the Iu-PS respectively. Both of these interfaces were based on the A interface, but the Iu-CS was for circuit-switched access, while the Iu-PS was for packet-switched connections. This represented a fundamental change in thinking for the interface between the mobile terminal and the core network. For GSM, the interface handling the circuit-switched calls and the interface handling the packet-switched access were very different. For UTRAN, it was decided to have one common way to access the core network, with only small differences for the circuit-switched and packet-switched connections. A high-level view of the architecture of this date, around 1999, is shown in Figure 1.1 (to be completely accurate, the Iu-CS interface was split into two parts, but we will disregard that for now in order not to make this description too complex).

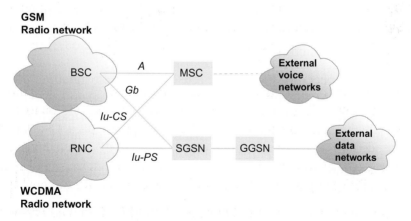

Figure 1.1: High-level View of The 3GPP Mobile Network Architecture.

The packet core network for GSM/GPRS and WCDMA/HSPA forms the basis for the evolution towards EPC. As a result, it is worthwhile taking the time for a brief review of the technology. Again, do not be put off by the number of acronyms. Parts II and III provide more details.

The packet core architecture was designed around a tunneling protocol named GTP (GPRS Tunneling Protocol) developed within ETSI and then continued within 3GPP after its creation. GTP is a fundamental part of 3GPP packet core, running between the two core network entities, the SGSN and the GGSN. GTP runs over IP and provides mobility, Quality of Service (QoS), and policy control within the protocol itself. As GTP was created for use by the mobile community, it has inherent properties that make it suitable for robust and time-critical systems such as mobile networks. Since GTP is developed and maintained within 3GPP, it also readily facilitates the addition of the special requirements of a 3GPP network such as the use of the Protocol Configuration Option (PCO) field between the terminal and the core network. PCO carries special information between the terminal and the core network, allowing for flexible, efficient running and management of the mobile networks.

GTP has from time to time faced criticism, however, from parts of the communication industry outside 3GPP. This has mainly been due to the fact that it was not developed in the IETF community, the traditional forum for standardization of Internet and IP technologies. GTP is instead a unique solution for 3GPP packet data services and was therefore not automatically a good choice for other access technologies. GTP was instead tailor-made to suit the needs of 3GPP mobile networks. Whether the criticism is justified or not is largely dependent on the viewpoint of the individual.

Regardless, GTP is today a globally deployed protocol for 3GPP packet access technologies such as HSPA, which has emerged as the leading mobile broadband access technology deployed prior to LTE. Due to the number of subscribers using GSM and WCDMA packet data networks, now numbering billions in total for both circuit- and packet-switched systems, GTP has been proven to scale very well and to fulfill the purposes for which it has been designed.

Another significant aspect of GPRS is that it uses SS7-based signaling protocols such as MAP (Mobile Application Part) and CAP (CAMEL Application Part), both inherited from the circuit-switched core network. MAP is used for user data management and authentication and authorization procedures, and CAP is used for CAMEL-based online charging purposes. Further details on CAMEL (Customized Applications for Mobile networks Enhanced Logic) are beyond the scope of this book. For our purposes, it is enough to understand that CAMEL is a concept designed to develop non-IP-based services in mobile networks. The use of SS7-based protocols can be seen as a drawback for a packet network created for delivering Internet connections and IP-based services.

The 3GPP packet core uses a network-based mobility scheme for handling user and terminal mobility, relying on mechanisms in the network to track movements of end-user devices and to handle mobility. Another aspect that was to become a target for optimization at a later date was the fact that it has two entities (i.e. SGSN and GGSN) through which user data traffic is carried. With the increased data volumes experienced as a result of WCDMA/HSPA, an optimization became necessary and was addressed in 3GPP Release 7, completed in early 2007 with the enhancement of the packet core architecture to support a mode of operation known as "direct tunnel" where the SGSN is not used for the user plane traffic. Instead, the radio network controller connects directly to the GGSN via the Iu-user plane (based on GTP). This solution, however, only applies to non-roaming cases, and also requires packet data charging functions to reside in the GGSN instead of the SGSN.

For further details on the packet core domain prior to SAE/EPC, please refer to 3GPP Technical Specification TS 23.060 (see References section for full details).

1.2.2 3GPP2 Radio Access Technologies

In North America, another set of radio access technology standards was developed. This was developed within the standards body called 3GPP2, under the umbrella of ANSI/TIA/EIA-41, which includes North American and Asian interests, towards developing global standards for those RAN technologies supported by ANSI/TIA/EIA-41.

3GPP2 developed the radio access technologies cdma2000®[1], providing 1xRTT and HRPD (High Rate Packet Data) services. cdma2000 1xRTT is an evolution of the older IS-95 cdma technology, increasing the capacity and supporting higher data speeds. HRPD defines a packet-only architecture with capabilities similar to the 3GPP WCDMA technology. The set of standards for the packet core network developed within 3GPP2 followed a different track to 3GPP, namely the reuse of protocols directly from the IETF, such as the Mobile IP family of protocols, as well as a simpler version of IP connectivity known as Simple IP, over a PPP link. The main packet core entities in this system are known as PDSN (Packet Data Serving Node) and HA (Home Agent), where terminal-based mobility concepts from IETF are used, in conjunction with mechanisms developed by 3GPP2 developed own mechanisms. It also uses Radius-based AAA infrastructure for its user data management, authentication, and authorization and accounting.

1.2.3 SAE – Building Bridges Between Different Networks

During the development of EPC, many operators using CDMA technologies specified by 3GPP2 became interested in the evolution of the core network ongoing in 3GPP

---

[1] cdma2000® is the trademark for the technical nomenclature for certain specifications and standards of the Organizational Partners (OPs) of 3GPP2. Geographically (and as of the date of publication), cdma2000® is a registered trademark of the Telecommunications Industry Association (TIA-USA) in the United States.

as they wished to join the LTE ecosystem and the development of the common packet core work under the umbrella of the SAE work item. As a result, work in both 3GPP and 3GPP2 was established to ensure that the EPS could support interworking towards 3GPP2 networks. EPS then needed to support the evolution of two very different types of core network and that created the framework of SAE work in 3GPP. SAE was therefore designed to both improve and build a bridge between two very distinct packet core networks.

The existing packet core networks were developed to serve certain market and operator requirements. These requirements have not changed with the evolution to EPS. Rather, with the evolution towards new radio networks and also the need to deliver new types of services across the core network, the EPS is instead required to support extra requirements on top of the old ones.

IETF-based protocols naturally play a key role in EPS. 3GPP developed both the IMS and PCC (Policy and Charging Control) Systems, where all the protocols are built on IETF-developed base protocols and then enhanced within the IETF as per 3GPP's requirements. This was not new or unchartered territory for 3GPP member companies, since 3GPP already had contributed extensively to the development within IETF of SIP, AAA, Diameter, and various security-related protocols.

The most contentious area of protocol selection was related to mobility management, where there were a few competing proposals in the IETF and progress was slow. The IETF settled on PMIP (Proxy Mobile IP) as the network-based mobility protocol, while both GTP and PMIP are compatible with the 3GPP standards.

1.3 A Shifting Value Chain

As we have shown, the telecoms industry has been defined by voice services for several decades. The manner in which services have been defined, implemented, and reached by consumers has been subject to a well-established, well-understood set of logics. An operator purchased network equipment from a network vendor and handset manufacturer. They then sold the handset on to the consumers and were the sole supplier of services and also content for end-users. Consumers, meanwhile, were limited to the selection of services that their particular operator provided for them. Operator networks were not particularly "open" – it was not possible for developers to create, install, and run software easily on a mobile network. In order to develop an application or service for a mobile network, it was necessary to undergo rigorous testing, comply with a large number of network standards, and establish contracts with an operator.

As discussed in the previous section, smartphones and other "connected devices" have begun to provide content outside the traditional operator network. End-users are now able to select and control the services and applications that they use on their mobile devices. Mobile broadband therefore does not just change the nature of the services delivered to end-users, it is also reshaping the value chain for data services in the mobile industry.

Mobile broadband pushes far beyond the traditional operator and network vendor value chain that has stayed nearly the same since the development of GSM in the early 1990s. Within data services, the ownership of the service and the ownership of the subscriber were typically the same; with so called app stores this changes as while operators retain ownership of the subscriber, it is not the same for services. End-users are able to instead access a wide variety of content available to all mobile subscribers, not just those subscribers on one particular operator network. This openness has increased demand for mobile broadband services in a feedback loop.

These services create large changes within the industrial structure, simultaneously generating opportunities for innovation, while fragmenting the industrial structure and creating a complex value chain for service development and delivery. The good coverage of mobile applications in turn increased sales volumes of mobile broadband technology, which has reduced the cost of many related technologies, e.g. Machine-to-Machine (M2M). This means that new services built around M2M, mobile broadband connectivity, and cloud storage have become economically viable in the last few years.

1.4 Terminology Used in This Book

As you progress through the chapters in this book, you will notice that there are several different acronyms used to describe different aspects of core network evolution. You will also notice these acronyms being used extensively in industry as well, so here we have included a brief description of their meanings and how we have used these terms in this book.

The common or everyday terminology used in the industry is not necessarily the same as the terms that have been used in standardization. On the contrary, there is something of a mismatch between the most commonly used terms in the mobile industry and the terms actually used in 3GPP specification work.

Below is a list of terms describing some of the most common acronyms in this book.

EPC: The new Packet Core architecture itself as defined in 3GPP Release 8 and onwards.
SAE: The work item, or standardization activity, within 3GPP that was responsible for defining the EPC specifications.
EPS: A 3GPP term that refers to a complete end-to-end system, i.e. the combination of the User Equipment (UE), E-UTRAN (and UTRAN and GERAN connected via EPC), and the Packet Core Network itself (EPC).

LTE/EPC: A term previously used to refer to the complete network; it is more commonly used outside of 3GPP instead of EPS. In this book we have used the term EPS instead of the term LTE/EPC.

E-UTRAN: Evolved UTRAN, the 3GPP term denoting the RAN that implements the LTE radio interface technology.

UTRAN: The RAN for WCDMA/HSPA.

GERAN: The GSM RAN.

LTE: Formally the name of the 3GPP work item (Long-Term Evolution) that developed the radio access technology and E-UTRAN, but in daily talk it is used more commonly instead of E-UTRAN itself. In the book we use LTE for the radio interface technology. In the overall descriptions we have allowed ourselves to use LTE for both the RAN and the radio interface technology. In the more technical detailed chapters (Parts III and IV) of the book we strictly use the terms E-UTRAN for the RAN and LTE for the radio interface technology.

2G/3G: A common term for both the GSM and WCDMA/HSPA radio access and the core networks. In a 3GPP2-based network 2G/3G refers to the complete network supporting CDMA/HRPD.

GSM: 2G RAN. In this book, the term does not include the core network.

GSM/GPRS: 2G RAN and the GPRS core network for packet data.

WCDMA: The air interface technology used for the 3G UMTS standard. WCDMA is also commonly used to refer to the whole 3G RAN, which is formally known as UTRAN.

WCDMA/HSPA: 3G RAN and the enhancements of the 3G RAN to high-speed packet services. Commonly also used to refer to a UTRAN that is upgraded to support HSPA.

WLAN/WiFi: WLAN refers to a specific access based on the IEEE 802.11 series, e.g. 802.11g. WiFi, meanwhile, refers to all of the wireless technologies that comply with the IEEE 802.11 series.

GSM/WCDMA: Both the second- and third-generation radio access technologies and RAN.

HSPA: A term that covers both HSDPA (High-Speed Downlink Packet Access) and Enhanced Uplink together. HSPA introduces several concepts into WCDMA, allowing for the provision of high-speed downlink and uplink data rates.

CDMA: For the purposes of this book, CDMA refers to the system and standards defined by 3GPP2; in the context of this book, it is used as a short form for cdma2000®, referring to the access and core networks for both circuit-switched services and packet data.

HRPD: High Rate Packet Data, the high-speed CDMA-based wireless data technology. For EPC, HRPD has been enhanced further to connect to EPS and support handover to and from LTE. Thus, we also refer to eHRPD, the evolved HRPD network, which supports interworking with EPS.

We also want to focus attention on the use of UE, Terminal, and Mobile Device in this book. These terms are used interchangeably in the book and all refer to the actual device communicating with the network.

Also, we use the word "interface" to refer to both the reference points and the actual interfaces.

PART II

Overview of EPS

CHAPTER 2

Architecture Overview

This chapter introduces the EPS architecture, mainly presenting a high-level perspective of the complete system as defined in the 3GPP SAE work item. In subsequent sections, we introduce the logical nodes and functions in the network. By the end of this chapter, the main parts of the EPS architecture should be understandable and readers will be prepared for the full discussion about each function and interface, as well as all applicable signaling flows that follow in Parts III and IV.

2.1 EPS Architecture

There are several domains in EPS, each one a grouping of logical nodes that interwork to provide a specific set of functions in the network.

A network implementing 3GPP specifications is illustrated in Figure 2.1. On the left of the diagram are four clouds that represent different RAN domains that can connect to the EPC, including the second and third generations of mobile access networks specified by 3GPP, more commonly known as GSM and WCDMA respectively. LTE is of course the latest mobile broadband radio access as defined by 3GPP. Finally, there is the domain called "non-3GPP access networks". This denotes any packet data access network that is not defined by 3GPP standardization processes, for example

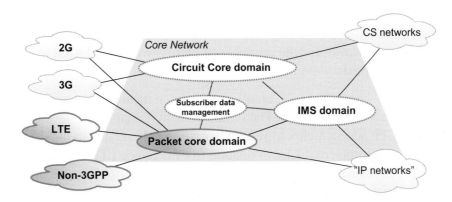

Figure 2.1: 3GPP Architecture Domains.

EPC and 4G Packet Networks.
DOI: http://dx.doi.org/10.1016/B978-0-12-394595-2.00002-5

eHRPD, WLAN, fixed network accesses, or some combination of these. This also means that 3GPP does not specify the details about these access technologies – these specifications are instead handled by other standardization forums, such as 3GPP2, IEEE, or Broadband Forum. Interworking with these access domains will be covered in more detail in later chapters of the book.

The Core Network is divided into multiple domains (Circuit Core, Packet Core, and IMS), as illustrated above. As can also be seen, these domains interwork with each other over a number of well-defined interfaces. The subscriber data management domain provides coordinated subscriber information and supports roaming and mobility between and within the different domains.

The Circuit Core domain consists of nodes and functions that provide support for circuit-switched services over GSM and WCDMA.

Correspondingly, the Packet Core domain consists of nodes and functions that provide support for packet-switched services (primarily IP connectivity) over GSM, WCDMA, and HSPA. Furthermore, the Packet Core domain also provides support for packet-switched services over LTE and non-3GPP access networks that in general have no relation to the Circuit Core (except for some specific features needed for voice handovers in relation to LTE). The Packet Core domain also provides functions for management and enforcement of service- and bearer-level policies such as QoS.

The IMS domain consists of nodes and functions that provide support for multimedia sessions based on SIP (Session Initiation Protocol), and utilizes the IP connectivity provided by the functions in the Packet Core domain.

In the middle of all of this, there is also a subscriber data management domain, where the handling of the data related to the subscribers utilizing the services of the other domains resides. Formally, in the 3GPP specifications, it is not a separate domain in and of itself. Instead, there are subscriber and user data management functions in the Circuit Core, Packet Core, and IMS domains interacting with subscriber data bases defined by 3GPP. However, for the purposes of clarity, we have elected to show this as a domain in and of itself.

The main emphasis here is the EPC architecture, which means the evolution of the Packet Core domain and Subscriber Data Management domain. The development of LTE as a new 3GPP access technology is of course closely related to the design of EPC. Due to the importance of LTE in relation to EPC (since LTE only connects via the Packet Core domain) we also provide a brief description of LTE on a high level. For a deeper insight into the interesting area of advanced radio communications, we recommend Dahlman (2011). The Circuit Core and the IMS domains are described in Chapter 5, where we look further into the topic of voice services.

We will now leave the high-level view of the 3GPP network architecture and turn our attention to the evolution of the Packet Core domain, or EPC.

While the logical architecture may look quite complex to anyone not familiar with the detailed functions of EPC, do not be put off. The EPS architecture consists of a few extra new functional entities in comparison to the previous core network architectures, with a large number of additional new functions and many new interfaces where common protocols are used. We will address the need for these additions and the perceived complexity by investigating all of these step by step.

Figure 2.2 illustrates the logical architecture developed for EPS, together with the Packet Core domain defined prior to EPC. It also shows how the connection to this "legacy" 3GPP packet core is designed (in fact, this specific connection comes in two flavors itself, a fact that adds to the complexity of the diagram, but more about that later).

Note here that Figure 2.2 illustrates the complete architecture diagram, including support for interconnection of just about any packet data access network one can think of. It is unlikely that any single network operator would make use of all these logical nodes and interfaces; this means that deployment options and interconnect options are somewhat simplified.

What is not visible in Figure 2.2 is the "pure" IP infrastructure supporting the logical nodes as physical components of a real network. These functions are contained in the underlying transport network supporting the functions needed to run IP networks, specifically IP connectivity and routing between the entities, DNS functions supporting selection and discovery of different network elements within and between operators networks, support for both IPv4 and IPv6 in the transport and application layer (the layers are more clearly visible when we go into the details in Parts III and IV).

Be aware that all nodes and interfaces described in this chapter (and, in fact, throughout the complete book) are logical nodes and interfaces – that is, in a real network implementation, some of these functions may reside on the same physical piece of infrastructure equipment; different vendors may have different implementations. In essence, different functions may be implemented in software and connect with one another via an internal interface, rather than via an actual cable. Also, the physical implementation of a particular interface may not run directly between two nodes; it may be routed via another physical site. Naturally, interfaces may also share transmission links.

One example is the X2 interface, connecting two eNodeBs (which will be explained in more detail later), that may physically be routed from eNodeB A together with the S1 interface (which connects an eNodeB to an MME in the core network) to a site

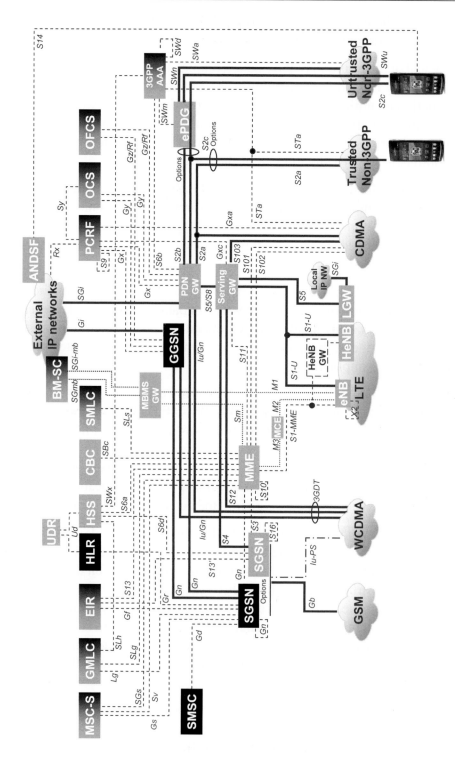

Figure 2.2: Architecture Overview.

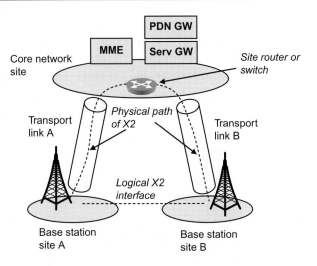

Figure 2.3: Logical and Physical Interfaces.

in the network with core network equipment. From this site, it would be routed back onto the radio access and finally to eNodeB B. This is illustrated in Figure 2.3.

2.1.1 Basic IP Connectivity Over LTE Access

At the core of the EPC architecture is the function required to support basic IP connectivity over LTE access. The plain vanilla EPC architecture, which one cannot live without when deploying LTE, appears as in Figure 2.4.

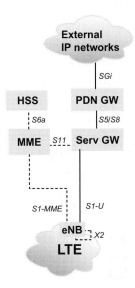

Figure 2.4: Basic EPC Architecture for LTE.

Two main principles have guided the design of the architecture. First of all, the strong wish to optimize the handling of the user data traffic itself, through designing a "flat" architecture. A flat architecture in this context means that as few nodes as possible are involved in processing the user data traffic. The primary motivation for this was to allow a cost-efficient scaling of the infrastructure operating on the user data traffic itself, an argument increasingly important as mobile data traffic volumes are growing quickly and are expected to grow even faster in the future with the introduction of new services relying on IP as well as new powerful access technologies such as LTE.

The second guiding principle was to separate the handling of the control signaling (shown as dotted lines) from the user data traffic. This was motivated by several factors. The need to allow independent scaling of control and user plane functions was seen as important since control data signaling tends to scale with the number of users, while user data volumes may scale more depending on new services and applications, as well as the capabilities in terms of the device (screen size, supported codecs, etc.). Allowing for both the control signaling functionality and the user data functionality to be implemented in optimized ways was another rationale for separating these functions in the logical architecture. A third important factor guiding the decision was that the split between control signaling and user data functions enabled more flexibility in terms of network deployment, as it allowed for the freedom of locating infrastructure equipment handling user data functions in a more distributed way in the networks, while at the same time allowing for a centralized deployment of the equipment handling the control signaling. This was in order to save valuable transmission resources and minimize delays between two parties connected and utilizing a real-time service such as voice or gaming. In addition to this, the split between the control signaling and user data functions allows for optimized operational costs through having these functions at separate physical locations in the network; by separating this functionality, the network nodes are more scalable, in particular when it comes to supporting high-bandwidth traffic. Only those nodes that are associated with end-user traffic need to be scaled for the high throughput, rather than both the traffic and signaling nodes as would have been the case previously. Finally, the chosen architecture was similar to the existing packet core architecture for evolved HSPA, allowing the possibility of a smooth migration and co-location of functionality supporting both LTE and HSPA.

Looking at the architecture itself, let us start with the radio network. First of all, in the LTE radio network there is at least one eNodeB (the LTE base station). The functionality of the eNodeB includes all features needed to realize the actual wireless connections between user devices and the network. The features of the LTE eNodeB will be described in Section 15.1.1.

In a reasonably sized network scenario, there may be several thousand eNodeBs in the network; many of these may be interconnected via the X2 interface in order to allow for efficient handovers.

All eNodeBs are connected to at least one MME (short for "Mobility Management Entity") over the S1-MME logical interface. The MME handles all LTE-related control plane signaling, including mobility and security functions for devices and terminals attaching over the LTE RAN. The MME also manages all terminals that are in *idle mode*, including support for Tracking Area management and paging. Idle modes will be further described in Section 6.4.

The MME relies on the existence of subscription-related user data for all users trying to establish IP connectivity over the LTE RAN. For this purpose, the MME is connected to the HSS (the Home Subscriber Server) over the S6a interface. The HSS manages user data and related user management logic for users accessing over the LTE RAN. Subscription data includes credentials for authentication and access authorization, and the HSS also supports mobility management within LTE as well as between LTE and other access networks (more about this later). The HSS and Subscriber Data Management will be further described in Chapter 10.

The user data payload – the IP packets flowing to and from the mobile devices – are handled by two logical nodes called the Serving Gateway (Serving GW) and the PDN Gateway (PDN GW), where PDN is "Packet Data Network".

The Serving GW and PDN GW are connected over an interface called either S5 (if the user is not roaming, i.e. the user is attached to the home network) or S8 (if the user is roaming, i.e. attached to a visited LTE network).

The Serving GW terminates the S1-U user plane interface towards the base stations (eNodeBs), and constitutes the anchor point for intra-LTE mobility, as well as (optionally) for mobility between GSM/GPRS, WCDMA/HSPA and LTE. The Serving GW also buffers downlink IP packets destined for terminals that happen to be in idle mode. For roaming users, the Serving GW always resides in the visited network, and supports accounting functions for inter-operator charging and billing settlements.

The PDN GW is the point of interconnection to external IP networks through the SGi interface. The PDN GW includes functionality for IP address allocation, charging, packet filtering, and policy-based control of user-specific IP flows. The PDN GW also has a key role in supporting QoS for end-user IP services. For example, for the GTP-based variant of S5/S8 (more about the S5/S8 variants below) the PDN GW handles the packet bearer operations and supports transport-level QoS through marking IP packets with appropriate DiffServ code points based on the parameters associated with the corresponding packet bearer. (An in-depth description of how QoS works in the EPS architecture can be found in Section 8.1.)

Something unique with the EPC architecture is that one of the interfaces is specified in two different variants (you guessed correctly – it is S5/S8). One of these variants utilizes the GTP protocol over S5/S8 (more about this in Section 16.2), which is also used to provide IP connectivity over GSM/GPRS and WCDMA/HSPA networks. The other variant utilizes the IETF PMIPv6 protocol over S5/S8 (more about this in Sections 6.2 and 16.4), but this variant has been shown in real-life deployments to have very limited market traction. GTP is already today the de-facto protocol used to interconnect hundreds of mobile networks worldwide, allowing IP connectivity to be established just about anywhere in the world where there is GSM/GPRS or WCDMA/HSPA coverage.

Since PMIPv6 and GTP do not have exactly the same feature set, this means the functional split between the Serving GW and the PDN GW is somewhat different depending on what protocol is deployed over S5/S8. In fact, in theory nothing would prevent both variants from being used simultaneously in the same network. Also note that S5 in itself may not be in use at all in most non-roaming traffic cases. It is a quite possible scenario that many operators will choose to deploy equipment that can combine Serving GW and PDN GW functionalities whenever needed, in theory reducing the amount of hardware needed to process the user data plane by up to 50% (depending on network dimensioning and assumed traffic load).

In some traffic cases, the S5 interface is very much needed, however, resulting in a division of the Serving GW and PDN GW functionality between two physical pieces of infrastructure equipment (two "Gateways"). Note that for a single user/terminal point of view, there can only be a single Serving GW active at any given time.

In addition to the roaming use case based on the S8 interface connecting the Serving GW in the visited network and the PDN GW in the home network, the split GW deployment within an operator network using the S5 interface may be used in three cases:

1. When a user wants to connect to more than one external data network at the same time, and not all of these can be served from the same PDN GW. All user data relating to the specific user will then always pass the same Serving GW, but more than one PDN GW.
2. When an operator's deployment scenario causes the operator to have their PDN GWs in a central location whereas the Serving GWs are distributed closer to the LTE radio base stations (eNodeBs).
3. When a user moves between two LTE radio base stations that does not belong to the same *service area*, the Serving GW needs to be changed, while the PDN GW will be retained in order not to break the IP connectivity. (The concepts of service areas and pooling will be described in detail in Section 6.6.)

The roaming case and the three intra-network cases are illustrated in Figure 2.5.

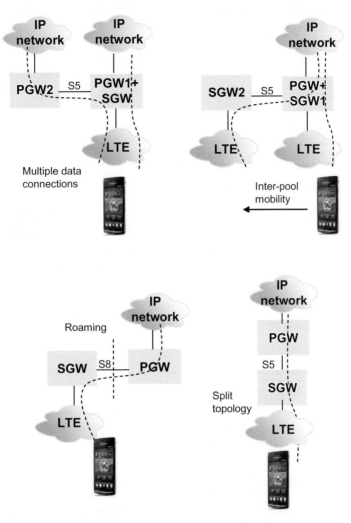

Figure 2.5: Use Cases Requiring a Split of SGW and PGW.

Control plane signaling between the MME and the Serving GW is exchanged over the S11 interface, which is one of the key interfaces in the EPC architecture. Among other things, this interface is used to establish IP connectivity for LTE users through connecting Gateways and radio base stations, as well as to provide support for mobility when users and their devices move between LTE radio base stations.

2.1.2 Adding More Advanced Functionality for LTE Access

Expanding somewhat on the basic architecture described above means introducing some more interfaces and some additional advanced features targeting the control of end-user IP flows; these additional features are covered in the following section.

For the purposes of this section, an "IP flow" can normally be thought of as all IP packets flowing through the network associated with a specific application in use, e.g. a web browsing session or a TV stream.

See the architecture diagram in Figure 2.6, which shows a few more details than the previous illustration. Three new logical nodes and associated interfaces are added – the PCRF, the OCS, and the OFCS.

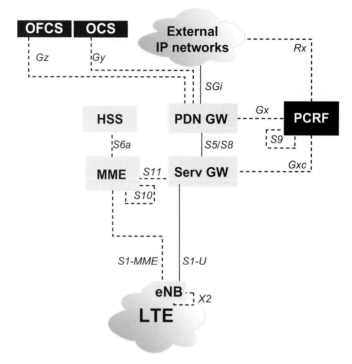

Figure 2.6: Adding Policy Control and Charging Support to the Basic EPC Architecture.

The PCRF (Policy and Charging Rules Function) makes up a key part of a concept in the EPC architecture (and in the 3GPP packet core architecture in general) called PCC (Policy and Charging Control). The PCC concept is designed to enable flow-based charging, including, for example, online credit control, as well as policy control, which includes support for service authorization and QoS management.

What, then, is a "policy" in the 3GPP architecture context? Think of it as a rule for what treatment a specific IP flow will receive in the network, for example how the data will be charged for or what QoS will be awarded to this service. Both the charging and the policy control functions rely on all IP flows being classified (in the PDN GW/ Serving GW) using unique packet filters that operate in real time on the IP data flows.

The PCRF contains policy control decisions and flow-based charging control functionalities. It terminates an interface called Rx, over which external application servers can send service information, including resource requirements and IP flow-related parameters, to the PCRF. The PCRF interfaces the PDN GW over the Gx interface and for the case where PMIPv6 and not GTP is used on S5, the PCRF would also interface the Serving GW over an interface called Gxc.

In the roaming case, a PCRF in the home network controls the policies to be applied. This is done via a PCRF in the visited network over the S9 interface, which hence is a roaming interface between PCRFs.

OFCS is short for Offline Charging System while OCS is short for Online Charging System. Both systems interface the PDN GW (through the Gz and Gy interfaces respectively) and support various features related to charging of end-users based on a number of different parameters such as time, volume, event, etc. Section 8.3 contains a description of the charging support in the EPC architecture.

Also shown in Figure 2.6 is an interface called S10. It connects MMEs together, and is used when the MME that is serving a user has to be changed for one reason or another, due to maintenance, to a node failure, or the most obvious usage, when a terminal moves between two *pools*. As stated above, the pooling concept will be described in detail in Section 6.6.

2.1.3 *Interworking Between LTE and GSM/GPRS or WCDMA/HSPA*

Given the fact that any new radio network is normally brought into service well before complete radio coverage is achieved (if that ever happens), the ability to allow for continuous service coverage through interworking with other radio networks is a key feature in any mobile network architecture. In many markets, LTE is deployed in frequency bands around 2 GHz or higher. While the data capacity normally increases as one moves into a higher frequency band (as there is more spectrum available), the ability to cover a given geographical area with a given base station output power quickly decreases with higher frequencies. Simply put, the gain in increased data capacity is unfortunately paid for by much less coverage.

For LTE deployment, interworking with existing access networks supporting IP connectivity hence becomes crucial. The EPS architecture addresses this need with two different solutions. One addresses GSM/GPRS and WCDMA/HSPA network operators, while the other solution is designed to allow LTE interworking with CDMA access technologies (1xRTT and eHRPD). Section 6.4 includes more details on inter-system mobility support in EPC.

For interworking between LTE and GSM/GPRS or WCDMA/HSPA networks, 3GPP has made the solution somewhat more complex than one would think necessary. 3GPP has in fact defined two *different* options for how to interconnect LTE and WCDMA/HSPA or GSM/GPRS. We describe both below.

2.1.3.1 Interworking Based on Gn-SGSN

The SGSN has been part of the packet core architecture since the first GSM/GPRS specification release in 1997. Back in those days, it was introduced to support the brand new service called GPRS, which was and still is the packet data connectivity service of GSM. In 1999, the ability to also serve IP connectivity over WCDMA networks was added to the SGSN. Note that the IP connectivity service over WCDMA was greatly enhanced in 2005 through the definition of HSPA, which however has no real impact on the packet core architecture itself. HSPA is mainly an enhancement of the WCDMA radio access technology.

In the GPRS architecture, an SGSN connects to a GGSN, which acts as the point of interconnect to external IP networks for all packet data sessions over GSM/GPRS and WCDMA/HSPA. In fact, it is the SGSN that selects which GGSN to use for a specific terminal. Subscriber data for GSM/GPRS and WCDMA/HSPA packet data access is stored in the HLR, which is connected to the SGSN (see Figure 2.7).

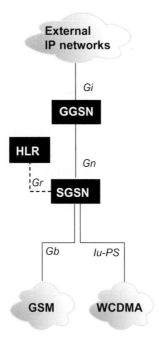

Figure 2.7: Packet Core Network for GSM/GPRS and WCDMA/HSPA.

When a user is moving between two networks that happen to be served by two different SGSNs, these two SGSNs interact over an interface (quite illogically also called Gn) to support IP session continuity; that is, the IP address, and all other data associated with the IP session itself, is maintained through keeping the GGSN unchanged when changing from one access network to the other.

If we disregard physical packet data equipment that may or may not have smooth migration paths to support the EPC architecture and features, the *logical* SGSN node has a key role to play also for LTE/EPC, while that is not the case for the *logical* GGSN node. Existing SGSNs and GGSNs keep serving non-LTE users as before, but the SGSNs are also utilized by multi-RAT LTE devices when out of LTE coverage.

The legacy packet core architecture and control signaling procedures form the base for the first solution for interworking between LTE and GSM/GPRS or WCDMA/HSPA described here. It was actually the second solution defined by 3GPP but it is the most straightforward to understand.

This solution includes GSM and WCDMA radio networks attaching to SGSN as today, but then includes the MME and the PDN GW acting towards the SGSN as another SGSN and a GGSN respectively. The MME and PDN GW are in fact replicating the signaling needed for movements between GSM/GPRS and WCDMA/HSPA to also apply for mobility with LTE. The MMEs and the PDN GWs act towards the SGSN as SGSNs and GGSNs respectively (see Figure 2.8).

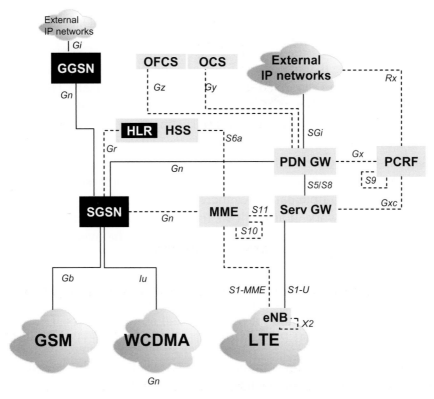

Figure 2.8: Interworking Between LTE and GSM/GPRS or WCDMA/HSPA Using Gn.

This includes both the MME and the PDN GW interfacing the SGSN over the standard packet core Gn interface. It may even be a Gn interface with an older date, that was specified and in operation before EPC was designed. This latter case is referred to as a pre-Rel8-SGSN.

Traditionally, the SGSN has interfaced a logical node called HLR (Home Location Register), which is the main database for user data in GSM and WCDMA. The interface between SGSN and HLR is called Gr. The MME instead interfaces the HSS (Home Subscriber Server) as described above. When moving between GSM/ WCDMA and LTE, there must not be inconsistent information in the network about, for example, to what radio network a specific terminal is currently attached. This means that the HLR and HSS need either to share a single set of data, or to ensure consistency through other means such as close interaction between the two network functions. In Release 8, 3GPP did not specify any detailed solution to this problem. In fact, the 3GPP specifications partly avoid the problem by defining HLR as a subset of HSS in later versions of the standards, but also outline interworking between legacy HLR and EPC nodes. This is further described in Chapter 10. As for the actual solution of ensuring this data consistency, this approach means that it may vary between different vendors of network infrastructure equipment.

3GPP Release 9 addresses this issue through creating the User Data Convergence (UDC) architecture (see Figure 2.9).

In the UDC architecture the processing logic and the different interfaces towards the network infrastructure are separated from the user data base itself. The processing logic is contained in a number of "front-ends", logical nodes that contain all functionality for processing user data and interfacing external nodes. The user data itself is stored in a separate logical node called UDR (User Data Repository), and this data is shared across all front-ends, accessible over the Ud interface. This architecture ensures data consistency and simplified data provisioning for the operator – instead of multiple databases there is a single user database that serves all needs within the network. It also allows for high availability solutions in that loss of functionality in one front-end network node can easily be overcome by use of a different front-end.

An example of applying the UDC architecture to a multi-access network also serving LTE is shown in Figure 2.10, where a common set of user data for all users is stored in the UDR database. This is valid for both LTE and non-LTE users. The SGSN and the MME then interface the HLR-Front-End (HLR-FE) and the HSS Front-End (HSS-FE) respectively.

UDC is described in more detail in Chapter 10.

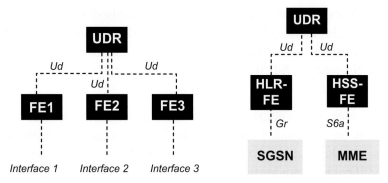

Figure 2.9: UDC Architecture.

Figure 2.10: UDC Architecture Applied to a 3GPP Multi-Access Network.

2.1.3.2 *Interworking Based on S4-SGSN*

However, the Gn-SGSN solution is not the only option for interworking between LTE and GSM/GPRS/WCDMA/HSPA. In fact, 3GPP first defined another solution referred to as the S4-SGSN solution, which is part of EPC. This is described in Figure 2.11.

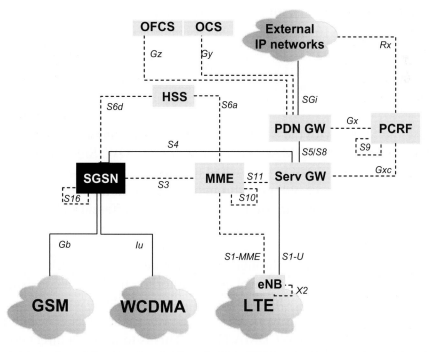

Figure 2.11: Interworking Between LTE and GSM/GPRS or WCDMA/HSPA Using S3/S4.

Just like the Gn-SGSN interworking architecture described, the S4-SGSN solution naturally also includes an SGSN interfacing the GSM/GPRS and WCDMA/HSPA radio networks using the Gb and Iu-PS interfaces. So far there are no differences. In fact, the Gn-SGSN and the S4-SGSN solutions are completely transparent to radio networks and terminals.

The SGSN, however, implements some new interfaces. Three of these (called S3, S4, and S16) rely on an updated version of the GTP protocol, the protocol that has been used since the early days of GPRS in the late 1990s, and which forms a core part of the 3GPP packet core architecture. All three are used instead of the different variants of the Gn interface present in the "legacy" packet core architecture. A fourth new interface is S6d, which mimics the MME S6a interface, towards the HSS for retrieving subscriber data from the HSS, but for the SGSN it is naturally data related to GSM and/or WCDMA, not to LTE. Just as with S6a, the IETF Diameter protocol is used over S6d, eliminating the need for the SGSN to support SS7/MAP signaling towards the HLR, and also allowing for usage of a common set of subscription data for GSM/WCDMA and LTE. Since the Subscriber Data Management part of a network is quite complex to migrate, 3GPP also describes the option of keeping the Gr interface instead of S6d when deploying S4-SGSNs in order to facilitate migration. This is further described in Chapter 10.

S3 is a signaling-only interface. It is used between the SGSN and the MME to support inter-system mobility. S16 is the SGSN–SGSN interface, while S4 connects the SGSN and the Serving GW. Note that there is a difference compared to the Gn-SGSN solution where the SGSN interfaces the PDN GW and treats this like a GGSN. S4 contains both a user plane and a control plane part, where the user plane part of GTP is not changed and hence is identical between the two solutions.

Connecting the SGSN with the Serving GW creates a common anchor point for LTE, GSM/GPRS, and WCDMA/HSPA in the Serving GW. Since the Serving GW for all roamers is located in the visited network, this means that all user traffic related to one roaming user will pass through this point in the network, regardless of which radio network is being used. This is new and different to how roaming is handled in the Gn-SGSN solution, where the SGSN itself implements the roaming interface for GSM and WCDMA and the Serving GW only for LTE. With all roaming traffic instead passing through a single point in the network, it allows for the visited network operator to control and monitor the traffic in a consistent way, potentially based on policies. One potential drawback is that user traffic needs to pass through one additional network node on its way to the PDN GW, but there is a solution to that, at least for WCDMA/HSPA. This is to utilize a direct connection between the Radio Network Controller (RNC) in the WCDMA radio network and the Serving GW. This interface is called S12 and is optional; if used, it means that the SGSN will only handle the control signaling for WCDMA/HSPA. The primary driver for this is that the network then does not have to be scaled in terms of SGSN user capacity, important due to the large increase of data sent over wireless networks (see Figure 2.12).

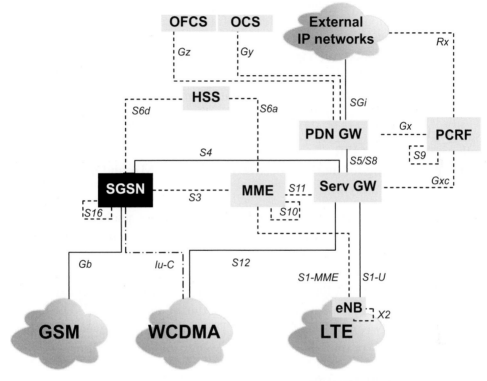

Figure 2.12: Direct Tunnel Support for WCDMA/HSPA.

It should be noted that this ability to let the user data bypass the SGSN is in fact also possible with the Gn-SGSN solution. This means that the WCDMA RNC would directly interface the PDN GW for the user traffic connections. A difference is that this would not work for roamers though, since, as stated above, roaming traffic always passes through a Gn-SGSN.

A further difference when utilizing S4-SGSN instead of Gn-SGSN is that the 3GPP specifications then allow for optimization of the signaling load for all terminals in idle mode. This concept is called ISR (short for Idle Mode Signaling Reduction). In short it means that terminals that are in idle mode (no traffic ongoing, no radio bearers established) are allowed to move between radio access networks without having to register to the network. This decreases mobility signaling and lowers the battery consumption in the terminal. The main drawback of ISR is that paging of terminals will consume more network resources and that ISR, if used, may cause additional delay at call setup for voice-over-LTE (VoLTE). ISR will be explained in detail in Section 6.4.3.

Summarizing, the main arguments for selecting a network architecture based on S4-SGSN are:

- It allows for harmonized signaling across all accesses based on Diameter
- It enables a common roaming architecture for all accesses where radio access changes (which may be frequent) are not necessarily exposed to the home network, and LTE-GSM/WCDMA mobility signaling is kept in the visited network instead of being carried to the home network
- It allows for using a common subscriber data profile for all accesses
- It simplifies the usage of QoS differentiation when moving between access technologies
- It allows for usage of ISR
- It supports signaling for mobility between GSM/WCDMA and non-3GPP access technologies such as WiFi.

In order to further optimize the packet handover performance, packet forwarding may be used (but it is optional). This means that any packets destined for the user device that may have happened to have been sent "downwards" from the PDN GW to either the SGSN or the Serving GW may be forwarded to the corresponding node in the target system. This is not absolutely required, but may improve the user experience of a handover, since in theory no data need to be lost during the handover. The case of packet forwarding between LTE and GSM/WCDMA is supported over Gn/S4 between SGSN and Serving GW or directly between the source and target radio network nodes.

2.1.4 Support for 3GPP Voice Services

In addition to the primary focus on enabling an efficient Mobile Broadband solution, the support for voice services was also given high priority in the 3GPP EPC specification work. From the start, it was agreed that LTE is a packet-only access network, allowing an optimization for packet services. Since the voice services so far have been realized using circuit-switched (CS) technologies, this meant that specific mechanisms were introduced to also allow for voice services in addition to the packet data services offered over the LTE access. Section 5.2 explains the subject of voice for LTE in more depth. In this chapter, we will highlight the applicable parts of the architecture.

In brief, two main solutions for voice are available. One solution is to use IMS (IP Multimedia Subsystem) mechanisms and realize voice services using the 3GPP MultiMedia Telephony (MMTel) framework that is using voice-over-IP. The other solution is to stick to the "old" circuit-switched way of providing voice services. The first option is used in the GSMA VoLTE (voice-over-LTE) solution and will be further described in Chapters 5 and 11. The second option in the 3GPP specifications is called CS Fallback and is realized through users temporarily leaving LTE to perform voice calls over GSM or WCDMA, and then returning when the voice call is finished. This may not be the most elegant of solutions, but can be seen primarily as a gap-filler in cases where the IMS infrastructure is not deployed.

2.1.4.1 *VoLTE Services and SRVCC*

VoLTE voice services means that voice is carried as IP packets. The solution is based on IMS/MMTel functionality combined with LTE/EPC capabilities to ensure proper establishment and management of voice bearers, including emergency sessions.

When a user engaged in a VoLTE voice call moves around, it is not unlikely that the user device may find that the LTE radio coverage is being lost. After all, this is a mobile system, and this may happen more or less frequently depending on how users move around and how complete the operator LTE coverage is. For this purpose, 3GPP has specified mechanisms to hand over an ongoing voice call in IMS/MMTel in LTE to another system (GSM or WCDMA) with better radio coverage. What happens then depends on whether the new (target) system can support IMS/MMTel or not. If this is the case, this will be solved through a packet handover procedure (see Section 17.4) and the IMS/MMTel session will continue after the handover (it is no longer called VoLTE for obvious reasons, since it is no longer in LTE). The case of a packet handover to GSM/WCDMA is supported using handover signaling over the S3 (or Gn) interface between the MME and the SGSN.

If IMS/MMTel-based voice services cannot be used in the target system, the VoLTE session will instead be handed over to a circuit-switched call in GSM or WCDMA. This procedure is called SRVCC (Single Radio Voice Call Continuity). To achieve a smooth handover, the SRVCC procedure involves pre-registration of the terminal in the target system CS domain (i.e. the system that the terminal will be attached to instead of LTE after the handover) and efficient handover signaling. The MME communicates with the MSC over the Sv interface for this purpose. If the target system supports simultaneous voice and data (which often is the case with WCDMA), data bearers can be moved from LTE in parallel with setting up the CS voice call. Data bearer mobility signaling is then carried over either the S3 or Gn interface between MME and SGSN. If simultaneous voice and data is not supported in the target system, the data session is suspended for the duration of the voice call.

2.1.4.2 *Voice Services Based on CS Fallback*

If no IMS is present in the network, LTE users have to temporarily leave LTE during voice calls and instead use GSM or WCDMA. The MME interacts with the GSM/WCDMA core network infrastructure over the SGs interface to allow for pre-registration in the CS domain and to establish the CS voice bearer at the time of call setup. If the target system supports simultaneous voice and data (which often is the case with WCDMA), data bearers can be moved from LTE in parallel with setting up the CS voice call. Data bearer mobility signaling is then carried over either the S3 or Gn interface between MME and SGSN. If simultaneous voice and data is not supported in the target system, the data session is suspended for the duration of the voice call.

Figure 2.13 highlights the parts of the architecture that apply to voice handovers. Note that in order to simplify this diagram, it includes the GSM/WCDMA MSC and its interfaces to the GSM and WCDMA radio networks but does not show the SGSN, which may be used to serve simultaneous data services. The diagram is also simplified in the sense that only an MSC is shown. In most modern deployments, the MSC is replaced by an MSC Server handling the signaling and a Media GW handling the media part of the voice call.

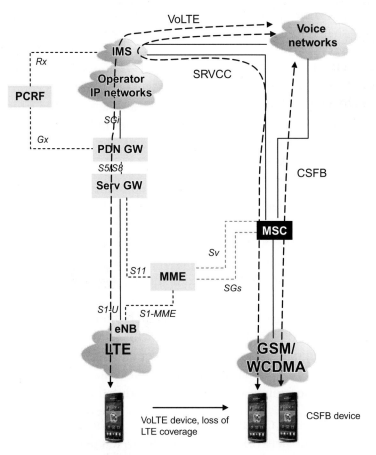

Figure 2.13: 3GPP Voice Solution Overview.

Also note that detailed descriptions of the mechanisms used in the MSC and the IMS system to enable SRVCC are beyond the scope of this book.

2.1.5 *Interworking Between LTE and CDMA Networks*

As stated above, an important part of the objectives behind the creation of EPC has been to allow efficient interworking with legacy mobile network infrastructure in order to allow for wide area service coverage. As there was significant interest towards using a common packet core network, strong efforts were devoted to also designing a solution linking LTE and the CDMA technologies defined by 3GPP2, allowing for efficient and smooth handovers between the different radio access technologies.

Interworking between LTE and CDMA networks covers two different aspects:

- Interworking with CDMA 1xRTT networks for voice services
- Interworking with CDMA eHRPD networks for data services.

eHRPD ("Evolved High Rate Packet Data") has evolved from CDMA EV-DO with the aim to offer more advanced services over CDMA and to connect into the 3GPP EPC architecture. It also supports handovers between LTE and CDMA access.

The combined LTE and eHRPD architecture is shown in Figure 2.14. A key part of the architecture is the S2a interface between the eHRPD HSGW (HRPD Serving GW) and the PDN GW. Connecting the HSGW to the PDN GW creates a common

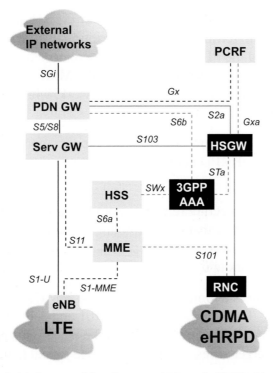

Figure 2.14: Interworking Between LTE and eHRPD Networks.

IP anchor point for both LTE and CDMA access. S2a is used for signaling and data transport in a similar way as S5 can be used between a Serving GW and a PDN GW for 3GPP access technologies if these GWs are physically separated. One difference is, however, that for S2a GTP is not an option when used for eHRPD. Instead this is a PMIP-based interface.

In order to share a common set of subscriber data in the HSS, the HSGW connects to a 3GPP AAA Server over the Diameter-based STa interface, and the 3GPP AAA Server connects to the HSS over SWx.

A fundamental function for allowing efficient interworking between two access technologies is to enable a common set of subscription data to be used – both for authentication purposes as well as for keeping track of which access network the user is currently attached to. The core of the solution is to allow for the HSS to act as a common database for all subscription data. This allows the HSGW to execute authentication of SIM-based terminals attaching over CDMA eHRPD access. For this purpose, the HSGW in the eHRPD network is connected to the EPC architecture over the STa interface. STa is terminated in a logical node called the 3GPP AAA Server, which in real-life implementations either may be a software feature inside the HSS or stand-alone AAA equipment interfacing the HSS over the Diameter-based SWx interface.

The PDN GW interfaces the 3GPP AAA Server over the S6b interface. This interface is used by the PDN GW to retrieve certain subscription data. It is also used to store the information regarding the PDN GW the user is connected to, in order to facilitate that, when the user moves and attaches over LTE, the MME will be able to select the same PDN GW as was used for the eHRPD network, which is a pre-requisite for maintaining the IP session also after the handover.

Finally, the EPC architecture allows for a common policy controller (the PCRF) to apply policies also in the eHRPD network. This is done over the Gxa interface to the HSGW.

An optional part of the architecture relates to optimizing the handover performance from LTE to CDMA eHRPD – that is, minimizing the time that data transmission is interrupted during a handover. There are two main solutions for this, which have different impacts on network and terminals.

The first solution introduces an interface between the MME serving LTE and the CDMA access network. This interface is called S101 and is used when a packet data handover from LTE to an eHRPD network is to take place. This procedure involves a pre-registration in the target access network as well as actual handover signaling, both carried over the S101 interface between the MME and the eHRPD access network. This functionality is described in more detail in Section 6.4.5.

To further optimize the packet data handover performance between LTE and eHRPD, there is also an S103 interface specified. This is an interface used to forward any IP data packets destined for the terminal that happened to end up in the Serving GW while the user terminal was executing the handover to eHRPD. These packets can then be forwarded to the HSGW in the eHRPD network, thus achieving close to a true lossless handover performance. The value of this packet forwarding can be questioned since the move from LTE to eHRPD anyway means a significant drop in peak data rates in most cases and would anyway cause data protocols such as TCP to reduce the transmission rate. The actual value hence depends on the application in use. Packet forwarding using S103 can be seen as an optional optimization that may add value in some use cases.

The second solution for optimizing the handover performance relies on the LTE + CDMA-capable device having support for simultaneous communication with the LTE and eHRPD access networks. This means a more complicated terminal but a simpler network solution without the need for S101 and related procedures.

For LTE + CDMA-capable mobile phones, voice service support is of course a must. There are multiple solutions possible, but assuming that voice is not carried over eHRPD, there are two main options:

- Use IMS-based Voice-over-LTE (VoLTE) when in LTE coverage but handover ongoing calls to CDMA 1xRTT voice services when LTE coverage is lost. This procedure is known as Single-Radio Voice Call Continuity (SRVCC).
- Always use the 1xRTT infrastructure for voice calls, even if within LTE coverage. This solution does not require any IMS solution, but leads to interruption of the data service during voice calls. This procedure is known as Circuit-Switched Fallback (CSFB).

Both SRVCC and CSFB are realized through procedures executed over the S102 interface between the MME and the 1xRTT MSC.

Figure 2.15 illustrates the two options for using voice services in a combined LTE + CDMA 1xRTT network, either through using an IMS-based VoLTE voice service and then only utilizing 1xRTT voice when out of LTE coverage (SRVCC), or through always relying on 1xRTT (CSFB). Note that in the SRVCC case, the call is handled in the IMS system even if the 1xRTT access is used. The details on MSC-IMS interworking are beyond the scope of this book though.

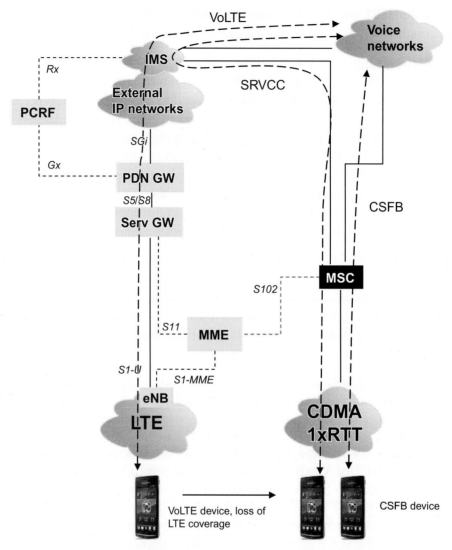

Figure 2.15: Voice Solution Overview for LTE + CDMA Operators.

2.1.6 *Interworking Between 3GPP Access Technologies and Non-3GPP Access Technologies*

The EPC architecture has been designed to allow interconnection with just about any access technology. This creates a common way of treating access to a PDN regardless of the access technology used, meaning that, for example, a terminal's IP address assignment, access to general IP services, as well as network features like user subscription management, security, charging, policy control and VPN connections, can be made independent of the access technology – be it wireless or fixed (Figure 2.16).

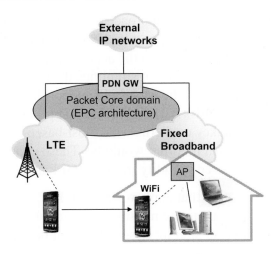

External IP networks

PDN GW

Packet Core domain
(EPC architecture)

LTE

Fixed Broadband

AP

WiFi

Figure 2.16: Interworking Between 3GPP Access and Non-3GPP Access Technologies.

Besides the common and access-independent feature set for service treatment, the architecture also allows for interworking between 3GPP technologies (i.e. GSM/ GPRS, WCDMA/HSPA, and LTE) and non-3GPP technologies (e.g. CDMA as described above, WiFi, or a fixed access technology).

Think of this as a use case: you carry a device that can access, among other technologies, LTE and WiFi. You are connected to the LTE/EPC network and move indoors, into your house. There you have a fixed broadband connection connected to a WiFi-capable home router. Depending on preferences, the device may in this situation switch access from LTE to WiFi. The EPS network then includes features to maintain the sessions also during this handover between two quite different access technologies.

The key functionality desired is support for mobility in the PDN GW. Mobile IP was designed in the 1990s by the IETF to provide IP host mobility, which is the ability of a portable computer to connect to a visited IP network, and establish a connection to the home IP network through tunneling of IP packets. To all corresponding hosts, this computer would appear as still being in the home network. Mobile IP technology has since been used to provide mobility for packet data services in mobile networks based on CDMA technology.

Due to the diversity of the requirements when specifying EPC, this part of the architecture has (somewhat unfortunately) come out with quite a few options. First of all, either "client/host-based" or "network-based" Mobile IP can be used. Host-based means that the Mobile IP client resides in the terminal, and that IP tunnels are established between the terminal and the PDN GW across the access network. Network-based means that there are functions in the access network that act on behalf of the terminal, and provide mobility support.

The major advantage with a host-based approach is that it may work over any access network, as long as there is adequate support in the terminal itself. This function may be totally transparent to the functionality in the access network. The advantage with a network-based approach is the opposite – it simplifies the terminal client application, but instead requires that there is specific Mobile IP support in the network itself. CDMA eHRPD is one access network where the latter approach has been chosen, as described above. One of the key concerns with "host-based mobility" was how secure, trusted, and efficient such mobility would be. These concerns partly drove the development of the network-based mobility track in 3GPP and in IETF.

The parts of the EPC architecture that apply to "non-3GPP" access support are shown in Figure 2.17. As can be seen, there are multiple ways to interconnect to a non-3GPP access network. We will make an attempt to sort these out below.

First of all, there are two ways to distinguish between the available options:

1. Is it a connection to a "trusted" or an "untrusted" network?
2. Are "network-based" or "host-based" mobility mechanisms used?

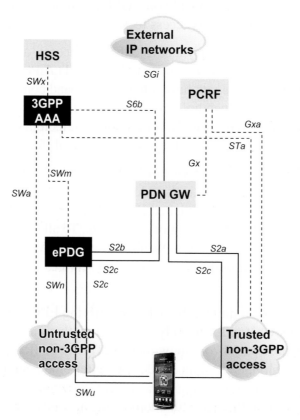

Figure 2.17: EPC Architecture for Non-3GPP Access.

Network-based and host-based concepts were described above, so what is a "trusted" and an "untrusted" access network? Simply put, this is really an indicator whether the 3GPP operator (owning the PDN GW and the HSS) trusts the security of the non-3GPP access network. A typical "trusted" network may be an eHRPD network or a fixed network owned by the operator and reachable over WiFi access, while an "untrusted" network may be, for example, usage of WiFi in a public café and connecting to the PDN GW over the public Internet.

The S2a, STa, and Gxa interfaces were described above for eHRPD and also apply in this context in the same manner. STa and Gxa apply to any trusted non-3GPP access network, and are used for user data management and policy control respectively. S2a is used for data connectivity when network-based mobility schemes are used in combination with trusted networks.

The corresponding interfaces for untrusted networks are S2b, SWm and SWa (both used for AAA purposes), and Gxb. There is, however, a major difference here. Since the operator may not trust the non-3GPP access network that is used by the device when attaching, the S2b, SWm, and Gxb interfaces do not interface the access network itself as S2a, STa, and GXa do, but instead interface a new logical node called the ePDG (evolved Packet Data Gateway). This is an evolution of the PDG that is specified in earlier versions of the 3GPP standards to allow interconnection (but not inter-access mobility) of WiFi access to a 3GPP network. Typically, the ePDG belongs to the mobile operator. More information on this can be found in Section 7.3. SWa is used for access authentication in the untrusted access network.

In the untrusted case, encrypted tunnels are established between the user devices and the ePDG using IPsec; this is to ensure that each device can communicate with the network in a secure way. This creates a logical association between each mobile terminal and the ePDG, referred to as the SWu interface, which carries signaling needed for management of the tunnel itself, as well as user data. The ePDG connects to the PDN GW and data as well, as signaling is transferred using the S2b interface between these two nodes.

The interface between the untrusted access network and the ePDG is called SWn. It carries all signaling and data between the two networks – the untrusted access and the operator network to which the ePDG belongs. SWu traffic and signaling is hence always routed over the SWn interface.

One final interface to understand in this non-3GPP access solution is SWm. This is a signaling interface only, and connects the ePDG to the 3GPP AAA Server. It is used to transport AAA-related parameters between the 3GPP AAA Server (which itself may get the data from the HSS over SWx) and the ePDG, for establishment and authentication of the necessary IPsec tunnels between the ePDG and the terminal.

In addition to the network-based mobility solutions for trusted and untrusted networks, there is also the host-based mobility solution, which relies on the S2c interface between the mobile device and the PDN GW. This means that an overlay solution is created that does not require any specific support from the underlying non-3GPP access network. It can be used over either trusted or untrusted access networks.

Host-based and network-based mobility solutions are further discussed in Sections 6.4.6 and 17.7.

2.1.7 Support for Broadcasting in Cellular Networks

In the network architecture there are specific features, nodes, and interfaces defined to support broadcasting of content to multiple users simultaneously; this saves network resources when the same content is to be received by many users. This is based on a technology called eMBMS (Enhanced Multimedia Broadcast and Multicast Service) and is defined for WCDMA and LTE. The architecture for supporting broadcasting over LTE is shown in Figure 2.18.

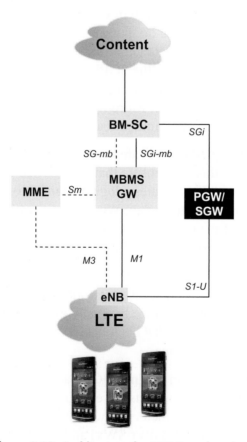

Figure 2.18: Architecture for LTE Broadcasting.

The BM-SC is a node controlling the broadcasting sessions and interacting with media sources and the end-user devices (via the PDN GW).

The MBMS GW is a logical node responsible for transmitting session data downstream to the base stations over the M1 interface, as well as for invoking the MME in MBMS signaling to control the broadcast sessions. The MBMS GW interfaces the BM-SC for both signaling and data transfer via the SG-mb and SGi-mb interfaces respectively.

The MME is connected to the MBMS GW over the Sm interface, and communicates with the LTE RAN over the M3 interface, relaying session control information received from the MBMS GW.

The MBMS architecture within LTE RAN consists of two logical entities – eNodeB (the base station) and MCE (Multi-cell/Multicast Coordination Entity) – interconnected using the RAN-internal M2 interface. The MCE is not shown explicitly in Figure 2.18. These entities are further described in Chapter 12.

The eNB receives control signaling from the MME (via the MCE) and data from the MBMS GW, and uses dedicated MBMS radio channels for broadcasting of control information and data respectively.

2.1.8 Positioning Services

The network architecture further includes support for determining the geographical position of a specific end-user device. This is useful both for commercial use, such as for instance triggering sending of location-based information or advertising, and for emergency situations when it is important to accurately locate a user in need of support. The 3GPP solution is referred to as LCS (LoCation Services). The architecture related to positioning services for LTE access is shown in Figure 2.19.

There are two basic ways of providing positioning information:

* User-plane-based methods, when the user device determines its position and communicates to a server in the network over the IP connection.
* Control-plane-based methods, when the position is determined using either only mechanisms in the network or a combination of mechanisms in the device and the network, and the position is made available to the GMLC (Gateway Mobile Location Center).

There are various methods of actually determining the position. The accuracy of these vary, and also whether they can be used in all environments, such as for example indoors.

* CID and E-CID – Different variants when the device may utilize the cell identities, signal strengths, and timing difference of signals from different base stations to calculate the position.

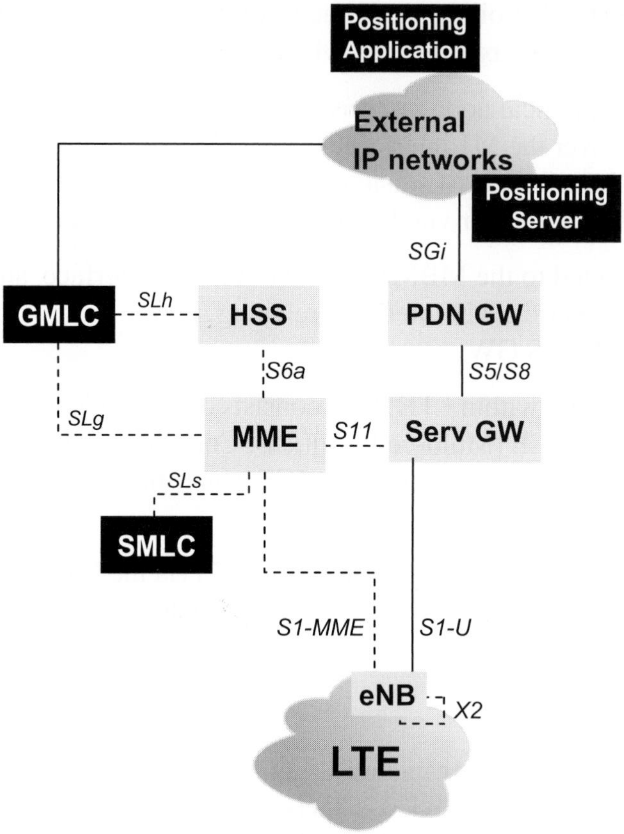

Positioning
Application

External
IP networks

Positioning
Server

SGi

GMLC — *SLh* - - - HSS — PDN GW

S6a   *S5/S8*

SLg

MME *S11* Serv GW

SLs

SMLC

S1-MME   *S1-U*

eNB *X2*

LTE

Figure 2.19: Architecture for Positioning over LTE.

- OTDOA – Observed Time Difference of Arrival. The device measures the differences between time stamps of signals sent out as reference signals from sites with known geographical locations.
- A-GNSS – Relies on the existence of a receiver for satellite-based positioning, for example GPS, in the device. The time it takes to determine the position can be shortened through utilizing reference information available from the network.

Control-plane-based positioning methods rely on the existence of an E-SMLC (Enhanced Serving Mobile Location Center) that calculates the position of the device on a request from the GMLC. Positioning signaling between the E-SMLC and the eNB, the device, and the GMLC is conveyed via the MME over the SLg and SLs interfaces, in addition to the existing S1-MME interface towards the eNB.

Positioning mechanisms for LTE are further described in Chapter 13.

2.1.9 Optimizations for Small Cells and Local Access

In order to allow for access to content that is physically located close to the base station, alternative solutions for allowing local IP connectivity have been specified in the architecture. One example is a small-sized base station located on a corporation's premises and being accessed by employees who primarily want to access the corporate Intranet. The 3GPP architecture specifies multiple options for how to realize this, which will be further described in Chapter 14. One option, referred to as LIPA (Local IP Access), means that the base station has a built-in IP GW functionality referred to as L-GW (Local GW), which is a simplified variant of a combined Serving GW + PDN GW. The LIPA feature developed in Release 10 is geared more towards end-user premise function where traffic towards a local network (printers, file storage, etc.) can be sent to/received from. The architecture is shown in Figure 2.20.

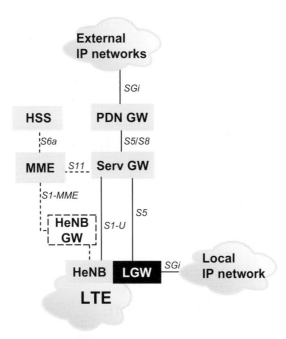

Figure 2.20: Network Architecture for Home eNB Support.

Note that the LTE base stations for corporate or home usage are called HeNB – "Home eNB" – in order to distinguish these from the eNBs serving the macro network. Regardless of whether LIPA is used or not, there is an optional HeNB GW in the architecture between the MME and the HeNBs. This GW appears to the MME as one single eNB, and as an MME towards all HeNBs. The HeNB GW relays messages between the MME and the HeNBs while protecting the MME from signaling due to, for example, HeNBs being powered on and off. It also allows for an efficient scaling of the number of HeNBs that can be connected to an MME. In order to protect the signaling and user traffic between the HeNBs and the HeNB GW and/or the MME and SGW, a Security GW is deployed (not shown in the figure).

The concept of "Closed Subscriber Groups" (CSG) is used to control access to specific cells, typically in home or enterprise locations. CSG identifies a group of subscribers as "members" who are permitted to access one or more CSG cells of the PLMN. A CSG manager can add, delete, or modify CSG subscription for a user.

2.1.10 Miscellaneous Features

We are almost through this overview of the 3GPP EPS architecture developed under the SAE work umbrella, including all the logical network nodes and the interfaces. As a final part, we will look at a few distinct features in the architecture that may be seen as somewhat outside the core of the architecture (Figure 2.21).

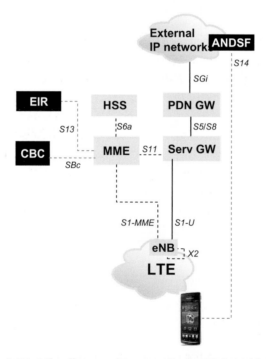

Figure 2.21: Miscellaneous Features in the EPC Architecture.

The first of these functions is called the Public Warning System (PWS) and is considered an important safety feature for countries endangered by nature catastrophes. It simply means that a warning is received by the Cell Broadcast Center

(CBC) from, say, a government agency monitoring seismic activity and predicting earthquakes. The CBC interfaces the MMEs in the network over the SBc interface. Since all terminals in the network must be reachable for this warning, the MMEs must convey the warning to all terminals that happen to be in idle mode, and whose location is only known with the accuracy of a Tracking Area, which may or may not contain lots of base stations and radio cells. PWS was defined in 3GPP Release 9, and is specified as an extension of the Release 8 feature called ETWS (Earthquake and Tsunami Warning System).

Another feature is the support for the Equipment Identity Register (the EIR), which is optionally used by the MME when a user attaches. The EIR is a database that contains information regarding whether the device used to attach to the network happens to be stolen or not. If that is the case, the MME can reject the attach attempt. The MME interfaces the EIR with the S13 interface. EIR is a function imported into EPC from the 2G/3G core network architecture, where the SS7-based Gf interface is used between SGSN and EIR.

The final function we intend to describe in this chapter is the ANDSF entity. ANDSF is short for Access Network Discovery and Selection Function, and put simply it is a function in the network that the operator can use to control how users and their devices prioritize between different access technologies if several non-3GPP access networks are available. The ANDSF can also assist end-user devices to discover available access networks. It is a means to give the network operator the possibility to control how users attach to the network, based on a number of criteria. The ANDSF logical entity interfaces the user device over an interface called S14. This is a logical interface carried over the user plane between the terminal and the ANDSF server, which is via the eNB and the Serving GW, or via other access technologies, and PDN GWs. The S14 interface will be further described in Section 15.11.

2.1.11 Summary of the Architecture Overview

The purpose of the description above was to make the overall EPS architecture more comprehensive and understandable. When putting all the pieces back together, we arrive at the complete architecture diagram as shown at the beginning of this chapter (Figure 2.22). Note that the figure does not necessarily represent completely all possible components within a 3GPP defined architecture, but, rather, focuses on the most relevant key components based on current deployment trends.

Large parts of the remainder of this book describe each of the network elements (or nodes), each of the interfaces, and each of the functions in greater detail.

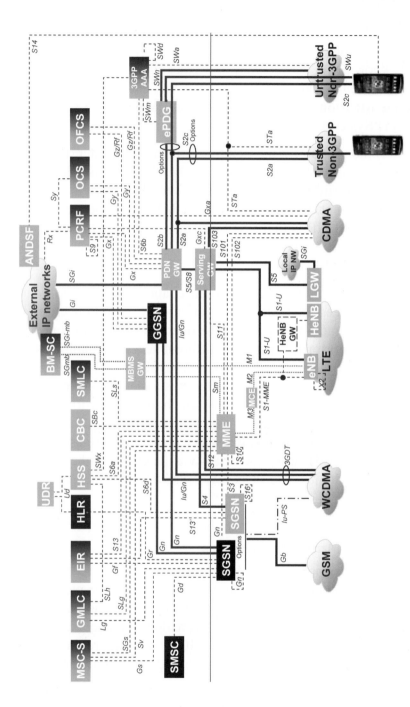

Figure 2.22: EPC Architecture Overview.

2.2 *Mobile Network Radio Technologies*

Even though the subject of this book is Packet Core Networks, it is important to understand the functionality of the supported radio access technologies. Readers with no interest at all in radio technologies may, however, choose to skip the remainder of this chapter.

LTE (also known as E-UTRAN) is the latest addition to the radio access technologies specified by 3GPP. LTE relies on EPC for core network functionality, and is naturally closely related to EPC through many interdependencies throughout the extensive standardization efforts. The concepts and functionality of LTE are described below. In addition, both GSM and WCDMA are briefly described.

However, we will start at a very basic level and describe the basic concepts of mobile radio networks in general.

2.2.1 *Overview of Radio Networks for Mobile Services*

Mobile networks (or cellular networks) consist of a number of base stations, each serving wireless transmission and reception of digital information in one or several "cells", where a cell refers to a specific portion of the overall geographical area the network serves. In most deployment cases one base station serves three cells through careful antenna configurations and frequency planning (Figure 2.23).

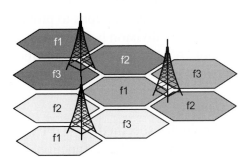

Figure 2.23: Cells and Base Stations.

The size and the outline of the cell are controlled by a number of factors, including base station and terminal power levels, frequency bands (radio signals using lower frequencies propagate over longer distances than radio signals using higher frequencies if the same power level is used), and antenna configurations. The radio wave propagation environment also has a significant effect on the cell size; there is a large difference depending on whether there are lots of buildings, mountains, hills, or forests in the area, compared to a surrounding area that is fairly flat and mostly uninhabited.

A fundamental ability of a cellular network is to allow the usage of the same frequency in multiple cells (see Figure 2.23, where f1 denotes a specific frequency). As can be seen, this frequency can be reused in multiple cells. This means that the total capacity of the network is greatly increased compared to the case where different frequencies would be needed for every site. The most intuitive way of allowing this frequency reuse is to make sure that base stations supporting cells using exactly the same subset of the available frequencies are geographically located sufficiently far apart to avoid radio signals from interfering with each other. However, GSM, WCDMA, and LTE have functionality that also allows adjacent cells to use the same frequency sets.

Base stations are located at sites that are carefully selected in order to optimize the overall capacity and coverage of the mobile services. This means that in areas where many users are present, for example in a city center, the capacity needs are met through locating the base station sites more closely to each other and hence allowing more (but smaller) cells, while in the countryside, where not so many users are present, the cells are normally made larger to cover a large area with as few base stations as possible.

Base stations are connected to other network nodes through transmission links, referred to as the RAN backhaul network. GSM and WCDMA radio networks include a centralized node (Base Station Controller (BSC) and Radio Network Controller (RNC) respectively) implementing some of the radio network functionality, while LTE radio networks rely solely on the base stations to provide the complete set of radio functions. The exact functional division between base stations and BSC/RNC is beyond the scope of this book. A brief overview of the most important functions of any digital cellular radio network is described in the following section.

2.2.2 Radio Network Functionality

Although the three radio technologies specified by 3GPP so far (GSM, WCDMA, and LTE) are different in several ways, the basic functionality of radio networks is common.

The most important functions of a cellular radio network include:

- Transmission and reception of data over radio carriers. This perhaps goes without saying – wireless transmission is naturally a key feature of a radio network. The characteristics of radio transmission are dependent on many parameters, such as distance from transmitter, used frequency, if any party is moving, transmission power that is used, height of antenna, and so on. Detailed fundamentals of electromagnetic wave propagation are, however, beyond the scope of this book.

- Modulation and demodulation of the radio carriers. This is a fundamental feature of both analog and digital wireless transmission. For a digital system, this means that the flow of bits relating to a specific service flow (e.g. a video stream), which may arrive at speed of, say, 2 Mbit/s, influences a high-frequency radio carrier through different means. This means that one or more fundamental characteristics of this radio carrier are changed (modulated) at constant time intervals depending on the next set of bit or bits – zeros or ones. This could be, for example, the phase, amplitude, or the frequency of the carrier. Every change corresponds to a "symbol", which may consist of one, two, or more bits. Today's most advanced mobile systems (HSPA and LTE) allow for the usage of up to six bit symbols, which means that there are 64 different symbols used. As an example, "001010" may mean a specific phase and a specific amplitude of the carrier, while "111011" may mean the same phase but a different amplitude (see Figure 2.24, where the two dark points represents these two symbols, the arrow represents the carrier amplitude, and the circles represent the phase of the carrier).

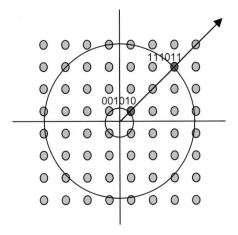

Figure 2.24: Modulation.

These more advanced symbol-modulation schemes place high requirements on the quality of the radio channel in order to avoid too frequent misinterpretations by the receiver that is converting the radio carrier changes into a flow of bits on the receiver side, the process known as demodulation.

- Scheduling of transmission of data from multiple users. This includes buffering of data from different users or applications while waiting for free radio capacity, and may include different priorities between queues of data, allowing different QoS to be applied to different flows of data. There are a number of different scheduling algorithms proposed for optimum sharing of the available transmission capacity based on the needs of services and users.

- Error correction schemes, in order to ensure that the number of bit errors, inevitably occurring over any transmission link, is minimized. There are two main approaches to error correction – FEC (Forward Error Correction) and ARQ (Automatic Repeat reQuest). FEC means that additional data bits (called redundancy bits) are added to the user data through which one- or multiple-bit errors may be detected and/or corrected. ARQ means that errors detected in a received block of data, for example by looking at a checksum, trigger a request to the other party to resend the data. FEC and ARQ are normally combined to increase the performance of the radio channel in that some errors can be corrected through FEC while larger errors need retransmission through ARQ. How to best balance FEC and ARQ mechanisms in order to maximize the performance is dependent on the requirements of the service and the characteristics of the actual radio channel, and normally also varies over time. Advanced radio communication systems here deploy adaptive coding, whereby the coding protection varies based on knowledge of the radio channel characteristics.
- Paging of idle terminals. This allows terminals to save battery power through entering what is called "idle mode". This means that the network no longer require the terminals to tell the network every time they move from one cell to another cell. Instead, terminals can move within a larger geographical area (defined by the network operator) without contacting the network so frequently and thus saving battery power. This is of course only possible when no services are active. If services are triggered from the network (e.g. through an incoming voice call), the terminal is told to reattach through the usage of broadcasting paging messages. If services are triggered from the end-user terminal (e.g. the user wants to make a voice call), the terminal is first triggered to leave idle mode, including notifying the network of exactly which cell it will be using for the upcoming voice call.
- Mobility (handover) support to allow for continuous service coverage even if user devices have to change base station or cell due to physical movements in the area. This is of course a very common scenario and very valuable to end-users. There is no need to stand still when using a mobile network service.
- Interference management in order to minimize the disturbance between multiple user devices or cells that share the same or neighboring frequency bands.
- Encryption of user data and signaling to protect the integrity and content of the user transmissions, as well as to protect the network from hostile attacks.
- Power control to efficiently utilize the available power, as well as to minimize interference between terminals.

2.2.3 GSM

GSM is the first generation of digital radio technologies specified by ETSI and later inherited by 3GPP. Since the first generation of cellular networks was based on analog transmission, GSM is normally referred to as a 2G (second-generation) technology. The development of GSM was driven by a wish to specify a standard that could be supported globally, and was supported by many operators and telecom equipment vendors. The work started during the 1980s and the first GSM networks came on air in 1991. Since then, the success and global adoption of GSM has been considerable. In April 2009, the number of GSM users around the world passed 2.3 billion, making GSM easily the most successful mobile technology to date.

Multiple users share the capacity of one GSM radio carrier that occupies 200 kHz spectrum in any of the supported frequency bands. The most common GSM bands are 900 and 1800 MHz, but 1900 and 850 MHz are also supported and used in some countries.

GSM is a TDMA (Time-Division Multiple Access) technology, meaning that the radio channel is divided into radio "frames". A radio frame is, simply put, an exact number of bits being transmitted on the channel. Each user is then allocated one or more time slots of each radio frame. When GSM is used for telephony, one time slot is needed for every voice call. One GSM radio channel consists of eight time slots per radio frame, meaning that in this case up to eight users can share one channel simultaneously. It is actually possible to squeeze up to 16 users into one GSM channel by utilizing what is called half-rate coding, at the expense of fewer bits being available per user and hence potentially a somewhat degraded voice quality (Figure 2.25).

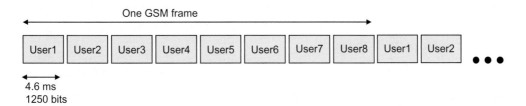

Figure 2.25: GSM Timeframes.

GSM also contains support for packet data services. These services are referred to as GPRS (General Packet Radio Service), and were specified as an add-on to GSM during the mid-1990s. The first GPRS services came on air in the late 1990s. Since the number of bits possible to transmit during one time slot is quite limited due to the small bandwidth of the GSM radio carrier, GPRS allows for one user to temporarily use more than one time slot in order to support higher data speeds. With the addition of EDGE technology (Enhanced Data rates for GSM Evolution), which adds more advanced signal processing technologies to the GSM channel, the peak rates of packet data services over GSM may reach above 400 kbits/s under favorable radio conditions and given that all eight time slots are allocated to a single user.

2.2.4 WCDMA

WCDMA is a third-generation (3G) radio technology, and the first version was specified by 3GPP during the late 1990s (it is normally referred to as Release 99 WCDMA).

WCDMA is specified for 5 MHz wide channels, significantly more than GSM. Besides the more advanced signaling processing deployed with WCDMA, this also means that higher data rates can be supported. Release 99 WCDMA could in theory support user downlink data rates of 2 Mbits/s, but in practice the limit in these networks is 384 kbits/s.

WCDMA is fundamentally different to GSM in that TDMA is not used as a means to separate the traffic from multiple users. Instead, the concept of CDMA is deployed, meaning that a specific code is allocated to each terminal. This code is combined with the data to be transmitted, and used to modulate the radio carrier. All terminals transmitt on the same 5 MHz channel and are separated due to the nature of the code instead of being allocated different frequencies or time slots. In order to allow communication with terminals further away from the base station site, the WCDMA concept includes advanced power control mechanisms that control the power levels of all terminals in the cell 1500 times a second.

Another feature of WCDMA is the ability to support soft handover and macro diversity, mechanisms allowing the terminal to communicate with more than one base station or cell simultaneously, enhancing the performance for the end-user, especially towards the outer parts of the cell.

From Release 5, 3GPP specified HSPA as an add-on technology to WCDMA, allowing more efficient usage of the available radio capacity for packet services, and allowing much higher data rates for end-users.

Through the introduction of more advanced modulation schemes, "MIMO" techniques (which are further described in the LTE section below), and multi-carrier support (utilizing more than one 5 MHz channel), HSPA peak data rates have been significantly increased in both downlink and uplink directions since the first version of HSPA in Release 5. The evolution is shown in Table 2.1.

2.2.5 LTE

Hand in hand with the work on 3GPP System Architecture Evolution, the work on the next generation RAN was carried out in the LTE study.

Table 2.1: HSPA Peak Data Rate Evolution

	Downlink		Uplink	
3GPP Release	*Peak Data Rate (Mbit/s)*	*Maximum Bandwidth (MHz)*	*Peak Data Rate (Mbit/s)*	*Maximum Bandwidth (MHz)*
Release 5	14	5	Not covered	5
Release 6	14	5	5.7	5
Release 7	28	5	11	5
Release 8	42	10	11	5
Release 9	84	10	22	10
Release 10	168	20	22	10
Release 11	336–672	40	70	10

In the same way that the outcome of the SAE work was specifications of EPC, the LTE work led to specifications of E-UTRAN, short for Evolved UTRAN.

However, names and terms that have been used for some time tend to stick in people's minds. LTE is now the official term for the radio access technology used for E-UTRAN.

The work on LTE started in late 2004 and early 2005 by defining a set of targets for the upcoming technical study and subsequent specification work. These targets can be found in the 3GPP technical report 25.913. A summary of the most important targets include:

- Downlink and uplink peak data rates of at least 100 and 50 Mbits/s respectively, assuming that a 20 MHz wide spectrum is used.
- The time it takes to change a user device from an idle to an active state shall not be more than 100 ms.
- The latency (delay) of user data shall not be more than 5 ms in the radio access network.
- Spectrum efficiency of 2–4× compared to a Release 6 3G network (spectrum efficiency is measured as the cell throughput in bits/s/MHz).
- Interruption time during a handover from LTE to GSM or WCDMA of maximum 300 or 500 ms for non-real-time and real-time services respectively.
- Support for both FDD and TDD multiplexing schemes with the same radio access technology (FDD means transmission and reception on different frequencies, while TDD utilizes the same frequency but with transmission and reception separated in the time domain).
- Support for a wide range of channel bandwidths, ranging from 1.4 to 20 MHz.

Implementations of LTE as defined in 3GPP Release 8 meet and in many respects even exceed these requirements. This is made possible by careful selection of technologies, including utilization of advanced signaling processing mechanisms.

The LTE radio network is connected to the EPC through the S1 interface, a key interface in the EPS architecture (Figure 2.26).

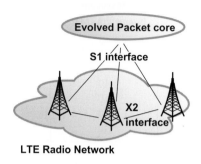

LTE Radio Network

Figure 2.26: LTE and Simplified Connection to EPC.

LTE base stations are optionally also interconnected to each other through the X2 interface. This interface is used to optimize performance, e.g. for handover between base stations or cells.

The S1 interface is divided into two parts:

1. S1-MME carries signaling messages between the base station and the MME. In addition it also carries signaling messages between the terminal and the MME that are relayed via the base station and piggybacked on radio interface signaling messages over the air interface.
2. S1-U carries user data between the base station and the Serving GW.

The S1 interface is shown in Figure 2.27 and will be further described in Sections 15.2.1 and 15.3.11 for the control plane and the user plane respectively.

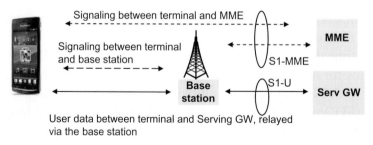

Figure 2.27: S1-MME and S1-U Interfaces.

A key technology for LTE is the OFDM (Orthogonal Frequency Division Multiplexing) transmission scheme, which is used for downlink transmission, i.e. from the base stations to the end-user devices. This is key to meeting the spectrum flexibility requirements. The basic concept of OFDM is that the total available channel spectrum (e.g. 10 MHz) is subdivided into a number of 15 kHz channels, each carrying one subcarrier. The available capacity (the usage of these subcarriers) can be controlled in both the time and frequency domains at the same time (Figure 2.28).

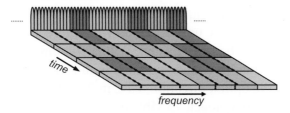

Figure 2.28: Downlink Channel-Dependent Scheduling in Time and Frequency Domains.

OFDM also has the benefit of being very robust against *multipath fading*, i.e. the variations in signal strength that are typical for mobile communications and are caused by the signal between transmitter and receiver propagating over multiple paths at the same time. Reflections of the radio waves in various objects mean that multiple copies of the signal arrive at the receiving antenna, since these are not synchronized in time due to slightly different propagation distances (Figure 2.29).

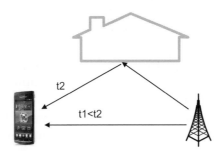

Figure 2.29: Multipath Propagation.

In the uplink direction, i.e. from the end-user device to the base station, a slightly different multiplexing scheme is deployed for LTE. As opposed to the LTE downlink, uplink transmission relies on only a single carrier. The key benefit of this is a lower peak-to-average ratio, i.e. the amplitude of the power used for transmission does not vary as much as in the OFDM case. This means a more efficient terminal power-amplifier operation in the end-user terminal, allowing for lower overall power consumption and hence longer battery life. In order to efficiently multiplex many users with the need to transmit at the same time, LTE allows for allocation of only a subset of the available uplink channel to each user. It is still a single carrier per user, as opposed to the downlink transmission scheme.

High peak rates in general rely on radio channels with good characteristics, i.e. with low noise and interference, as well as limited load on the cell that is being used. To take advantage of such channels, LTE allows for advanced modulation schemes, collectively referred to as HOM (Higher Order Modulation) schemes. 64QAM is one such scheme, allowing for six bits ($2^6 = 64$) to be transmitted for every symbol change on the radio carrier. Another technique is to simultaneously utilize multiple antennas in either the downlink or both the uplink and downlink directions. This technique is called MIMO, and further boosts capacity. With a combination of HOM and MIMO, really high peak rates are made possible with LTE – above 300 Mbits/s in the downlink direction. In the uplink direction, LTE in 3GPP Release 8 did not include MIMO and the peak rates may reach 75 Mbits/s.

Following the initial specification work of LTE in 3GPP Release 8, Release 10 added functionality and capabilities, turning LTE into what is popularly called "LTE-Advanced".

The background to the name *LTE-Advanced* is the ITU term *IMT-Advanced*, which includes a set of requirements for the next generation of mobile technologies. The IMT-Advanced requirements are specified in ITU-R M.2134 (see References) and include:

- Peak spectral efficiency of at least 15 bits/s/Hz DL
- Peak spectral efficiency of at least 6.75 bits/s/Hz UL
- Support for bandwidths of at least 40 MHz
- Transition time from idle to active state of less than 100 ms
- User-plane latency for small IP packets of less than 10 ms for both UL and DL.

The 3GPP design targets for Release 10 include full compliance to the IMT-Advanced requirements, and hence the term "LTE-Advanced" is frequently used for LTE Release 10. 3GPP, however, specified more aggressive requirements for LTE than the ITU requirements above. The 3GPP requirements are specified in TS 36.913 and include:

- Peak spectral efficiency of at least 30 bits/s/Hz DL
- Peak spectral efficiency of at least 15 bits/s/Hz UL
- Downlink peak data rates of at least 1 Gbit/s
- Uplink peak data rates of at least 500 Mbit/s
- Transition time from idle to connected state of less than 50 ms.
- Reduced user-plane latency compared to LTE Release 8
- Mobility at speeds up to 350 km/h
- Support for six new frequency bands.

Also important are the requirements on backwards compatibility that state that:

- LTE Release 8 terminals shall work in an LTE Release 10 network
- LTE Release 10 terminals shall work in an LTE Release 8 network.

In order to meet these requirements, LTE Release 10 includes a number of new and enhanced features:

- Carrier aggregation
- Enhanced multi-antenna support
- Improved heterogeneous network support
- LTE relaying.

Each of these key features is described below.

Carrier aggregation support is perhaps the most important difference from LTE Release 8. It allows combination of multiple carriers for communication with a single device. This can extend the available bandwidths far beyond the limitation of 20 MHz in LTE Release 8. A maximum of five individual carriers can be combined, enabling up to 100 MHz aggregated spectrum for an individual terminal.

This allows for much higher data rates than LTE Release 8. Since the carriers do not need to be contiguous in frequency, it also means that operators with a fragmented spectrum can achieve higher data rates than what is possible with LTE Release 8. Three variants of carrier aggregation are supported with LTE Release 8 (see Figure 2.30).

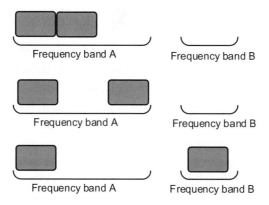

Figure 2.30: Carrier Aggregation Options.

Older LTE terminals that do not support carrier aggregation can still use the individual carriers, fulfilling the requirement on backwards compatibility.

Enhanced multi-antenna support is the other key feature added to support the peak data rate requirements of LTE Release 10. This is an extended variant of the MIMO technique as described above, realized through usage of multiple antennas in both directions. In Release 10, there can up to eight so-called *antenna ports*, allowing for up to eight parallel *transmission layers*, in the downlink direction, and up to four antenna ports, corresponding to up to four transmission layers, in the uplink direction. Although an antenna port is a logical concept, in most cases it can be seen as corresponding to a physical transmission antenna.

When combining the carrier aggregation and multiple antenna support features, the data rates of an LTE Release 10 system can reach up to 30 bits/s/Hz $\times$ 100 MHz = 3 Gbit/s in the downlink, and 15 bits/s/Hz $\times$ 100 MHz = 1.5 Gbit/s in the uplink.

Improved heterogeneous network support does not primarily address the achievable peak rates, but, rather, improves the management of inter-cell interference, which is a common problem when deploying combinations of small and large cells in the same geographical area, since these cells use different power levels and partly overlap each other (Figure 2.31).

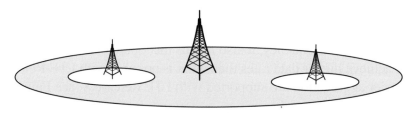

Figure 2.31: Heterogeneous Network Deployment.

LTE relaying is a technique introduced in Release 10 to allow for usage of LTE as a transport backhauling technology, interconnecting the LTE base stations with the EPC network (Figure 2.32). Operators with sufficient LTE spectrum can then allocate part of the available LTE capacity for the backhaul and the remaining part for end-user terminals. LTE relaying introduces the concept of Donor base stations and Relay base stations. Donor base stations have two tasks – they both serve mobile terminals as well as Relay base stations. Relay base stations serve mobile terminals and connect to the Donor base station for the backhaul link. Terminals will not see a difference between Donor base stations and Relay base stations; from a terminal perspective they are functionally identical, a prerequisite for the relaying solution to be backwards compatible.

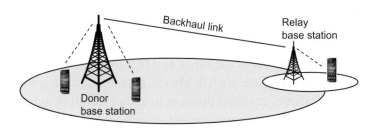

Figure 2.32: LTE Relaying.

LTE capabilities and performance were further enhanced in 3GPP Release 11, where the main feature addressing LTE-Advanced requirements is *Coordinated Multipoint Transmission and Reception*, abbreviated as CoMP. CoMP is a technology where transmission between the user device and the network can be handled via multiple reception/transmission points (e.g. sectors of a macro site or geographically separated nodes). In the downlink direction from the network to the device, either the scheduling and beam-forming of the individual points are dynamically coordinated to minimize interference levels, or the data transmission actually takes place over multiple transmission points simultaneously, leading to improved signal-over-interference and signal-over-noise levels and hence better performance (Figure 2.33).

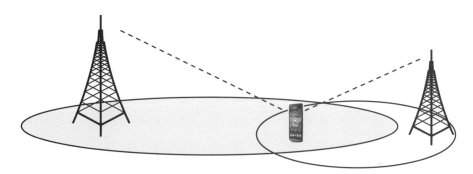

Figure 2.33: Multipoint Communication Using CoMP.

In the uplink direction, the CoMP scope of Release 11 is more limited, focusing on improving reception of multiple reference signals simultaneously and possibly scheduling enhancements.

CoMP can either be deployed between multiple points served by the same eNB, or between multiple points served by different eNBs, e.g. a macro and a pico base station. In both cases, requirements on the inter-site transmission links are very strict Win terms of low delay and high bandwidth due to the requirements of very precise scheduling across cells.

Terminals with LTE capabilities are classified into different categories depending on their capabilities in terms of which peak rates, modulation schemes, and MIMO variants they support. Table 2.2 shows the different terminal categories and the maximum peak rates supported.

Table 2.2: LTE Device Categories

	3GPP Release							
	Rel. 8/9/10/11				*Rel. 10/11 Only*			
Category	1	2	3	4	5	6	7	8
Downlink peak rate (Mbit/s)	10	50	100	150	300	300	300	3000
Uplink peak rate (Mbit/s)	5	25	50	50	75	50	150	1500

A more detailed description of LTE and LTE-Advanced is beyond the scope of this book, but an excellent source of information of advanced 3GPP radio technologies with a focus on LTE is Dahlman (2011).

CHAPTER 3

EPS Deployment Scenarios and Operator Cases

Deployment of an EPS network architecture is naturally coupled to an operator's plan to offer new or enhanced services to users, be they corporate users or private consumers. Since most services and applications would be supported also over third-generation packet data access technologies like HSPA, the primary motivation for an LTE investment is the enhanced characteristics – for the individual end-user service in terms of higher data rates and lower delay, as well as for the overall network capacity – and to enable a converged service offering based on IP technology over time.

As spectrum suitable for LTE becomes available either through opening up new frequency bands or through refarming of already existing frequency bands where other mobile access technologies are deployed, new opportunities are opening up to serve the market with mobile data services. In many countries this opportunity attracts both incumbent operators as well as potential newcomers to the market (the "Greenfielders"). The rules for awarding LTE spectrum differ between countries, and range from pure commercial processes (spectrum auctions) to more authority-controlled ways of allocating or reallocating spectrum, in many cases intended to increase the competitive situation in the local market.

LTE has been specified to support a wide range of different frequency bands, which simplifies global deployment and facilitates the emergence of large numbers of end-user devices. A challenge is, however, that the large flexibility creates a situation where international inter-operator roaming is not always straightforward since different regions of the world use partly different LTE spectrum. There is, after all, a limit to how many different frequency bands a single LTE device can support.

Table 3.1 is taken from the 3GPP specification 36.101, and shows the different frequency bands for which LTE operation is specified.

EPC is very much an evolution of the existing Packet Core Network deployed for data services over GSM/GPRS or WCDMA/HSPA, but the case for upgrading an existing Packet Core Network to EPC without deploying LTE is not obvious. Hence the most

EPC and 4G Packet Networks.
DOI: http://dx.doi.org/10.1016/B978-0-12-394595-2.00003-7

65

Table 3.1: Different Frequency Bands for LTE Operation

E-UTRA Operating Band	Uplink (UL) Operating Band BS Receive UE Transmit $F_{UL\_low}-F_{UL\_high}$	Downlink (DL) Operating Band BS Transmit UE Receive $F_{DL\_low}-F_{DL\_high}$	Duplex Mode
1	1920–1980 MHz	2110–2170 MHz	FDD
2	1850–1910 MHz	1930–1990 MHz	FDD
3	1710–1785 MHz	1805–1880 MHz	FDD
4	1710–1755 MHz	2110–2155 MHz	FDD
5	824–849 MHz	869–894 MHz	FDD
6[1]	830–840 MHz	875–885 MHz	FDD
7	2500–2570 MHz	2620–2690 MHz	FDD
8	880–915 MHz	925–960 MHz	FDD
9	1749.9–1784.9 MHz	1844.9–1879.9 MHz	FDD
10	1710–1770 MHz	2110–2170 MHz	FDD
11	1427.9–1447.9 MHz	1475.9–1495.9 MHz	FDD
12	699–716 MHz	729–746 MHz	FDD
13	777–787 MHz	746–756 MHz	FDD
14	788–798 MHz	758–768 MHz	FDD
17	704–716 MHz	734–746 MHz	FDD
18	815–830 MHz	860–875 MHz	FDD
19	830–845 MHz	875–890 MHz	FDD
20	832–862 MHz	791–821 MHz	FDD
21	1447.9–1462.9 MHz	1495.9–1510.9 MHz	FDD
22	3410–3490 MHz	3510–3590 MHz	FDD
23	2000–2020 MHz	2180–2200 MHz	FDD
24	1626.5–1660.5 MHz	1525–1559 MHz	FDD
25	1850–1915 MHz	1930–1995 MHz	FDD
33	1900–1920 MHz	1900–1920 MHz	TDD
34	2010–2025 MHz	2010–2025 MHz	TDD
35	1850–1910 MHz	1850–1910 MHz	TDD
36	1930–1990 MHz	1930–1990 MHz	TDD
37	1910–1930 MHz	1910–1930 MHz	TDD
38	2570–2620 MHz	2570–2620 MHz	TDD
39	1880–1920 MHz	1880–1920 MHz	TDD
40	2300–2400 MHz	2300–2400 MHz	TDD
41	2496–2690 MHz	2496–2690 MHz	TDD
42	3400–3600 MHz	3400–3600 MHz	TDD
43	3600–3800 MHz	3600–3800 MHz	TDD

[1]Band 6 is not applicable.

important operator scenarios to look at are those where existing or new operators are deploying LTE. We separate these into three main cases:

1. Existing GSM/GPRS or WCDMA/HSPA operators deploying LTE
2. Existing CDMA operators deploying LTE
3. Other operators deploying LTE ("Greenfielders").

Each of these scenarios will be discussed below. It is important to note that each of the scenarios is intended to be a realistic example, but in many areas there are options and variants that may be executed differently between operators.

3.1 Scenario 1: Existing GSM/GPRS and/or WCDMA/HSPA Operators Deploying LTE/EPC

This is a very common scenario where an operator has an existing GSM and/or WCDMA radio network and corresponding core network infrastructure. Normally the operator will have support for HSPA data services over WCDMA as part of a Mobile Broadband service offering.

The focus will be on getting a cost-efficient deployment of LTE and EPC by expanding the installed infrastructure as much as possible while ensuring that the existing customer base is not negatively affected in any way. The main thinking is normally to strive for a common service offering over a common core network solution, and a combination of two or three radio access technologies. Different subscription models and data plans then provide a means to differentiate the service offering across the customer base.

In the initial phase of LTE rollout, one can assume that the LTE coverage is quite limited in comparison to GSM or WCDMA coverage. The coverage will of course depend on which frequency band is used for LTE deployment. An example of a possible coverage situation is shown in Figure 3.1.

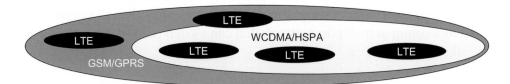

Figure 3.1: Overlapping Radio Coverage.

3.1.1 First Phase – Initial EPC Deployment

3.1.1.1 Physical Deployment

It can be assumed that EPC equipment is initially deployed as new, separate, network nodes in order to not affect the GSM/WCDMA infrastructure, which typically supports revenue-generating services for millions of users at the time of LTE introduction. Of course, proper dimensioning of the new EPC network and the individual nodes needs to take place, as well as detailed planning of how to integrate the nodes into the operator IP infrastructure. DNS configurations also need to be updated to serve LTE access needs.

The split of the control-plane and the user-plane functions in MMEs and PGW/SGWs respectively allows for flexibility in the physical deployment of the nodes. It could be that in order to optimize the transport resource usage, the PGW/SGWs may be located topologically closer to the LTE base stations than the MMEs. When even higher data rates are achieved over the air interface, even more careful design of the physical transport and overall network topology is needed. This is in order to avoid unacceptable delays in the transmission path, which would impact the data rates achievable to the devices.

A further selection to be made is whether to physically split the SGW and PGW when deploying, or to use combined SGW/PGW nodes in the network. In most use cases it makes sense to combine the functions in one physical node in order to simplify management and minimize the amount of hardware to be deployed. There may, however, be reasons to split the nodes in some specific operator cases such as when the operator's service network is located in a central location based on regions (e.g. east/west coast location based separation) and the end-user traffic needs to be routed to a PGW via such a central network. The EPC architecture allows for both variants.

It should be noted that even if combined SGW/PGW nodes are deployed, for individual users connected to a specific Packet Data Network there may be a physical split of the GWs anyway depending on specific use cases, as discussed in Section 2.1.1.

3.1.1.2 Resource Pooling

A natural part of the initial deployment phase of LTE/EPC is to utilize the pooling capabilities that are built into the architecture design from the start. In practice this means that a single LTE base station can be connected to multiple MMEs and SGWs and distribute the end-user devices over the available nodes, either uniformly or non-uniformly using arbitrary "weight factors". This requires that all sites are connected to a common transport network that can switch (L2) or route (L3) traffic between all sites (see Figure 3.2).

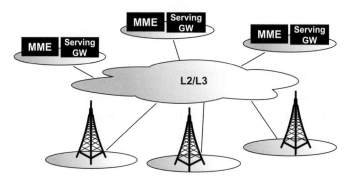

Figure 3.2: Pooling of EPC Resources.

Deploying EPC in pooled configurations has a number of benefits:

• It allows for load distribution over multiple nodes, ultimately leading to an optimized capacity utilization.
• It allows for very high service availability, since there are always backup nodes that can be utilized in case one of the EPC nodes is out of service.
• It allows for controlled maintenance windows with no impact on service availability. Through blocking of traffic to, for example, an MME and redirection of traffic to other MMEs in the pool, the blocked MME can be taken out of service for maintenance purposes with no end-user impact.
• It allows for easy capacity expansion – just add nodes to a live EPC pool and take them into service once configured and ready.

3.1.1.3 Multi-Access Support

While the first trial deployments of LTE may be single-access technology only, a commercial deployment would likely include multi-access-capable devices. This gives users the opportunity to use, for example, HSPA access when out of LTE radio coverage.

A first step in deployment of LTE/EPC services would be to offer common subscriptions. This means that while the infrastructure for the GSM/WCDMA network is logically and physically separated from the LTE/EPC infrastructure, end-user devices with multi-access capabilities can connect to any of the networks. The preferences regarding which network to prioritize when the device has coverage from multiple access networks is then typically controlled from settings in the radio network and is dependent on received signal strengths, while configurations in the device software can be used to select between, for example, LTE only or LTE/WCDMA modes of operation.

Since a single UICC/SIM card (which must be a USIM) is used in the multi-access-capable device, authentication and authorization of the user attaching will be done using a single set of user credentials and a single IMSI, regardless of whether the UE is attaching over LTE, GSM/GPRS, or WCDMA/HSPA.

This requires subscriber data provision in the HSS and parts of the User Data Management infrastructure to enable USIMs to be used to access also over LTE. Provision of consistent subscriber data in the HLR and HSS parts of the User Data Management infrastructure can either be done by utilizing a single database as described in Section 2.1.3, or through "Double Provisioning", where the necessary user credentials are stored both in the HSS and HLR. This is illustrated in Figure 3.3.

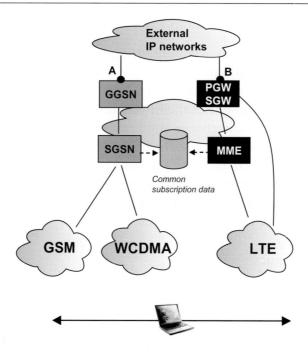

Figure 3.3: Common Subscriber Data for Multiple Access Technologies.

When attaching over LTE, the user device is served by the MME and the payload is handled by an SGW/PGW. When attaching over GSM or WCDMA and GPRS, the user device is instead served by an SGSN and payload is handled by a GGSN. The user session needs to be re-established when changing to and from LTE, and the IP address allocated to the end-user is changed.

The change of access network may be more or less automated or hidden for the user, but since the IP address changes with the change of radio network, this provides quite a basic level of mobility support, and it means that services and applications running on the user device may not be usable when the device has moved; some of these will require an application to re-establish the session. This is because when the user has moved and wants to use the new access network, the network views this as a new data connection instead of the movement of an existing data connection. The device is normally given a new IP address from the network, which then may or may not cause problems for the applications in use in the device. Furthermore, there is normally quite a long service interruption between loss of coverage and connection over network A and the establishment of IP connectivity over network B.

This simplest form of interworking basically only requires a single set of subscription data, allowing the end-user to attach over either of the networks as well as the user carrying a portable device that includes support for both radio technologies. There is in fact no specific support required in the network itself.

Another deployment aspect that needs to be decided is whether or nor dynamic policy control, QoS differentiation, and policy-based charging will be used for services over LTE. This will then require the deployment of one or several PCRFs interconnected to the PGWs.

3.1.2 Second Phase – Integration with Existing Packet Core

3.1.2.1 Inter-System Mobility

The next step to improve the service offering is to provide "session continuity", which ensures that end-user IP sessions established over any access networks will survive movements to and from LTE. This is done by interconnecting the Packet Core network used to serve GSM and WCDMA with the EPC network used to serve LTE. This allows LTE-capable devices to move in and out of LTE coverage while retaining the IP session and corresponding IP address. This support is provided without affecting the end-user services of non-LTE-capable devices, which at this point is quite likely still used by the vast majority of subscribers.

Session continuity is realized by retaining a stable "IP anchor point" in the network, which means it is not necessary to change the IP address of the device at all irrespective of any moves between radio access networks. In theory, applications and services will not be dependent on the access network that is in use or on any possible movements between them. This is, of course, only partly true. Some services may rely on very high data speeds or very low network delay, criteria that may not possibly be met for all of the networks due to limited radio coverage or by limitations in the access technologies themselves. Remember that the wireless data performance and capabilities offered by LTE may be superior to HSPA, which in turn is vastly superior to GPRS, the IP connectivity service offered on GSM networks.

In practice, inter-system mobility is added to the deployed network by interconnecting the SGSNs serving GSM and WCDMA with the MME and the PGW in the EPC part of the network (see Figure 3.4).

In this step, the existing SGSNs and GGSNs keep serving non-LTE users as before, but the SGSNs are also utilized by LTE devices when they are out of LTE coverage. As described in Chapter 2, there are two methods of interconnecting

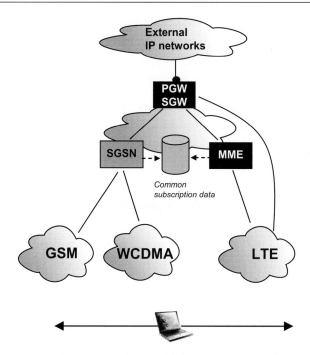

Figure 3.4: Interconnecting EPC and the Packet Core Network for GSM/WCDMA.

the legacy SGSN with the EPC network. In this phase of network deployment, we assume the Gn interface is used in order to minimize the impact on the legacy network.

This solution assumes that there is no change in the way GSM and WCDMA radio networks attach to the SGSNs, but in parallel with the GGSNs serving GSM/GPRS and WCDMA/HSPA services, the new PGWs incorporate GGSN functions and are connected to the SGSNs over the Gn interface. For control signaling, the MMEs connect to the SGSN as well.

This means that the MMEs and the PDN GWs act towards the SGSN as SGSNs and GGSNs respectively. In this case the EPC system (MMEs and PGWs) adapts to the GSM/WCDMA Packet Core system, which is not affected.

In order to support session continuity, the SGSNs in the network must be able to distinguish between a terminal that attaches over GSM/GPRS or WCDMA/HSPA but is not capable of moving to LTE, from a terminal that in fact can connect to LTE but is currently attaching to GSM/GPRS or WCDMA/HSPA due to lack of LTE radio coverage. In order to achieve session continuity, the latter terminal must always be using a PDN GW as the anchor point and never a GGSN, since there is no logical connection between the LTE radio network and a GGSN. If an incorrect choice of

IP anchor point were to be made by the SGSN, IP sessions would be dropped when changing access network to LTE.

Consider the example in Figure 3.5; terminal A has GSM/WCDMA support but is not capable of utilizing LTE access, while terminal B can use all three radio access networks. The simplest case is when terminal B attaches to the LTE radio network. It is then served by the MME, which will select a PDN GW and Serving GW (shown in the figure as a combined PGW/SGW node).

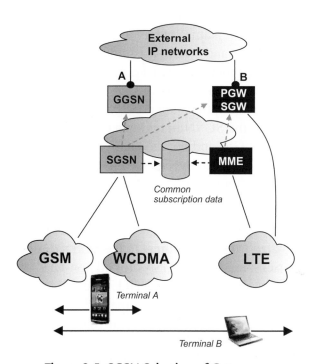

Figure 3.5: SGSN Selection of Gateways.

When either of the terminals attach over GSM or WCDMA radio it is served by an SGSN. For terminal B this may happen when there is no LTE coverage, while terminal A does not have LTE support so an SGSN is always serving this terminal.

There are various ways to make sure the SGSNs select the correct GW nodes (GGSN or PGWs). Note that the existing subscriber base, which so far has been served by GGSNs, can equally well be served by PGWs – the functionality for these users is identical in practice. A PGW can serve users over all access technologies.

The simplest way to provide session continuity for LTE users is hence to either make sure that all GGSNs in the network are upgraded to also support PGW functionality, or to replace the GGSNs with new PGW nodes. Either of these options makes the selection by the SGSN irrelevant – it will always select a correct IP anchor point since they are all PGWs. It should be noted that any upgrade of GGSNs to PDN GWs (including GGSN functionality) may affect charging and policy control systems, which may need to be updated to support Release 8 functionality.

If this method is not possible, another option is to make the SGSN distinguish between GGSNs that are still deployed in the network to serve terminals with GSM/WCDMA capabilities and the PGWs that can serve users over any access.

The SGSN may use different ways of achieving this. The most obvious way is to utilize the "APN" (Access Point Name), which is a part of the configuration data related to a user's subscription, and ensure it is pointing at the preferred external network. Since only terminals that include LTE radio access support (terminal B in the example) may ever move and attach to an LTE RAN, the simplest solution is to make sure that only these subscriptions are configured with an APN that is associated with a PDN GW. This helps the SGSN in taking a correct decision and ensuring that terminal B is using the PDN GW and not the GGSN as the IP anchor point. This solution is completely transparent to the SGSNs; in fact the SGSNs act as if selecting between two GGSNs with different APNs, it does, however, affect the configuration of the operator's DNS system.

Another solution is to base the selection on knowledge about the terminal capabilities. This is information signaled from the terminal to the SGSN during network attachment, and can be used by the SGSN to select GGSNs for all non-LTE-capable terminals (terminal A in the example) and PDN GWs to all LTE-capable terminals (terminal B in the example). This allows for the usage of the same APN for all terminals (something that is often preferred by operators), but requires the SGSN to have support for this selection mechanism that was defined in 3GPP Release 8. It means that the SGSN is no longer fully transparent to the introduction of LTE/EPC in the network.

Deployment of inter-system mobility support also requires upgrade of the GSM or WCDMA network that the terminals fall back on when losing LTE coverage. The required support is to broadcast system information on neighboring LTE cells in order for the terminals to find their way back to LTE when in coverage again.

3.1.2.2 LTE Roaming

It can also be assumed that when LTE roaming is deployed by the operator then the existing GPRS/HSPA roaming solution cannot be reused. There are two main differences:

- Instead of providing a roaming interface from the SGSN, LTE roaming relies on interconnecting the SGW in the visited network (VPLMN) and the PGW in the home network (HPLMN) over the GTPv2-based S8 interface. This is functionally similar to Gp, which is used for GPRS/HSPA roaming but requires a different network configuration since the SGW is involved.

- Instead of utilizing an MAP/SS7-based interface to interconnect the SGSN in the VPLMN and the HLR in the HPLMN, LTE relies on Diameter-based signaling. The MME needs to connect to the HSS in the VPLMN over the S6a interface. The GSMA technical document IR.88 recommends the use of a Diameter Agent in both the VLPMN and HPLMN to provide redundancy, improved scalability, and security for the inter-operator Diameter connections. More information on Diameter Agents can be found in Chapter 16.

Figure 3.6 shows a typical LTE roaming configuration for the most common case, where traffic is tunneled to the home network.

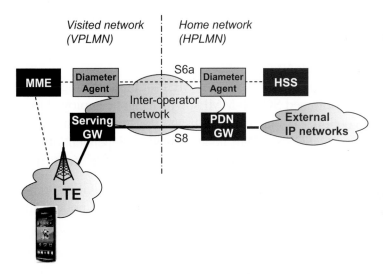

Figure 3.6: LTE Roaming Architecture.

The IP connection between the operators carries Diameter signaling (S6a) as well as GTP signaling and payload (S8). The actual connection between the operators may be realized either through a direct peer-to-peer connection or by routing via a third-party roaming service provider, a GRX/IPX operator.

The case of local breakout in the visited network is, however, also a valid scenario, and is controlled by subscription data in the home operator HSS. This scenario is not shown, but relies on the use of a PDN GW in the visited network.

3.1.2.3 Capacity Considerations
Since LTE/EPC is a "flat" network architecture, directly connecting the individual LTE base stations to the MMEs and the SGWs in the EPC network, control signaling load

levels in EPC will need to be carefully monitored and dimensioned for. Different types of devices and different end-user applications behave differently in the signaling plane, and combined with operator decisions on how often to move devices from active to idle mode, this will directly impact the signaling capacity dimensioning for the EPC nodes. This will also change over time when more LTE users and new devices come onto the market, so continuous monitoring of signaling load levels and subsequent network tuning is important. For the user plane the dimensioning is similar to how nodes are dimensioned for HSPA traffic, but also taking into account that user traffic triggers signaling, especially if advanced packet inspection mechanisms combined with dynamic policy control are used.

3.1.3 Third Phase – Further Optimizations Towards the Common Core

Once an integrated core network solution has been achieved by interconnecting the SGSNs with EPC, there are multiple measures that can be taken to optimize the EPC solution.

3.1.3.1 Multi-Access-Capable Nodes

Given the very similar roles of SGSN and MMEs for the signaling part, a natural product offering is combined SGSN–MMEs, which can be deployed to simplify network operations and to optimize overall capacity usage. This may mean that allocation of processing power in the SGSN–MME nodes is made independent of whichever access the users happen to be using – the total node capacity is made access independent.

A natural step for the operator is then also to upgrade all GGSNs to SGW/PGWs, or to phase out old GGSNs and replace them with SGW/PGW nodes. This brings at least three benefits:

- It removes the need for the SGSNs selecting different GW nodes for different types of terminals.
- It makes the total GW node capacity available for any access the users may be attaching over, reducing the need for over-dimensioning of GW capacity.
- It simplifies operations since all GW nodes are then identical.

3.1.3.2 Deployment of S4-SGSNs

Upgrading the Gn-SGSNs to S4-SGSNs provides further benefits for the operator by enabling harmonization of the complete Packet Core solution. Instead of operating different solutions for SGSN and MME signaling, deploying the S4-SGSN aligns the SGSN and MME signaling over all Diameter and GTPv2 interfaces and furthermore uses the EPC roaming architecture over the S8 interface for all accesses, not only LTE, also allowing all HSPA traffic to completely bypass the SGSN. It should be noted that radio networks and terminals are not affected by the migration to S4-SGSNs, which simplifies the migration considerably.

3.1.3.3 Evolution of the Subscriber Data Management Solution

Beyond the initial deployment of a common set of subscriber data for all accesses, deploying the S4-SGSN also allows the use of the HSS for handling subscriber data for all accesses, not only LTE. This means that EPS data used for LTE can also be applied to GSM/GPRS and WCDMA/HSPA access. The interconnection of the SGSN and HSS is then performed over the S6d interface, which in practice is identical to the S6a interface used between MME and HSS.

It should be noted that in order to simplify the migration to S4-SGSNs, Gr can still be used by S4-SGSNs until a migrated Subscriber Data Management solution supporting S6d is in place, a process that may take longer due to the migration of subscriber data for millions of users. A deployment initially relying on UDC architecture splitting the database and the front-end simplifies this migration.

3.1.3.4 Performance Optimizations

Additional features can be introduced in LTE/EPC to enhance the performance of the network, especially to reduce the time a device needs to move between two access networks when, for example, losing coverage.

An example is a terminal moving from LTE to WCDMA. Instead of the device being triggered to move from LTE and then having to read all system information from the WCDMA broadcast channel, some of the system information can be provided to the device already before leaving LTE. This requires that the LTE base station has the information available for the applicable WCDMA cell. This can be provided by the LTE base station subscribing to any changes in the WCDMA system information, and by the SGSN and MME forwarding the applicable information from the RNC to the applicable LTE base stations.

Further optimization of the inter-access mobility case is provided by the use of Packet Handovers, where the target cell is prepared in advance for the move, including pre-establishing necessary bearer resources, with handover signaling taking place between the two access networks via the Packet Core. Packet Handovers are further discussed in Chapters 6 and 17.

Voice services are discussed further in Chapter 11.

3.2 Scenario 2: Existing CDMA Operators Deploying LTE/EPC

The LTE/EPC specifications include various mechanisms to facilitate the introduction of LTE for an operator using CDMA technology. Depending on the strategies and plans for further investments into CDMA technology, the migration plans for LTE can be assumed to vary substantially across the CDMA operator community.

In practice this means an initial deployment is implemented as an overlay solution, similar to what was described in Section 3.1.1 for GSM/WCDMA operators. This includes deploying LTE base stations, MMEs, SGW/PGWs, an HSS, and the necessary IP infrastructure equipment including DNS servers. Deployments of PCRFs are optional but are required for several use cases as discussed above.

One difference here is, of course, that the terminals need to be dual-mode with CDMA + LTE support if any multi-access capabilities are to be offered. Another difference is that the terminals will not be served by an SGSN when out of LTE coverage, but instead by a PDSN or HSGW that is part of the CDMA Packet Core infrastructure.

There are different ambition levels available for CDMA operators deploying LTE. One option is to rely on Dual-Radio devices, terminals that can be attached to CDMA for voice services and simultaneously to LTE for data services. This means very limited impact on the CDMA network solution, at the expense of a somewhat more complex terminal with higher battery consumption.

In order to support multi-access support with a single subscription based on a USIM used in the CDMA + LTE device, the CDMA infrastructure needs to be upgraded to eHRPD capabilities. This includes the PDSN being upgraded to an HSGW and the AAA infrastructure being upgraded to support the use of USIMs over CDMA. The AAA may interface the HSS node used for LTE access in order to share a consistent set of subscriber data.

If a common Gateway for LTE and CDMA access is requested, the HSGW needs to be interconnected to the PGW over the S2a interface. This creates a single IP anchor point for the session, and allows for movements between LTE and CDMA access networks, including mobility of existing data sessions on the loss of LTE coverage.

For improved performance of the mobility case, i.e. a reduced data service interruption time, there are two possible ways forward:

- Dual-Radio devices are used that pre-register in the target access network before source system radio coverage is lost. This reduces the time needed to establish the session over the target access system.
- Single-Radio devices are used, and the CDMA and LTE/EPC parts of the network are interconnected through the S101 interface, which allows for handover signaling between the access networks. While reducing the service interruption time further and simplifying the dual-access CDMA + LTE devices compared to the Dual-Radio option, this variant has a larger impact on the investments needed, since it requires additional support in the CDMA and LTE/EPC infrastructure.

LTE roaming for a CDMA + LTE operator would be supported in the same way as for a GSM/WCDMA operator deploying LTE, but the case of inter-system mobility during roaming may need specific considerations since it is less likely that CDMA is available in all countries to which the operator's customers may travel. If, instead, HSPA is the primary fallback technology when LTE coverage is lost in these countries, this may either call for deployment of an HSPA roaming solution in the CDMA + LTE operator network, i.e. the HPLMN, or that access over HSPA is disabled for the roamers.

Voice services are discussed further in Chapter 11.

3.3 Scenario 3: New Operators Deploying LTE/EPC

Mobile Greenfield operators, i.e. operators with no prior mobile network deployments, can have different backgrounds. Examples include:

- Fixed broadband access operators who want to broaden the service offering by offering access over LTE in selected areas, possibly as a complement to using WiFi in the device and accessing over a fixed broadband connection when at home or in Enterprise environments.
- Cable TV operators expanding their offering.
- Fixed Wireless access operators, offering broadband access services over some other radio technology for stationary devices but not for mobile terminals.
- Completely new entrants in the mobile marketplace.

The fact that an operator is a Greenfielder means that there are no legacy installations of equipment for mobile services to relate to, but of course physical sites, IP transport networks, and spectrum (in the case of Fixed Wireless operators) may be useful assets when deploying a new LTE/EPC network.

As for all other installations, the network solution will need to include LTE base stations, MMEs, SGW/PGWs, an HSS, and necessary IP infrastructure equipment, including DNS servers. Deployment of PCRFs is again optional but is required for several use cases, as discussed above.

A Greenfield operator, like other operators, would have a fairly spotty LTE coverage for a considerable time. This may be fine for a Fixed Wireless operator who can build coverage only where services are to be offered, but for operators intending to offer mobile data services, there is also a need to provide services outside of the LTE coverage areas. Since there is no fallback technology operated by the Greenfielder, there is a need to instead partner with an existing mobile operator to provide subscribers with access to, for example, HSPA outside of LTE coverage. This would require establishment of solutions for national roaming, in practice very similar to the international roaming case.

CHAPTER 4

Data Services in EPS

EPS is a faster, lower latency mobile broadband solution. In contrast to previous generations of core networks of mobile broadband, LTE and EPS were designed primarily with IP connectivity and data services, rather than just voice services in mind. There were several reasons behind this change. This chapter covers these changes from the perspective of data services, while Chapter 5 covers voice services.

First, when GPRS and WCDMA were originally developed during the 1990s, it was not clear what data services end-users would want or that would be successful on a mobile operator's network. Uptake of data services developed on GPRS actually took a very long time – Wireless Access Protocol, or WAP, is one example of an IP technology that the mobile industry experimented with that had very low success rates. HSPA, meanwhile, with its much higher bandwidth for mobile broadband, as well as Wi-Fi hotspots, gave a good indication that there was a high demand for mobile broadband for people on the move to connect to corporate systems.

The launch of the first "smartphone" in 2007 – the iPhone – changed everything: Mobile Internet became an intuitive service for end-users and, for the first time, content on mobile devices was decoupled from operator networks. Mobile data traffic has increased rapidly, surpassing mobile voice in Q4 2009 (Ericsson, 2011). By Q1 2011, data was responsible for twice the amount of traffic on operator networks than voice, as shown in Figure 4.1.

In addition, an increasing number of devices are now significantly more than just phones – tablets, eBook readers, and other "connected devices" that benefit from having mobile connectivity have exploded in popularity. These new devices have created an application-centric ecosystem, which places the end-user at the center of the content value chain and are fueling changes in mobile network usage. Smartphone users are browse the Internet at many different points during the day. They also use a multitude of chat applications online, play games, and check emails almost constantly – often even before getting out of bed. These changes in usage patterns require network operators and vendors alike to re-think network design, operations, and management.

EPC and 4G Packet Networks.
DOI: http://dx.doi.org/10.1016/B978-0-12-394595-2.00004-9

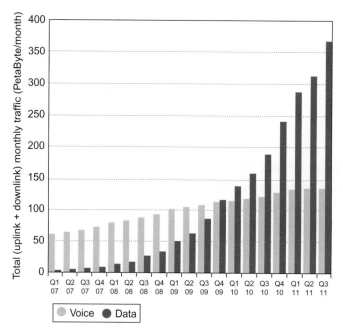

Figure 4.1: Global Traffic in Mobile Networks 2007–2011.
Source: Ericsson (2012).

The use of IP technology within EPS, therefore, is about significantly more than reducing the cost of delivering voice services; it is about the mobile broadband platform emerging to take a far greater role within the global economy and our society. Smartphones, tablets, and app stores are just the beginning of the changes that mobile broadband will deliver through data services – connectivity has become as essential as roads and electricity in end-users' lives.

The shift to an all-IP mobile operator environment therefore represents a fundamental shift in both the types of services that run on an operator's network and the nature of the value chain that forms the basis of the industrial structure. We are now at a tipping point, as the majority of services accessed via mobile technologies become data services, rather than traditional voice or SMS services.

The following sections cover two aspects of data services on EPC – messaging services and Machine-to-Machine communication.

4.1 Messaging Services

The ability to send messages to users of mobile devices has become immensely popular since the introduction of the Short Messaging Service (SMS) in GSM. The introduction of more advanced messaging services such as the Multimedia Messaging

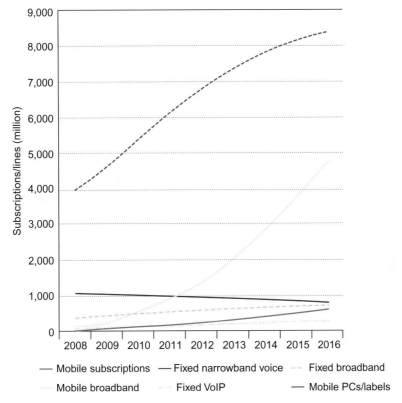

Figure 4.2: Fixed and Mobile Subscriptions, 2008–2016.
Source: Ericsson (2011).

Service (MMS) has offered the ability to also include photos, graphics, and sound in addition to text in the messages. Instant messaging and chat-like services have also been introduced as a means to further enhance the messaging experience for users.

Just as for voice, there are two fundamentally different ways of realizing messaging support with EPC – either using an IP-based solution (like IMS-based messaging or SMS-over-IP) or using the circuit-switched infrastructure that is normally used to deliver SMS messages over GSM and WCDMA. The fact that LTE is a packet-only radio access calls for some specific mechanisms to be included in the latter case.

For the case where messages are sent based on IP, there are no specific features needed in the EPC. Messages are sent transparently through the network from a messaging server to the client, and are treated just like any IP packet by the EPC. How the messaging application as such is realized is independent of EPC (as long as IP is used as the transport technology) and is beyond the scope of this book. Any sort of media (text, video, sound, graphics, etc.) can be included in messages sent using IP.

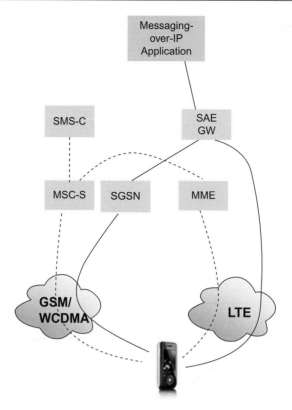

Figure 4.3: Options for Messaging Services.

When the circuit-switched infrastructure is used for delivering messages, the MME interacts with the MSC Server. The MSC Server is normally connected to a messaging center for delivery of SMS messages over control channels in, for example, GSM and WCDMA, and via the interaction with MME, this solution can also be used for LTE. Messages are then included in NAS signaling messages between MME and the mobile device. This solution supports only SMS text messaging, meaning that other types of messages (e.g. MMS) need to be based on IP, just as for GSM and WCDMA.

The two variants of messaging transmission are illustrated in Figure 4.3, where the dotted lines denote SMS transmission using signaling interfaces and the solid lines denote Messaging-over-IP transmission.

4.2 Machine Type Communication

As broadband speeds increase and the cost of semiconductors has dropped, the possibility of using sensor networks and other Machine-to-Machine (M2M) technologies in conjunction with mobile broadband has begun to be explored. There are multiple use cases of M2M within a mobile broadband context. M2M and similar technologies will drive further demand for mobile broadband over the coming years. This section explores a few such use cases.

In addition, the traffic generated for M2M devices is predicted to grow 22-fold from 2011 to 2016. Cisco, for example, predicts that M2M technologies will account for nearly 5% of total mobile data traffic in 2016 compared to 4% in 2011, as shown in Figure 4.4.

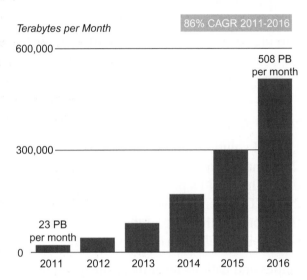

Figure 4.4: Traffic Growth Forecast for M2M Devices.
Source: Cisco VNI Mobile (2012).

These technologies will be useful in several domains, from corporations to cities; many different economic actors are set to benefit from such technologies.

4.2.1 *Industrial and Corporate Uses*

One of the most obvious benefits of M2M communication is within companies and industrial complexes. Supply chain management, for example, is often improved with an increased level of understanding about the exact location and environmental conditions that the goods being delivered are in. Alternatively, a truck can be given a new delivery route in real time if, say, a traffic jam occurs on the original route. In fact, it is quite possible that the use of M2M and similar technologies will become a matter of competitive advantage for the companies involved. Some use cases of benefit to industry are:

- Built into private cars for communicating service needs, the car's position (retrieved using GPS), as well as receiving up-to-date traffic data for traffic guidance systems

- Built into water or electricity meters for remote control and/or remote meter reading
- Built into street-side vending machines for communicating when goods are out of stock or when enough coins are present to justify a visit for emptying
- Built into taxi cars for validating credit cards
- Built into delivery cars for fleet management, including optimization of delivery routes and confirming deliveries
- Built into ambulances for sending life-critical medicine data to the hospital prior to arrival in order to increase chances of successful treatments
- Built into surveillance cameras for home or corporate security purposes.

These industrial uses are just the start of the use of mobile broadband, however.

4.2.2 Societal – M2M and Sustainable Development

In addition to all this, M2M solutions and services have a wider role to play in the future of our world. 2007 marked a landmark year for the world: for the first time in history, more than 50% of the world's population were living in cities, rather than rural areas (UNPD, 2009). This trend shows no signs of reversing. By 2016, it is predicted that users living on less than 1% of the Earth's total land area are set to generate around 60% of mobile traffic (Ericsson Consumer Labs).

The infrastructure of cities and nations must therefore adapt accordingly, from roads, lighting, metro/commuter trains, and pipelines to name just a few (HM Treasury, 2011). Much of this infrastructure will be instrumented with sensors and actuators for more efficient management, and all these devices associated with infrastructure will be connected to large-scale data analysis and management systems, the data of which needs effective capture, analysis, and visualization in order to be applied effectively in the development of smart, sustainable societies and cities. In the UK alone, this market represents a significant investment by both the government and private sector alike. The use of M2M and ICT in assisting the delivery of economic, social, and environmental outcomes for nations and regions is rapidly becoming an area of concern for professionals working in this space (Broadband Commission, 2012).

In order for this transformation to be fulfilled, however, it is also necessary to build a data analytics systems that can handle the real-time nature of the flow of data from mobile and sensor networks. These technologies will allow for new forms of innovation on the data contained in mobile networks. As illustrated in Figure 4.5, the data provided from ICT implemented within a city context will provide innovation in enterprise, city, and government services.

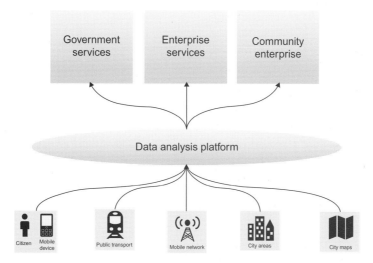

Figure 4.5: M2M and Data Analytics.

Here we see that the mobile broadband platform has the potential to become a nexus of contracts not just in the mobile telecommunications industry, but the wider global economy. EPS is the system that enables this transformation. In the following chapters, we will investigate the technology that is driving this mobile broadband revolution.

CHAPTER 5

Voice Services in EPS

Voice services have been the primary source of revenue for mobile operators since the dawn of basic mobile service offerings in the 1950s. The emergence of GSM technology in the early 1990s was the starting point for the unprecedented global adoption of mobile communication services. By the end of 2011 the number of mobile subscriptions in the world reached around 6 billion and is growing by millions each week. Despite tremendous growth in data-centric mobile Internet devices, the majority of terminals are still voice centric and, at the time of writing this book, voice services generated more than 60% of operators' revenues.

Voice traffic is still growing, in particular in emerging markets, and given the importance of voice services for billions of users, it is not surprising that while the majority of efforts behind designing the EPS architecture and procedures primarily targeted a high-quality IP access service, the importance of efficient voice support via EPS was also acknowledged right from the start of the work.

With the introduction of IP technologies, end-users now have a variety of options in addition to traditional voice services delivered by mobile network operators; two well-known examples by the so-called "Over The Top" (OTT) service providers are Skype or WhatsApp. Although these OTT voice solutions are growing, there is still a significant market demand for operator-delivered solutions, for LTE also.

There are two fundamentally different ways that voice services can be realized for LTE users: using Circuit-Switched Fallback or VoLTE/MMTel based on IP Multimedia Subsystem (IMS) technologies. These two different approaches and the differences between them are presented in the following sections. Chapter 11 discusses the implementation of voice services in EPS.

5.1 Realization of Voice Over LTE

The LTE radio access design is optimized for IP-based services. It provides a packet-only access with no connection to the circuit-switched mobile core network. This is in contrast to GSM, WCDMA, and CDMA, which support both circuit- and packet-switched services. Naturally, this difference affects the technical solution selection for delivery of voice services to LTE end-users.

EPC and 4G Packet Networks.
DOI: http://dx.doi.org/10.1016/B978-0-12-394595-2.00005-0

One of the main benefits of the previous generations of mobile voice services has been continuous service coverage, supported through handovers between radio cells and between base stations. Coverage availability for LTE, however, will be dependent on network operator build-out plans and the frequency bands that are allocated to LTE within a country or region. For some end-users, therefore, radio coverage can be assumed to be non-continuous or even spotty, in particular for initial LTE deployments.

For EPC, two basic approaches have been guiding the work in defining voice service support. Simply put, voice services on LTE are produced using the circuit-switched infrastructure used for voice calls in GSM, WCDMA, and CDMA, or the IMS and MMTel applications are used on the packet-switched infrastructure.

5.2 Voice Services Using IMS Technology

MMTel is the IMS-based service offering for voice calls, standardized by 3GPP. Since EPS is designed to efficiently carry IP flows between two IP hosts, MMTel is a natural choice for offering voice services when in LTE coverage. In addition to traditional voice services, MMTel offers end-users more possibilities than circuit-switched technologies; for example, video, text, or other media may be added to the voice component in order to enhance the communication experience and value. Moreover, MMTel allows for evolution from today's voice and video telephony networks to fully fledged multimedia communication and fallback to 2G and 3G circuit-switched telephony.

Voice over LTE (VoLTE), meanwhile, is GSMA profile IR.92 based on the MMTel standard. This profile covers every layer of the network, including IMS features, media requirements, bearer management, LTE radio requirements, and common functions, such as the IP version. It includes a subset of general IMS and MMTel service features, selected to provide an IP telephony service with similar user experience as today's 3G and GSM networks. It should be noted that the VoLTE profile features represent the minimum set of features that can act as a serving point for operators. More details of VoLTE are given in Chapter 11.

As discussed, operators cannot rely on the fact that LTE coverage is present everywhere where a user may want to make a voice call. Full voice service coverage therefore relies on the following:

- Other access networks are complementing the LTE access network in terms of coverage
- The device used to make the voice call (a traditional mobile phone or another device) also supports these access technologies and the technology used for voice calls in that technology (like circuit-switched procedures in, for example, GSM)
- Inter-system handovers are possible.

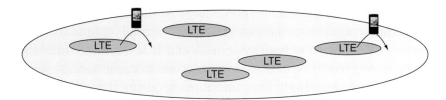

Figure 5.1: Voice Services and the Need for Mobility Support.

Figure 5.1 shows the areas that have LTE coverage (small dark areas in the figure), and those that have technology with much better radio coverage (large lighter area).

There are three different scenarios that need to be considered:

1. A voice call is established when in LTE coverage (dark area), and the user does not move outside the LTE coverage area for the duration of the call. For this use case, MMTel would be used to provide the voice service over LTE.
2. A voice call is established when outside the LTE coverage area (light area). The call would then instead be established using circuit-switched access over, for example, WCDMA. Depending on the solution, the call could be converted into a SIP-based call and handled by the IMS system, or it could be handled as a traditional circuit-switched call by the MSC.
3. A voice call is established when in LTE coverage (dark area) and, during the voice call, the user moves outside the LTE coverage area. If the system depicted as a light area can support IMS/MMTel voice services, this would be handled by a "Packet Handover" between LTE and the other system (e.g. WCDMA/HSPA or eHRPD) and the voice service would continuously be served as an IP-based service and handled by the IMS infrastructure. If this is not the case, specific measures are needed to secure service continuity when LTE coverage can no longer be maintained. The 3GPP solution for this is called Single-Radio Voice Call Continuity (SRVCC).

5.3 Single-Radio Voice Call Continuity (SRVCC)

SRVCC is designed to allow for the handover of a voice call between a system that supports the IMS/MMTel voice service and a second system where there is insufficient radio access support for carrying the MMTel service. This could be, for instance, due to insufficient bandwidth for IP services, or insufficient QoS support in the network.

SRVCC hence defines a solution for how an IP-based voice call in "system A" (dark gray area) can be handed over to "system B" (light gray area), which serves the voice call using circuit-switched procedures.

So why is this called a "Single Radio" procedure? Additional complexity of this handover procedure comes from the fact that most "normal" terminals (the end-user devices) cannot be connected to both systems A and B at the same time. Instead, a very quick handover has to be executed in order not to cause serious service degradation such as an annoyingly long interruption during the voice call. This is because the end-user device would require more complex and expensive radio filters, antennas, and signal processing if simultaneous connections to two systems would need to be maintained. This is where the "Single Radio" comes in. It extends the 3GPP Release 7 VCC solution, which allows for handovers between an IMS-based service over WiFi and circuit-switched services over, for example, GSM. In Release 7 VCC, the assumption is, however, that dual-radio is used, i.e. the terminal is simultaneously connected to both WiFi and GSM at the same time. This is possible due to the difference between a system with local coverage and low transmitting power (WiFi) and a system with wide coverage and relatively high transmitting power (GSM).

3GPP has specified the following combinations for SRVCC (system A to system B):

- LTE to GSM
- LTE to WCDMA
- WCDMA (HSPA) to GSM
- WCDMA to WCDMA
- LTE to 1xRTT.

In 3GPP Release 11, features were added to allow the possibility to start a CS call on GSM or WCDMA and transfer to MMtel Service on LTE. This is known as return SRVCC or rSRVCC. The following rSRVCC combinations are specified (system A to system B):

- GSM to LTE
- WCDMA to LTE.

The solution is based on the principle that IMS is kept as the system serving the user for the complete duration of the call (it is the "service engine" for the voice call), and also when the user is served by system B. SRVCC includes interaction between the MME of the EPC core network and the MSC Server of the circuit-switched core network, as well as an IMS VCC Domain Transfer Function.

SRVCC is described in more detail in Chapter 11.

5.4 Circuit-Switched Fallback

Circuit-Switched Fallback (CSFB) is an alternative solution to using IMS and SRVCC in order to provide voice services to users of LTE. The fundamental differences are that IMS is not part of the solution, and in fact voice calls are never served over LTE at all. Instead, CSFB relies on a temporary inter-system change (aka Fallback) that moves the UE from LTE to a system with 2G and/or 3G radio access and where circuit-switched voice calls can be served.

The solution relies on the CSFB LTE terminals being "registered" not only in EPS but also in the circuit-switched domain when powered up and attaching to LTE. This dual-domain registration is handled by the network through an interaction between the MME and the MSC Server in the circuit-switched network domain.

There are then two use cases to consider – voice calls initiated by the mobile user or voice calls received by the mobile user:

1. If the user is to make a voice call, the terminal switches from LTE (system A) to a system with circuit-switched voice support (system B). Any packet-based services that happened to be active on the end-user device at this time are either handed over and continue to run in system B but may be on lower data speeds or are suspended until the voice call is terminated and the terminal switches back to LTE again and the packet services are resumed. Which of these cases applies will depend on the capabilities of system B.
2. If there is an incoming voice call to a user who is currently attached to LTE, the MSC Server requests paging in LTE for the specific user. This is done via the interface between the MSC Server and the MME. The terminal receives the page via LTE, and temporarily switches from LTE to system B, where the voice call is received. Once the voice call is terminated, the terminal may switch back to LTE.

The details of CSFB are described in Chapter 11.

5.5 Comparing MMTel/SRVCC and CSFB

The two approaches on how to offer voice services to LTE users are fundamentally different.

The main strengths of IMS/MMTel and SRVCC include:

• Allowing for simultaneous usage of high-speed packet services over LTE and voice calls.
• MMTel offers an enhanced experience to the end-user, enabling the addition of additional media components within the voice call itself.

The main strengths of CSFB include:

- No need to rely on deployment of IMS infrastructure and services before offering voice as a service to LTE users.
- Same feature and service set offered for voice services when in LTE access as when in a system supporting circuit-switched voice calls. The circuit-switched core network infrastructure can also be utilized for LTE users.

As discussed above, both approaches rely on the end-user device used for the voice call being capable of supporting access not only to LTE but also systems with presumed wider radio coverage (e.g. GSM), as well as the capability to execute circuit-switched voice calls.

It should also be noted that both solutions can be simultaneously supported in the same network, and it can be assumed that operators initially deploying CSFB may over time migrate towards MMTel/VoLTE solutions.

5.6 IMS Emergency Calls and Priority Services

To deploy a new first-line telephony service, operators need to ensure that the service complies with local regulations. The regulations vary across the globe but one of the basic requirements for a telephony service is support for basic emergency calls. In addition to the basic emergency calls an operator may need to support unauthenticated IMS emergency calls (e.g. IMS emergency calls from UEs without a SIM or without a valid subscription).

While emergency calls or emergency services typically refer to a citizen-to-authority communication, Priority Services is a service that allows representatives from authorities to communicate with each other during extraordinary situations where they may receive prioritized service access over other users. Priority services may, for example, be used in disaster situations to ensure that, say, the fire brigade, police, and medical staff can communicate even if the telephony networks are overloaded.

3GPP introduced basic IMS emergency calls in 3GPP Release 7 and introduced 3GPP radio access and EPC support for emergency bearer services during Release 9. 3GPP specifications have since then gradually added more and more features to support the regional requirements for IMS emergency calls such as support for other media and priority services. More details on these aspects are described in Chapter 11.

PART III

Key Concepts and Services

CHAPTER 6

Session Management and Mobility

6.1 IP Connectivity and Session Management

LTE and Enhanced UTRAN introduced significant improvements in the amount of bandwidth provided for wireless transmission using cellular technology, truly paving the way for mobile broadband. In order to provide support for the type of applications and traffic that these access networks would enable, it was necessary to prepare the core network. In conjunction with the evolution of the access networks provided with LTE, the convergence of technologies at the IMS level meant that it was now also possible to provide a common packet core with appropriate policy, security, charging, and mobility. This provides end-users with ubiquitous access to network services across different access networks and also provides them with session continuity across various access technologies. This section covers the IP connectivity and management of the sessions within EPC. A high-level view of the steps involved is shown in Figure 6.1.

6.1.1 The IP Connection

6.1.1.1 General

The most fundamental task of the EPC is to provide IP connectivity to the terminal for both data and voice services. This is perhaps an even more significant task than in GSM/WCDMA systems, where the circuit-switched domain is also available. With EPS, only the IP-based packet-switched domain is available. Even though there are different possibilities for interworking towards the circuit-switched domain, as described in Chapter 5, only the IP-based packet-switched domain is natively supported when using E-UTRAN.

IP connectivity will have certain properties and characteristics depending on the scenario and type of services that the user wants to access. First, IP connectivity will be provided towards a certain IP network. This would in many cases be the Internet, but it could also be a specific IP network where the telecom operator provides certain services, for example based on IMS. Then, IP connectivity would be provided using

EPC and 4G Packet Networks.
DOI: http://dx.doi.org/10.1016/B978-0-12-394595-2.00006-2

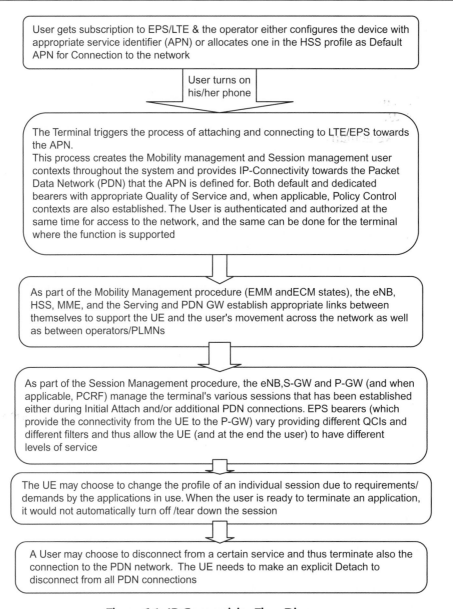

User gets subscription to EPS/LTE & the operator either configures the device with appropriate service identifier (APN) or allocates one in the HSS profile as Default APN for Connection to the network

User turns on his/her phone

The Terminal triggers the process of attaching and connecting to LTE/EPS towards the APN.
This process creates the Mobility management and Session management user contexts throughout the system and provides IP-Connectivity towards the Packet Data Network (PDN) that the APN is defined for. Both default and dedicated bearers with appropriate Quality of Service and, when applicable, Policy Control contexts are also established. The User is authenticated and authorized at the same time for access to the network, and the same can be done for the terminal where the function is supported

As part of the Mobility Management procedure (EMM andECM states), the eNB, HSS, MME, and the Serving and PDN GW establish appropriate links between themselves to support the UE and the user's movement across the network as well as between operators/PLMNs

As part of the Session Management procedure, the eNB,S-GW and P-GW (and when applicable, PCRF) manage the terminal's various sessions that has been established either during Initial Attach and/or additional PDN connections. EPS bearers (which provide the connectivity from the UE to the P-GW) vary providing different QCIs and different filters and thus allow the UE (and at the end the user) to have different levels of service

The UE may choose to change the profile of an individual session due to requirements/demands by the applications in use. When the user is ready to terminate an application, it would not automatically turn off /tear down the session

A User may choose to disconnect from a certain service and thus terminate also the connection to the PDN network. The UE needs to make an explicit Detach to disconnect from all PDN connections

Figure 6.1: IP Connectivity Flow Diagram.

one or both of the available IP versions, IPv4 and/or IPv6. Additionally, the IP connectivity should fulfill certain QoS requirements depending on the service being accessed. For example, the connection may need to provide a certain guaranteed bit rate or allow a prioritized treatment over other connections. The following sections discuss how the above concepts are solved in EPS, both for the 3GPP family of accesses and for other accesses connected to EPC.

6.1.1.2 The PDN Connectivity Service

As mentioned above, the EPS provides the user with connectivity to an IP network. In EPS, as well as in 2G/3G Packet Core, this IP network is called a "Packet Data Network" (PDN). In EPS, the IP connection to this PDN is called a "PDN connection". Why do we use fancy names such as "Packet Data Network" and "PDN connection"? Why not just simply say "IP network" and "IP connection"? After all, isn't that what we actually mean? There are two reasons for this terminology: the first is technical and the second is historical. First of all, the PDN connection comprises more than just basic IP access; QoS, charging, and mobility aspects are all parts of a PDN connection. Also, when GPRS was originally specified, there was a desire to support different Packet Data Protocol (PDP) types besides IP, such as Point-to-Point Protocol (PPP; with support for different network-layer protocols) and X.25. Since GPRS could provide access to PDN types other than just IP networks, it made sense at that time to refer to them as PDNs rather than just IP networks. In reality, however, IP-based PDNs became the most commonly used variant in the vast majority of deployments on the market. It has therefore been decided that EPS only supports access to IP-based PDNs using IPv4 and/or IPv6. The name "Packet Data Network", however, remains.

The operator may provide access to different PDNs with different services. One PDN could, for example, be the public Internet. If the user establishes a PDN connection to this "Internet PDN", the user can browse websites on the Internet or access other services available on the Internet. Another PDN could be a specific IP network set up by the telecom operator to provide operator-specific services, for example based on IMS. In summary, if the user establishes a PDN connection to a specific PDN, he or she would only get access to the services provided on that PDN. It is of course possible to provide multiple services in a single PDN. The operator chooses how to configure its networks and services.

A terminal may access a single PDN at a time or it may have multiple PDN connections open simultaneously, for example to access the Internet and the IMS services simultaneously if those services happen to be deployed on different PDNs. In the latter case, the terminal would have multiple IP addresses, one (or two if both IPv4 and IPv6 are used) for each PDN connection. Each PDN connection represents a unique IP connection, with its own IP address (or pair of IPv4 address and IPv6 prefix); see Figure 6.2.

One PDN connection is always established when the terminal attaches to the EPS (refer to Chapters 6 and 17 for a high-level description of the attachment procedure for the different accesses). During the attachment procedure, the terminal may provide information about the PDN that the user wants to access. The information is carried

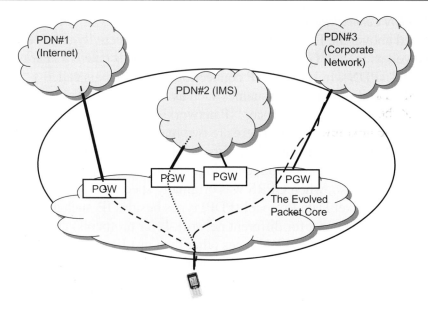

Figure 6.2: UE with Multiple Simultaneous PDN Connections.

in a parameter called "Access Point Name" (APN). The APN is a character string that contains a reference to the PDN where the desired services are available. The network uses the APN when selecting the PDN for which to set up the PDN connection; the operator defines what APNs (and corresponding PDNs) are available to a user as part of their subscription. If the terminal does not provide any APN during the attachment procedure, the network will use a default APN defined as part of the user's subscription profile in the HSS in order to establish the PDN connection. It should be noted that the APN is used to select not only PDN but also the PDN GW providing access to that PDN. These selection functions are described in more detail in Chapter 9.

Additional PDN connections may be established when the terminal is attached to EPS. In this case the terminal sends a request to the network to open a new PDN connection. The request must always contain an APN to inform the network about what PDN the user wants to access.

The terminal may at any time close a PDN connection.

6.1.1.3 Relation Between EPC, Application, and Transport Layers

The PDN connection provides the user with an IP connection to a PDN. When a PDN connection is established, context data representing the connection is created in the UE, the PDN GW and, depending on access technology used, also in other core network nodes in between, for example the MME and Serving GW for E-UTRAN access. The EPS is concerned with this "PDN connection layer" and associated functions such as IP address management, QoS, mobility, charging, security, policy control, etc.

The PDN connection is a logical connection between a specific IPv4 address and/or IPv6 prefix allocated to a UE and a particular PDN. The user data belonging to the PDN connection is transported between the terminal and the base station over the underlying radio connection. An underlying transport network between the network entities in the EPC also carries the user data. At the same time, the application (or service) that the user may be running is transported on top of the PDN connection. Here, and in the next few sections, we use the term "application" in a generic manner, including protocol layers on top of IP.

The transport network in the EPC provides IP transport that can be deployed using different technologies such as MPLS, Ethernet, wireless point-to-point links, etc. Over the radio interface, the PDN connection is transported on top of a radio connection between the UE and the base station. Figure 6.3 provides an illustration of the relation between application layer, PDN connection "layer", and transport layer. The IP transport layer entities in the backbone network, such as IP routers and layer 2 switches, are not aware of the PDN connections as such. In fact, these entities are typically not aware of per-user aspects at all. Instead, they operate on traffic aggregates and if any traffic differentiation is needed, it is typically based on Differentiated Services (DiffServ) and techniques operating on traffic aggregates.

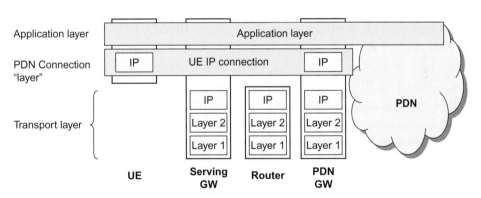

Figure 6.3: Schematic View of Application, PDN Connection, and Transport Layers for 3GPP Family of Accesses.

Figure 6.3 provides a schematic illustration of the application layer, PDN connection "layer", and transport layer for EPS when the UE is connected over a GERAN, UTRAN, or E-UTRAN access. It may be noted that the user IP connection (the PDN connection) is separate from the IP connection between the EPC nodes (the transport layer). This is a common feature in mobile networks where the user plane is tunneled over a transport network in order to provide per-user security, mobility, charging,

QoS, etc. This distinction between PDN connection and transport layer is also important when we consider aspects related to IP address allocation, QoS, etc. below.

Figure 6.3 also illustrates the different layers in the case where the UE is connected to EPC over other accesses. In general, an IP-based transport layer below the PDN connection layer is present in all cases. It should, however, be noted that the details regarding the protocol layers below the PDN connection layer depend on what mobility protocol is used and what other protocols are used when accessing over a non-3GPP access.

6.1.1.4 IP Addresses

A key task of the EPS is to provide IP connectivity between a UE and a PDN. The IP address allocated to the UE comes from the PDN where the UE is accessing. It should be noted that this IP address, and the IP address domain of the PDN, is different from the IP network (or backbone) that provides the IP transport between nodes within the EPC. The backbone providing the IP transport in the EPC can be a purely private IP network used solely for the transport of PDN connections and other IP-based signaling in the EPC, either for a single operator in non-roaming cases or between operators in roaming scenarios. The PDN is, however, an IP network where a user gains access and is provided services, for example the Internet. This section is only concerned with the IP addresses allocated to the UE.

Each PDN may provide services using IPv4 and/or IPv6. A PDN connection must thus provide connectivity using the appropriate IP version. Currently, the majority of the IP networks where end-users gain access, for example using GPRS or fixed broadband accesses, is based on IPv4. That is, the user will be assigned an IPv4 address and access IPv4-based services. Also, most services available on the Internet are IPv4 based. It is likely, however, that IPv6 deployments will become more common in the EPS/E-UTRAN time frame and it is thus important that EPS provides efficient support for both IPv4 and IPv6, for example to allow easy migration and coexistence.

Usage of IPv6 instead of IPv4 is primarily motivated by the vast number of IPv6 addresses available for allocation to devices and terminals. Some operators are already experiencing or will soon experience shortages of IPv4 addresses to varying degrees. The allocation of IPv4 addresses across the world and between organizations differs greatly. IPv6 does not have this problem since the addresses are 128 bits long, in theory providing 2^{128} addresses. In comparison to the 32 bits used for IPv4, IPv6 therefore provides significantly more addresses.

However, since the IP infrastructure and applications on both private networks and the Internet are still mostly based on IPv4, the introduction of IPv6 is a great challenge in terms of migration and smooth introduction. This is because IPv4 and IPv6 are not interoperable protocols; IPv6 implements a new packet header format, designed to reduce the amount of processing an IP header requires. Due to this fundamental difference in headers, workarounds are needed to enable them to function on the same network. Multiple mechanisms exist allowing, for example, devices using IPv6 to communicate with applications based on IPv4 (as they are on the Internet), as well as transporting IPv6 packets over IPv4 infrastructure. These solutions all have their pros and cons, but specific details are beyond the scope of this book. Interested readers are referred to the many excellent books on IPv6 available; some examples are Li et al. (2006) and Blanchet (2006), but many others exist.

Originally, in the 2G/3G core network, each PDN connection (i.e. PDP context) supported one IP address only. In order for a terminal to request both an IPv4 and an IPv6 address/prefix, it had to activate two PDN connections (two "primary" PDP contexts), one for each IP version. With EPS, this has changed. A terminal activating a PDN connection in EPS may request an IPv4 address, an IPv6 prefix, or both for that PDN connection. That is, EPS supports three types of PDN connections: IPv4 only, IPv6 only, as well as dual-stack IPv4/IPv6. The solution defined for EPS was also added to the 2G/3G GPRS architecture in Release 9. This means that support for dual-stack PDP context is now also available in the 2G/3G GPRS specifications.

EPS supports different ways to allocate an IP address. The IP address may be assigned using different protocols depending on the access used. The detailed procedure for allocating an IP address also depends on deployment aspects as well as the IP version (v4 or v6). This is explained in more detail in the following sections.

IP Address Allocation in 3GPP Accesses
The methods used to allocate IPv4 addresses and IPv6 prefixes are different. Below we will describe how IPv4 addresses and IPv6 prefixes are allocated in EPS. There are two main options for allocating an IPv4 address to the UE in 3GPP access:

1. One alternative is to assign the IPv4 address to the UE during the attach procedure (E-UTRAN) or PDP context activation procedure (GERAN/UTRAN). In this case the IPv4 address is sent to the UE as part of the attach accept message (E-UTRAN) or Activate PDP Context Accept message (GERAN/UTRAN). This is a 3GPP-specific method of assigning an IP address and this is the way it works in most 2G/3G networks. The terminal will also receive other parameters needed for the IP stack to function correctly (e.g. DNS address) during the attachment (E-UTRAN) or during PDP context activation (GERAN/UTRAN). These parameters are transferred in the so-called Protocol Configurations Options (PCO) field.

2. The other alternative is to use DHCPv4 (often referred to as just DHCP). In this case the UE does not receive an IPv4 address during attachment or PDP context

activation. Instead, the UE uses DHCPv4 to request an IP address after attachment (E-UTRAN) and PDP context activation (for GERAN/UTRAN) are completed. This method to allocate IP addresses is similar to how it works, for example, in Ethernet and WLAN networks, where terminals use DHCP after the basic layer 2 connectivity has been set up. When DHCP is used, the additional parameters (e.g. DNS address) are also sent to the UE as part of the DHCP procedure.

Whether alternative 1 or 2 is used in a network depends on what is requested by the UE, as well as what is supported and allowed by the network. It should be noted that both these alternatives are supported already in 2G/3G core network standards, even though alternative 1 is used in the vast majority of existing 2G/3G networks. One difference is, however, that in 2G/3G core networks, the selected IPv4 allocation method (alternative 1 or 2) is configured per APN. This means that only one method is supported for each APN. In EPS, however, this has been made more flexible to allow both methods to coexist for the same APN. It is still possible to deploy an EPS network so that only one method per APN is used, if desired.

We now proceed to the IP address allocation procedure for IPv6. The primary method supported in EPS is stateless IPv6 address auto configuration (SLAAC). In addition, a IPv6 prefix delegation (PD) using DHCPv6 was also added in Release 10. For more details on basic IPv6 features, please refer to Hagen et al. (2006). When IPv6 is used, a /64 IPv6 prefix is allocated for each PDN connection and UE using SLAAC. The UE can utilize the full prefix and can construct the IPv6 address by adding an Interface Identifier to the IPv6 prefix. Since the full /64 prefix is allocated to the UE and the prefix is not shared with any other node, the UE does not need to perform Duplicate Address Detection (DAD) to verify that no one else is using the same IPv6 address.

With stateless IPv6 address auto configuration, attachment and (for GERAN/UTRAN) PDP context activation are completed first. The GW then sends an IPv6 Router Advertisement (RA) to the UE after attachment and PDP context activation are completed. The RA contains the IPv6 prefix that is allocated to this PDN connection. The RA is sent over the already established PDN connection and is therefore sent only to a specific terminal. This is different compared to some non-3GPP access networks, where many terminals share the same layer 2 link (e.g. Ethernet). In these networks the RA is sent as a broadcast to all connected terminals. After completing the IPv6 stateless address auto configuration, the terminal can use stateless DHCPv6 to request other necessary parameters, for example DNS address. The option to allocate IPv6 prefix(es) using stateful DHCPv6 is currently not supported by EPS.

After the UE has been allocated a /64 prefix using SLAAC, the option to allocate additional prefix(es) to the UE is supported using IPv6 prefix delegation. This option is, for example, useful if the UE is in fact an IPv6 router and the network needs to configure this router with a prefix that is used in the site of the UE. One can see this delegation mechanism as a way to automate the process of configuring UEs operating as routers with appropriate prefixes at the end-user site. With IPv6 Prefix Delegation (PD), the UE first establishes a PDN connection and is configured with a /64 prefix using SLAAC, as described above. Later, the UE may ask for additional prefix(es) using the DHCPv6 protocol and IPv6 prefix options. The prefixes assigned using DHCPv6 PD may have shorter length than /64 and thus represent a larger network than the prefix assigned using SLAAC. One specific aspect of the DHCPv6 PD solution for EPC is that the PDN connection must still be associated with only a single IPv6 prefix, even if DHCPv6 is used. If PD is not used, this prefix would typically be a /64 prefix but if PD is supported, the prefix needs to be shorter than /64. When allocating prefixes to the UE using PD, those prefixes thus need to be taken from the shorter prefix associated with the PDN connection. The reason for associating each PDN connection with a single prefix is that the PCC system, charging systems, etc. currently require that each PDN connection is associated only with a single IPv6 prefix. It can be noted that IPv6 Prefix Delegation in E-UTRAN and other 3GPP accesses is currently only supported for the architecture variant using GTP-based S5/S8.

IP Address Allocation in Other Accesses

The way by which IPv4 addresses and/or IPv6 prefixes are assigned in other accesses differs depending on what access is used and what mobility protocol (GTPv2, PMIPv6, MIPv4, or DSMIPv6) is used.

When the terminal attaches from a trusted non-3GPP access and PMIPv6 or GTP is used on the S2a interface, the address allocation is quite similar to how it works in 3GPP accesses. (Note that support of GTP on S2a has been limited to the case when the non-3GPP access is WLAN.) An access may, for example, have specific means to deliver the IPv4 address to the UE, or DHCPv4 may be used. For IPv6, stateless IPv6 address auto configuration is typically supported. The IP layers are illustrated in Figure 6.4. For more details on the attachment and detachment procedures when using S2a, see Chapter 17.

When the terminal attaches in an untrusted non-3GPP access and PMIPv6 or GTP is used on the S2b interface, the terminal receives the IP address from the PDN during the IKEv2-based authentication with the ePDG. It may be noted that before IKEv2 is performed and the IPSec tunnel is set up, an additional IP address is involved. The reason is that the terminal needs local IP connectivity from the untrusted non-3GPP access in order to communicate with the ePDG. This local IP address, however, does not come from a PDN but is only used to set up the IPSec tunnel. This is illustrated in Figure 6.5. For more details on the attachment and detachment procedures when using S2b, see Chapter 17.

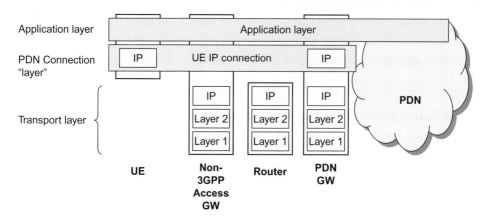

Figure 6.4: Schematic View of Application, PDN Connection, and Transport Layers for Trusted Non-3GPP Accesses When PMIP or GTP Is Used on S2a.

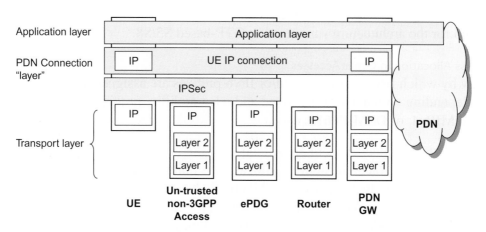

Figure 6.5: Schematic View of Application, PDN Connection, and Transport Layers for Untrusted Non-3GPP Accesses When PMIP or GTP Is Used on S2b.

From an IP address allocation point of view, the situation is somewhat similar when DSMIPv6 is used. The terminal receives its IP address (also referred to as Home Address in Mobile IP terminology) during DSMIPv6 bootstrapping with the PDN GW. However, the terminal first needs to acquire a local IP address to be used as Care-of Address. Therefore, the UE has two IP addresses, one for the local connection (Care-of Address) and one for the PDN connection (Home Address). This is illustrated in Figure 6.6. In addition to an IPv6 Home Address, a UE using DSMIPv6 in non-3GPP accesses may also ask for additional IPv6 prefix(es) using DHCPv6

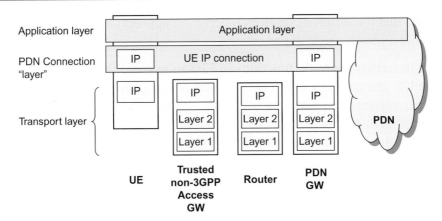

Figure 6.6: Schematic View of Application, PDN Connection, and Transport Layers for Trusted Non-3GPP Accesses When DSMIPv6 (S2c) Is Used.

Prefix Delegation signaling. For more details on the attachment and detachment procedures when using S2c, see Chapter 17.

The most involved case, which is not shown in the figure, is when DSMIPv6 is used over untrusted non-3GPP accesses. In this case the terminal uses three IP addresses: a local IP address to establish the IPSec tunnel towards an ePDG, the IP address received from ePDG which is used as a Care-of Address, and then finally the IP address for the PDN connection received during DSMIPv6 bootstrapping with the PDN GW.

For more details on DSMIPv6, see Sections 16.3, 17.6, and 17.7.

6.2 Session Management, Bearers, and QoS Aspects

6.2.1 General

Providing PDN connectivity is not just about getting an IP address; it is also about transporting the IP packets between the UE and the PDN in such a way that the user is provided with a good experience of the service being accessed. Depending on whether the service is a voice call using Voice-over-IP, a video streaming service, a file download, a chat application, etc., the QoS requirements for the IP packet transport are different. The services have different requirements on bit rates, delay, jitter, etc. Furthermore, since radio and transport network resources are limited and many users may share the same available bandwidth, efficient mechanisms must be available to partition the available (radio) resources between the applications and the users. The EPS needs to ensure that all these different service requirements are supported and that the different services receive the appropriate QoS treatment in order to enable a positive user experience.

This section describes the basic functions in EPS to manage the user-plane path between the UE and the PDN GW. One key task of the session management features is to provide a transmission path of a well-defined QoS. The basic principles around session management are covered in this chapter, but the more detailed QoS aspects can be found in the subsequent chapters on QoS.

In the subsections below we introduce the "EPS bearer", which is a central concept in E-UTRAN and EPS both for providing the IP connection as such and for enabling QoS. We also look at several aspects related to how GERAN and UTRAN accesses are connected to EPS, but without going into the same level of detail regarding session management for GERAN/UTRAN. Finally, we look at similar aspects for other accesses connected to EPS.

6.2.2 The EPS Bearer for E-UTRAN Access

For E-UTRAN access in EPS, one basic tool to handle QoS is the "EPS bearer". In fact, the PDN connectivity service described above is always provided by one or more EPS bearers (also denoted as "bearer" for simplicity). The EPS bearer provides a logical transport channel between the UE and the PDN for transporting IP traffic. Each EPS bearer is associated with a set of QoS parameters that describe the properties of the transport channel, for example bit rates, delay and bit error rate, scheduling policy in the radio base station, etc. All conforming traffic sent over the same EPS bearer will receive the same QoS treatment. In order to provide different QoS treatment to two IP packet flows, they need to be sent over different EPS bearers. All EPS bearers belonging to one PDN connection share the same UE IP address. The QoS aspects and their relation to the EPS bearers will be discussed in more detail in Section 8.1.

6.2.2.1 Default and Dedicated Bearers

A PDN connection has at least one EPS bearer but it may also have multiple EPS bearers in order to provide QoS differentiation to the transported IP traffic. The first EPS bearer that is activated when a PDN connection is established in LTE is called the "default bearer". This bearer remains established during the lifetime of the PDN connection. Even though it is possible to have an enhanced QoS for this bearer, in most cases the default bearer will be associated with a default type of QoS and will be used for IP traffic that does not require any specific QoS treatment. Additional EPS bearers that may be activated for a PDN connection are called "dedicated bearers". This type of bearer may be activated on demand, for example when an application is started that requires a specific guaranteed bit rate or prioritized scheduling. Since dedicated bearers are only set up when they are needed, they may also be deactivated when the need for them no longer exists, for example when an application that needs specific QoS treatment is no longer running.

6.2.2.2 User-Plane Aspects

The UE and the PDN GW (for GTP-based S5/S8) or Serving GW (for PMIP-based S5/S8) use packet filters to map IP traffic onto the different bearers. Each EPS bearer is associated with a so-called Traffic Flow Template (TFT) that includes the packet filters for the bearer. These TFTs may contain packet filters for uplink traffic (UL TFT) and/or downlink traffic (DL TFT). The TFTs are typically created when a new EPS bearer is established, and they can then be modified during the lifetime of the EPS bearer. For example, when a user starts a new service, the traffic filters corresponding to that service can be added to the TFT of the EPS bearer that will carry the user plane for the service session. The filter content may come either from the UE or from the PCRF (see Section 8.2 on PCC for more details).

The TFTs contain packet filter information that allows the UE and PDN GW to identify the packets belonging to a certain IP packet flow aggregate. This packet filter information is typically an IP 5-tuple defining the source and destination IP addresses, source and destination port, as well as protocol identifier (e.g. UDP or TCP). It is also possible to define other types of packet filters based on other parameters related to an IP flow. The filter information may contain the following attributes:

- Remote IP Address and Subnet Mask
- Protocol Number (IPv4)/Next Header (IPv6)
- Local Address and Mask (introduced in Release 11)
- Local Port Range
- Remote Port Range
- IPSec Security Parameter Index (SPI)
- Type of Service (TOS) (IPv4)/Traffic Class (IPv6)
- Flow Label (IPv6).

The word "remote" refers to the entity on the external PDN with which the UE is communicating, while "local" refers to the UE itself. The UE IP address was, until Release 11, not included in the TFT since it was understood that the UE is only assigned a single IP address, or possibly a single IP address of each IP version, per PDN connection. The local UE IP address would not be needed in the TFT. This is true for IPv4, but with IPv6, as we described above, the UE is assigned a full /64 prefix and can configure any IPv6 address from that prefix. In addition to that, with IPv6 prefix delegation, as supported from Release 10, a UE can request additional (and smaller than /64) prefixes. If a UE acts as a router for other devices in a local network, these other devices would use IPv6 addresses from the prefix(es) assigned by EPC. In order for the UE to provide correct mapping of uplink traffic onto bearers there is in this case a need to also include the specific UE IP address used by the IP flows. Therefore, the local IP address and mask was added in Release 11.

Some of the above-listed attributes may coexist in a packet filter while others mutually exclude each other. Table 6.1 lists the different packet filter attributes and possible combinations. Each packet filter in a TFT is associated with a precedence value that determines in which order the filters shall be tested for a match.

Table 6.1: Valid Packet Filter Attribute Combinations

Packet Filter Attribute	Valid Combination Types		
	I	II	III
Remote address and subnet mask	X	X	X
Protocol number (IPv4)/next header (IPv6)	X	X	
Local address and mask	X	X	X
Local port range	X		
Remote port range	X		
IPSec SPI		X	
TOS (IPv4)/traffic class (IPv6) and mask	X	X	X
Flow label (IPv6)			X

An example of how a TFT is used could be that the UE starts an application that connects to a media server in the PDN. For this service session, a new EPS bearer may be set up with the appropriate QoS parameters and bit rates. At the same time, packet filters are installed in the UE and the PDN GW that directs all traffic for the corresponding media onto that newly established EPS bearer. The Policy and Charging Control (PCC) system may be used at service establishment to ensure that the right QoS and TFT are provided.

When an EPS bearer is established, a bearer context is created in all EPS nodes that need to handle the user plane and identify each bearer. For E-UTRAN and a GTP-based S5/S8 interface between Serving GW and PDN GW, the UE, eNodeB, MME, Serving GW, and PDN GW will all have bearer context. The exact details of the bearer context will differ somewhat between the nodes since the same bearer parameters are not relevant in all nodes. Furthermore, as will be seen further below, when PMIP-based S5/S8 is used, the PDN GW will not be aware of the EPS bearers.

Between the core network nodes in EPC, the user-plane traffic belonging to a bearer is transported using an encapsulation header (tunnel header) that identifies the bearer. The encapsulation protocol is GTP-U. When E-UTRAN is used, GTP-U is used on S1-U and can also be used on S5/S8. The other alternative, to use PMIP on S5/S8, is described further below. The GTP-U header contains a field that allows the receiving node to identify the bearer the packet belongs to. Figure 6.7 illustrates two EPS bearers in a GTP-based system. A user-plane packet encapsulated using GTP-U is illustrated in Figure 6.8. For more information on GTP, see Section 16.2.

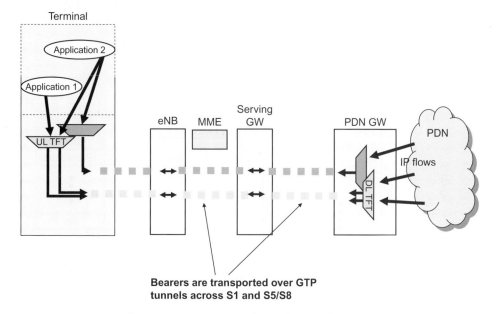

Figure 6.7: EPS Bearer for GTP-Based System.

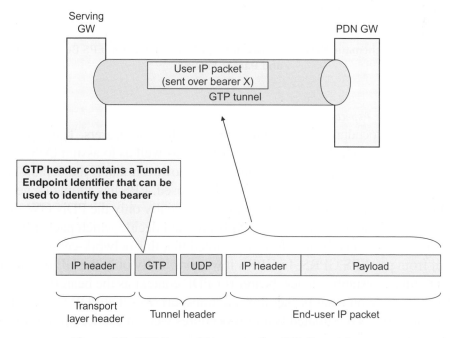

Figure 6.8: EPS Bearer Transport for GTP-Based System.

6.2.2.3 PDN Connections, EPS Bearers, TFTs, and Packet Filters – Bringing It All Together

In this chapter we have so far described several different concepts used in EPS and E-UTRAN to provide IP connectivity to a PDN and provide appropriate packet transport: PDN connections, EPS bearers, TFTs, and packet filters. Before going into more detail on bearer procedures in EPS, it may be useful to see how they all relate to each other. Figure 6.9 provides an illustration of how the UE, PDN connection, EPS bearer, TFT, and packet filters within the TFT relate to each other.

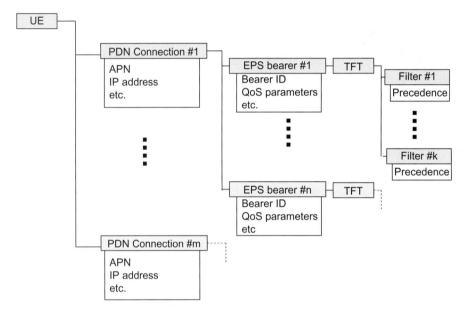

Figure 6.9: Schematic Relation Between UE, PDN Connection, EPS Bearer, TFT, and Packet Filters.

6.2.2.4 Control-Plane Aspects

There are several procedures available in EPS to control the bearers. These procedures are used to activate, modify, and deactivate bearers, as well as to assign QoS parameters, packet filters, etc., to the bearer. Note, however, that if the default bearer is deactivated the whole PDN connection will be closed. EPS has adopted a network-centric QoS control paradigm, meaning that it is basically only the PDN GW that can activate, modify, and deactivate an EPS bearer and decide which packet flows are transported over which bearer. It may be noted that this network-centric approach is different from pre-EPS GPRS. In GPRS it was originally only the UE that would take the initiative to establish a new bearer (or PDP context as the bearers are called in GPRS) and decide on what packet flows to transport over that PDP context. In 3GPP Release 7, the NW-initiated bearer procedures were introduced in GPRS by specifying a new procedure with the long name "network-requested secondary PDP context activation procedure". In this procedure it is the GGSN that takes the initiative to create a "dedicated bearer", known as secondary PDP context in 2G/3G packet cores, as well as to assign packet filters. The move towards a network-centric approach has been taken one step further with EPS, since it is now only the PDN GW that can activate a new bearer and decide which packet flows are transported over which bearer. For more information on the network-centric QoS control paradigm, see Section 8.1.

It should be noted that when an EPS bearer is established or modified, the state in the radio access may also be modified to provide an appropriate radio layer transport for each active EPS bearer. More information about this can be found in Dahlman et al. (2011) and Section 8.1.

6.2.2.5 Bearers in PMIP- and GTP-Based Deployments

In the illustration for bearers in GTP-based systems above, the EPS bearers extend between the UE and the PDN GW. This is how it works when GTP is used between Serving GW and PDN GW. As explained in Chapter 2, it is also possible to deploy EPS with PMIP between Serving GW and PDN GW. While GTP is designed to support all functionality required to handle the bearer signaling, as well as the user-plane transport, PMIP was designed by IETF to only handle functions for mobility and forwarding of the user plane. PMIP thus had no built-in features to bearers or QoS-related signaling. During 2007 there were long discussions in 3GPP whether PMIP should be extended to support bearer-related signaling, as well as to allow user-plane marking to identify the EPS bearers, similar to GTP. It was eventually decided, however, that the PMIP-based reference points would not be aware of the EPS bearers. This means that it is not possible for the bearers to extend all the way between UE and PDN GW. Instead, the bearers are only defined between UE and Serving GW when PMIP-based S5/S8 is used. Consequently, it is not sufficient to have the packet filters only in UE and PDN GW. Without bearer markings between S-GW and PDN GW, also the Serving GW would need to know the packet filters in order to map the downlink traffic onto the appropriate bearer towards the UE. This is illustrated in Figure 6.10.

The observant reader will have noticed from Figure 6.10 that the PDN GW still uses the packet filters, just as with the GTP-based S5/S8. Why is this so? Isn't it enough that the Serving GW has the packet filters in this case? It is true that the PDN GW does not need the packet filters to do the bearer mapping of downlink traffic, since this is instead done by the Serving GW when PMIP-based S5/S8 is used. The PDN GW, however, still performs important functionality such as bit rate enforcement and charging for the different IP flows. This is not directly related to the EPS bearers but it is the reason why the PDN GW also has packet filter knowledge for PMIP-based S5/S8. These functions of the PDN GW, common to both PMIP-based S5/S8 and GTP-based S5/S8, are further described in Section 8.2 on PCC.

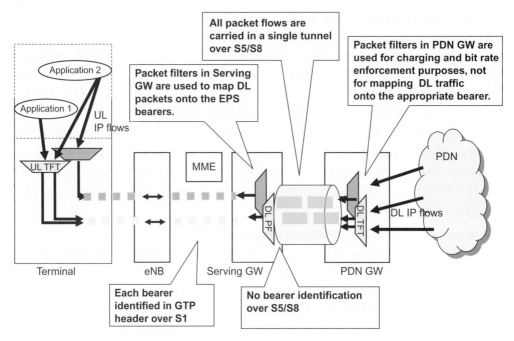

Figure 6.10: EPS Bearer When PMIP-based S5/S8 Is Used.

6.2.3 Session Management for EPS and GERAN/UTRAN Accesses

The 2G/3G Core Network uses the concept of PDP contexts to provide PDN connectivity and QoS management in the core network. The PDP context is defined between the UE and the GGSN and defines all the information used for a certain connection, including PDP address, QoS class, etc. The PDP context corresponds to EPS bearers for E-UTRAN.

The PDP context concept is partially maintained also when 2G/3G accesses are connected to EPC. In principle it would have been possible to replace the PDP context procedures for 2G/3G with the EPS bearer procedures, but it was considered preferable to maintain the PDP context procedures, at least between UE and the SGSN, in order to limit the impact on the UE. Within the EPC, however, the EPS bearer procedures are also used when the UE is in 2G/3G access. The SGSN provides the mapping between PDP context and EPS bearer procedures, and maintains a one-to-one mapping between PDP contexts and EPS bearers. There are several reasons why this architecture is preferable:

- By using PDP context procedures between UE and SGSN, the UE can use similar ways to connect when the 2G/3G access connects to EPC as when it connects to the GPRS architecture.

- By using the EPS bearer in the EPC also for 2G/3G access, it is easier for the PDN GW to handle mobility between E-UTRAN and 2G/3G. Since there is a one-to-one mapping between PDP context and EPS bearer, the handover between 2G/3G access and E-UTRAN access is simplified.

The interfaces where PDP context procedures are used and where EPS bearer procedures are used respectively are illustrated in Figure 6.11.

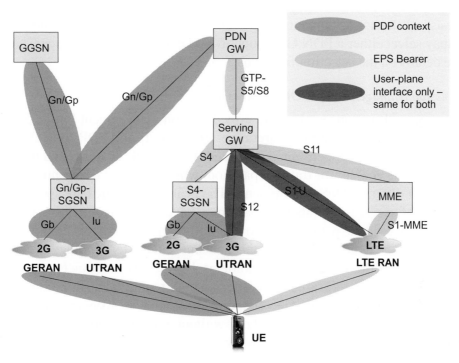

Figure 6.11: Usage of PDP Context Procedures and EPS Bearer Procedures for EPS as Well as for EPS Interworking with Gn/Gp-Based SGSNs.

An SGSN using S4 maps the UE-initiated PDP context procedures (activation/ modification/deactivation) over GERAN/UTRAN into corresponding EPS bearer procedures towards the Serving GW. One aspect that complicates this mapping is the fact that PDP contexts are controlled by the UE while EPS bearers are controlled by the NW. One consequence is that the PDN GW needs to be aware that the UE is using 2G/3G and adjust its behavior accordingly. For example, when the UE is using GERAN/UTRAN and has requested activation of a secondary PDP context, the PDN GW must activate a new EPS bearer corresponding to the PDP context. This is not the case if the UE is using E-UTRAN. In this case the UE cannot directly request activation of an EPS bearer, the UE can only make a request for certain resources, and the PDN GW can decide whether or not to activate a new EPS bearer or modify an existing EPS bearer. The signaling towards the UE does not change when SGSN is in the Gn/Gp or S4 mode of operation, i.e. support of Gn/Gp or S4 architecture variant is invisible to the UE. S4-SGSN architecture enables the common UTRAN/LTE network.

The SGSN also needs to map parameters provided from the UE (e.g. QoS parameters defined according to pre-Release 8 GPRS) into corresponding EPS parameters towards the Serving GW. See Section 8.1 for further details on QoS.

Another aspect that is worth mentioning in this context is that an SGSN supporting Gn/Gp may select either a PDN GW (supporting Gn/Gp) or a GGSN. See Chapter 9 for more details on the selection functions.

6.2.4 Session Management for Other Accesses

In the previous subsections we have described session management for the 3GPP family of accesses and the use of bearers (i.e. EPS bearers and PDP contexts) to handle the user-plane path between the UE and the network. It was found that the bearer is the basic enabler for traffic separation, which provides differential treatment for traffic with differing QoS requirements. The bearer procedures are specific to the 3GPP family of accesses but other accesses may have similar features and procedures to manage the user-plane path and to provide traffic separation between different types of traffic. The details of the QoS mechanisms and the terminology used may differ between the different access technologies, but the key function to provide differentiated treatment for traffic with differing QoS requirements is common to all accesses that are able to provide QoS support. There are also accesses that basically only support best effort delivery of packets, without any differentiated treatment. Most vanilla IEEE 802.11b WLAN networks fall into this category.

We will not describe access-specific QoS "bearer" capabilities and procedures that may be supported by the different accesses outside the 3GPP family of accesses. We will, however, look at how QoS is managed using the PCC architecture when these accesses interwork with EPS as described in Section 8.2.

6.3 Subscriber Identifiers and Corresponding Legacy Identities

Permanent and temporary subscriber identities are constructed to identify not only a particular subscriber, but also the network entities where the permanent and temporary subscriber records are stored.

6.3.1 Permanent Subscriber Identifiers

Subscriptions are identified with an IMSI (International Mobile Subscriber Identity). Each subscription is assigned a unique IMSI. The IMSI is an E.164 number (basically a string of digits like a phone number) with a maximum length of 15 digits. The IMSI is constructed by an MCC (Mobile Country Code), an MNC (Mobile Network Code), and an MSIN (Mobile Subscriber Identity) (Figure 6.12).

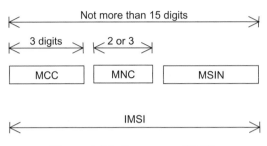

Figure 6.12: Structure of IMSI.

The MCC identifies the country and the MNC identifies the network within the country. The MSIN in turn is a unique number for each subscriber within a particular network.

The IMSI is the permanent subscription identifier and it is used as a master key in the subscriber database (HSS). The IMSI is also stored in the USIM (an application running on the smartcard provided by the operator). By its construction the IMSI allows any network in the world to find the home operator of the subscriber; specifically it provides a mechanism to find the HSS in the home operator network.

The IMSI is also used in 2G/3G networks and there is no change of the purpose or format of the IMSI for SAE/LTE.

6.3.2 Temporary Subscriber Identifiers

Temporary subscriber identifiers are used for several purposes. They provide a level of privacy since the permanent identity does not need to be sent over the radio interface. But more importantly they provide a mechanism to find the resources where the subscriber's temporary context is stored. The temporary context for the subscriber is stored in an MME (or SGSN in the 2G/3G case) and, for example, the eNodeB needs to be able to send signaling from a UE to the correct MME where the subscriber's context resides.

Pooling of MMEs was an integral part of SAE/LTE design (as opposed to 2G/3G where pooling was a feature added a few years after the original design). Hence, the temporary identities in SAE/LTE could be designed with pooling in mind. This has resulted in a cleaner design of the temporary identities for SAE/LTE.

Figure 6.13 illustrates the temporary identifiers. The GUTI (Globally Unique Temporary ID) is a worldwide unique identity that points to a specific subscriber

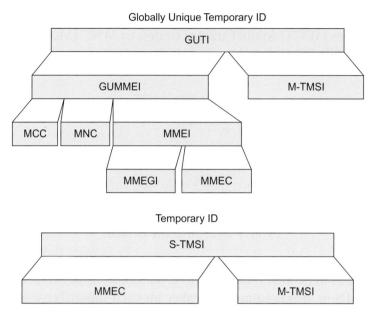

Figure 6.13: Structure of GUTI and S-TMSI.

context in a specific MME. The S-TMSI is unique within a particular area of a single network. The UE can use the S-TMSI when communicating with the network as long as it stays within a TA that is part of the TA list it has received.

The GUTI consists of two main components: (1) the GUMMEI (Globally Unique MME Identifier) that uniquely identifies the MME that allocated the GUTI; and (2) the M-TMSI (MME Temporary Subscriber Identity) that identifies the subscriber within the MME.

The GUMMEI is in turn constructed from MCC (country), MNC (network), and MMEI (MME identifier, the MME within the network). The MMEI is constructed from an MMEGI (MME Group ID) and an MMEC (MME Code).

The GUTI is a long identifier and to save radio resources, a shorter version of the GUTI is used whenever possible. The shorter version is called S-TMSI and it is unique only within a group of MMEs. It is constructed from the MMEC and the M-TMSI. The S-TMSI is used for, as an example, paging of UEs and service request.

One can imagine the different temporary identities, a set of pointers, to network resources in EPS (Figure 6.14).

6.3.3 Relation to Subscription Identifiers in 2G/3G

Why is there a need for a relation between the EPS and the 2G/3G identifiers? The main reason is mobility between GSM/WCDMA and LTE. When the UE moves from

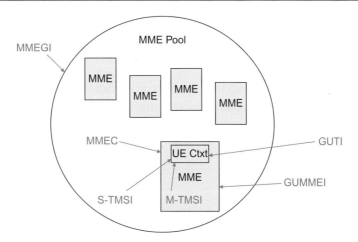

Figure 6.14: Identifiers as Pointers.

one access to the other, it should be possible to locate the node where the UE context is stored; for example, when moving to LTE from MME there is a need to find the UE context in the SGSN and vice versa. Another reason is that the implementation of combined SGSN and MME nodes will be available and it is preferable to map the temporary identifier used on LTE and the temporary identifier used on GSM/WCDMA so that they point to the same combined node.

The IMSI – the common permanent subscription identifier across GSM, UMTS, and EPS – and the same subscription can be used to access all technologies.

The temporary identities in 2G/3G look a little bit different, primarily since the original design of GSM/UMTS assumed a strictly hierarchical system where an RA was controlled by a single SGSN node. Hence the GSM/UMTS original design did not explicitly include any SGSN pool and SGSN node identifiers. When pooling was added in the GSM/UMTS system, these node identifiers were encoded inside the temporary identifier.

In GPRS the P-TMSI is the identifier that identifies the subscriber context inside an SGSN. The globally unique identifier in GPRS has no explicit name but by combining the RA ID (RAI) and the P-TMSI you get a globally unique identifier. The RAI is constructed by MCC, MNC, LAC, and RAC.

Pooling is discussed in another chapter but the basic principle for pooling in 2G/3G is that the P-TMSI range is divided among the SGSNs in a pool. There are a set of bits inside the P-TMSI called NRI (Network Resource Identifier) that point to specific SGSNs, and each SGSN in a pool is assigned one (or more) unique NRIs. The NRI in 2G/3G hence corresponds to the MME code and the NRI identifies an SGSN in a pool just as the MME code identifies an MME in an MME pool.

The operator needs to ensure that the MMEC is unique within the MME pool area, and if overlapping pool areas are in use, they should be unique within the area of overlapping MME pools.

The GUTI is used to support subscriber identity confidentiality, and in the shortened S-TMSI form to enable more efficient radio signaling procedures (e.g. paging and Service Request).

6.4 Mobility Principles

6.4.1 General

In the early days of GSM, the system supported a single radio access technology (GSM) and there was no mobility to/from other technologies. Since then 3GPP has developed WCDMA and LTE. 3GPP2 has developed CDMA (1xRTT and HRPD) and in addition other forums have developed access technologies, such as WLAN and fixed access, such as xDSL, Passive Optical Networks (PON), and Cable. With such a range of access technologies available to users to pick from and also for operators to select as their preferred system of choice, mobility has become quite complex. There is a need and desire to find a "common" set of tools allowing the end-user's devices to converge towards supporting a core set of mobility mechanisms.

With EPS, 3GPP aimed to provide not only a common core network for all access technologies, but also *mobility* between heterogeneous access technologies. EPS is the first complete realization of multi-access convergence: a packet core network that supports full mobility management, access network discovery, and selection for any type of access network.

In this section we will give an overview of the mobility functionality in EPS, starting from the mobility functionality for LTE, WCDMA, and GSM, then adding on HRPD, WLAN, and other access technologies.

Mobility is the core feature of mobile systems, and many of the major system design decisions for EPC are derived directly from the need to support mobility. The functionality of mobility management is required to ensure the following:

- That the network can "reach" the user, for example in order to notify the terminal about incoming calls
- That a user can initiate communication towards other users or services such as Internet access, and
- That ongoing sessions can be maintained as the user moves, within or between access technologies.

Associated functionality also ensures the authenticity and validity of the user's access to the system. It authenticates and authorizes the subscription and prepares the network and the user's device (i.e. the UE) with subscription information and security credentials.

6.4.2 Mobility within 3GPP Family of Accesses

6.4.2.1 Cellular Idle-Mode Mobility Management

Idle-mode mobility management in cellular systems like LTE, GSM/WCDMA, and CDMA is built on similar concepts. To be able to reach the UE, the network tracks the UE, or rather the UE updates the network about its location, on a regular basis. The radio networks are built by cells that range in size from tens and hundreds of meters to tens of kilometers. It would not be practical and would cause a lot of signaling to keep track of a UE in idle mode every time it moves between different cells. It is also not practical to search for the UE in the whole network for every terminating event (e.g. an incoming call). Hence the cells are grouped together into "registration areas" (Figure 6.15).

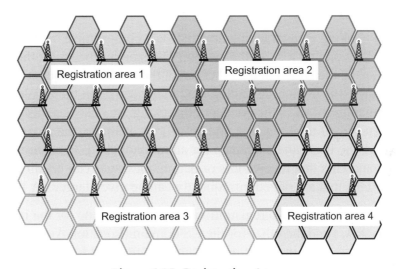

Figure 6.15: Registration Areas.

The base stations broadcast registration area information and the UE compares the broadcasted registration area information with the information it has previously stored. If the registration area information does not match the information stored in the UE, it starts an update procedure towards the network to inform that it is now in a different registration area.

For example, when a UE that was previously in registration area 1 moves into a cell in registration area 2, it will notice that the broadcast information includes a different registration area identity. This difference in the stored and broadcast information triggers the UE to perform a registration update procedure towards the NW. In the registration update procedure, the UE informs the network about the new registration area it has entered. Once the network has accepted the registration update, the UE will store the new registration area.

In EPS the registration areas are called Tracking Areas (TAs). In order to distribute the registration update signaling, the concept of tracking area lists was introduced in EPS. The concept allows a UE to belong to a list of different TAs. Different UEs can be allocated to different lists of tracking areas. As long as the UE moves within its list of allocated TAs, it does not have to perform a tracking area update. By allocating different lists of tracking areas to different UEs, the operator can give UEs different registration area borders and so reduce peaks in registration update signaling, for example when a train passes a TA border.

In addition to the registration updates performed when passing a border to a TA where the UE is not registered, there is also a concept of periodic updates. Periodic updates are used to clear resources in the network for UEs that are out of coverage or have been turned off.

The size of the tracking areas/tracking area lists is a compromise between registration update load and the paging load in the system. The smaller the areas, the fewer the cells needed to page the UEs. On the other hand, there will be frequent TA updates. The larger the area, the higher the paging load in the cells, but there is less tracking area update signaling. In LTE, the concept of tracking area lists can also be used to reduce the frequency of tracking area updates. If, for example, the movement of UEs can be predicted, the lists can be adapted for an individual UE to ensure that they pass fewer borders, and UEs that receive lots of paging messages can be allocated smaller TA lists, while UEs that are paged infrequently can be given larger TA lists.

In GSM/WCDMA there are two registration area concepts: one for the PS domain (Routing Areas, RAs) and the other for the CS domain (Location Areas, LAs). GSM and WCDMA cells may be included in the same Routing and Location Areas, allowing the UE to move between technologies without performing Routing Area Updates (RAUs). The Routing Areas are a subset of the Location Areas and can only contain cells from the same LA. There is no support for lists of Routing or Location Areas in GSM/WCDMA and hence all UEs share the same RA/LA borders. There is, however, another optimization that has been introduced in GSM/WCDMA, since the RA is a subset of the LA; the UE can perform combined RAU/LAU where the UE is tracked on an RA basis. Since the RA is a subset of the LA, the network also knows in which LA the UE is. The UE can hence perform combined updates when crossing the RA borders and no extra LAU procedures are needed. This optimization does, however, require support in both the UE and the NW. The combined update procedure is gaining momentum in GSM deployments, while so far it has not been broadly deployed in WCDMA networks. The registration areas concept are summarised in Table 6.2.

Table 6.2: Registration Area Representation for 3GPP Radio Accesses

Generic Concept	EPS	GSM/WCDMA GPRS	GSM/WCDMA CS
Registration area	List of tracking areas (TA list)	Routing Area (RA)	Location Area (LA)
Registration area update procedure	TA Update procedure	RA Update procedure	LA Update procedure

A summary of the idle mobility procedure in EPS is:

• A TA consists of a set of cells
• The registration area in EPS is a list of one or more TAs
• The UE performs TA Update when moving outside its TA list
• The UE also performs TA Update when the periodic TA Update timer expires. An outline of the Tracking Area Update procedure is shown in Figure 6.16 and it contains the following steps:

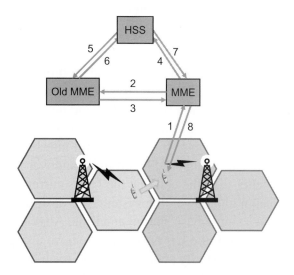

Figure 6.16: TAU Procedure.

1. When the UE reselects a new cell and realizes that the broadcast TA ID is not in their list of TAs, the UE initiates a TAU procedure to the network. The first action is to send a TA update message to the MME.
2. Upon receipt of the TA message from the UE, the MME checks if a context for that particular UE is available; if not it checks the UE's temporary identity to determine which node keeps the UE context. Once this is determined the MME asks the old MME for the UE context.
3. The old MME transfers the UE context to the new MME.

4. Once the MME has received the old context, it informs the HSS that the UE context has now moved to a new MME.

5–6. The HSS cancels the UE context in the old MME.

7. The HSS acknowledges the new MME and inserts new subscriber data in the MME.

8. The MME informs the UE that the TAU was successful and as the MME was changed it supplies a new GUTI (where the MME code points back to the new MME).

6.4.2.2 Paging

Paging is used to search for Idle UEs and establish a signaling connection. Paging is, for example, triggered by downlink packets arriving to the Serving GW. When the Serving GW receives a downlink packet destined for an Idle UE, it does not have an eNodeB address to which it can send the packet. The Serving GW instead informs the MME that a downlink packet has arrived. The MME knows in which TA the UE is roaming and it sends a paging request to the eNodeBs within the TA lists. Upon receipt of the paging message, the UE responds to the MME and the bearers are activated so that the downlink packet may be forwarded to the UE.

6.4.2.3 Cellular Active-Mode Mobility

Great effort has been put into optimized active-mode mobility for cellular systems. The basic concept is somewhat similar across the different technologies with some variations in the functional distribution between UE and networks. While in active mode, the UE has an active signaling connection and one or more active bearers, and data transmission may be ongoing. In order to limit interference and provide the UE with a good bearer, the UE changes cells through handover when there is a cell that is considered to be better than the cell that the UE is currently using. To save on complexity in the UE design and power, the systems are designed to ensure that the UE only needs to listen to a single base station at a time. Also, for inter-RAT handover (e.g. E-UTRAN to UTRAN HO) the UE only needs to have a single radio technology active at a time. It may need to rapidly switch back and forth between the different technologies, but at any instance of time only one of the radio technologies is active.

To determine when to perform handover the UE measures the signal strength on neighboring cells regularly or when instructed by the network. As the UE cannot send or receive data at the same time as it measures neighboring cells, it receives instruction from the network on suitable neighboring cells that are available and on which the UE should measure. The network (eNodeB) creates measurement time gaps where no data is sent or received to/from the UE. The measurement gaps are used by the UE to tune the receiver to other cells and measure the signal strength. If the signal strength is significantly stronger on another cell, the handover procedure may be initiated.

In E-UTRAN the eNodeBs can perform direct handover via the direct interface (known as X2 interface) between eNodeBs. In the X2-based HO procedure, the source eNodeB and the target eNodeB prepare and execute the HO procedure. At the end of the HO execution, the target eNodeB requests the MME to switch the downlink data path from the source eNodeB to the target eNodeB. The MME in turn requests the Serving GW to switch the data path towards the new eNodeB.

If downlink packets are sent before the Serving GW has switched the path towards the new eNodeB, the source eNodeB will forward the packet over the X2 interface.

If the X2 interface is not available between eNodeBs, the eNodeB can initiate a handover involving signaling via the core network. This is called S1-based handover. The S1-based HO procedure sends the signal via the MME and may include change of MME and/or SGW.

6.4.3 Idle-Mode Signaling Reduction (ISR)

Idle-mode signaling reduction is a feature that allows the UE to move between LTE and 2G/3G without performing Tracking Area (TA) or Routing Area (RA) updates once ISR has been activated. As idle-mode mobility between RATs may be rather common, especially in deployments with spotty coverage, ISR can be used to limit the signaling between the UE and the network, as well as signaling within the network.

In 2G/3G there is also similar functionality that allows the UE to move between GERAN and UTRAN cells in idle mode without performing any signaling. This is implemented by using a common SGSN and common RAs; that is, the GERAN and UTRAN cells belong to the same RA and can be paged for terminating events in both GERAN and UTRAN cells that belong to the RA.

It would have been possible to use a similar concept for SAE but that would have required a combined SGSN/MME node. Due to this difference in architecture (SGSN vs. MME) and the area concepts (RAs vs. list of TAs), it was decided to implement the ISR functionality in a different way between E-UTRAN and GERAN/UTRAN than between GERAN and UTRAN.

The ISR feature enables signaling reduction with separate SGSN and MME, and also with independent TAs and RAs. The dependency between EPC and 2G/3G is minimized at the cost of ISR-specific node and interface functionality.

The idea behind the ISR feature is that the UE can be registered in a GERAN/UTRAN RA at the same time as it is registered in an E-UTRAN TA (or list of TAs). The UE keeps the two registrations in parallel and runs periodic timers for both registrations independently. Similarly, the network keeps both registrations in parallel and it also ensures that the UE can be paged in both the RA and the TA(s) it is registered in.

6.4.3.1 ISR Activation

A prerequisite for ISR activation is that the UE, SGSN, MME, Serving GW, and HSS support ISR. The ISR support is specified to be mandatory for the UE but optional for network entities. ISR also requires an S4-SGSN and is not supported with a Gn/Gp-SGSN.

On the first attachment to the network, ISR is not activated. ISR can only be activated when the UE has first been registered in an RA on 2G/3G and then registers in a TA or vice versa.

If the UE first registers on GERAN/UTRAN and then moves into an LTE cell, the UE will initiate a TA update procedure. In the TA update procedure, the SGSN, MME, and Serving GW will communicate their capabilities to support ISR, and if all nodes support ISR, the MME will indicate to the UE that ISR is activated in the TAU accept message.

A simplified example of a TA update procedure with ISR activation is shown in Figure 6.17. Before the TAU procedure is performed, the UE has attached in GERAN/UTRAN and has an active registration with an SGSN. There is hence an active MM context in UE, SGSN, and HSS, and the SGSN has a control connection with the Serving GW:

1. The UE initiates a TAU procedure.
2. The MME requests the UE context and indicates to the SGSN that it is ISR capable. The SGSN responds with the UE context and indicates that it supports ISR.
3. The Serving GW is informed that the UE is registered with the MME and that ISR is activated.

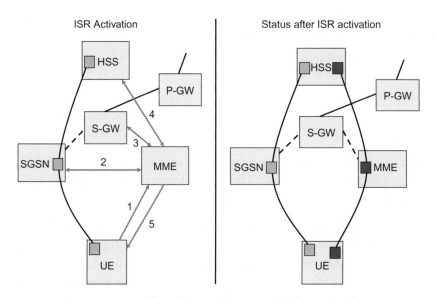

Figure 6.17: Outline of ISR and ISR Activation Procedure.

4. The HSS may be updated with the MME address. The update type indicates that HSS will not cancel the SGSN location.
5. The MME informs the UE that the TAU procedure was successful and that ISR is activated.

As shown in Figure 6.17, when ISR is activated, the UE is registered with both MME and SGSN. Both the SGSN and the MME have a control connection with the Serving GW, and the MME and SGSN are both registered in HSS. The UE stores MM parameters from SGSN (e.g. P-TMSI and RA) and from MME (e.g. GUTI and TA(s)) and the UE stores session management (bearer) contexts that are common for E-UTRAN and GERAN/UTRAN accesses. SGSN and MME store each other's address when ISR is activated.

When ISR is activated, a UE in idle state can reselect between E-UTRAN and GERAN/UTRAN (within the registered RAs and TAs) without any need to signal to the network.

6.4.3.2 Paging
When the UE is in idle mode and ISR is active and downlink data arrives in the Serving GW, it will send a downlink data notification to both the MME and the SGSN. The MME will then initiate paging in the TA where the UE is registered, and the SGSN will initiate paging in the RA where the UE is registered. When the UE receives the paging message, it will perform a service request procedure on the RAT it is currently camping on. As part of the service request procedure, the Serving GW will be requested to establish a downlink data connection towards an eNodeB if the UE was camping on E-UTRAN and towards the SGSN/RNC if the UE was camping on GERAN/UTRAN. When the Serving GW receives this request to establish the downlink data connection, it will also inform the SGSN or MME to stop paging on the other RAT.

A simplified example of the paging procedure, when ISR is active and the UE camps on E-UTRAN, is outlined in Figure 6.18.

6.4.3.3 ISR Deactivation
ISR activation has to be refreshed at every RAU and TAU procedure and the UE will deactivate ISR if it does not receive ISR active indication in the RAU accept and TAU accept messages.

The UE and the network run independent periodic update timers for GERAN/UTRAN and for E-UTRAN. The UE will perform an RA update if it camps on GERAN/UTRAN when the periodic RAU timer expires and it will perform a TA update if it camps on E-UTRAN when the periodic TAU timer expires. If the UE camps on a different RAT when the periodic timers expires, it will not perform the

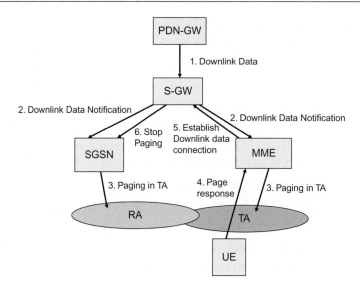

Figure 6.18: Simplified Example of the Paging Procedure When ISR Is Active.

update procedure; for example, if it camps on E-UTRAN when the periodic RAU timer expires, it will remain camping on E-UTRAN and not perform an RAU update.

When the MME or SGSN do not receive periodic updates, the MME and SGSN may decide to implicitly detach the UE. The implicit detachment removes session management (bearer) contexts from the node performing the implicit detachment and it also removes the related control connection from the Serving GW. Implicit detachment by one CN node (SGSN or MME) deactivates ISR in the network.

When the UE cannot perform periodic updates in time, it starts a Deactivate ISR timer. When this timer expires and the UE is not able to perform the required update procedure, the UE locally deactivates ISR.

ISR will be deactivated by MME or SGSN, which is done by omitting the signaling of "ISR activation" when there is:

• A CN node change resulting in context transfer between the same type of CN nodes (SGSN to SGSN or MME to MME)
• A Serving GW change.

There are also situations where the UE needs to deactivate ISR locally. For example:

• Modification or activation of additional bearers
• After updating either MME or SGSN about the change of the DRX parameters or UE capabilities
• E-UTRAN selection by a UTRAN-connected UE (e.g. when in URA_PCH to release Iu on the UTRAN side).

6.4.4 *Closed Subscriber Group*

The Closed Subscriber Group (CSG) allows for control of access by the UE to a cell or a group of cells identified by a CSG Identifier within a PLMN. For the HeNB subsystem, CSG ID is mandatory for the operation of functions as it controls the UE's access to the HeNB during Mobility procedures (Attach, Service Request, TAU, etc.). There are three modes of operation: Closed, Hybrid, and Open access modes.

- When the HeNB is configured for Open access mode, HeNB is able to provide services to subscribers of any PLMN, subject to roaming agreement.
- When the HeNB is configured for Hybrid access mode, the HeNB is able to provide services to its associated CSG members, and subscribers of any PLMN not belonging to its associated CSG, subject to roaming agreement.
- When the HeNB is configured for Closed access mode, only users that belong to its associated CSG are able to obtain services from it.

CSG information is downloaded to MME as part of the subscription data from HSS when provided by a CSG manager for a subscriber. The CSG subscription data in HSS includes the following information:

- The CSG Subscription Data is a list of CSG IDs (up to 50) per PLMN and for each CSG ID optionally an associated expiration date indicates the point in time when the subscription to the CSG ID expires; an absent expiration date indicates unlimited subscription.
- For a CSG ID that can be used to access specific PDNs via Local IP Access, the CSG ID entry includes the corresponding APN(s).

The Local IP Access information is used to indicate support for the feature LIPA, as described in Chapter 14.

The CSG mode of operation is enforced in HeNB and MME via different functions and these are specified in 3GPP TS 23.401 and TS 36.300:

- CSG subscription handling function stores and updates the user's CSG subscription data at the UE and the network (e.g. HSS, CSS and MME/SGSN).
- For Closed mode, CSG access control function ensures a UE has valid subscription at a CSG where it performs an access or a handover.
- As part of access control, MME checks that the CSG ID for the CSG cell is part of the user's subscription data and that the expiration timer, if present, is valid before allowing access. If the subscription is not valid or the timer has expired, then the request is rejected with an appropriate error code (i.e. Not authorised for this CSG). This error code then would trigger an action in the UE that it needs to remove the CSD ID from its Allowed CSG List if it is present in the list.

- Admission and rate control function is used to provide different admission and rate control for CSG and non-CSG members for a hybrid CSG cell, as explained above.
- A paging optimization function is optionally used to filter paging messages based on TAI List, user's CSG subscription data, and CSG access mode in order to avoid paging at CSG cells where the UE is not allowed access.
- The VPLMN Autonomous CSG roaming function is optionally supported whereby a VPLMN, if allowed by the HPLMN, stores and manages VPLMN-specific CSG subscription information for roaming UEs without interaction with the HSS. Note that a HPLMN operator may explicitly request disallowing of this function for its subscribers. The CSG Subscriber Server (CSS) in the VPLMN stores the necessary information for a roaming user's CSG access and MME/SGSN verifies when this function is supported in the VPLMN, for each roaming subscriber requesting access in a CSG where the function is supported and allowed. This function is available from Release 11 onwards.

CSG provision by a CSG manager is done by managing the list (e.g. add/modify/remove and handle supervision of length of time allowed access) of subscribers per CSG and supporting functions on how to store/manage CSG information in the UE and in the network. By handling all CSG members for a specific CSG ID in a PLMN via a single CSG list, management of the subscribers is greatly improved. In the UE, the Allowed and Operator CSG lists are installed/provided via OMA Device Management as specified in 3GPP TS 24.285 or Over The Air (OTA) Activation procedures as specified in 3GPP TS 31.102 via the CSG List Server, which ensures access to the authorized CSG by a UE. Figure 6.19 shows the interfaces and relationship of the CSG provision in 3GPP as specified in 3GPP TS 23.002.

An expired CSG subscription should not be removed from the HSS subscription data before it is removed from the UE's Allowed CSG list or Operator CSG list. When a CSG subscription is canceled it should be handled as an expired subscription in HSS subscription data to allow execution of the appropriate procedure so it is removed from UE's Allowed CSG list or Operator CSG list first. 3GPP TS 23.008 describes CSG expiration handling as follows: when a CSG-Id expires, or the expiration date is changed (added or modified) to an expired date, the CSG-Id should be removed from the UE (e.g. by OMA DM or Over The Air (OTA) update). After successful removal of the CSG-Id from the UE, the HLR/HSS should delete the CSG-Id and, if applicable, update the MME. The two operations (i.e. the removal from the UE and the deletion/update of the MME by the HSS) may not be correlated in the sense that they may be performed independently by different systems. CSG List Server and HSS interaction is not specified in 3GPP. For the UEs in connected mode whose

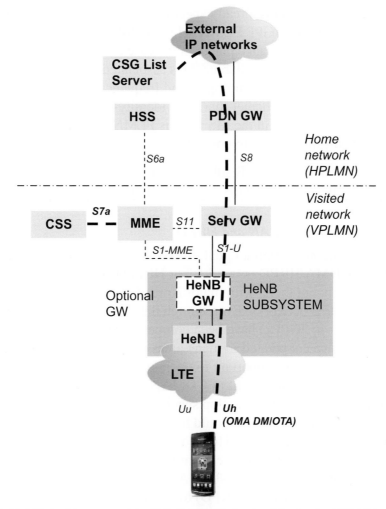

Figure 6.19: HeNB Architecture with CSG Configuration Entities in the HPLMN and VPLMN, as Applicable.

CSG subscription has expired, MME informs the HeNB and appropriate actions may be taken to either remove the UE from the CSG by handing over to a suitable cell or releasing the radio link so the UE is moved to an Idle state.

Chapters 2 and 7 provide further information on the architectural and security aspects of the HeNB subsystem, which utilizes CSG ID for its operation.

There are two modes of CSG selection that affect the UE registration procedure in a 3GPP system that takes place as part of the PLMN selection process: Automatic mode and Manual mode. This function is specified in 3GPP TS 23.122. For the Automatic mode of selection, the CSG ID is already specified in the Allowed or Operator CSG list in the UE and the normal 3GPP PLMN selection followed by UE registration and tracking area update procedures take place. In the case of Manual mode, the UE needs to indicate to the user the list of available CSGs associated with the PLMNs across supported 3GPP Radio Access Technologies and frequencies. For each entry in the list, an indication is provided whether that CSG identity is in the Allowed CSG list or in the Operator CSG list stored in the MS for this PLMN. The user at this point needs to make a selection for the process to continue.

For further information on the HeNB subsystem and CSG management, see "3GPP Femtocells: Architecture and Protocols" (Horn, 2010).

6.4.5 *Mobility Between E-UTRAN and HRPD*

As already mentioned, one important goal of EPS is to support efficient interworking and mobility with CDMA/HRPD networks. For HRPD networks, there is a significant subscriber base already out there, with major North American and Asian operators operating CDMA/HRPD networks. Even though the two technologies (one developed in 3GPP and the other in 3GPP2) have been competing over the last 20 years, the two bodies have also cooperated in many areas in order to develop common standards that are strategically important to operators; examples of these are IMS and PCC development. For the benefit of CDMA, operators and other companies cooperated extensively in order to develop special optimized HO procedures between E-UTRAN and HRPD access, which would have efficient performance and reduced service interruption during handover. This work was brought into the mainstream 3GPP standards under the SAE work item umbrella and thus produced the so-called Optimized Handover between E-UTRAN and HRPD. HRPD networks then became known as evolved HRPD (eHRPD) to highlight the changes required for interoperability and connectivity with EPC and E-UTRAN.

Mobility between E-UTRAN and HRPD has been specified to allow an efficient handover with minimal interruption time also for those terminals that can only operate a single radio at a time. There is thus no need for the terminal to operate both HRPD and E-UTRAN interfaces simultaneously. Despite the fact that these terminals support multiple radio technologies, this property of being able to operate only one radio at a time is sometimes referred to as "single radio capability".

Note that in the early deployment of E-UTRAN in an existing HRPD network, it is considered more prevalent, and thus more important, to support E-UTRAN to HRPD handover than the reverse direction, since it is assumed that the HRPD networks would have sufficient coverage to keep a user within the HRPD system. In order to support a network-controlled handover from E-UTRAN to HRPD, the eNodeB can be configured with HRPD system information that is sent over E-UTRAN in order to assist the terminal in preparation for cell reselection or handover from the E-UTRAN to the HRPD system. The terminal also makes appropriate measurements on HRPD cells while being connected to E-UTRAN. Similarly to the active-mode mobility for 3GPP accesses described in the previous section, measurement time gaps have to be provided to the terminal in order to allow the terminal to use only a single radio at a time, that is either E-UTRAN or HRPD. The measurements are reported to the eNodeB to allow the E-UTRAN to make the appropriate handover decisions.

The purpose of the optimized procedures is to minimize the total service interruption time experienced at the UE, by having the UE prepare the target access system before actually leaving the source access system. The preparation in the target access system is done by enabling the UE to exchange access-specific signaling with the target access over the source access. The S101 interface between the MME and HRPD Access Network is used to tunnel the signaling between UE and target access system. A benefit of letting the UE prepare the target access via a tunnel over the source access is that the direct exchange of UE context between the different access networks can be minimized. The impact on access networks can be minimized since neither access needs to adapt its signaling towards another access technology.

The handover between E-UTRAN and HRPD is performed in two phases:

1. A pre-registration (or preparation) phase where the target access and specific core network entity for the specific access (MME for E-UTRAN and HRPD S-GW or HSGW for eHRPD access) is prepared ahead of the actual handover.
2. A handover (or execution) phase where the actual access network change occurs.

In the pre-registration/preparation phase for E-UTRAN to HRPD handovers, the UE communicates with the HRPD access network via the E-UTRAN access and the MME. The HRPD signaling is forwarded transparently by the E-UTRAN and MME between the UE and the HRPD RNC. This is illustrated in Figure 6.20. In the E-UTRAN to HRPD direction, there is no time limit of how long a UE may be pre-registered in the HRPD system before the handover takes place. Therefore, pre-registration may take place well in advance of the actual handover. When the decision to hand over from E-UTRAN to HRPD is taken, the handover phase is executed. During this phase, some additional preparation of HRPD target access is performed before the actual user-plane path switch takes place.

The E-UTRAN may redirect the UE to HRPD using RRC Connection Release with Redirection Information set according to 3GPP TS 36.331. If pre-registration has

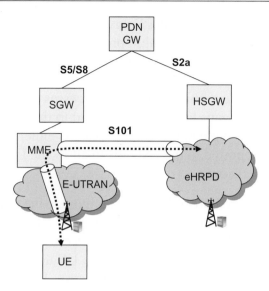

Figure 6.20: Terminal Performing Pre-Registration in HRPD While Being Connected to E-UTRAN.

The dashed line illustrates HRPD-specific signaling transparently forwarded by E-UTRAN and MME.

not been performed successfully, at the reception of the redirection message, the UE acquires the HRPD channel and performs the non-optimized handover by triggering access authentication and authorization followed by an attachment procedure towards the HRPD access network using HRPD-specific procedures and then the Access Network prepares connection towards EPC using S2a/PMIPv6.

If pre-registration is successful the UE, upon receiving the redirection message, follows the RRC Connection Release with Redirection procedure to reselect the HRPD cell according to TS 36.331 and then performs the idle-mode optimized handover procedure starting from informing the HRPD about the Inter-Technology Idle mode mobility event according to HRPD procedures and then tuning into HRPD and continuing connection to the network.

The handover from HRPD to E-UTRAN is also performed with a preparation phase where E-UTRAN-specific signaling is exchanged between UE and MME via the S101 interface. This time the signaling is transported transparently via the HRPD access network. A difference with handover from HRPD to E-UTRAN is that the pre-registration/preparation takes place immediately prior to the terminal switches from HRPD to E-UTRAN radio, i.e. just before the actual handover is performed. There is thus no long-lived pre-registration state in target access in this case. For more details on the HO procedures between E-UTRAN and HRPD, see Figure 6.21, Figure 6.22 and Section 17.7.

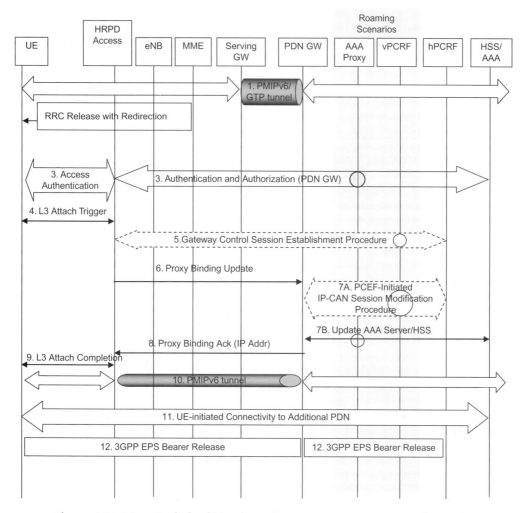

Figure 6.21: Non-Optimized Handover from E-UTRAN to HRPD Idle Mode.

6.4.6 Generic Mobility Between 3GPP and Non-3GPP Accesses

6.4.6.1 General IP Session Continuity

Generic mobility between heterogeneous access types, such as between a 3GPP access and WLAN or WiMAX, are also supported by EPS. Generic mobility is, as the name suggests, generic in the sense that the mobility procedures are not specifically adapted to any particular access technology. Instead, the procedures are generic enough to be applicable to any non-3GPP access technology, such as WLAN and WiMAX, as long as the accesses support some basic requirements. The generic mobility procedures are, for example, also applicable for HRPD if the optimized mechanisms described in the previous section are not deployed.

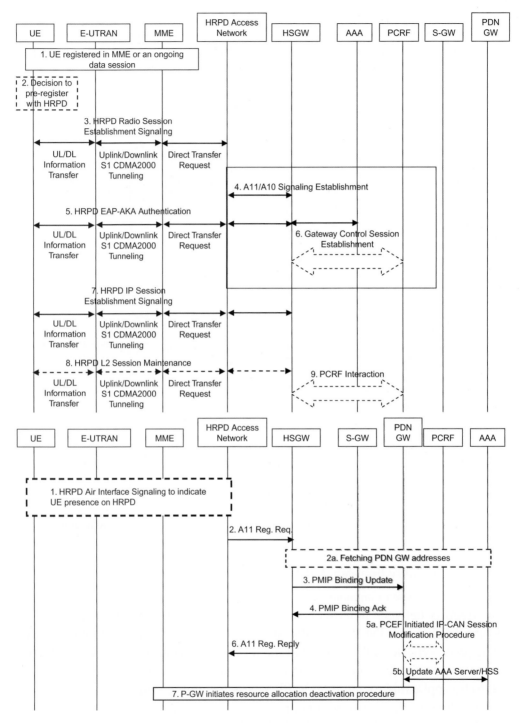

Figure 6.22: Pre-Registration Followed by Optimized Idle-Mode Handover from E-UTRAN to HRPD.

Since the generic mobility procedures are supposed to work with any access, they are also not optimized towards any specific access technology. In 3GPP and other standards forums, generic mobility is also referred to as non-optimized handover. However, a goal has still been to provide efficient handover procedures also in this generic case. In particular, for a terminal that is able to operate in multiple access technologies simultaneously – often referred to as "dual-radio capable" terminals – it is possible to prepare the target access before performing the actual handover. The terminal can, for example, perform authentication procedures in target access while still using the source access to transfer user-plane data. This is, in a sense, similar to the pre-registration described for HRPD with the difference that the "pre-registration" is done in the actual target access instead of via the source access.

A key difference compared to the mobility mechanisms within and between 3GPP accesses and the optimized interworking between LTE and HRPD is that generic mobility does not assume any interaction between the two access networks. Instead, the source and target access networks are fully decoupled.

The generic handover is always triggered by the terminal – that is, there are no measurements of cells in the target access being reported to the radio access network where the UE is attached. There are also no handover commands from the source access to trigger the handover. Instead, it is up to the terminal to decide when to initiate the handover. The terminal may, for example, base its decision on measurement of signaling strength of available access networks. The operator may also use the Access Network Discovery and Selection Function (ANDSF) to provide the terminal with information about access networks as well as policies for access network selection. The ANDSF is described further in a subsequent section.

The EPS supports two different mobility concepts for generic mobility between 3GPP and non-3GPP accesses: host-based and network-based mobility. Host-based mobility is a term often used to denote a mobility scheme where the terminal (or host) is directly involved in movement detection and mobility signaling. Mobile IP, defined by IETF, is one example of such a mobility protocol. In this case the terminal has IP mobility client software.

Another type of mobility protocol and mobility scheme is the network-based mobility management scheme. In this case the network can provide mobility services for a terminal that is not explicitly exchanging mobility signaling with the network. It is a task of the network to keep track of the terminal's movements and ensure that the appropriate mobility signaling is executed in the core network in order for the terminal to maintain its session while moving. GPRS Tunneling Protocol (GTP) is an example of a network-based mobility protocol. Proxy Mobile IPv6 (PMIPv6) is another example of a network-based protocol that is used to support mobility.

EPS supports multiple mobility protocol options using host- and network-based mobility protocols. For 3GPP accesses, only network-based mobility protocols are used: either GTP or PMIPv6. Over non-3GPP accesses the network-based mobility protocols GTP and PMIPv6 are also supported. EPS supports two host-based mobility schemes: Dual-Stack Mobile IPv6 (DSMIPv6) and Mobile IPv4 (MIPv4) over non-3GPP accesses. Over 3GPP accesses host-based mobility is not used. Instead, it is always assumed that the 3GPP access is the "home link" in the host-based Mobile IP sense. See Section 16.3 for further descriptions of this and other Mobile IP concepts.

For a more detailed description of the mobility protocols and mobility mechanisms, see Chapter 16 for protocol descriptions and Chapter 17 for details of the handover procedures.

Host- and network-based mobility schemes have different properties. A host-based scheme requires support for the mobility protocol in the terminal. Host-based mobility is often generic in the sense that it assumes nothing or very little about any particular access network. It can therefore be used also over those access networks not supporting mobility at all. A consequence is that basic host-based mobility is not optimized for any particular access network.

Network-based mobility, on the other hand, requires that the access network supports the mobility protocol. GTP, for example, is developed by 3GPP to support mobility between 3GPP accesses and includes support for context transfer and other features needed to provide a seamless handover within and between 3GPP accesses. GTP has also been specified for use with WLAN networks and for generic non-3GPP accesses.

Even though the division between host- and network-based mobility schemes often provides a useful classification, it should be noted that the distinction is not always completely clear. For example, with network-based mobility protocols, the terminals often need to be mobility aware and handle mobility even if they do not explicitly participate in the mobility signaling. Therefore, even though the division into host- and network-based mobility is a useful categorization on a high level and when discussing a specific protocol, it is also important to remember that in reality the full solution typically contains parts of both. EPS supports different mobility protocols in different accesses and there may be a change in mobility protocol when moving between access networks. The terminal may, for example, move from a non-3GPP access where DSMIPv6 is used to a 3GPP access where GTP or PMIPv6 is used. It is the task of the PDN GW to ensure that IP session continuity is also provided when different mobility protocols are used in the different accesses.

6.4.6.2 Simultaneous Multi-Access
In 3GPP Release 10, the support for simultaneous multi-access was introduced where the UE is connected to both 3GPP and non-3GPP access simultaneously. These

features were partly driven by the desire to offload the 3GPP access by using WLAN instead. The following flavors of simultaneous connectivity in multiple accesses are provided:

- Multi-access PDN Connectivity (MAPCON): The support for having one (or more) PDN connection(s) in 3GPP (2G/3G/LTE) access and one (or more) PDN connection(s) in a non-3GPP access. Mobility of each PDN connection between 3GPP and non-3GPP access is also supported. MAPCON is supported for both the host- and network-based mobility schemes.
- IP flow mobility (IFOM): The support for having one PDN connection over both 3GPP access and WLAN access simultaneously and choosing which access to route traffic over on a per-IP-flow basis. Seamless movement of IP flows between 3GPP and WLAN access is supported. IFOM is only supported for the DSMIPv6-based mobility solution and is an extension of that solution.
- Non-seamless WLAN offload (NSWO): The support for routing traffic over WLAN without traversing the Evolved Packet Core. Mobility (IP session continuity) with 3GPP access is not supported. For traffic using NSWO there is no IP mobility anchor in the EPC. Mobility (IP session continuity) between WLAN and 3GPP access is thus not supported.

More details on the simultaneous multi-access solutions can be found in Chapter 14.

6.4.6.3 IP Mobility Mode Selection

The EPS specifications allow both host- and network-based mobility; different operators may make different choices about which of the two to deploy in their networks. An operator may also choose to deploy both. In 3GPP accesses, there is only a choice between two network-based mobility protocols (PMIPv6 or GTP) and the selection of one protocol over the other has no impact on the terminal, since the network choice of protocol is transparent to the UE. It should be noted that even if multiple mobility protocols are supported in a network deployment, only a single protocol is used at a time for a given UE and access type.

Terminals may support different mobility mechanisms. Some terminals may support host-based mobility and thus have a Mobile IP client installed (Dual-Stack Mobile IPv6 and/or Mobile IPv4 client). Other terminals may support IP-level session continuity where network-based mobility protocols are used. There may also be terminals that support both mechanisms. In addition, some terminals may have neither a Mobile IP client nor support IP session continuity using network-based mechanisms. These terminals will not support IP-level session continuity but could still attach to EPC using different access technologies.

There is thus a need for selecting the right mobility mechanism when a terminal attaches to the network or is making a handover. If network-based mobility protocols are selected, there also needs to be a decision whether session continuity (i.e. IP address preservation between the accesses) shall take place.

EPS has defined different means for how this selection can be done. The rules and mechanism for selecting the appropriate mobility protocol is referred to as IP Mobility Mode Selection (IPMS).

One option with IPMS is to statically configure the mobility mechanism to use in the network and the terminal. This is possible, for example, if the operator is only supporting a single mobility mechanism and if the operator can assume that the terminals used in its network support the deployed mobility mechanism. If a user switches to another terminal not supporting the mobility protocol deployed by the operator, IP-level session continuity may not be possible.

The other option is to have a more dynamic selection where the decision to use either network- or host-based mobility is made as part of attachment or HO procedures. It should be noted that over 3GPP accesses only network-based mobility is supported using either PMIPv6 or GTP. Therefore, mobility mode selection is only needed when the terminal is using a non-3GPP access.

IPMS is performed when the user attaches in a non-3GPP access or sets up an IPSec tunnel towards an ePDG, before an IP address is provided to the terminal. The terminal may provide an indication about its supported mobility schemes during network access authentication, by using an attribute in the EAP-AKA and EAP-AKA' protocols. The indication from the UE informs the network if the terminal supports host-based mobility (DSMIPv6 or MIPv4) and/or if it supports IP session continuity using network-based mobility (using GTP or PMIPv6 in the network). The network may also learn about the terminal capabilities using other mechanisms. For example, if a terminal has attached in 3GPP access and performed bootstrapping for DSMIPv6, the network implicitly understands based on the already performed bootstrapping procedure that the terminal is capable of using DSMIPv6. Based on its knowledge about the terminal and the capabilities of the network, the network decides which mobility mechanism to use for the particular terminal. If the network has no knowledge about a terminal's capabilities, the default is to use network-based mobility protocols. Host-based mobility can only be selected by the network if it knows that the terminal supports the appropriate mobility protocol (DSMIPv6 or MIPv4).

6.4.7 Access Network Discovery and Selection

The Access Network Discovery and Selection Function (ANDSF) is a function that was defined in 3GPP TS 23.402. The ANDSF is a function in the network that provides the UE with policies and network selection information for influencing how users and their devices prioritize between different access technologies if several non-3GPP access networks are available. ANDSF thus provides the operator with a tool for steering terminal access selection. The policies for access technology selection can also be preconfigured in the terminals but dynamically updated or modified by the ANDSF.

As depicted in Figure 6.23, the architecture for ANDSF contains a reference point S14 to the UE to provide the UE with access network selection policies and/or access discovery hints. The protocol on S14 between UE and ANDSF is IP based and utilizes Open Mobile Alliance (OMA) Device Management (DM). The UE can thus connect to the ANDSF via any IP-based access. Authentication and communication security between the UE and ANDSF is based on the Generic Bootstrapping Architecture (GBA), where GBA is used for authentication based on SIM-card credentials with IMSI as the user identity.

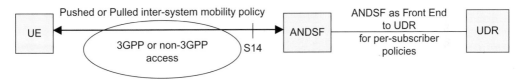

Figure 6.23: Non-Roaming Architecture for ANDSF.

The solution supports both "pull" and "push", i.e. the UE can either request the information or the ANDSF can initiate the data transfer to the UE. It should, however, be noted that ANDSF is (by design) not suitable for frequent/dynamic updates of policy. ANDSF is thus not intended as a very dynamic access selection mechanism controlling access selection in real time. Furthermore, the detailed UE behavior when using info/policies received from ANDSF is mostly implementation dependent.

The UE may provide information to the ANDSF that assists the ANDSF when generating discovery and selection information. The UE will, for example, provide its identity (IMSI) but may also provide its location (such as GPS coordinates or cell identities of nearby radio base stations), as well as information about what types of ANDSF information are requested by the UE. The ANDSF may use this information to provide the UE with subscriber-specific network discovery and selection information that applies in the vicinity of the UE.

The ANDSF can provide the UE with three types of information:

- Access Network Discovery and Selection Information
- Inter-System Mobility Policies (ISMPs)
- Inter-System Routing Policies (ISRPs).

The Access Network Discovery and Selection Information includes a list of access networks available in the vicinity of the UE. The list may include both access types from the 3GPP family of accesses as well as other access types such as WLAN. This will help the UE when discovering access networks and simplify and speed up the scanning needed in the UE. The information may include the access technology types (e.g. WLAN), network identifiers (e.g. SSID in the case of WLAN) as well as validity conditions (e.g. in what locations and at what time of day the discovery information is valid). Figure 6.24 illustrates the content of the discovery object.

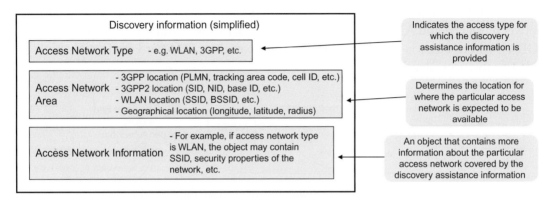

Figure 6.24: Overview of Discovery Information That May be Provided by ANDSF.
Not all information elements are shown.

The Inter-System Mobility Policies (ISMPs) are a set of operator-defined rules and preferences that affect the inter-system mobility decisions taken by the UE, i.e. decisions about whether to use 3GPP or non-3GPP access, and what type of non-3GPP access (e.g. WLAN). The ISMPs do not contain policies that influence the choice of 3GPP access (e.g. GERAN, UTRAN, or E-UTRAN). Such access selection between 3GPP radio technologies is handled by other methods not related to ANDSF. The UE uses the inter-system mobility policy when it can use only a single radio access interface at a given time. (When the UE is capable of using multiple radio interfaces at a time, the ISRPs are as described below.) The ISMPs may indicate when inter-system mobility is allowed or restricted, as well as the most preferable access technology type or access network that should be used to access EPC. For example, an inter-system mobility policy may indicate that inter-system handover from E-UTRAN access to WiMAX access is not allowed. It may also indicate, for example, that WLAN access is preferable to WiMAX access. The ISMPs can also indicate if a specific access network identifier is preferable to another (e.g. WLAN SSID A is preferable to WLAN SSID B). Similar to the discovery information, the ISMPs also include validity conditions, i.e. conditions indicating when a policy is valid (a time duration, a location area, etc.). Figure 6.25 illustrates the content of the ISMP object.

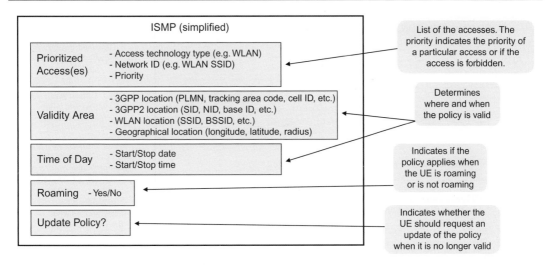

Figure 6.25: Overview of ANDSF ISMP Object.
Not all information elements are shown.

The Inter-System Routing Policies (ISRPs) are a set of operator-defined rules and preferences that affect the routing decisions taken by the UE, i.e. decisions about whether to route traffic via 3GPP or a particular type of non-3GPP access. The ISRPs also affect the decisions taken by the UE for what traffic to offload in WLAN access without forwarding it via EPC and a PDN GW. Similar to ISMPs, the ISRPs do not affect the selection between different 3GPP accesses (GERAN, UTRAN, or E-UTRAN). The UE uses the inter-system routing policies when it can route IP traffic simultaneously over multiple radio access interfaces. The ISPRs include three types of routing policies:

- Routing policies for a specific APN. This type of ISRP is used by a UE that can establish multiple PDN connections simultaneously in different accesses. For this purpose the ISRP includes filter rules that identify a prioritized list of access technologies and/or networks that should be used by the UE to route PDN connections to specific APNs. A filter rule also identifies which radio accesses are restricted for PDN connections to specific APNs (e.g. WLAN is not allowed for PDN connection to APN-x). Figure 6.26 illustrates the content of this type of ISRP object.
- Routing policies for IP flow mobility. This type of ISRP is used by a UE that uses DSMIPv6 together with IP flow mobility for routing separate IP flows belonging to a single PDN connection over both 3GPP access and WLAN access simultaneously. In this case the ISRPs include filter rules identifying a prioritized list of access technologies and/or networks that should be used by the UE to route traffic that matches specific IP traffic filters. A filter rule may identify traffic based on, for example, destination and/or source IP address, transport protocol,

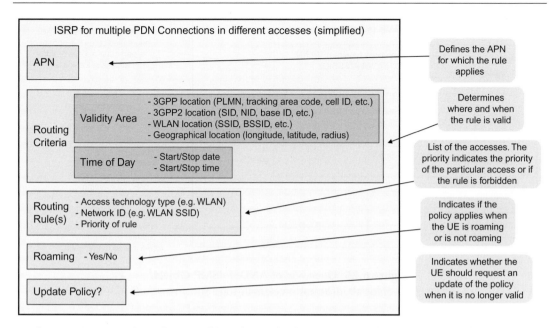

Figure 6.26: Overview of ISRP Object for Multiple PDN Connections in Different Accesses.
Not all information elements are shown.

destination/source port numbers, DSCP, destination domain name, and application identity. A filter rule can also identify which radio accesses are restricted for traffic that matches specific IP traffic filters (e.g. WLAN is not allowed for RTP/RTCP traffic flows on APN-x). Figure 6.27 illustrates the content of this type of ISRP object.

• Routing policies for WLAN offload. This type of ISRP is used by a UE that uses the local IP address allocated in a WLAN network for traffic that does not need to traverse the EPC and a PDN GW. These rules contain the same type of information as the ISRPs for IP flow mobility but the purpose is different. The ISRPs for WLAN offload indicate what traffic shall or shall not be non-seamlessly offloaded to a WLAN. Figure 6.27 illustrates the content of this type of ISRP object.

For more details on the ANDSF MO, see the text on ANDSF-related interfaces in Chapter 15.

The ISMPs and ISRPs may be provided in the UE and can also be updated by the ANDSF based on network triggers or after receiving a request from a UE for network discovery and selection information.

The ANDSF Managed Object (MO) used with OMA-DM to carry the information between the UE and ANDSF is specified by 3GPP and is described in further detail in Chapter 15.

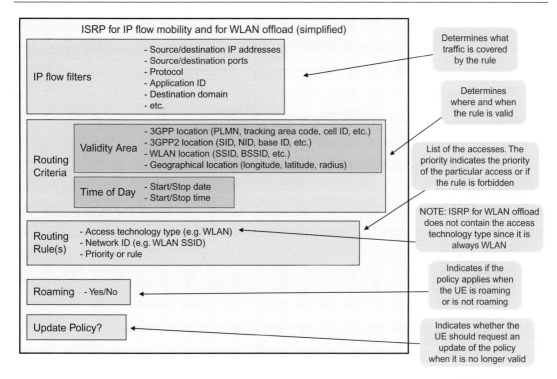

Figure 6.27: Overview of ISRP Object for IP Flow Mobility and for WLAN Offload.
Not all information elements are shown.

In Release 8 only the non-roaming architecture was specified but in Release 9 ANDSF has been extended with a roaming architecture, where a roaming UE can connect to both an ANDSF in the home network (H-ANDSF) and an ANDSF in the visited network (V-ANDSF), see Figure 6.28. The H-ANDSF provides the UE with discovery information only for access networks that provide connectivity to the home network, while the V-ANDSF provides the UE with discovery information only for access networks that provide connectivity to the particular visited network. Slightly different principles apply for the mobility and routing policies. The H-ANDSF provides mobility and routing policies that apply generally, while the V-ANDSF provides mobility and routing policies that will be valid only within the particular visited network. Since the policies from H-ANDSF and V-ANDSF in some cases may overlap, the UE needs to resolve any conflicts that may exist between the policies provided by the H-ANDSF and the policies provided by the V-ANDSF. In general the policies provided by the serving network (i.e. the V-ANDSF in roaming scenarios) take precedence.

The information and policies provided by the ANDSF to the UE may also depend on the UE's subscription data. From Release 9 onwards, the ANDSF can act as a

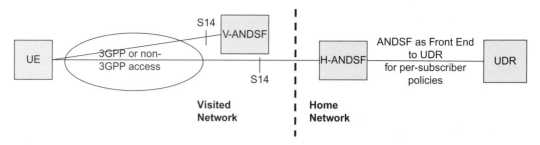

Figure 6.28: Roaming Architecture for ANDSF.

front-end in the UDC architecture in order to get subscription data for individual subscribers (see Chapter 10 for more information on the UDC architecture). The ANDSF (H-ANDSF in case of roaming) may take the subscription data, e.g. the list of access networks or access technology types that the UE is authorized to use, into account when generating the inter-system mobility policies.

6.5 Interworking with Managed WLAN Networks

WLAN (often also referred to as WiFi for IEEE 802.11-based WLAN) has recently become an important technology for mobile data. A key factor in this development is the strong growth of handsets supporting WLAN. From being a technology that was only included in a few of the most advanced handsets, WLAN is now supported by all advanced handsets and is also available in more basic devices. In addition to the growth of WLAN-enabled handsets, the explosion of mobile broadband traffic has also created an increased interest among mobile operators for WLAN technologies. The manifold increase of mobile data traffic has resulted in mobile operators looking at WLAN as a possibility to offload some of their 3GPP networks using unlicensed WLAN spectrum. However, there is a huge installed base of WLAN networks, e.g. in residential, enterprise, and public environments, and these deployments differ considerably in terms of security properties, performance, access methods, etc. It is therefore a challenge for mobile operators to capture the benefits of WLAN and incorporate them into mobile operator environments.

Even though non-3GPP accesses have been discussed above it was at a more generic level (with the exception for HRPD). The importance and increased interest of WLAN, however, warrants a dedicated section on managed WLAN.

WLAN integration with 3GPP networks has been defined in different variants since 3GPP Release 6. Some of the general aspects about connecting over a non-3GPP access to EPC, as defined from Release 8 onwards, have been described in the previous sections. However, connectivity to the 3GPP domain over WLAN access has actually been supported already since 3GPP Release 6 based on the Interworking WLAN (I-WLAN) specifications. All these solutions have been standardized as "overlay" solutions using IP-based tunneling over the WLAN network. As mentioned in the previous sections on generic non-3GPP access, protocols like IPSec and DSMIPv6 are used to connect to the EPC. A reason for using overlay solutions has been to avoid any impact on the WLAN networks. They do, however, require terminal support, including client software for, say, IPSec and DSMIPv6. However, there has been a lack of terminals supporting these solutions and take-up in real deployments has been very limited.

In 3GPP Release 11, new solutions for connecting the user via WLAN to EPC have been included. These solutions rely on WLAN being treated as a "trusted non-3GPP access" and are targeting managed WLAN deployments. The intent is to improve the user experience when using WLAN and to give operators a greater possibility for tighter coupling between WLAN networks and the EPC domain. One key aspect that has made these developments more realistic is the increased number of WLAN-enabled handsets that support 3GPP-based access authentication, i.e. the capability to smoothly connect and authenticate to WLAN networks using SIM cards with no or minimal user involvement. These solutions will be the topic in the rest of this section.

In 3GPP Release 11, a new architecture for managed WLAN has been defined. The solution treats the WLAN access as a "trusted non-3GPP access" and uses the S2a interface (see above and Chapter 3) for connecting the UE to EPC. Even though connectivity via S2a for trusted non-3GPP access in general has been supported since Release 8, it mostly targeted other mobile networks such as CDMA and WiMAX. However, with Release 11 the support for trusted WLAN has been explicitly added to the specifications. In addition, the S2a interface that was defined in Release 11 to use PMIPv6 only has been enhanced to also support GTP.

The Release 11 solution for managed WLAN access does not assume any new functionality in the UEs. The solution will work with standard WLAN terminals as long as they support SIM-based access authentication based on IEEE 802.11 and IEEE 802.1X. There is no need for additional software for connecting to EPC.

The general description of session management and IP address allocation using S2a for trusted non-3GPP accesses in previous sections still applies, but additional details for WLAN have been worked out. The basic architecture for connecting Trusted WLAN Access Networks (TWANs) to EPC is shown in Figure 6.29. The internal structure of the TWAN is not defined by 3GPP, but 3GPP does place requirements on

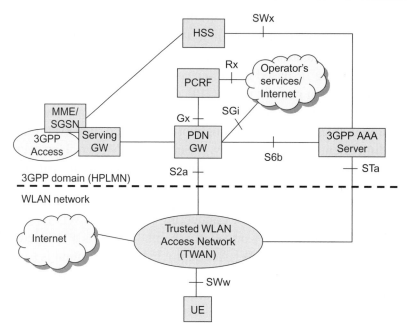

Figure 6.29: Architecture for Trusted WLAN Access with S2a-Based Access to EPC (Non-Roaming).

functionality that need to be supported. The functions of the TWAN are illustrated in Figure 6.30 and briefly described below:

- A WLAN Access Network (WLAN AN) includes collection of one or more WLAN access points.
- A Trusted WLAN Access Gateway (TWAG) terminates S2a towards the PDN GW. It also terminates a logical per-UE point-to-point (p2p) link between the UE and the TWAG.
- A Trusted WLAN AAA Peer (TWAP) terminates STa and relays the AAA information between the WLAN Access Network and the 3GPP AAA Server or Proxy in the case of roaming.

The per-UE point-to-point link between the UE and the TWAG is required when traffic for that UE is routed via S2a, since the UEs sharing the same WLAN AP may be connected to different PDN GWs and be assigned IP addresses from different subnets, possibly with overlapping address spaces. There should be no direct communication among these UEs and since WLAN is a shared medium the S2a solution requires that the p2p link exists between each UE and the TWAG. In particular, it is assumed that the WLAN AN enforces upstream and downstream forced forwarding between the UE's WLAN radio link and the TWAG.

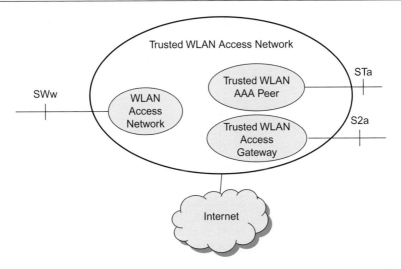

Figure 6.30: Functionality of Trusted WLAN Access Network (TWAN).

For more details on the attachment and detachment procedures when a UE connects via a trusted WLAN and the S2a connection to EPC is established, see Chapter 17.

Due to the assumption in Release 11 that there is no impact on UEs, some limitations of the current solution exist. For example, the Release 11 solution does not support handover with IP address preservation when moving between 3GPP access and TWAN. Furthermore, there is only support for connecting to the default APN (as stored in HSS) and for a single PDN connection. Discussions are ongoing to extend the solution after Release 11. For example, 3GPP may add support for IP address preservation at handover to/from TWAN, support for non-default APNs from the UE, and support for multiple simultaneous PDN connections over TWAN. These features will require enhanced support not only in the network but also in the UE.

6.6 Pooling, Overload Protection, and Congestion Control

EPC has been designed from the start with pooling of network elements as a foundation of the system, in contrast to 2G and 3G systems, where pooling was added as an afterthought. The pooling mechanisms in EPC are efficient for an operator's network by allowing them to centralize and pool a group of their signaling nodes. It should be noted that the network elements are stateful; as a result the UE context is stored in every node involved in handling the UE.

In 2G and 3G networks, the core network was designed as a hierarchical system; when a terminal or UE was located in a particular cell, it was only able to connect to one Base Station (BS). While there were one-to-many relationships between the

SGSN and Base Station Controller (BSC) and also between the BSC and BS, in practice the UE was only able to be connected to one BS, BSC, and therefore only one SGSN (Figure 6.31).

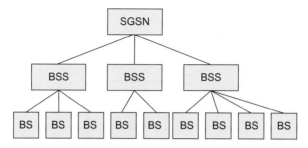

Figure 6.31: Original Hierarchical System Design for GSM/GPRS.

The same hierarchical structure applied in WCDMA networks for the Radio Network Controller (RNC) and node B. Naturally, this does not utilize the capacity of the nodes to their full potential; in the strictly hierarchical systems of the 2G and 3G networks, the capacity of the nodes can never be perfectly balanced, for example when users move in and out of cities. The network must be dimensioned according to the peak load in a particular area, which may be much higher than the *average* load in that area. Through applying a pooling mechanism, it is possible to dimension the whole pooled capacity to the peak rates of a much larger region; for example, it is now possible to create a pool of network nodes for the entire London region, rather than the need to split the network up into many smaller regions.

So, how is this implemented in an actual network configuration?

As described in Section 6.3, an MME has several different identifiers associated with it, which help to manage the pooling mechanism. As a brief review, the *MME Group* identity refers to the name of the pool that the MME belongs to, while the *MMEC* identifies the actual node within a group; the *MMEI* is therefore formed by combining the MMEGI and MMEC together. Figure 6.32 shows an example of an MME pool.

In LTE, the eNodeB knows which MMEs it can communicate with. When a terminal that is already registered enters a cell, it sends a service request to the network, which contains a GUTI, within which the MMEI is encapsulated. The eNodeB checks the GUTI code and if the MMEI is within the pool of MMEs for which it has a connection, it simply uses the MMEI to send it to the correct one. When a GUTI is available, therefore, it makes everything reasonably simple – the eNodeB can always route you to the correct MME. If the GUTI is absent, however, or the GUTI points

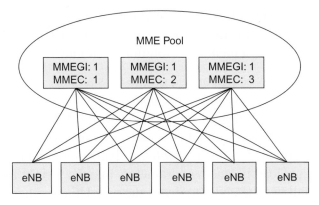

Figure 6.32: MME Pool with eNBs.

to an MME that is not within the pool associated with that particular eNodeB, the eNodeB selects a new MME and forwards the service request to the new MME. The MME then returns a GUTI to the eNodeB, which naturally includes its own MMEI.

The eNodeB cannot just randomly choose any MME in the pool to send the service request to, however, because all of the MME nodes may have different capacities. Randomly allocating UEs to them thus may overload nodes with lower capacity. When configured, therefore, the MME is assigned a "weight factor", which indicates its capacity. The MME then informs all of the eNodeBs in its pool what this weight factor is. The eNodeBs can therefore distribute UEs across the pool of MMEs accordingly.

There are certain situations where it may be useful to manipulate the weighting factor of an MME. Take the example of an already established pool of eNodeBs and MMEs that you wish to insert a new MME into. Initially that particular MME can inform the eNodeBs that is has a higher weight factor than it actual will during "normal" operations. This means that the particular MME will then be allocated some UEs by eNodeBs. Once the new MME reaches a suitable load, it will update its weight factor and inform the eNodeBs of this change.

The implementation of pooling in the core network architecture also implies some changes with regard to network dimensioning. In the hierarchical systems, when you dimension your system, you will generally apply some general formula, for example peak load plus an additional 20% capacity, in order to handle extra load. This is dimensioned across an area, however, not one particular region as with a pooled network architecture. This means that the numbers of subscribers that you are dealing with when calculating excess capacity are significantly different.

In addition to load balancing and overload control of MMEs, the tremendous growth of smart phones and data traffic has also created a need for an improved mechanism for possible load rebalancing of MME nodes in a pool, as well as other overload and congestion control mechanisms in the EPS. These mechanisms should complement possible vendor-specific implementation solutions that may be already in place.

MME load rebalancing can be performed on the UEs already registered, which can then be moved to another MME within the pool area. This would typically be done by triggering O&M operation to move users from a specific MME. There are a number of ways to utilize this function without adversely affecting users' experience or network/system performance. Some possible mechanism to manage load rebalancing best would be to adjust load balancing parameters like the weight factor in the eNBs ahead of time. This allows the eNBs to divert the new UEs to other MMEs in the pool and thus reduces the number of UEs being moved from the MME itself. Of course, whenever time load rebalancing occurs, it is important to consider the effect of such load redistribution towards other MMEs in the pool and to ensure that these nodes are not overwhelmed by the incoming new UEs. So it is clear that load rebalancing is not the means of controlling already overloaded nodes but rather to ensure that the MMEs in a pool are properly balanced during normal conditions. As the UEs are being moved from one MME to another, MME load rebalancing needs signaling towards the UE depending on whether the UE is active and connected to the network or whether the UE is in an idle state. In the case of connected UEs, S1 release procedures with "load balancing TAU required" are used to trigger the UE to release its connection and perform TA update procedures without any parameters that are associated with the previous MME (GUMMEI, etc.). Other mechanisms may be used as well, such as S1 Connection Inactivity, which causes the UEs to be released. If the UE is in an idle state, the MME may either wait for attachment or TA update procedures to be completed before triggering the S1 release procedure or MME may first page the UE and bring it to the connected mode and then trigger the S1 release procedure. Regardless of the mechanism used, care must be taken so that the load rebalancing process is moving UEs gradually into other MMEs in the pool area.

Even with load rebalancing/load management, additional mechanisms are available in the 3GPP specification to ensure that MMEs can manage and control situations that would prevent/minimize overload conditions in the network. MME is the first point of entry to the core network and it is crucial to maintain the network's integrity that MMEs are operating in a managed environment so the number of users able to access the operator's network is maximized. The control mechanisms an MME may employ to control its load include informing eNBs via an S1 interface overload protection mechanism to reduce a certain amount of traffic by an eNB. The eNodeB may reduce, say, non-emergency and non-high-priority originating calls by rejecting the requests in the eNB. MME implementation may use other internal event or trigger information to determine and manage the load situation by invoking overload signaling towards its serving eNBs. When MMEs deem it necessary, the signaling is triggered to start the overload supervision in an eNB (see Figure 6.33).

Figure 6.33: Overload Start.

The message indicates the amount of overload and a list of GUMMEI(s) indicating the MMEs affected by the request.

Once the overload condition in the MME(s) no longer exists, a message is sent to the appropriate eNBs stopping the process (see Figure 6.34).

Figure 6.34: Overload Stop.

The eNB(s) then removes the restrictions according to the provided list of GUMMEI(s).

Additionally, the 3GPP system provides various responses (known as Cause Code Values) to UEs in order to control certain aspects of the system behavior such as congestion, requesting EPS bearer activation request, etc. But these were not sufficient to manage the tremendous growth that is being experienced by mobile broadband deployment (e.g. HSPA, LTE).

As 3GPP architecture is being prepared for the eventuality of billions of connected devices for services like Machine Type communications, concerns on how to manage these devices – which may exhibit special characteristics like low mobility and reduced traffic activity as well as low volume of traffic compared to high-mobility/ high-volume traffic activity smart phones – became prominent in the standards work. Since the MME is responsible for managing access of the number of UEs and type of UEs into the EPC and MME is also the node with the responsibility for keeping UEs connected and UE network connections and user-plane connections to various PDNs, it is apparent that the system needed to be fitted with tools to enable MME to coordinate and control the flow of access into the EPC, as well as the volume of signaling and data traffic generated from both originating and terminating attempts to/ from UEs.

So far we have mainly focused on the interaction between MME and eNB/UE, but MME load may also be increased due to unusual events arising from the user-plane nodes such as the Serving GW or PDN GW. These may be due to problems along the SGi interface or other issues that could be controlled by managing the signaling that can possibly trigger load in the MME from the Serving GW and/or PDN GW.

Mechanisms have been introduced to allow the MME to control the amount of download data notification triggers received from a Serving GW, which causes possible paging of the Idle UEs and increase of load in the network. Based on the MME internal conditions (such as load threshold), an MME can request a Serving GW to reduce the number of downlink data notifications for low priority traffic, which is usually determined in the Serving GW by the ARP value of the associated bearer as well as configured operator policy. The MME provides the Serving GW with a throttling factor and a throttling delay per MME, and the Serving GW always resets these values with the latest received data from the same MME. The Serving GW resumes normal operating status once the delay time has expired for that MME throttling setup. 3GPP TS 29.274 specifies the throttling related parameters shown in Table 6.3 in the Downlink Data Notification Ack message.

Table 6.3: Throttling-Related Information Elements in a Downlink Data Notification Acknowledgement

Information Elements	Condition/Comment	IE Type
Data Notification Delay	The MME/SGSN shall include an adaptive delay indication to the SGW to delay the number of Data Notification indications, if the rate of Downlink Data Notification event occurrence in the MME/SGSN becomes significant (as configured by the operator) and the MME/SGSN's load exceeds an operator configured value.	Delay Value
DL Low-Priority Traffic Throttling	The MME/SGSN may send this IE to the SGW to request the SGW to reduce the number of Downlink Data Notification requests it sends for downlink low-priority traffic received for UEs in idle mode served by that MME/SGSN in proportion to the Throttling Factor and during the Throttling Delay. For instance, if the DL low priority traffic Throttling IE indicates a Throttling Factor of 40% and a Throttling Delay of 180 seconds, the SGW drops by 40% the number of Downlink Data Notification requests it sends for downlink low-priority traffic received for UEs in idle mode served by that MME/SGSN, during a period of 180 seconds.	Throttling

The PDN GW has the ability to control the number of PDN connections towards an APN served by a PDN GW that may be experiencing some form of unusual disturbances towards the PDN served by a specific troubled APN. When PDN GW starts to reject PDN connection or additional bearer activation requests for a specific APN due to congestion or overload or other detected problem conditions related to that APN, the GW may request the MME to use additional backoff time for the UEs associated with that identified APN. The MME should then reject any PDN connection requests related to that APN and it may additionally provide the UEs with a backoff timer for session management-related procedures such as PDN connection or EPS bearer activation requests. UEs then should not trigger any further session management procedures towards that specific APN until the timer has expired, even if the UE has changed Cell/TA/PLMN or Radio Access Type. In severe conditions (if so deemed by the MME), the MME can use the session backoff timer with PDN deactivation procedures to reduce load in that specific APN. Armed with this type of information, an MME can control its own paging load as well as the number of UEs allowed to connect to a possible stressed APN/PDN GW.

In addition to these mechanisms, further congestion management is possible within an MME by equipping the MME with mobility management and session management and general congestion control backoff timers to the UEs as part of various mobility and session management procedures. These backoff timers may be applied in a general manner or they may be applied in conjunction with a specific APN, which may be facing problems during a certain unforeseen event that may be outside the control of the mobile network operator. The use of the general mobility management congestion control mechanism allows MME to reject access requests from UEs and also allows the MME to control repeated attempts by UEs to try to re-establish connections via returning backoff timers for all mobility management procedures except detachment and high-priority/emergency accesses (assuming that the UEs respect the backoff timer rules). This can be very important in cases where certain network conditions may be made worse by aggressive terminals' repeated attempt to restore network connections. This backoff timer continues to run unless the UE is being paged or the UE has moved into a new PLMN that is not part of the UE equivalent PLMN list. For a connected UE, Tracking Area Update and Handover commands should be followed while this backoff timer is running. In the APN-based mobility or session management congestion control process, the backoff timers and related actions are associated with a specific APN. The mobility management procedures affected would be when the MME rejects UEs' Attach Requests, which may include the backoff timer, and any subsequent requests are also rejected as long as the backoff timer is running by the MME, which also maintains the timer associated with each UE and the APN in effect. The session management backoff timers are applied to procedures like PDN connection and bearer activation/ modification, and may be triggered by conditions within an MME or triggered by situations and events triggered by PDN GW. In addition to rejecting session management procedures, the MME may initiate PDN connection deactivation associated with the APN in question.

A combination of the above-mentioned procedures and tools used smartly in conjunction with clever implementation choices should provide some powerful options for operators to manage/control network resources. But as many of these tools rely on the appropriate actions and respecting the rules of behavior/engagement towards the networks by the terminals, their success and failure will be greatly influenced by their compliance to the specifications. More in-depth details such as specific cause codes, error conditions, and UE behaviors can be found in 3GPP specifications TS 23.401, 24.301, and 29.274.

CHAPTER 7

Security

7.1 Introduction

Providing security is one of the key aspects of mobile networks. One of the more obvious reasons is that the wireless communication can be intercepted by anyone within a certain range of the transmitter and with the technical skills and equipment to decode the signaling. There is thus a risk that the data that is transferred can be eavesdropped, or even manipulated, by third parties. There are also other threats; for example, an attacker may trace a user's movement between radio cells in the network or discover the whereabouts of a specific user. This may constitute a significant threat to users' privacy. Apart from security aspects directly related to end-users, there are also security aspects related to the network operator and the service providers, as well as security between network operators in roaming scenarios. For instance, there should be no doubt regarding which user and roaming partner were involved in generating certain traffic in order to assure correct and fair charging of subscribers.

There are also regulatory requirements related to security and these may differ between countries and regions. The regulations can, for example, be related to exceptional situations where law enforcement agencies can request information about the activities of a terminal as well as intercept the telecommunications traffic. The framework in a telecommunications system for supporting this is called "lawful intercept". There may also be regulations to ensure that end-users' privacy is protected when using mobile networks. Requirements like these are generally captured in the national and/or regional laws and regulations by the responsible authorities for that specific nation/region.

Below we discuss different aspects of security in mobile networks, starting with a brief discussion on key security concepts and security domains. Then security aspects relating to end-users as well as within and between network entities are discussed. We conclude this chapter with a description of the framework for lawful intercept.

EPC and 4G Packet Networks.
DOI: http://dx.doi.org/10.1016/B978-0-12-394595-2.00007-4

7.2 Security Services

Before we go into the actual security mechanisms of EPS, it may be useful to briefly go through some basic security concepts that are important in cellular networks.

Before a user is granted access to a network, *authentication* in general has to be performed. During authentication the user proves that he or she is who he/she claims to be. Typically, *mutual authentication* is desired, where the network authenticates the user and the user authenticates the network. Authentication is generally done via a procedure where each party proves that it has access to a secret known only to the participating parties, for example a password or a secret key.

The network also verifies that the subscriber is *authorized* to access the requested service, for example to get access to EPS using a particular access network. This means that the user must have the right privileges (i.e. a subscription) for the type of services that are requested. Authorization for an access network is often done at the same time as authentication. It should be noted that different kinds of authorization may be required in different parts of the network and at different instances during an IP session. The network may, for example, authorize the use of a certain access technology, a certain QoS profile, a certain bit rate, access to certain services, etc.

Once the user has been granted access, there is a desire to protect the signaling traffic and user-plane traffic between the UE and the network, and between different entities within the network. *Ciphering* and/or *integrity protection* may be applied for this purpose. With ciphering (i.e. encryption and decryption) we ensure that the information transmitted is only readable to the intended recipients. To accomplish this, the traffic is modified so that it becomes unreadable to anyone who manages to intercept it, except for the entities that have access to the correct cryptographic keys. Integrity protection, on the other hand, is a means of detecting whether traffic that reaches the intended recipient has not been modified, for example by an attacker between the sender and the receiver. If the traffic has been modified, integrity protection ensures that the receiver is able to detect it. Ciphering and integrity protection serve different purposes and the need for ciphering and/or integrity protection differs depending on what traffic it is. Furthermore, the data protection may be done on different layers in the protocol stack and, as we will see, EPS supports data protection features on both protocol layers 2 and 3 depending on the scenario.

In order to encrypt/decrypt as well as to perform integrity protection, the sending and receiving entities need *cryptographic keys*. It may seem tempting to use the same key for all purposes, including authentication, ciphering, integrity protection, etc. However, using the same key for several purposes should generally be avoided. One reason is that if the same key is used for authentication and traffic protection, an attacker that manages to recover the ciphering key by breaking, for example, the encryption algorithm would at the same time learn the key used also for authentication and integrity protection. Furthermore, the keys used in one access should not be the same as the keys used in another access. If they were to be the same, the keys recovered by an attacker in one access with weak security features could be reused to break accesses with stronger security features. The weakness of one algorithm or access thus spreads to other procedures or accesses. To avoid this, keys used for different purposes and in different accesses should be distinct, and an attacker who manages to recover one of the keys should not be able to learn anything useful about the other keys. This property is called *key separation* and, as we will see, this is an important aspect of EPS security design. In order to achieve key separation, the UE and the EPC derives distinct keys that are used for different purposes. The keys may be derived during the authentication process, at mobility events, and when the UE moves to a connected state.

By *privacy protection* we here mean the features that are available to ensure that information about a subscriber does not become available to others. For example, it may include mechanisms to ensure that the permanent user ID is not sent unnecessarily often in clear text over the air link. If done, this would mean that an eavesdropper could detect the movements and travel patterns of a particular user.

Laws and directives of individual nations and regional institutions (e.g. the European Union) typically define a need to intercept telecommunications traffic and related information. This is referred to as *lawful intercept* and may be used by law enforcement agencies in accordance with the laws and regulations.

7.2.1 Security Domains

In order to describe the different security features of EPS it is useful to divide the complete security architecture into different security domains. Each domain may have its own set of security threats and security solutions. 3GPP TS 33.401 divides the security architecture into different groups or domains:

1. Network access security
2. Network domain security
3. User domain security
4. Application domain security
5. Visibility and configurability of security.

The first group is specific to each access technology (E-UTRAN, GERAN, UTRAN, etc.), whereas the others are common for all accesses. Figure 7.1 provides a schematic illustration of different security domains.

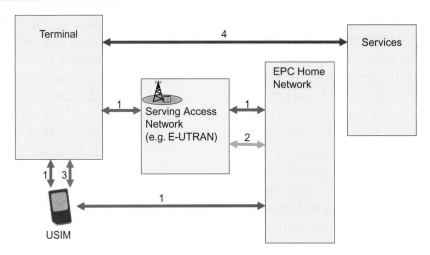

Figure 7.1: Schematic Diagram of Different Security Domains.

7.2.1.1 Network Access Security

By network access security we mean the security features that provide a user with a secure access to the EPS. This includes mutual authentication as well as privacy features. In addition, protection of signaling traffic and user-plane traffic in the particular access is also included. This protection may provide confidentiality and/or integrity protection of the traffic. Network access security is generally access specific – that is, the detailed solutions, algorithms, etc. differ between access technologies. Further details for different types of accesses are provided later in this chapter.

7.2.1.2 Network Domain Security

Mobile networks contain many network entities and reference points between these entities. Network domain security refers to the features that allow these network nodes to securely exchange data and protect against attacks on the network between the nodes.

7.2.1.3 User Domain Security

User domain security refers to the set of security features that secure the physical access to terminals. For example, the user may need to enter a PIN code before being able to access the terminal.

7.2.1.4 Application Domain Security

Application domain security is the security features used by applications such as HTTP (for web access) or IMS.

Application domain security is generally end to end between the application in the terminal and the peer entity providing the service. This is in contrast to the previous security features listed that provide hop-by-hop security – that is, they apply to a single link in the network only. If each link (and node) in the chain that requires security is protected, the whole end-to-end chain can be considered secure.

Since application-level security traverses on top of the user-plane transport provided by EPS, and as such is more or less transparent to EPS, it will not be discussed further in this book. For more information on IMS security, see for example Camarillo and Garcia-Martin (2008).

7.2.1.5 Visibility and Configurability of Security

This is the set of features that allows the user to learn whether a security feature is in operation or not and whether the use and provision of services will depend on the security feature. In most cases the security features are transparent to the user and the user is unaware that they are in operation. For some security features the user should, however, be informed about the operational status. For example, use of encryption in E-UTRAN depends on operator configuration and it should be possible for the user to know whether it is used or not, for example using a symbol on the terminal display. Configurability is the property where the user can configure whether the use or provision of a service will depend on whether a security feature is in operation.

7.3 Network Access Security

As mentioned previously, network access security is in many aspects specific to each access. Below we go into some detail on access security in different types of accesses such as E-UTRAN, HRPD, and a WLAN hotspot. With these three examples we describe the different possibilities of gaining access to the EPS. We also describe additional aspects for the case where DSMIPv6 is used.

Common to all cases is the use of USIM.

7.3.1 Access Security in E-UTRAN

It was clear from the start of the standardization process that E-UTRAN should provide a security level at least as high as that of UTRAN. Access security in E-UTRAN therefore consists of different components, similar to those that can be found in UTRAN:

- Mutual authentication between UE and network
- Key derivation to establish separate keys for ciphering and integrity protection
- Ciphering, integrity, and replay protection of NAS signaling between UE and MME
- Ciphering, integrity, and replay protection of RRC signaling between UE and eNB
- Ciphering of the user plane. The user plane is ciphered between UE and eNB
- Use of temporary identities in order to avoid sending the permanent user identity (IMSI) over the radio link.

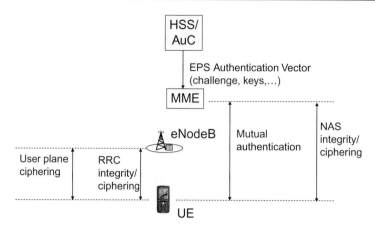

Figure 7.2: Security Features for E-UTRAN.

Figure 7.2 illustrates some of these components in the network.

Below we will discuss in detail how each of these components has been facilitated.

The authentication procedure in E-UTRAN is in many ways similar to the authentication procedure in GERAN and UTRAN, but there are also differences. To understand the reason behind these differences, it is useful to first briefly look at the security features of GERAN and UTRAN systems. As with all security features in communication systems, what was considered sufficiently secure at one point in time may not turn out to be sufficient years later when attack methods and computing power have developed further. This is also true for 3GPP radio accesses. When GERAN was developed, some limitations were purposely accepted. For example, mutual authentication is not performed in GERAN where it is only the network that authenticates the terminal. It was thought that there was no need for the UE to authenticate the network, since it was unlikely that anyone would be able to set up a rogue GERAN network. When UTRAN/UMTS was developed, enhancements were made to avoid some of the limitations of GERAN. For example, mutual authentication was introduced. These new security procedures are one reason why a new type of SIM card was needed for UMTS: the so-called UMTS SIM (or USIM for short). With the introduction of E-UTRAN, further improvement is taking place. One important aspect is, however, that it has been agreed that the use of USIM in the terminal will be sufficient to access E-UTRAN – that is, no new type of SIM card is needed. The new features are instead supported by software in the terminal and the network.

Mutual authentication in E-UTRAN is based on the fact that both the USIM card and the network have access to the same secret key K. This is a permanent key that is stored on the USIM and in the HSS/AuC in the home operator's network. Once configured, the key K never leaves the USIM or the HSS/AuC. The key K is thus not used directly to protect any traffic and it is also not visible to the end-user or even the terminal. During the authentication procedure, other keys are generated from the key K in the terminal and in the network that are used for ciphering and integrity protection of user-plane and control-plane traffic. For example, one of the derived keys is used to protect the user plane, while another key is used to protect NAS signaling. One reason why several keys are produced like this is to provide key separation and to protect the underlying shared secret K. In UTRAN and GERAN, the same keys are used for ciphering of control signaling and user traffic, and hence this is also an enhancement compared to these earlier standards. This is, however, not the only key management enhancement, as will be discussed below.

The mechanism for authentication as well as session key generation in E-UTRAN is called EPS Authentication and Key Agreement (EPS AKA). Mutual authentication with EPS AKA is done in the same manner as for UMTS AKA, but as we will see when we go through the procedure, there are a few differences when it comes to key derivation.

EPS AKA is performed when the user attaches to EPS via E-UTRAN access. Once the MME knows the user's IMSI, the MME can request an EPS authentication vector (AV) from the HSS/AuC, as shown in Figure 7.3. Based on the IMSI, the HSS/AuC looks up the key K and a sequence number (SQN) associated with that IMSI. The AuC increases the SQN and generates a random challenge (RAND). Taking these parameters and the master key K as input to cryptographic functions, the HSS/AuC generates the UMTS AV. This AV consists of five parameters: an expected result (XRES), a network authentication token (AUTN), two keys (CK and IK), and the RAND. This is illustrated in Figure 7.3. Readers familiar with UMTS will recognize this Authentication Vector as the parameter that the HSS/AuC would send to the SGSN for access authentication in UTRAN. For E-UTRAN, however, the CK and IK are not sent to the MME. Instead, the HSS/AuC generates a new key, K_{ASME}, based on the CK and IK and other parameters such as the serving network identity (SN ID). The SN ID includes the Mobile Country Code (MCC) and Mobile Network Code (MNC) of the serving network. A reason for including SN ID is to provide a better key separation between different serving networks to prevent a key derived for one serving network being (mis)used in a different serving network. Key separation is illustrated in Figure 7.4.

K_{ASME}, together with XRES, AUTN, and RAND, constitutes the EPS AV that is sent to the MME. The CK and IK never leave the HSS/AuC when E-UTRAN is used. In order to distinguish the different AVs, the AUTN contains a special bit called the "separation bit" indicating whether the AV will be used for E-UTRAN or for UTRAN/GERAN. A reason for going through this extra step with the new key K_{ASME}, instead of using CK and IK for ciphering and integrity protection as in UTRAN, is to

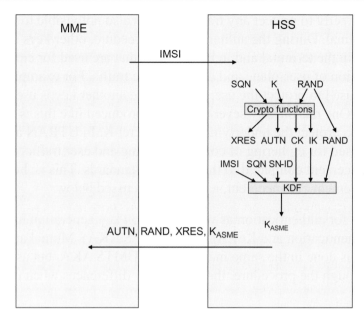

Figure 7.3: MME Fetching the EPS Authentication Vector from HSS/AuC.

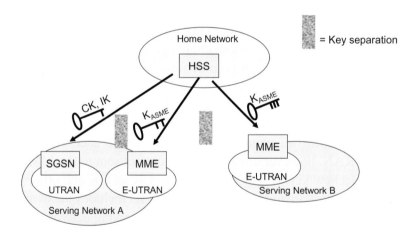

Figure 7.4: Key Separation Between 3GPP Accesses and Serving Networks.

provide strong key separation for legacy GERAN/UTRAN systems. For more details on the generation of the EPS AV, see 3GPP TS 33.401.

Mutual authentication in E-UTRAN is performed using the parameters RAND, AUTN, and XRES. The MME keeps K_{ASME} and XRES but forwards RAND and AUTN to the terminal shown in Figure 7.5. Both RAND and AUTN are sent to the USIM. AUTN is a parameter calculated by the HSS/AuC based on the secret key K and the SQN. The USIM now calculates its own version of AUTN using its own key K and SQN, and compares it with the AUTN received from the MME. If they are consistent, the USIM authenticates the network. Then the USIM calculates a response RES using cryptographic functions with the key K and the challenge RAND as input parameters. The USIM also computes CK and IK in the same way as when UTRAN is used (it is, after all, a regular UMTS SIM card). When the terminal receives RES, CK, and IK from the USIM, it sends the RES back to the MME. The MME authenticates the terminal by verifying that the RES is equal to XRES. This completes the mutual authentication. The UE then uses the CK and IK to compute K_{ASME} in the same way as HSS/AuC did. If everything has worked out, the UE and network have authenticated each other and both UE and MME now have the same key K_{ASME} (note that none of the keys K, CK, IK, or K_{ASME} was ever sent between UE and the network).

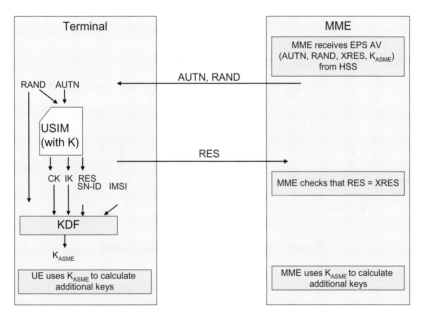

Figure 7.5: EPS AKA Between UE and MME.

Now all that remains is to calculate the keys to be used for protecting traffic. As mentioned above, the following type of traffic is protected between UE and E-UTRAN:

- NAS signaling between UE and MME
- RRC signaling between UE and eNB
- User-plane traffic between UE and eNB.

Different keys are used for each set of procedures above, and also different ciphering and integrity protection keys are used. The key K_{ASME} is used by UE and MME to derive the keys for ciphering and integrity protection of NAS signaling (K_{NASenc} and K_{NASint}). In addition, the MME also derives a key that is sent to the eNB (the K_{eNB}). This key is used by the eNB to derive keys for ciphering of the user plane (K_{UPenc}) as well as ciphering and integrity protection of the RRC signaling between UE and eNB (K_{RRCenc} and K_{RRCint}). The UE derives the same keys as eNB. The "family tree" of keys is typically referred to as a *key hierarchy*. The key hierarchy of E-UTRAN in EPS is illustrated in Figure 7.6.

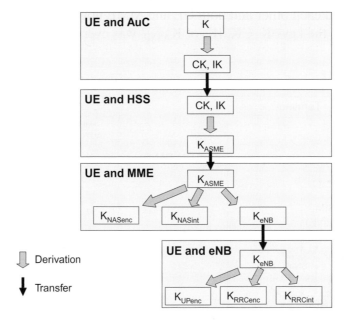

Figure 7.6: Key Hierarchy for E-UTRAN.

Once the keys have been established in the UE and the network, it is possible to start ciphering and integrity protection of the signaling and user data. The standard allows use of different cryptographic algorithms for this, and the UE and the NW need to agree on which algorithm to use for a particular connection. The EPS encryption algorithms (EEA) currently supported for NAS, RRC, and UP ciphering are shown in Table 7.1. EEA0, 128-EEA1, and 128-EEA2 are mandatory to support in the UE, eNB, and MME, while 128-EEA3 is optional to support. The EPS integrity protection algorithms (EIA) currently supported for RRC and NAS signaling integrity protection are shown in Table 7.2. The algorithms 128-EIA1 and 128-EIA2 are mandatory to support in the UE, eNB, and MME, while 128-EIA3 is optional to support. The Null integrity protection algorithm EIA0 is only used for unauthenticated emergency calls.

Table 7.1: Ciphering Algorithms for LTE

Name	Algorithm	Comment
EEA0	Null ciphering algorithm	When this algorithm is selected, there is no ciphering of the messages. Supported from Release 8.
128-EEA1	SNOW 3G-based algorithm	Supported from Release 8
128-EEA2	AES-based algorithm	Supported from Release 8
128-EEA3	ZUC-based algorithm	Added in Release 11

Table 7.2: Integrity Protection Algorithms for LTE

Name	Algorithm	Comment
EIA0	Null integrity protection algorithm	When this algorithm is selected, there is no integrity protection of the messages. Added in 3GPP Release 9 to support unauthenticated emergency calls
128-EIA1	SNOW 3G-based algorithm	Supported from Release 8
128-EIA2	AES-based algorithm	Supported from Release 8
128-EIA3	ZUC-based algorithm	Added in Release 11

For more details on the ciphering and integrity algorithms supported with E-UTRAN, see 3GPP TS 33.401.

The final aspect that should be mentioned is identity protection. In order to protect the permanent subscriber identity (i.e. IMSI) from being exposed in clear text over the radio interface, temporary identities are used whenever possible in a similar way to what is done in UTRAN. See the identities section in Chapter 6 for a description on how temporary identities are used in E-UTRAN.

A main enhancement in E-UTRAN as compared to UTRAN is, as was discussed above, the strong key separation between networks and key usage. A few other enhancements are also worth briefly mentioning:

- Larger key sizes. E-UTRAN supports not only 128-bit keys but can (in future deployments) also use 256-bit keys.
- Additional protection against compromised base stations. Due to the flattened architecture in E-UTRAN, additional measures were added to protect against a potentially compromised "malicious" radio base station. One of the most important features is the added forward/backward security: each time the UE changes its point of attachment (due to mobility) or when the UE changes from the Idle to the Connected state, the air interface keys are updated according to a sophisticated procedure. This means that even in the unlikely event that the keys used so far have been compromised, security can be restored.

7.3.2 Interworking with GERAN/UTRAN

In this book we will not describe the security features applicable to GERAN and UTRAN in any detail. The interested reader is instead referred to books dedicated to GERAN and UTRAN, see for example Kaaranen (2005). However, the interworking between GERAN/UTRAN and E-UTRAN will be discussed below.

When a UE moves between GERAN/UTRAN and E-UTRAN, there are different possibilities in order to establish the security context to be used in the target access. One possibility would be to perform a new authentication and key agreement procedure every time the UE enters a new access. In order to reduce the delays during handover between GERAN/UTRAN and E-UTRAN, however, this may not be desirable. Instead, handovers can be based on *native* or *mapped* security contexts. If the UE has previously established a native security context in E-UTRAN access by running EPS AKA, then moved to GERAN/UTRAN and later returns to E-UTRAN, the UE and network may have cached a native security context for E-UTRAN, including a native K_{ASME}, from the previous time the UE was in E-UTRAN. In this way a full AKA procedure in the target access is not needed during the inter-RAT handover. If a native context is not available, it is instead possible to map the security context used in the source access to a security context for the target access. This security context mapping is supported when moving between different 3GPP accesses. When mapping is performed, the UE and MME derive keys applicable to the target access (e.g. K_{ASME} for E-UTRAN) based on the keys used in the source access (e.g. CK, IK for UTRAN). The mapping is based on a cryptographic key derivation function (KDF) having the property that it protects the source context from the mapped target context. This assures that if attackers compromise a mapped context they get no information about the context from which it was mapped. An example of such a mapping is illustrated in Figure 7.7.

There are, however, a few important aspects related to such a mapping of security context since the protection is only one way. If the source context has already been compromised, then the mapped context will inherit this property. Also, as we have already mentioned above, the level of security is not the same in all accesses. This is where key separation becomes important. Unless the different accesses are kept separate from a security point of view, the vulnerability of one access may spread into other accesses not susceptible to the same vulnerabilities. Therefore, if a security context from, for example, GERAN has been mapped to a security context for E-UTRAN, it is highly recommended that a cached native security context is activated, or if no such context exists, a full EPS AKA run is performed as soon as possible after entering E-UTRAN to establish a fresh, native, E-UTRAN security context.

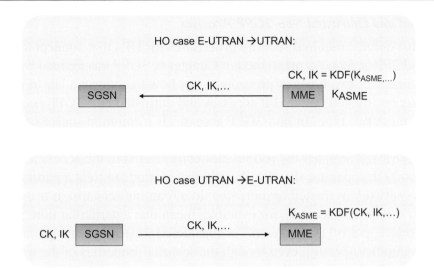

Figure 7.7: Examples of Mapping Security Context in Handover Between E-UTRAN and UTRAN.

7.3.3 Special Consideration for IMS Emergency Calls

IMS emergency calls are in some cases, from a security point of view, treated in a special way. Normally, the UE is required to authenticate to the network as described above in order to get access to the network services. Such a UE can make authenticated IMS emergency calls, assuming that IMS emergency is supported by the UE and the network. However, depending on regulatory requirements the operator may be required to provide emergency calls also for terminals where authentication fails. As a side note, in some jurisdictions the opposite is true: unauthenticated emergency calls are explicitly forbidden by regulators.

There may be several reasons why authentication of the UE fails. There may, for example, be situations where the USIM is missing from the UE or the USIM does not correspond to a valid subscription. There may also be network failures that prevent, say, the MME or SGSN from obtaining authentication vectors. In roaming situations it may also happen that the serving network does not have a roaming agreement with the subscriber's home network. In all of these cases authentication will fail and the security procedures described in previous sections cannot be applied.

As mentioned above, support for unauthenticated emergency calls may need to be supported, depending on regulatory requirements. Other non-IMS emergency services are, however, not allowed for unauthenticated UEs. If the network is configured to allow unauthenticated IMS emergency calls, such calls will be handled without the confidentiality and integrity protection services described in previous sections.

7.3.4 Trusted and Untrusted Non-3GPP Accesses

3GPP has also defined required security procedures for UEs that connect to the EPC using a non-3GPP access. As mentioned in Chapter 6, 3GPP has defined two classes of accesses, or rather two types of procedures, for how to connect a UE to EPC via a non-3GPP access: trusted non-3GPP accesses and untrusted non-3GPP accesses. The definition of these two types of non-3GPP accesses is a common source of confusion. It should, however, be noted that whether a specific non-3GPP access network is considered as trusted or untrusted is only indirectly related to the access technology itself. It is rather the operator that decides whether it wants to treat a particular non-3GPP access network as trusted or untrusted. In a roaming scenario, it is the home operator that decides. This could, for example, mean that a particular non-3GPP access network (e.g. a WLAN network) is considered trusted by one operator but untrusted by another operator, even though the security properties of the network are the same for both operators. It may instead be that the operators have different preferences when it comes to how a 3GPP UE should connect to EPC via that network. As described in Chapter 6, connectivity solutions using IPsec tunnels are used in untrusted non-3GPP networks, while connectivity solutions for trusted non-3GPP networks, rely on the connectivity solutions native to the particular access technology without additional secure tunneling from the UE.

The description for when a non-3GPP access is considered as trusted was recently updated and is described in TS 33.402 as: "When all of the security feature groups provided by the non-3GPP access network are considered sufficiently secure by the home operator, the non-3GPP access may be identified as a trusted non-3GPP access for that operator. However, this policy decision may additionally be based on reasons not related to security feature groups." The description of when to consider a non-3GPP access as untrusted is described in the same specification as: "When one or more of the security feature groups provided by the non-3GPP access network are considered not sufficiently secure by the home operator, the non-3GPP access may be identified as an untrusted non-3GPP access for that operator. However, this policy decision may additionally be based on reasons not related to security feature groups."

In the following sections we will look more closely at the access security in trusted and untrusted non-3GPP accesses.

7.3.5 Access Security in Trusted Non-3GPP Accesses

As mentioned above, the criteria for treating a non-3GPP access as trusted are not based on the access technology type but it is, rather, an operator decision. However, before moving into the technical details of access security in trusted non-3GPP accesses we will look at two examples of access technologies that an operator may decide to treat as trusted.

One example of a cellular technology not specified by 3GPP, but that can be used to provide access to EPC, is evolved HRPD (eHRPD). The security features of HRPD that are specified by 3GPP2 are thus not under the control of 3GPP. Still, HRPD has the capabilities to provide a strong access control, mutual authentication, as well as protection of signaling and user-plane traffic sent over the HRPD radio link. Even though not specified by 3GPP, it is reasonable that these security features are sufficient for providing access to EPC. Typically, HRPD would be connected directly to EPC using the S2a or S2c reference points.

Another access technology that has the capability to provide strong access control and protection over the radio link is IEEE 802.11 WLAN, assuming that the latest IEEE 802.11 security standards are applied. Assuming that the WLAN network is deployed with such security features in place, it is reasonable that an operator may also want to treat WLAN as a trusted non-3GPP access.

Access authentication in trusted non-3GPP accesses is based on EAP-AKA or, to be more precise, 3GPP has agreed to use a revision of EAP-AKA called EAP-AKA′, but more about this below. On a high level, EAP-AKA (and EAP-AKA′) is a method to perform AKA-based authentication over an access even if there is not native support for EPS AKA or UMTS AKA in that particular access. This makes it possible to perform 3GPP-based access authentication using the same credentials – the shared secret key K located in USIM and HSS/AuC – as for 3GPP accesses. EAP-AKA runs between the UE and the 3GPP AAA Server (Figure 7.8). In order to perform the AKA-based authentication, the 3GPP AAA Server downloads the Authentication Vector from HSS/AuC.

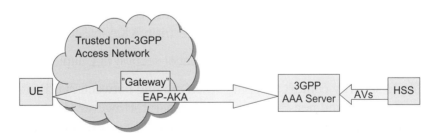

Figure 7.8: EAP-AKA-Based Access Authentication for a Trusted Non-3GPP Access Network.

It should, however, be noted that the key hierarchy and details regarding key derivation differ somewhat between EPS AKA and EAP-AKA. With EAP-AKA, the Authentication Vector from HSS/AuC is a starting point for the authentication procedure, just as in EPS AKA. Then, a Master Key is derived by the UE and 3GPP AAA Server based on the CK and IK; in a way this is conceptually similar to how

K_{ASME} is derived from CK and IK with EPS AKA. This Master Key is used to derive further keys, for example keys that are used to protect user-plane and control-plane traffic in the particular trusted non-3GPP access. In this book we will not go into further detail regarding security in specific non-3GPP access technologies such as eHRPD or WLAN.

As already mentioned above, EAP-AKA′ is used for access authentication in trusted non-3GPP accesses. EAP-AKA′, specified in IETF RFC 5448, is a small revision of EAP-AKA, defined in IETF RFC 4187. The revision made in EAP-AKA′ is the introduction of a new key derivation function that binds the keys derived within EAP-AKA′ to the identity of the access network. In practice, this means that the access network identity is taken into account in the key derivation schemes. The procedure is thus more aligned with EPS AKA and strengthens the key separation.

Figure 7.9 provides an example of an authentication using EAP-AKA/EAP-AKA′. At the level of detail shown in Figure 7.9, the difference between EAP-AKA and EAP-AKA′ is only that a few parameters associated with the serving access network name are included when using EAP-AKA′. For example, with EAP-AKA′ the access network identity is included in the EAP-Request/Identity. The access network identity is used by the AAA Server and the UE in the key derivation functions. Although not explicitly illustrated in the figure, the EAP messages between the EAP Peer and the EAP Authenticator are carried over an underlying protocol specific to the type of access. The EAP messages between the EAP Authenticator and the EAP Server are carried in an AAA protocol. For trusted non-3GPP access the AAA protocol is based on Diameter. The interested reader may want to compare the EAP-AKA′ message exchange in Figure 7.9 with the EPS AKA message exchange over E-UTRAN described in Section 7.3.1. The EPS AKA over E-UTRAN and the EAP-AKA′ for accesses supporting EAP are two ways to perform AKA-based authentication.

For more details on EAP, see the EAP protocol section in Chapter 16.

There is no security context mapping when moving between 3GPP and eHRPD, or between 3GPP and non-3GPP access in general. However, despite the lack of context mapping, there are means to optimize the security procedures during handover between 3GPP access and non-3GPP access. If the UE is capable of communicating in both 3GPP access and non-3GPP access at the same time, as for example is typically possible in 3GPP and WLAN accesses, the UE can perform the EAP-AKA′ procedure in the target non-3GPP access already before leaving the source 3GPP access. Once the handover takes place, the security context is already established in the target non-3GPP access. Another example that does not require simultaneous communication on both 3GPP and non-3GPP accesses but still allows for optimizations is the optimized eHRPD interworking described in Chapter 6.

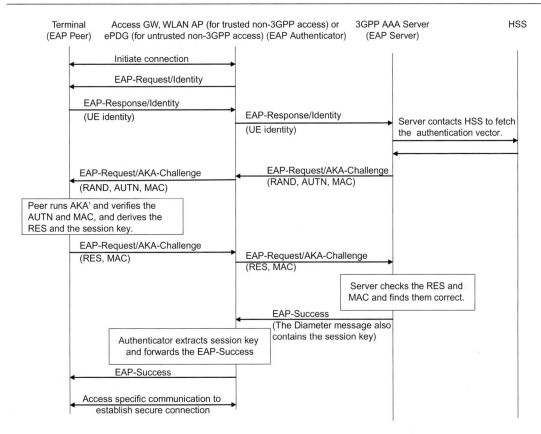

Figure 7.9: EAP-AKA′/EAP-AKA Call Flow.
Only a subset of the information elements is shown.

When a UE is active in E-UTRAN access and the dedicated signaling connection with eHRPD (via S101) is set up, the UE can perform the EAP-AKA′ procedure for eHRPD before actually handing over to eHRDP access. Also, in the other direction – that is, handover from eHRPD to E-UTRAN – the EPS AKA security procedures for E-UTRAN can be performed via the S101 signaling connection.

For privacy protection, EAP-AKA supports means to use temporary identities (so-called pseudonyms) in a similar way as for E-UTRAN access. The EAP-AKA pseudonyms are, however, of a different format than the temporary identities used in 3GPP accesses.

7.3.6 Access Security in Untrusted Non-3GPP Access

Even though WLAN was mentioned as an example of a possible trusted non-3GPP access in the previous section, there may also be cases when an operator prefers to treat a WLAN network as an untrusted non-3GPP access. WLAN can be used in many scenarios, for example in corporate environments, in the home, or in public location such as airports and coffee shops. The level of security provided by a WLAN access also differs between deployments. For home and corporate use, WLAN security solutions such as WEP and WPA are often used. In public places it is, however, still common to turn off WLAN security completely. Instead, access control is provided by means of a web page where the user can enter a username and a password. The user may, for example, have received a username and password for a temporary subscription at the same time he or she bought a cup of coffee. Once the user has entered the credentials on the web page, Internet access is provided. The WLAN access in this case does not provide any encryption or integrity protection of the user plane and is vulnerable to many types of attacks. In such deployments, WLAN access to EPS will therefore most likely be handled as an untrusted non-3GPP access. In other deployments with WLAN security turned on, WLAN might be treated as a trusted non-3GPP access as described in the previous section.

Now the user wants to get access to the services provided by his or her operator via EPS. In general, the coffee shop where the end-user is located and wants to connect does not have any agreement to connect to the EPC directly. Equally important, since no or very limited security is provided, providing direct access to EPC would make EPC vulnerable to attacks. The solution to this problem, as defined by EPS, is to set up an IPsec tunnel between the UE and a so-called ePDG (evolved Packet Data Gateway) inside the operator's network when the access is untrusted (Figure 7.10). The ePDG acts as a secure entry point into the EPC.

In order to set up the IPsec tunnel, the UE must first perform mutual authentication towards the ePDG and operator network, as well as establish keys for the IPsec security association. This is done using the IKEv2 protocol. Once the UE has connected to the WLAN and discovered the IP address of the ePDG (this can be done using DNS) it starts the IKEv2 procedure. As part of IKEv2, public key-based authentication with certificates is used to authenticate the ePDG. The UE, on the other hand, is authenticated in a similar manner as for E-UTRAN – that is, based on the credentials on the USIM. Within IKEv2, EAP-AKA is run to perform the AKA-based authentication and key agreement. Therefore, the USIM-based authentication described in the previous sections can also be performed when the UE accesses over a generic WLAN hotspot. EAP-AKA is the same protocol as was described in Section 7.3.5. The difference is that now EAP-AKA is run as part of the IKEv2 procedure, while in Section 7.3.5 EAP-AKA' is run as part of the attachment procedure in a trusted non-3GPP access such as eHRPD. Similar properties regarding key generation and privacy protections as mentioned about EAP-AKA' in Section 7.3.5 apply to

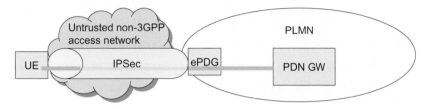

(a) Authentication and tunnel setup using IKEv2 and EAP-AKA

(b) User plane traversing IPSec tunnel between UE and ePDG

Figure 7.10: (a) Authentication Using IKEv2 and EAP-AKA Between UE and ePDG. (b) User-Plane Traffic is Protected in IPsec Tunnel Between UE and ePDG.

EAP-AKA in this scenario as well. An example call flow of EAP-AKA is shown in Figure 7.9.

It should be noted that the WLAN hotspot has been used in this section as an example scenario where connection using IKEv2 and IPsec to an ePDG is suitable. The ePDG may also be used for any access that can provide IP connectivity, for example from the DSL connection at home, independent of security properties of the underlying access.

7.3.7 Special Considerations for Host-Based Mobility DSMIPv6

In the previous sections we have described the security features for two main scenarios, a user attaching to an access network that provides a high level of security, for example E-UTRAN and eHRDP, and a user attaching to an access network where additional security protection is needed (IPsec tunnel over an unprotected WLAN access).

However, the choice of mobility protocol also affects the security features. As has been described in earlier chapters, specifically the mobility section of Chapter 6, there are two main methods for providing access in EPS, either using network-based mobility (GTP or PMIP) or host-based mobility (DSMIPv6 or MIPv4). When network-based mobility is used, the access security described in the previous chapters provides the necessary security between UE and the EPC. However, when host-based mobility is used, there is a need to also provide security for the host-based mobility protocol between UE and PDN GW.

When DSMIPv6 is used, the DSMIPv6 signaling between UE and PDN GW is integrity protected using IPsec. In order to establish the IPsec security association for DSMIPv6 signaling, the user is first authenticated using IKEv2 and EAP-AKA, as illustrated in Figure 7.11. This EAP-AKA-based authentication for DSMIPv6, as well as IPsec protection of the DSMIPv6 signaling, is done in addition to any access-level authentication and user-plane protection that may be performed (as described in previous chapters). This means that the DSMIPv6 signaling may be protected twice, first using the general user-plane protection on access level and then using IPsec between UE and PDN GW. Additionally, in case there is an ePDG on the path, there is an IPsec tunnel between UE and ePDG.

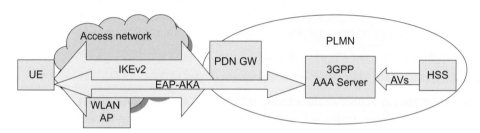

Figure 7.11: IKEv2- and EAP-AKA-Based Authentication and Key Agreement for DSMIPv6.

Furthermore, DSMIPv6 also allows the IPsec Security Association (SA) established during DSMIPv6 authentication to be used for user-plane traffic. To enable user-plane protection, the UE and the Home Agent (PDN GW) creates a separate so-called Child SA for the user plane. In this case the user plane between UE and Home Agent (PDN GW) is tunneled in IPsec. This enhanced security for the user plane when using DSMIPv6 is an optional feature that was introduced in 3GPP Release 10.

On a high level, the basic security features for the MIPv4 control plane are similar to those of DSMIPv6 – that is, the MIPv4 signaling needs to be integrity protected to ensure that only authenticated UEs can send MIPv4 signaling messages to the PDN GW. However, the details of the security solution for MIPv4 are quite different compared to DSMIPv6. MIPv4 performs integrity protection of signaling using a special authentication element in the signaling messages. The messages are thus not protected using IPsec between UE and the network. The MIPv4 protocol also does not support enhanced security of the user-plane tunnel. For more details on the MIPv4 security solution, see 3GPP TS 33.402.

7.4 Network Domain Security

When GSM/GERAN was developed, no solution was specified for how to protect the traffic in the core network. This was perceived not to be a problem, since the

GSM networks typically were controlled by a small number of large institutions. Furthermore, the original GSM networks were only running circuit-switched traffic. These networks used protocols and interfaces specific for circuit-switched voice traffic and typically only accessible to large telecom operators. With the introduction of GPRS as well as IP transport in general, the signaling and user-plane transport in 3GPP networks now runs over networks and protocols that are more open and accessible to others than the major institutions in the telecom community. This brings a need to provide enhanced protection also to traffic running over core network interfaces. For example, the core network interfaces may traverse third-party IP transport networks, or the interfaces may cross operator boundaries as in roaming cases. 3GPP has therefore developed specifications for how IP-based traffic is to be secured also in the core network and/or between a core network and some other (core) network. On the other hand, it should be noted that also today, if the core network interfaces run over trusted networks, for example a transport network owned by the operator that is physically protected, there would be little need for this additional protection.

The specifications for how to protect the IP-based control-plane traffic is called Network Domain Security for IP-based control planes (NDS/IP) and are available in 3GPP TS 33.210. This specification introduces the concept of security domains. The security domains are networks that are managed by a single administrative authority. Hence, the level of security and the available security services are expected to be the same within a security domain. An example of a security domain could be the network of a single telecom operator, but it is also possible that a single operator divides its network into multiple security domains. On the border of the security domains, the network operator places Security Gateways (SEGs) to protect the control-plane traffic that passes in and out of the domain. All NDS/IP traffic from network entities of one security domain is routed via an SEG before exiting that domain towards another security domain. The traffic between the SEGs is protected using IPsec, or to be more precise, using IPsec Encapsulated Security Payload (ESP) in tunnel mode. The Internet Key Exchange (IKE) protocol, either IKEv1 or IKEv2, is used between the SEGs to set up the IPsec security associations. An example scenario is illustrated in Figure 7.12.

A case of special relevance to EPS and E-UTRAN is the S1-U interface between EPC and the E-UTRAN. This interface needs to be properly protected (physically and/ or by NDS/IP) since the user-plane data protection would otherwise be terminated in the eNB, potentially exposing sensitive data on S1. The S1-U interface is hence an exception in that it is a user-plane interface that is protected by IPsec in 3GPP networks. The S1-U interface has additional exceptional features compared to other IPsec protected interfaces in 3GPP networks, e.g. it is mandatory to provide confidentiality protection.

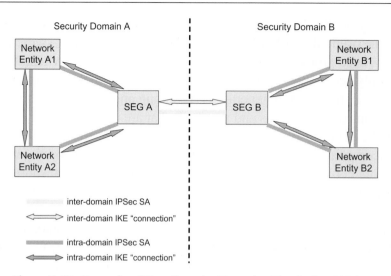

Figure 7.12: Example of Two Security Domains Employing NDS/IP.

Although NDS/IP was initially intended mainly for the protection of control-plane signaling only, it is possible to use similar mechanisms to protect the user-plane traffic. Indeed, for the aforementioned case of user data over S1, NDS/IP will be used in deployments that require it.

Also within a security domain – that is, between different network entities or between a network entity and an SEG – the operator may choose to protect the traffic using IPsec. The end-to-end path between two network entities in two security domains is thus protected in a hop-by-hop manner.

7.5 User Domain Security

The most common security feature in this user domain context is the secure access to the USIM. Access to the USIM will be blocked until the USIM has authenticated the user. Authentication is in this case based on a shared secret (the PIN code) that is stored inside the USIM. When the user enters the PIN code on the terminal, it is passed on to the USIM. If the user provided the right PIN code, the USIM allows access from the terminal/user, for example to perform the AKA-based access authentication.

7.6 Security Aspects of Home eNBs and Home NBs

Home eNBs and Home NBs are small base stations that typically are placed in customer premises such as residential or enterprise areas. This creates a number of challenges for the security area that are different compared to regular macro base stations.

In this section we will briefly describe a few of the challenges with Home (e)NBs and solutions that have been defined in 3GPP. (We will use the term H(e)NB when referring to both Home NB (HNB) and Home eNB (HeNB).) We will look at a few different aspects relating to H(e)NB security:

- Closed Subscriber Groups
- Device authentication
- Hosting party authentication
- IPsec tunnel establishment for backhaul link security
- Location verification
- Security procedures internal to the H(e)NB.

The security procedures internal to the H(e)NB are also important since H(e)NBs may be placed in public or residential areas, where it is easy to get physical access to the H(e)NB. The H(e)NB should therefore support a Trusted Environment (TrE), which is a logical entity within the H(e)NB that provides a trustworthy environment for the execution of sensitive functions and the storage of sensitive data.

7.6.1 H(e)NB Security Architecture

For a general description of the architecture for an H(e)NB subsystem, see Chapters 2 and 6. In this section we will focus on the security aspects of the H(e)NB subsystem. Figure 7.13 shows the security aspects of the H(e)NB system architecture. The H(e)NBs are typically located in customer premises, e.g. in the home of end-users, and the backhaul between the H(e)NB and the mobile operator's core network typically traverses a regular fixed broadband connection, which in most cases cannot be considered secure. This places special requirements on the protection of that traffic passing this backhaul link. For this reason the H(e)NB accesses the operator's security domain via a Security Gateway (SeGW).

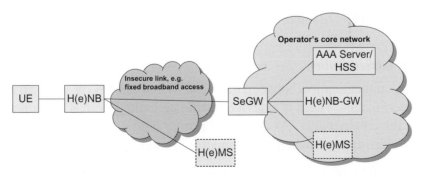

Figure 7.13: System Architecture for H(e)NB (Security Aspects Only).

The SeGW is a network element at the border of a security domain of the mobile operator. The SeGW is located between the H(e)NB and H(e)NB-GW, if an H(e) NB-GW is deployed. Otherwise the SeGW is located at the edge of the core network. After successful mutual authentication between the H(e)NB and the SeGW, as will be described in further detail below, a security tunnel is established between H(e)NB and the SeGW to protect information transmitted over the backhaul link. Any connection between the H(e)NB and the H(e)NB-GW or EPC is tunneled through the SeGW. Note that the SeGW for H(e)NB is a different logical entity from the Security GWs, or SEGs, used with NDS/IP.

7.6.2 Closed Subscriber Groups

Closed Subscriber Groups (CSG) is a concept that was introducce in Release 9, together with support for Home (e)NBs. With CSGs an operator is able to configure a cell with a certain access mode, where the access mode may be open, hybrid, or closed. This is done to control the access for HeNBs that are located, for example, in corporate or residential environments. When the cell is configured for open access mode, the cell is able to provide services as usual. This means that the cell is open to any subscriber of the same PLMN or to roaming subscribers of other PLMNs assuming there is a suitable roaming agreement. When the cell is configured for hybrid access mode, the cell provides services to both its associated CSG members and to subscribers not belonging to its associated CSG. When the cell is configured for closed access mode, only users that belong to its associated CSG are able to obtain services. The CSG concept is thus a way to allow the operator to restrict authorization to use certain cells to only the members of a Closed Subscriber Group.

An eNB that supports CSG and is configured to serve as a certain CSG cell is broadcasting a CSG Indicator and a specific CSG identity. A hybrid cell serves both CSG members and non-members. Such a cell is not broadcasting a CSG Indicator but is broadcasting a CSG identity.

For more details on HeNBs and CSG handling in general, see Chapter 6.

7.6.3 Device Authentication

Device mutual authentication between the H(e)NB and the SeGW is based on IKEv2 with digital certificates. There is a device certificate in the H(e)NB and a network certificate in the core network. As described further below, the authentication leads to an IPsec tunnel being established between H(e)NB and SeGW.

An example call flow of device authentication is provided in Figure 7.14.

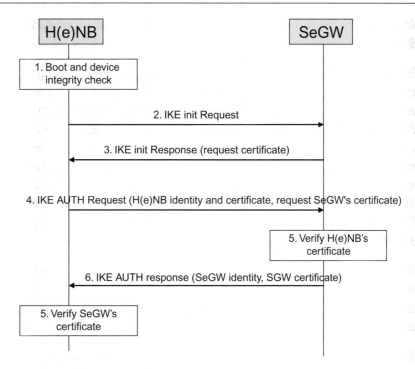

Figure 7.14: Device Authentication.

7.6.4 Hosting Party Authentication

In addition to authentication of the H(e)NB device towards the core network, there may also be an optional identification and authentication of the hosting party. The hosting party is the party that has a contractual relationship with the mobile operator for access via H(e)NB. In this case there is a so-called Hosting Party Module, provided by means of a USIM application hosted on a smartcard. The smartcards used in 3GPP systems are Universal Integrated Circuit Cards (UICCs). The UICC is distinct from the H(e)NB physical equipment but is inserted in the H(e)NB. The USIM on the UICC is used for identification and authentication of the Hosting Party towards the MNO.

The hosting party mutual authentication is optional and, if supported, is performed following successful device mutual authentication. The authentication is based on EAP-AKA between the H(e)NB/USIM and the AAA Server and HLR/HSS. The EAP-AKA procedure has already been described above for access authentication for non-3GPP accesses. When EAP-AKA is used for H(e)NB hosting party authentication, the SeGW acts as EAP authenticator and forwards the EAP protocol messages to the AAA Server to retrieve an authentication vector from AuC via HSS/HLR. As usual with USIMs and EAP-AKA, the user identifier is in the format of an IMSI. These IMSIs are available as subscription records in HLR/HSS as regular subscriptions, but should be marked there as used for H(e)NBs, e.g. by allocating dedicated ranges or by adding specific attributes to avoid misuse of these IMSIs for ordinary UEs.

An example call flow of device authentication followed by hosting party authentication is provided in Figure 7.15.

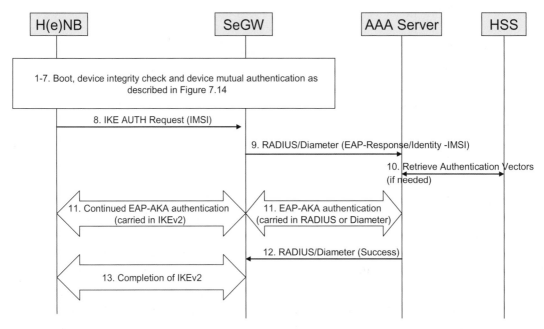

Figure 7.15: Hosting Party Authentication.

7.6.5 Backhaul Link Security

In order to protect the traffic between the H(e)NB and the H(e)NB-GW or core network, a secure tunnel between H(e)NB and the SeGW is established. The H(e)NB and SeGW use the IKEv2 authentication to set up an IPsec tunnel where all signaling, user, and management plane traffic over the interface between H(e)NB and SeGW is carried. The Encapsulating Security Payload (ESP) protocol is used for the IPsec tunnel.

IPsec use for the backhaul link is mandatory to implement but optional to use based on an operator policy. If the operator chooses not to use IPsec, mutual authentication between the H(e)NB device and the SeGW will be performed and the interface between the H(e)NB and SeGW will be secured with a mechanism that provides security for confidentiality and integrity protection of communications. The details for such non-IPsec-based solutions are, however, not specified by 3GPP.

In addition to the IPsec protection of the link between H(e)NB and H(e)NB-GW, secure communication between H(e)NB and the H(e)NB Management System using TLS has also been specified.

7.6.6 Location Verification

Even though an H(e)NB is typically deployed in customer premises, it still uses licensed radio frequencies belonging to the operator providing the H(e)NB. The H(e)NB also needs to fulfill the regulatory requirements for the country or region where it is located. It may therefore be a problem if an end-user moves the H(e)NB to another region or another country where, for example, the service providers operating in the given frequency bands and/or the regulations are different from where the H(e)NB was intended to be located. For these reasons 3GPP has also included Location Verification support in the H(e)NB specifications. The specification of the location verification feature is, however, not very detailed and different solutions are also allowed to fit different implementations and markets. One reason for this is that the requirements for location verification may vary between regulatory domains. Furthermore, it is a difficult task to find a single solution that fulfills all requirements and works in all conditions. For example, placing a GPS receiver in the H(e)NB and providing the geographic coordinates to a verifying entity in the core network may be a very accurate method but is not always reliable, e.g. in indoor locations where there is no GPS reception. There are also other possibilities, e.g. to verify the IP address and/or fixed access line identity used by the H(e)NB. We will not go further into the details of location verification. The interested reader may consult 3GPP TS 33.320.

7.7 Lawful Intercept

Lawful Interception (LI) is one of the regulatory requirements operators must satisfy as a legal obligation toward the Law Enforcement Agencies (LEA) and Government Authorities in most countries where they are operating their businesses. Within 3GPP standards, this is currently defined as: *Laws of individual nations and regional institutions (e.g. European Union), and sometimes licensing and operating conditions define a need to intercept telecommunications traffic and related information in modern telecommunications systems. It has to be noted that lawful interception shall always be done in accordance with the applicable national or regional laws and technical regulations* (as per 3GPP TS 33.106 "Lawful Interception Requirements"). LI allows appropriate authorities to perform interception of communication traffic for specific user(s) and this includes activation (requiring a legal document such as a warrant), deactivation, interrogation, and invocation procedures. A single user (i.e. interception subject) may be involved where interception is being performed by different LEAs. In such scenarios, it must be possible to maintain strict separation of these interception measures. The Intercept Function is only accessible by authorized personnel. As LI has regional jurisdiction, national regulations may define specific requirements on how to handle the user's location and interception across boundaries. As a necessary part of mobile communications systems, handover is a basic process in EPS. Interception is also carried out when handover has taken place, when required by national regulations.

This subsection deals with this aspect on a brief and high level in order to complete the overall EPS functionalities; it is intended as a description of the 3GPP LI standards and not of any function implemented in Ericsson's or other vendors' nodes. The LI function in itself does not place requirements on how a system should be built but, rather, requires that provisions be made for legal authorities to be able to get the necessary information from the networks via legal means, according to specific security requirements, without disruption of the normal mode of operations and without jeopardizing the privacy of communications not to be intercepted. Note that LI functions must operate without being detected by the person(s) whose information is being intercepted and other unauthorized person(s). As this is the standard practice for any communications networks already operating today around the world, EPS is no exception.

The process of collection of information is done by means of adding specific functions into the network entities where certain trigger conditions will then cause these network elements to send data in a secure manner to a specific network entity responsible for such a role. Moreover, specific entities provide administration and delivery of intercepted data to Law Enforcement in the required format. As an example, Figure 7.16 shows the LI architecture for some of the EPS nodes.

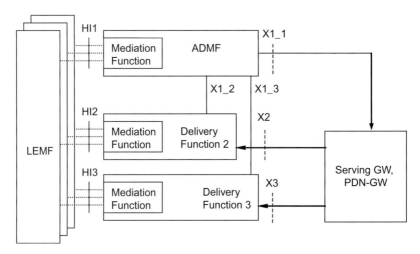

Figure 7.16: High-Level EPS-LI Architecture for S-GW/PDN-GW.

Intercept-related information (also referred to as Events) are triggered by activities detected at the network element. Some events applicable to the MME are:

- Attachment
- Detachment
- Tracking Area Update
- UE requested PDN connectivity
- UE requested PDN disconnection.

Events are triggered for E-UTRAN access if the following user-plane-related activities are detected at the Serving GW and at the PDN GW:

- Bearer activation (valid for both Default and Dedicated bearers)
- Bearer modification
- Bearer deactivation
- Start of intercept with bearer active
- UE requested bearer resource modification.

Depending on national regulations, intercept-related information collected may also be reported by the HSS.

Local regulations may allow for operators to charge for the services rendered towards the LI requesting party. The charging data collection process may include some or all of the following situations being supported:

- Use of network resources
- Activation and deactivation of the target
- Every intercept invocation
- Flat rate.

This brief overview represents high-level functions supported in EPS in order to fulfill the LI requirements. This does not in any way show the complete possibilities or aspects of this function, since it is deemed to be unrelated to the overall architecture aspects of the new system, but is, rather, shown for completeness of the system in itself.

CHAPTER 8

Quality of Service, Charging, and Policy Control

8.1 Quality of Service

Many mobile broadband operators aim to provide multiple services (Internet, voice, video) across their packet-switched access networks. These services will share the radio and core network resources with best effort services such as Internet browsing and e-mail download, and they all have different Quality of Service (QoS) requirements in terms of required bit rates as well as acceptable packet delays and packet loss rates. Furthermore, with mobile broadband subscriptions offering flat-rate charging, high-bandwidth services such as file sharing also become more common in cellular systems. In such a multi-service scenario, it is important that EPS provides an efficient QoS solution that ensures that the user experience of each service running over the shared radio links is acceptable. Simply solving these issues through over-provisioning is not economical; the available radio spectrum is limited and the costs of transmission capacity, including both spectrum allocations and backhaul links to potentially remote base stations, are important factors to an operator.

In addition to service differentiation, an important aspect is subscriber differentiation. The operator may provide differentiated treatment of the IP traffic for the same service depending on the type of subscription the user has. These subscriber groups can be defined in any way suitable to the operator, for example corporate vs. private subscribers, post-paid vs. pre-paid subscribers, and roaming vs. non-roaming subscribers, as illustrated in Figure 8.1.

The conclusion is that there is a need to standardize simple and effective QoS mechanisms for multi-vendor mobile broadband deployments. Such QoS mechanisms should allow the operator to enable service and subscriber differentiation and to control the performance experienced by the packet traffic of a certain service and subscriber group.

8.1.1 QoS in E-UTRAN

Before going into detail regarding the QoS parameters and mechanisms for E-UTRAN and EPS, we will put the EPS bearer QoS concept into a wider context.

EPC and 4G Packet Networks.
DOI: http://dx.doi.org/10.1016/B978-0-12-394595-2.00008-6

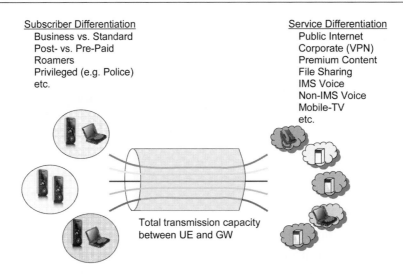

Subscriber Differentiation
Business vs. Standard
Post- vs. Pre-Paid
Roamers
Privileged (e.g. Police)
etc.

Service Differentiation
Public Internet
Corporate (VPN)
Premium Content
File Sharing
IMS Voice
Non-IMS Voice
Mobile-TV
etc.

Total transmission capacity
between UE and GW

Figure 8.1: Service and Subscriber Differentiation.

The EPS only covers QoS requirements for the traffic within the EPS – that is, between UE and PDN GW. If the service extends beyond that, QoS is maintained by other mechanisms that, for example, depend on operator deployments and service level agreements (SLAs) between network operators. This book will not go further into these aspects.

The EPS bearer service has been introduced in Chapter 6. The EPS bearer represents the level of granularity for QoS control in E-UTRAN/EPS and provides a logical transmission path with well-defined QoS properties between UE and the network. The QoS concepts of the EPS bearer is then mapped to the QoS concepts of the underlying transport. For example, over the E-UTRAN radio interfaces, the EPS bearer QoS characteristics are implemented using E-UTRAN-specific traffic handling mechanisms. Each EPS bearer is transported over an E-UTRAN radio bearer with the corresponding QoS characteristics. In the "backbone" network between eNB, Serving GW, and PDN GW, the EPS bearer QoS may be mapped to IP transport layer QoS, for example using DiffServ. In this book we will only very briefly touch upon the lower layer QoS mechanisms. The interested reader is referred to Dahlman (2011) for details regarding QoS mechanisms in the E-UTRAN radio layer.

8.1.1.1 Differences Compared to QoS for pre-EPS GERAN/UTRAN
The QoS solutions for E-UTRAN have a few differences compared to the QoS solutions defined for GERAN/UTRAN. The two most prominent differences are described below.

Bearer Control Paradigm
The bearer control paradigm has changed since GPRS Release 6 and earlier, as it was only the UE that could initiate a new bearer (i.e. a PDP context). The UE also

controlled the traffic-to-bearer mapping information. Release 7 was when the system was amended with an NW-initiated procedure to establish bearers and traffic mapping information. EPS and E-UTRAN, on the other hand, implement a fully network-controlled bearer concept. The UE may request resources, but it is always the network that controls the EPS bearer state and the traffic-to-bearer mapping.

QoS Parameters of a Bearer

E-UTRAN also simplifies the QoS parameters that are associated with each bearer. As we will describe in more detail in later subsections, the EPS bearer is associated with two QoS parameters, a QoS class and an allocation and retention parameter. Certain EPS bearers also have associated bit rates resulting in a total of four QoS parameters for those EPS bearers.

Pre-EPS GPRS, on the other hand, defines a QoS concept for GERAN/UTRAN with four Traffic Classes and 13 different bearer QoS Attributes. In Release 9 a 14th bearer QoS attribute, the Evolved ARP, was added to be used for GERAN/UTRAN, but we will go into the reasons for that briefly later. This QoS concept is often referred to as "Release 99 QoS" since it was introduced for Release 99 GPRS (completed in 2000). Each PDP context is assigned one of the four Traffic Classes together with values of the associated QoS Attributes. The QoS Attributes specify bit rates supported by the PDP context, the priority of the traffic, error rates, maximum transfer delay, etc. This generates a complex system and many of the QoS Attributes are not used in practice. We will not go into any further details on Release 99 QoS in this book. We will only discuss Release 99 QoS on a high level to understand how GERAN/UTRAN is used together with EPS. Readers interested in more information on GRPS and Release 99 QoS are instead referred to a book on GPRS, see for example Kaaranen (2005).

Subscribed QoS Parameters

In pre-EPS GPRS the subscribed QoS profile is the maximum QoS that can be allocated for each PDP context. A terminal activating multiple PDP contexts would thus have a subscribed QoS for each of them. In EPS, however, the subscribed QoS stored in HSS only applies to the default bearer. There is no such concept as subscribed QoS for a dedicated bearer. Instead, it is the PDN GW that determines the QoS of the dedicated bearer based on the authorized QoS received from the PCRF (Policy and Charging Rules Function). There is thus no need to have specific subscription parameters for dedicated bearers in the HSS. If the terminal is allowed to access a certain service, the PCRF will authorize the resources in the network.

Below we discuss the EPS bearer QoS in more detail. We then return to Release 99 QoS and GERAN/UTRAN to describe how QoS for GERAN/UTRAN works when GERAN/UTRAN is connected to the EPC.

8.1.1.2 QoS Parameters of an EPS Bearer

The basic aspects of the EPS bearer and its use for QoS purposes have already been introduced (Section 6.2). In this section we discuss the QoS parameters of the EPS bearer in more detail.

Each EPS bearer has two QoS parameters associated with it: the QoS Class Identifier (QCI) and the Allocation and Retention Priority (ARP). As we will see below, the QCI determines what user-plane treatment the IP packets transported on a given bearer should receive, while the ARP specifies the control-plane treatment a bearer should receive. Some EPS bearers also have associated bit rate parameters to support allocation of a guaranteed bit rate (GBR) when establishing the bearer.

As we will see below, the bearer concept and the associated QoS mechanisms provide two essential features: traffic separation and resource-based admission control. The class-based QoS concept using QCI allows the network to separate between bearers carrying real-time and bearers carrying non-real-time traffic. The network can then provide the appropriate forwarding treatment for each QoS class. For the support of services that require a certain GBR, the network reserves a GBR when establishing the corresponding bearer. These bearers are subject to admission control to ensure that sufficient resources are available before allowing the GBR bearer to be set up. These mechanisms are described further below.

QoS Class Identifier

The EPS uses a class-based QoS concept where each EPS bearer is assigned a QCI. This QCI is a number and the numerical value in itself does not represent any QoS property. The QCI is just a pointer, or reference, to node-specific parameters, which define what packet forwarding treatment a particular bearer should receive when processed in a node (scheduling weights, admission thresholds, queue management thresholds, link layer protocol configuration, etc.). The node-specific parameters for each QCI have been preconfigured by the vendor designing the node or by the operator owning the node (e.g. eNodeB).

Allocation and Retention Priority

The ARP is used to indicate a priority for the allocation and retention of bearers. It is typically used by the network to decide whether a bearer establishment or modification can be accepted or needs to be rejected due to resource limitations.

EPS supports 15 different ARP values. In the 2G/3G core network until Release 9, the packet core network supported only three ARP values, while 15 ARP values were supported by the GERAN and UTRAN radio networks. The three values on the packet core network needed to be mapped to the 15 values in the RAN. Providing only three values from the core network was considered sufficient since emergency calls run over circuit-switched voice services only and the ARP mechanism in the packet-switched domain could thus be used for commercial purposes only. However, with the packet-only EPS, emergency services also have to be supported in the packet-switched domain. Therefore, the definition of ARP for EPC has been more aligned with the ARP used with circuit-switched services and in GERAN/UTRAN radio networks. It will be noted that, starting from Release 9, a new Evolved ARP parameter with 15 values for use in the 2G/3G core network with GERAN/UTRAN has been introduced. In Release 9 and beyond, the definition of the ARP is thus aligned between RAN and Packet Core networks and between the different 3GPP radio accesses.

In situations where resources are scarce, the network can use the ARP to prioritize establishment and modification of bearers with a high ARP over bearers with a low ARP when performing admission control; note that bearers with high ARP are assigned low ARP *values* and vice versa. For example, a VoIP call for emergency services should have a higher chance of being accepted than a regular VoIP call, and should thus be assigned a high ARP. ARP also supports pre-emption of bearers. When there are exceptional resource limitations, the network can use the ARP to decide which bearers to drop. This could, for example, occur at handover situations. Another exceptional circumstance is disaster situations, where the ARP can be used to free up capacity by dropping low ARP bearers or giving access to emergency responders/ authorities dealing with the situations.

GBR and Non-GBR Bearers

One of the properties of a bearer is the bit rates it is associated with. We distinguish between two types of bearers: GBR bearers and non-GBR bearers. A GBR bearer has, in addition to the QoS parameters discussed above, associated bit rate allocations: the GBR and the Maximum Bit Rate (MBR). A non-GBR bearer does not have associated bit rate parameters.

A bearer with an associated GBR means that a certain amount of bandwidth is reserved for this bearer, independently of whether it is utilized or not. The GBR bearer thus always takes up resources over the radio link, even if no traffic is sent. The GBR bearer should not in normal cases experience any packet losses due to congestion in the network or radio link. This is ensured since GBR bearers are subject to admission control when they are set up. A GBR bearer is only allowed by the network if there are enough resources available. The MBR limits the bit rate that can be expected to be provided by a GBR bearer. Any traffic in excess of the MBR may be discarded by a rate shaping function. Up to Release 9, EPC only supports the case where the MBR and GBR are equal. However, from Release 10 onwards, this restriction is lifted and EPC now also supports MBR values greater than the GBR for a bearer. One of the reasons for such enhancements is to better support rate adaptive codecs where a minimum bit rate is guaranteed by the network (the GBR) while additional bandwidth may be allowed if available (the MBR). Additional system enhancements related to adaptive codecs were made by introducing support for Explicit Congestion Notification (ECN) to allow adjustment of the traffic when the network identifies a congestion situation. This had been introduced already in Release 9 for E-UTRAN access and in Release 10 for UTRAN access. ECN will be described in some more detail in Section 8.1.1.4 below.

A non-GBR bearer does not have a fixed bandwidth allocated and there is thus no guarantee of how much traffic it can carry. The non-GBR bearer may therefore experience packet loss in cases of congestion. The availability of radio resources for an existing non-GBR bearer thus depends on the total load of the cell as well as the QCI of the bearer. No transmission resources are reserved for non-GBR bearers. In EPS, non-GBR bearers are rate policed on an aggregate level instead of on a per-bearer (or PDP context) level. Consequently, even though the non-GBR bearers do not have any associated MBR, the operator may still police the utilized bandwidth of non-GBR bearers using the Aggregate Maximum Bit Rate (AMBR) as described further below.

Whether or not a bearer should be a GBR or a non-GBR bearer typically depends on what service is carried over that bearer. The GBR bearers are typically used for those services where it is better to block a service rather than degrade already admitted services in case resources are not available. Some services such as VoIP and streaming services benefit from a constant bandwidth and a GBR value may thus be needed to ensure a satisfactory user experience. If those resources are not available, it is better to block the service. Other services, such as Internet browsing, e-mail, and chat programs, normally do not require a constant fixed bandwidth. Those services would typically use non-GBR bearers. Whether to use GBR or non-GBR bearers for a certain service is dependent on operator configuration and can, for example, be controlled using the PCC framework. The choice depends to a large extent on the expected traffic load compared with the available capacity.

Standardized QCI Values and Corresponding Characteristics

Certain QCI values have been standardized to reference specific QoS characteristics. The QoS characteristics describe the packet forwarding treatment that the traffic for a bearer receives edge-to-edge between the UE and the GW in terms of certain performance characteristics such as priority, packet delay budget, and packet error loss rate. The standardized characteristics are not signaled on any interface; they should instead be understood as guidelines for the preconfiguration of node-specific parameters for each QCI. For example, the radio base station would need to be configured to ensure that traffic belonging to a bearer with a certain standardized QCI receives the appropriate QoS treatment. The goal of standardizing a QCI with corresponding characteristics is to ensure that applications/services mapped to that QCI receive the same minimum level of QoS in multi-vendor network deployments and in cases of roaming. The standardized QCI characteristics are defined in clause 6.1.7 in 3GPP TS 23.203. A simplified description can be found in Table 8.1.

Table 8.1: Standardized QCI Characteristics

QCI	Resource Type	Priority	Packet Delay Budget	Packet Error Loss Rate	Example Services
1	GBR	2	100 ms	10^{-2}	Conversational voice
2	GBR	4	150 ms	10^{-3}	Conversational video (live streaming)
3	GBR	3	50 ms	10^{-3}	Real-time gaming
4	GBR	5	300 ms	10^{-6}	Non-conversational video (buffered streaming)
5	Non-GBR	1	100 ms	10^{-6}	IMS signaling
6	Non-GBR	6	300 ms	10^{-6}	Video (buffered streaming), TCP-based (www, e-mail, chat, ftp, p2p file sharing, progressive video, etc.)
7	Non-GBR	7	100 ms	10^{-3}	Voice, video (live streaming) interactive gaming
8	Non-GBR	8	300 ms	10^{-6}	Video (buffered streaming), TCP-based (www, e-mail, chat, ftp, p2p filesharing, progressive video, etc.)
9	Non-GBR	9	300 ms	10^{-6}	Video (buffered streaming), TCP-based (www, e-mail, chat, ftp, p2p file sharing, progressive video, etc.)

The QCI values 1–4 are allocated for traffic that requires dedicated resource allocation for a GBR, while values 5–9 are not associated with GBR requirements. Each standardized QCI is associated with a priority level, where priority level 1 is the highest priority level. The Packet Delay Budget can be described as an upper bound for the time that a packet may be delayed between the UE and the PCEF (Policy and Charging Enforcement Function). The Packet Error Loss Rate can, in a simplified manner, be described as an upper bound for the rate of non-congestion-related packet losses.

Note that the description above gives a very simplified definition of the standardized QCIs, witholding many of the details. The purpose is to give the general reader a basic view of the topic. The interested reader should consult TS 23.203 for the complete definitions.

Apart from these standardized QCIs, non-standardized QCIs may also be used. In this case it is the operators and/or vendors who define what node-specific parameters are used for a given QCI.

8.1.1.3 APN-AMBR and UE-AMBR

In addition to the bit rate parameters associated with each GBR bearer, EPS also defines the AMBR parameters that are associated with non-GBR bearers. These parameters are not specific to each non-GBR bearer, but, rather, define a total bit rate that a subscriber is allowed to consume for an aggregate of non-GBR bearers. The bit rate consumed by the GBR bearers does not count towards the AMBR. Two variants of the AMBR are defined: the APN-AMBR and the UE-AMBR.

One reason why aggregate rate policing of non-GBR bearers is preferable over per-bearer policing is that network planning becomes easier. With a subscribed per-bearer MBR (as in GPRS) it is difficult to estimate the total bit rate that subscribers will use. Also, an AMBR can provide a more understandable subscription for the end-user compared to an MBR that is per bearer.

The APN-AMBR defines the total bit rate that is allowed to be used for all non-GBR bearers associated with a specific APN. This parameter is defined as part of a user's subscription but may be overridden by the PCRF. The APN-AMBR limits the total non-GBR traffic for an APN, independent of the number of PDN connections and non-GBR bearers that are opened for that APN. In other words, if a user has multiple PDN connections for the same APN, they all share the same APN-AMBR. For example, if an operator provides an APN for Internet access, the operator may then limit the total bandwidth for that APN and thus prevent the UE from increasing its accessible bandwidth by just opening new PDN connections to the same APN. This is different from the pre-Release 9 2G/3G core network where the subscribed QoS is defined per PDP context. The APN-AMBR is enforced by the PDN GW.

The UE-AMBR is defined per subscriber and defines the total bit rate allowed to be consumed for all non-GBR bearers of a UE. The subscription profile contains a subscribed UE-AMBR. However, the actual UE-AMBR value that is enforced by the network is set as the minimum of the subscribed UE-AMBR and the sum of the APN-AMBR of all active APNs (i.e. all APNs for which the UE has active PDN connections). The UE-AMBR is enforced by the base stations.

In 3GPP Release 8, APN-AMBR enforcement was introduced for EPC with enforcement by the PDN GW. For E-UTRAN access, UE-AMBR was introduced in the same release. Both APN-AMBR and UE-AMBR were also introduced for the 2G/3G core network in Release 9 with APN-AMBR enforcement in GGSN and with UE-AMBR enforcement in the 2G/3G radio networks.

Different AMBR values are defined for uplink and downlink directions. There are thus a total of four AMBR values defined: UL APN-AMBR, DL APN-AMBR, UL UE-AMBR, and DL UE-AMBR.

The UE-AMBR and APN-AMBR are not dependent on each other and an operator may choose to apply either UE-AMBR or APN-AMBR (or both). The enforcement of the UE-AMBR and APN-AMBR are two tools for the operator to realize the business model. The UE-AMBR may be used to put upper limits on a subscription or to limit the total amount of traffic in the network. The APN-AMBR, on the other hand, is more related to, for example, Service Level Agreements with external PDNs or subscriptions related to specific APNs.

8.1.1.4 User-Plane Handling

Some aspects of user-plane handling of EPS bearers were introduced in Chapter 6. In particular it was shown how packet filters in the UE and the GW are used to determine which IP flows should be carried over a certain EPS bearer. Now, after the description of the mechanisms available for QoS control in E-UTRAN, it is useful to look at user-plane handling and how the QoS functions and QoS parameters described above are allocated to different nodes in the network. Figure 8.2 indicates different user-plane QoS functions for E-UTRAN/EPS.

	UE	eNB	Transport network	PDN GW
Functions operating per EPS bearer				
Packet Filtering	X (uplink)			X (downlink)
GBR/ARP Admission		X		X
ARP Preemption		X		X
Rate Policing		X		X
Queue Management	X	X		
Uplink + Downlink Scheduling		X		
Configuring layer 1 and layer 2 protocols		X		
Map QCI to DSCP		X		X
Functions operating on transport layer, e.g. per DSCP				
Queue Management			X	
Uplink + Downlink Scheduling			X	

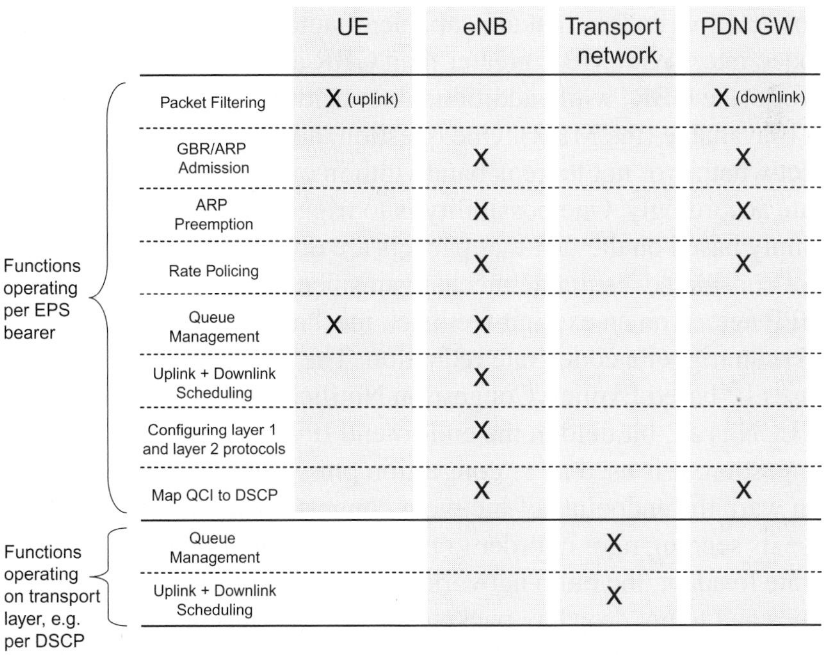

Figure 8.2: Overview of User-Plane QoS Functions for E-UTRAN/EPS.
The functional allocation for the PDN GW is for GTP-based S5/S8. For PMIP-based S5/S8, the bearer-related functions are moved to the Serving GW.

The UE and GW (PDN GW for GTP-based S5/S8 and Serving GW for PMIP-based S5/S8) carry out uplink and downlink packet filtering respectively in order to map the packet flows on to the intended bearer.

The GW and the eNB can implement functions related to admission control and pre-emption handling (i.e. congestion control) in order to allow these nodes to limit and control the load put on them. These functions can take the ARP value as an input in order to differentiate the treatment of different bearers in these functions, as described in the ARP section above.

The GW and eNB further implement functions related to rate policing. The goal of these functions is twofold: to protect the network from becoming overloaded and to ensure that the services are sending data in accordance with the specified maximum bit rates (AMBR and MBR). For non-GBR bearers, the PDN GW performs rate policing based on the APN-AMBR value(s) for both uplink and downlink traffic, while the eNB performs rate policing based on the UE-AMBR value for both uplink and downlink traffic. For GBR bearers, MBR policing is carried out in the GW for downlink traffic and in the eNB for uplink traffic.

In Release 10, support for GBR bearers with an MBR value greater than the GBR value was introduced. Prior to Release 10, only MBR equal to GBR was supported. As discussed above, a benefit with MBR greater than GBR is that it allows better utilization of adaptive codecs that are capable of automatically switching between different codec rates. With MBR greater than GBR a minimum bit rate is guaranteed by the network (the GBR) while additional bandwidth may be utilized by an application if available (the MBR). One question, however, is how an application should detect whether or not there is bandwidth in excess of the GBR available and adjust its rate accordingly. One possibility is to trigger codec rate reduction in the terminal simply based on the fact that packets are dropped by the network. This is, however, not considered a suitable mechanism since it may give poor user experience. Instead, 3GPP agreed on an explicit feedback mechanism where the network (e.g. the eNodeB) can trigger a codec rate reduction. The mechanism introduced in Release 9 uses IP-based Explicit Congestion Notification (ECN) specified in IETF RFC 3168. ECN is a 2-bit field in the end-to-end IP header that can be used to indicate congestion. It is used as a "congestion pre-warning scheme" by which the network can warn the endpoints of incipient congestion so that the sending endpoint can decrease its sending rate. In order to provide sufficient time for the end-to-end codec rate to adapt, the radio network should attempt to preserve the QoS characteristics and to not drop any packets on a bearer during a grace period after it has indicated congestion. The default grace period is defined to be 500 ms. After the grace period the radio network may maintain resource reservation for the GBR only.

By providing this ECN-based pre-warning and grace period before packet drops, the codec can more gracefully adjust the rate compared to scenarios where rate reduction is based only on dropped packets. In Release 9 the IP level ECN scheme is only applicable for E-UTRAN access and for voice media (Multimedia Telephony Service for IMS), while in Release 10 this feature was extended to also include UTRAN access as well as video services. It can also be noted that 3GPP has not specified any explicit feedback from the network to trigger a codec rate increase. Instead, the codec may stay at the GBR value for the rest of the call, or the endpoints may probe the connection to find out if a rate increase seems possible.

Figure 8.3 illustrates how the ECN handling is performed for uplink traffic. In step 1, the SIP session is negotiated with the full set of codec rates but without taking any network level congestion into account. After ECN has been successfully negotiated for a media stream the sender marks each IP packet as ECN-capable transport. This is shown in steps 2 and 3. The eNB may then mark the IP packets to indicate that congestion has been experienced (steps 4 and 5). In response to an indication of experienced congestion, the receiving side triggers a codec rate reduction (steps 6 and 7).

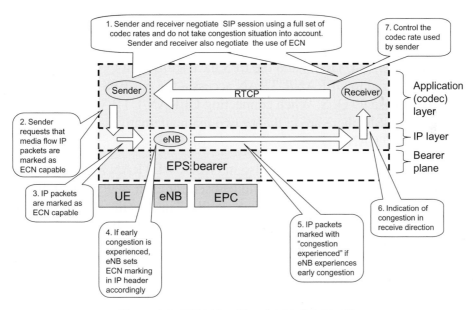

Figure 8.3: ECN Handling for Uplink Traffic.

In order to distribute radio network resources (radio and processing resources) between the established bearers, the eNB implements uplink and downlink scheduling functions. The scheduling function is, to a large extent, responsible for fulfilling the QoS characteristics associated with the different bearers.

The eNB is responsible for configuring the lower layer (layers 1 and 2) protocols of the radio connection of the bearer in accordance with the QoS characteristics associated with the bearer. Among other things, this includes configuring the error-control protocols (modulation, coding, and link layer retransmissions) so that the QoS characteristics, packet delay budget, and packet error loss are fulfilled. For more details on QoS handling in E-UTRAN, see Dahlman (2011).

On the transport level, i.e. the basic IP transport between EPC network entities (including intermediate transport entities in the packet cores such as regular IP routers), there is no awareness of the EPS bearers and instead queue management and packet forwarding treatment is done according to a transport layer mechanism such as DiffServ. The EPC entities map the QCI of the EPS bearer on to DiffServ Code Point (DSCP) values that are used by the transport network.

It should be noted that there are also service-aware QoS control functions operating on a finer granularity than the EPS bearer. These functions are defined as part of the Policy and Charging Control (PCC) architecture and are described in the PCC section of this chapter.

8.1.2 Interworking with GERAN/UTRAN

As mentioned above, E-UTRAN and the EPS have a different QoS control architecture compared to GERAN/UTRAN access – the pre-EPS QoS model is often referred to as Release 99 QoS. When connecting GERAN/UTRAN accesses to EPS, via an S4-based SGSN, there are in theory two main alternatives for how to handle QoS in GERAN/UTRAN:

1. Implement the EPS QoS solutions also in GERAN/UTRAN access. This would, for example, imply that the QCI would be used for each PDP context, instead of the GERAN/UTRAN (Release 99) QoS profile. This has the benefit that the 3GPP family of accesses in EPS uses the same QoS parameters. It should, however, be noted that the GERAN/UTRAN radio interface has to be backwards compatible with the Release 99 QoS scheme and PDP context procedures in order to allow pre-EPS terminals to connect. An EPS-based GERAN/UTRAN network would thus anyhow need to implement Release 99-based QoS.
2. Keep the existing Release 99 QoS and PDP context procedures for GERAN/ UTRAN. This would imply the fewest changes to the current GERAN/UTRAN radio interface, but there would be a need to specify a mapping to EPS-based QoS solutions and bearer procedures.

3GPP opted for the second alternative. The main motivation was that this was the simplest option that was also backwards compatible with pre-EPS terminals.

The bearer section of Chapter 6 provides a description of the interworking between PDP context and EPS bearer procedures. It describes interworking scenarios when the Gn/Gp interfaces are used between a SGSN and a PDN GW. In this section we look at the corresponding mapping between QoS parameters. Depending on scenario, different network entities need to perform mapping between QoS parameters associated with the EPS bearer and the QoS parameters associated with the PDP context. The detailed mapping between the EPS QoS parameters and the Release 99 QoS parameters is defined in Annex E of 3GPP TS 23.401 and briefly described below:

1. Up to Release 8, the ARP for a PDP context in the core network can take three possible values, while the ARP for an EPS bearer can take 15 possible values. The mapping between PDP context ARP values and EPS bearer ARP values is a one-to-many mapping. The exact scheme for which range of EPS bearer ARP values is mapped to an ARP value of a PDP context is not standardized but can be configured by each operator. In Release 9, however, a new QoS parameter for PDP contexts was introduced, the Evolved ARP, supporting 15 values. Therefore, from Release 9 onwards, the priority of the EPS bearer ARP can be mapped directly to/from the PDP Context Evolved ARP parameter, if the network supports this parameter.
2. The GBR and MBR values are mapped one-to-one (for GBR bearers only).
3. The MBR for PDP contexts without GBR is mapped to/from APN-AMBR.
4. The mapping between the standardized QCIs and the Release 99 Traffic Class and QoS Attributes is described in Annex E of 3GPP TS 23.401. Only a subset of the Release 99 QoS Attributes is specified by the mapping. The setting of the values of the other Release 99 QoS Attributes is not specified by the standard and is instead based on operator policy preconfigured in the SGSN.
5. In the first release of EPS (3GPP Release 8), UE-AMBR is only used in E-UTRAN and has no counterpart when using 2G/3G. In the next release (Release 9), UE-AMBR is also applied for 2G/3G accesses.

8.1.3 QoS Aspects When Interworking with Other Accesses

So far we have considered QoS aspects related to the 3GPP family of accesses. EPS does, however, support interworking and mobility with other accesses as well, defined by other standardization bodies. Each such access may have its own set of QoS mechanisms and QoS parameters, as defined by the relevant standardization body. It is not possible within this book to go through each access and discuss the access-specific QoS solutions. There are, however, a few aspects that are either independent of access or related to the interworking between EPS and the access-specific QoS mechanism.

One QoS parameter that is common to all accesses is the APN-AMBR as described in the APN-AMBR section above. The APN-AMBR is enforced by the PND GW and can be enforced independent of which access the UE may be using.

Other access-independent parameters are the QCI and ARP. As described above, the QCI and ARP are parameters of the EPS bearer when using 3GPP family of accesses. As we will see in the next section on PCC, the PCC architecture also uses QCI and ARP, but as access-independent parameters. When interworking with other accesses, these parameters are mapped by each individual access to access-specific parameters and mechanisms.

8.2 Policy and Charging Control

PCC provides operators with advanced tools for service-aware QoS and charging control. In wireless networks, where the bandwidth is typically limited by the radio network, it is important to ensure efficient utilization of the radio and transport network resources. Furthermore, different services have very different requirements on the QoS, which are needed for packet transport. Since a network generally carries many different services for different users simultaneously, it is important to ensure that the services can coexist and that each service is provided with an appropriate transport path. PCC also provides the means to control charging on a per-session or per-service basis.

PCC enables centralized control to ensure that the services are provided with the appropriate transport and charging, for example in terms of bandwidth, QoS treatment, and charging method. The PCC architecture enables control of the media plane for both the IP Multimedia Subsystem (IMS) and non-IMS services.

When in 3GPP access, bearer procedures are available for QoS management in the access. While the EPS bearer and PDP context procedures are specific to the 3GPP family of accesses, corresponding QoS procedures exist for many other accesses as well. In this section we focus on how the operator can *control* those QoS procedures and the charging mechanisms used for each service session.

When it comes to PCC, the term "bearer" is used in a more generic fashion to denote an IP transmission path with well-defined characteristics (e.g. capacity, delay, and bit error rate). This allows us to use the bearer terminology in an access-agnostic fashion, independent of the details of how this transmission path is created or how QoS is managed for each access technology.

The term "service session" is also important here. The bearer concept handles traffic aggregates – that is, all conformant traffic that is transported over the same bearer receives the same QoS treatment. This means that multiple service sessions transported over the same bearer will be treated as one aggregate. These bearer concepts still apply when PCC is used. As we will see, however, PCC adds a "service aware" QoS and charging control mechanism that in certain aspects is more fine grained – that is, it operates on a per-service session level rather than on a per-bearer level.

The PCC architecture for EPS is an evolution of the PCC architecture defined in 3GPP Release 7. PCC has nevertheless evolved significantly from 3GPP Release 7 in order to support new features in EPS, such as multiple access technologies, roaming, as well as other PCC enhancements. The goal in 3GPP has been to define an access-agnostic policy control framework and, as such, make it applicable to a number of accesses such as E-UTRAN, UTRAN, GERAN, HRPD, WLAN, and WiMAX. Furthermore, the introduction of a complete roaming model for PCC allows operators to have the same dynamic PCC, and provide the same access to services independently of whether a user is making this access through a gateway in their home or visited network. PCC has evolved and new features have been added since EPS Release 8. For example, support for using charging and spending limits as input to policy decisions has been added. Also, PCC support for application detection, reporting, and control based on (deep) packet inspection has been made more extensive.

It is also worth mentioning that standardization bodies standardizing other access technologies for fixed or wireless access have also created policy control specifications targeting their particular access technologies. When it comes to wireless accesses such as HRPD, an alignment towars a common policy control architecture based on 3GPP PCC has already materialized with the EPS. For the fixed accesses, in particular related to the standardization work being done in the Broadband Forum (BBF), the work towards alignment and/or interworking has also started. 3GPP Release 11 includes solutions for policy interworking between policy controllers in the 3GPP and BBF domains respectively. Corresponding solutions are being described in BBF specifications. Work is also starting in both 3GPP and BBF to specify a policy solution for the BBF domain based on 3GPP PCC, but since this work is still at a very early stage, it will not be covered in this book.

In this section we will introduce the architecture and functions of PCC. In Sections 8.2.1 and 8.2.2 we will introduce the PCC architecture, the basic functions of PCC, and how PCC works. In Section 8.2.3 we will look at two different principles for how QoS control can be achieved in EPS and how PCC handles this. In Sections 8.2.4–8.2.6 we will then look at additional features of PCC. In Section 8.2.4 we will go into PCC roaming. In Section 8.2.5 we will have a look at the PCC features that have been added in 3GPP Release 9 and beyond, such as policy control based on subscriber spending limits and PCC support for application detection and control. Finally, in Section 8.2.6 we will have a brief look at PCC support for Fixed Broadband Accesses.

Even though PCC is in general considered an optional part of the architecture, some key features now require mandatory use of PCC. Services like VoLTE (VoIP using IMS for LTE access) require PCC as determined by GSMA; Multimedia Priority Services (both IMS based and EPS bearer services and EPS bearer priority triggered by applications) require PCC usage and so do Multimedia IMS-based emergency services.

8.2.1 The PCC Architecture

The basic aspects of the EPS architecture, including the PCC aspects, were introduced in Chapter 2. In this section we give a more in-depth description, as well as describe the basic concepts and functions of PCC.

The reference network architecture for PCC in EPS is shown in Figure 8.4. The functional entities that are part of the PCC architecture are briefly described below. In addition to the reference points and network entities shown in Figure 8.4, an

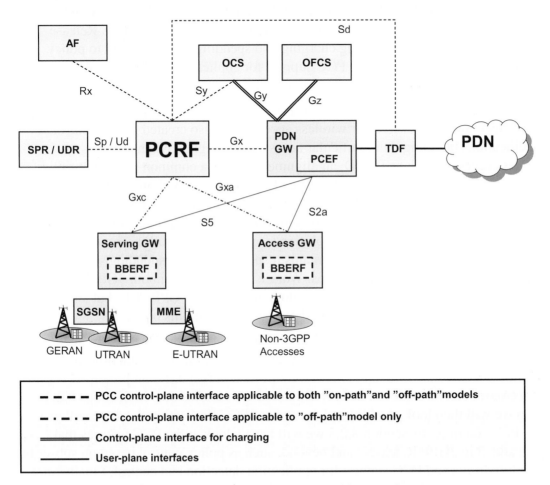

Figure 8.4: 3GPP PCC Non-Roaming Architecture for EPS.
Note that only a subset of the EPS reference points and EPS network entities is shown.

additional reference point, the S9a reference point, has been introduced to address policy interworking with Fixed Broadband Access networks. In order to not complicate the basic PCC architecture and description, the enhancements made to PCC to support Fixed Broadband policy interworking are described in a separate section below.

The Application Function (AF) interacts (or intervenes) with applications or services that require dynamic PCC. Typically the application level signaling for the service passes through, or is terminated in, the AF. The AF extracts session information from the application signaling and provides this to the PCRF over the Rx reference point. The AF can also subscribe to certain events that occur at the traffic-plane level (i.e. events detected by either PCEF or BBERF (Bearer Binding and Event Reporting Function)). Those traffic-plane events include events such as IP session termination or access technology type change. When the AF has subscribed to a traffic-plane event, the PCRF will inform the AF of its occurrence. The term "Application Function" is a generic term used by PCC for this entity, and in practice the AF functionality is contained within a specific network entity depending on the type of service. For the IMS, the AF corresponds to the P-CSCF. For a non-IMS service, the AF could, for example, be a video streaming server.

The PCRF has an interface to a database that contains subscription information, such as user-specific policies and data.

The Subscription Profile Repository (SPR) is the database that was defined as part of the PCC architecture in Release 7 and is maintained in later releases as a standalone logical entity containing PCC-related subscription data.

The User Data Repository (UDR) is another alternative for PCC-related subscription data. This option was introduced in 3GPP Release 10 in order to utilize the UDC architecture to store PCC-related subscription data. Further details on the UDC architecture are provided in Chapter 10.

The Online Charging System (OCS) is a credit management system for pre-paid charging. The PCEF interacts with the OCS to check out credit and report credit status. In 3GPP Release 11 an interface (Sy) between OCS and PCRF has also been introduced to enable policy control based on subscriber spending limits. This extension will be further described in Section 8.2.5.3.

The Offline Charging System (OFCS) is used for offline charging. It receives charging events from the PCEF and generates Charging Data Records (CDRs) that can be transferred to the billing system.

The PCRF is the policy control function of PCC. It receives session information over Rx as well as information from the access network via Gx. If a BBERF is used (see below), the PCRF also receives information via Gxa/Gxc reference points. (Gxb was defined between ePDG and PDN GW but is not used in the first releases of EPS.) The PCRF may also receive subscription information from the SPR. The PCRF takes the available information, as well as configured operator policies, into account and creates service-session-level policy decisions. The decisions are then provided to the

PCEF and the BBERF. Another task of the PCRF is to forward event reports between the BBERF, the PCEF, and the AF.

The PCEF enforces policy decisions (e.g. gating, maximum bit rate policing) received from the PCRF and also provides the PCRF with user- and access-specific information over the Gx reference point. The PCEF may also perform measurements of user-plane traffic (e.g. user-plane traffic volume and/or time duration of a session). It reports usage of resources to the OFCS and interacts with the OCS for credit management. The PCEF can also provide usage reports to the PCRF if policy decisions based on consumed volume are desired. In addition, the PCEF can support application detection and control (ADC) for specific applications and report this to the PCRF. The ADC functionality has been added in 3GPP Release 11 and will be further described in Section 8.2.5.1.

The Traffic Detection Function (TDF) is a functional entity that provides application detection and control (ADC) functionality using packet inspection as well as reporting of detected applications to the PCRF. The TDF has been introduced in the PCC architecture in Release 11 and will be described in further detail in Section 8.2.5.1.

In the PCC architecture for EPS, there are two main architecture alternatives for QoS control: with and without BBERF in the Access GW (e.g. Serving GW or HSGW). In common language the two alternatives are referred to as "off-path" and "on-path" models respectively. The BBERF supports a subset of the functions supported by the PCEF. The details regarding the BBERF and the two alternative architectures are discussed further below.

8.2.1.1 Multi-Access and the Off-Path PCC Model

As described in Section 6.4, EPS supports different mobility protocols depending on which access technology is used. For the 3GPP family of accesses (GERAN, UTRAN and E-UTRAN) the GTP or PMIPv6 may be used on the S5/S8 reference points. For connecting other accesses to EPC, it is possible to use GTP, PMIPv6, DSMIPv6, and Mobile IPv4 (MIPv4) on S2a/b/c reference points. These different protocols have different properties when it comes to how the EPS bearers are implemented. These differences result in different requirements on PCC.

When GTP is used between the Serving GW and the PDN GW, the bearers are terminated in the PDN GW and the PDN GW can thus use the bearer procedures to control the EPS bearers. We refer to this model as the "on-path" model because the QoS/bearer signaling takes place (using GTP) on the same "path" as the user plane. In this model, the PCRF controls the QoS by providing the QoS policy information to the PCEF via the Gx reference point. The BBERF and Gxa/Gxc have no role here and are thus not used at all in the "on-path" model.

When a Mobile IP-based protocol, such as PMIP or DSMIPv6, is used towards the PDN GW, the bearers and QoS reservation procedures are terminated closer to the (radio) access network and the PDN GW thus has no knowledge about bearers. For the 3GPP family of accesses, the bearers only extend between the UE and the Serving GW. Between the Serving GW and the PDN GW there is no notion of EPS bearers. See the bearer section in Chapter 6 for illustrations. For other accesses, the bearers and QoS reservation procedures (if they exist) extend between the UE and an "access GW" in the access network. In this case, the PDN GW only handles mobility signaling towards the access network and the UE, not any QoS signaling. Therefore, the PDN GW cannot control the QoS using bearer procedures and it is not sufficient for the PCRF to provide the QoS information to the PCEF. The PCRF has to provide the QoS information to the entity where the bearers are terminated. For this purpose, the BBERF and the Gxa/Gxc reference points are introduced.

When it comes to other functions of PCC, not related to where bearers are terminated, there is in most cases no difference between the "on-path" and "off-path" models. For example, the service-aware charging functionality is always located in the PCEF. Further details on the functional content of PCEF and BBERF can be found in later subsections.

For EPS, the PCEF is always located in the PDN GW. The BBERF location, however, depends on the particular access technology. For example, for the 3GPP family of accesses, the BBERF (if applicable) is located in the Serving GW, whereas for eHRPD access the BBERF is located in the HSGW. Since the PDN GW is the mobility anchor for the UE, the same PCEF is kept during the whole IP session. The BBERF allocated for a UE may, however, change due to the mobility of the UE. For example, the Serving GW may change as the UE moves within the 3GPP accesses. The BBERF location will also change when the UE moves between 3GPP access and other access technologies. Support for BBERF relocation is thus an inherent part of the off-path PCC architecture for EPS.

8.2.2 Basic PCC Concepts

As the name suggests, the purpose of PCC is policy and charging control.

Policy control is a very generic term and in a network there are many different policies that can be implemented, for example policies related to security, mobility, use of access technologies, etc. When discussing policies, it is thus important to understand the context of those policies. When it comes to PCC, policy control refers to the two functions gating control and QoS control:

1. Gating control is the capability to block or to allow IP packets belonging to IP flow(s) for a certain service. The PCRF makes the gating decisions that are then

enforced by the PCEF. The PCRF could, for example, make gating decisions based on session events (start/stop of service) reported by the AF via the Rx reference point.

2. QoS control allows the PCRF to provide the PCEF (and BBERF, if applicable) with the authorized QoS for the IP flow(s). The authorized QoS may, for example, include the authorized QoS class and the authorized bit rates. The PCEF or BBERF enforces the QoS control decisions by setting up the appropriate bearers. The PCEF also performs bit rate enforcement to ensure that a certain service session does not exceed its authorized QoS.

Charging control includes means for both offline and online charging. The PCRF makes the decision on whether online or offline charging will apply for a certain service session, and the PCEF enforces that decision by collecting charging data and interacting with the charging systems. The PCRF also controls what measurement method applies – that is, whether data volume, duration, combined volume/duration, or event-based measurement is used. Again, it is the PCEF that enforces the decision by performing the appropriate measurements on the IP traffic passing through the PCEF.

With online charging, the charging information can affect, in real time, the services being used and therefore a direct interaction of the charging mechanism with the control of network resource usage is required. Online credit management allows an operator to control access to services based on credit status. For example, there has to be enough credit left with the subscription in order for the service session to start or an ongoing service session to continue. The OCS may authorize access to individual services or to a group of services by granting credits for authorized IP flows. Usage of resources is granted in different forms. The OCS may, for example, grant credit in the form of certain amount of time, traffic volume, or chargeable events. If a user is not authorized to access a certain service, for example if the pre-paid account is empty, then the OCS may deny credit requests and additionally instruct the PCEF to redirect the service request to a specified destination that allows the user to refill the subscription.

PCC also incorporates service-based offline charging. With offline charging, the charging information is collected by the network for later processing and billing. The charging information therefore does not affect, in real time, the service being used. Since billing takes place after the service session is complete, for example via a monthly bill, this functionality does not provide any means for access control in itself. Instead, policy control must be used to restrict access and then service-specific usage may be reported using offline charging.

Online and offline charging may be used at the same time. For example, even for billed (offline charged) subscriptions, the online charging system may be used for functionality such as Advice of Charge. Conversely, for pre-paid subscribers, offline charging data generation may be used for accounting and statistics.

8.2.2.1 PCC Decisions, the PCC Rule, and the QoS Rule

The PCRF is the central entity in PCC making PCC decisions. The decisions can be based on input from a number of different sources, including:

- Operator configuration in the PCRF that defines the policies applied to given services
- Subscription information/policies for a given user, received from the SPR
- Information about the service received from the AF
- Information from the TDF or PCEF about detected applications
- Information from the charging system about subscriber spending limit status
- Information from the access network about what access technology is used, etc.

The PCRF provides its decisions in the form of so-called "PCC rules". The PCRF also provides a subset of the information in the so-called "QoS rules" to the BBERF if the "off-path" model is used. In this section we first describe the main content of the PCC rules and then the subset of information contained in a QoS rule.

A PCC rule contains a set of information that is used by the PCEF and the charging systems. First of all, it contains information (in a so-called "Service Data Flow (SDF) template") that allows the PCEF to identify the IP packets that belong to the service session. All IP packets matching the packet filters of an SDF template are designated an SDF. The filters in an SDF template contain a description of the IP flow and typically contain the source and destination IP addresses, and the protocol type used in the data portion of the IP packet, as well as the source and destination port numbers. These five parameters are often referred to as the IP 5-tuple. It is also possible to specify other parameters from the IP headers in the SDF template. The PCC rule also contains the gating status (open/closed), as well as QoS and charging-related information for the SDF. The QoS information for an SDF includes the QCI, MBR, GBR, and ARP. The definition of the QCI is the same as that described in the QoS section in this chapter for the EPS bearer and the reader is referred to that section for a more thorough description of those parameters. However, one important aspect of the QoS parameters in the PCC rule is that they have a different scope than the QoS parameters of the EPS bearer. The QoS and charging parameters in the PCC rule apply to the SDF. More precisely, the QCI, MBR, GBR, and ARP in the PCC rule apply to the IP flow described by the SDF template, while the QCI, MBR, GBR, and ARP discussed in Section 8.1 apply for the EPS bearer. A single EPS bearer may be used to carry traffic described by multiple PCC rules, as long as the bearer provides the appropriate QoS for the service data flows of those PCC rules. Below we will discuss further how PCC rules and SDFs are mapped to bearers. Table 8.2 lists a

Table 8.2: A Subset of the Elements That May be Included in a Dynamic PCC Rule

Type of Element	PCC Rule Element	Comment
Rule identification	Rule identifier	Used between PCRF and PCEF for referencing PCC rules
Information related to Service Data Flow detection in PCEF and BBERF	Service Data Flow Template	List of packet filters for the detection of the service data flow
	Precedence	Determines the order in which the service data flow templates are applied at PCEF
Information related to policy control (i.e. gating and QoS control)	Gate status	Indicates whether a SDF may pass (gate open) or will be discarded (gate closed)
	QoS Class Identifier (QCI)	Identifier that represents the packet forwarding behavior of a flow
	UL and DL maximum bit rates	The maximum uplink (UL) and downlink (DL) bit rates authorized for the service data flow
	UL and DL guaranteed bit rates	The guaranteed uplink (UL) and downlink (DL) bitrates authorized for the service data flow
	ARP	The Allocation and Retention Priority for the service data flow
Information related to charging control	Charging key	The charging system uses the charging key to determine the tariff to apply for the service data flow
	Charging method	Indicates the required charging method for the PCC rule. Values: online, offline, or no charging
	Measurement method	Indicates whether the SDF data volume, duration, combined volume/duration, or event will be measured
Usage Monitoring Control	Monitoring key	The PCRF uses the monitoring key to group services that share a common allowed usage. See Section 8.2.5.2 for further details

Text taken from 3GPP TS 23.203.

subset of the parameters that can be used in a PCC rule sent from PCRF to PCEF. For a full list of parameters, see 3GPP TS 23.203 and 3GPP TS 29.212.

The same standardized QCI values and QCI characteristics outlined in Section 8.1 apply when QCI is used in the PCC rule. The standardized QCI and corresponding characteristics are independent of the UE's current access. The access network receiving the PCC rule will thus map the QCI value of the PCC rule on to any access-specific QoS parameters that apply in that access. This is further elaborated below.

The discussion so far has assumed a case where the PCRF provides the PCC rules to the PCEF using Gx. These rules, which are dynamically provided by the PCRF, are denoted "dynamic PCC rules". There is, however, also a possibility for the operator to configure PCC rules directly into the PCEF. Such rules are referred to as "predefined

PCC rules". In this case the PCRF can instruct the PCEF to activate such predefined rules by referring to a PCC rule identifier. While the packet filters in a dynamic PCC rule are limited to the IP header parameters (the IP 5-tuple and other IP header parameters), filters of a PCC rule that is predefined in the PCEF may use parameters that extend the packet inspection beyond the IP 5-tuple. Such filters are sometimes referred to as Deep Packet Inspection (DPI) filters and they are typically used for charging control where more fine-grained flow detection is desired. The definition of filters for predefined rules is not standardized by 3GPP.

As described above, if the "off-path" model applies the PCRF needs to provide the QoS information to the BBERF via the Gxa/Gxc reference points. The QoS information provided to the BBERF is the same as is present in the corresponding PCC rule. However, since the BBERF only needs a subset of the information available in a PCC rule, the PCRF does not send the full PCC rule to the BBERF. Instead, the PCRF generates a so-called "QoS rule" with information from the corresponding PCC rule. The QoS rules contain the information needed for the BBERF to ensure that bearer binding (see below) can be performed. The QoS rules thus contain the information needed to detect the SDF (i.e. SDF template and precedence), as well as the QoS parameters (e.g. QCI and bit rates). QoS rules do not contain any charging-related information.

8.2.2.2 Use Cases

As a result of the interactions with the PCRF, the PCEF and BBERF perform several different functions. In this subsection we present two use cases in order to get an overview of the dynamics of PCC and how PCC interacts with the application level, as well as the access network level. Some of the aspects brought up in the use cases will be discussed in more detail later. The intention of placing the use cases first is that a basic overview of the procedures described in the use cases should simplify the understanding of the PCC aspects being discussed in the later subsections.

The first use case is intended to illustrate a service session set up using "on-path" PCC, network-initiated QoS control, and online charging (Figure 8.5).

The first use case is described below:

1. The subscriber initiates a service, for example an IMS voice call, and performs end-to-end application session signaling that is intercepted by the AF (P-CSCF in the IMS case). In the IMS case, the application signaling uses the Session Initiation Protocol (SIP). A description of the service is provided as part of the application signaling. In IMS, the Session Description Protocol (SDP) is used to describe the sessions.
2. Based on service description information contained in the application signaling, the AF provides the PCRF with the service-related information over the Rx

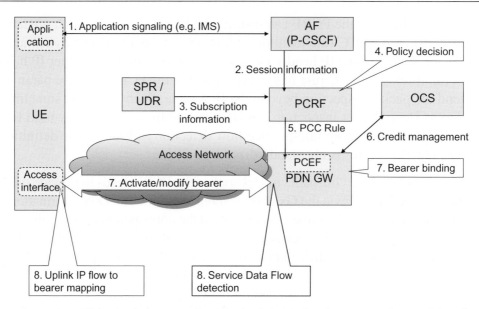

Figure 8.5: High-Level Use Case for PCC in the EPS for the "On-Path" Model and NW-Initiated Bearer Procedures.

interface. The session information is mapped at the AF from an SDP (e.g. SIP/SDP for IMS) into information elements in the Rx messages to the PCRF. This information typically includes QoS information (type of service, bit rate requirements) as well as traffic parameters (e.g. the IP 5-tuple) that allow identification of the IP flow(s) corresponding to this service session.

3. The PCRF may request subscription-related information from the SPR or UDR. (The PCRF may have requested subscription information earlier but it is shown at this step for illustrative purposes.)

4. The PCRF takes the session information, operator-defined service policies, subscription information, and other data into account when building policy decisions. The policy decisions are formulated as PCC rules.

5. The PCC rules are sent by the PCRF to the PCEF. The PCEF will enforce the policy decision according to the received PCC rule. All user-plane traffic for a given subscriber and IP connection pass through the network entity where the PCEF is located. For EPS, the PCEF is located in the PDN GW.

6. If the PCC rule specified that online charging should be used for this PCC rule, the PCEF contacts the OCS via the Gy reference point to request credit according to the measurement method specified in the PCC rule.

7. The PDN GW (PCEF) installs the PCC rules and performs bearer binding to ensure that the traffic for this service receives appropriate QoS. This may result in the establishment of a new bearer or modification of an existing bearer. More details on bearer binding are provided below.

8. The media for the service session are now being transported across the network and the PCEF performs SDF detection to detect the IP flow for this service. This IP flow is transported over the appropriate bearer. Further details on SDF detection can be found below.

The second example intends to illustrate the same basic use case but in a different network scenario using "off-path" PCC, UE-initiated QoS control, and offline charging. With UE-initiated QoS control, the UE and the network rely on UE-initiated triggers that start the bearer operations for this application. More discussion on UE- and network-initiated QoS control principles can be found below. Since offline charging is used, the PDN GW (PCEF) does not perform credit-based access control. The interactions with the charging system are therefore not shown in this use case.

It can be noted that the first three steps in the second use case are the same as in the first use case. These steps concern the application level signaling and the Rx signaling. This signaling is not dependent on access network properties, such as whether on-path or off-path PCC is used or whether UE-initiated or NW-initiated procedures are used. It is only the handling in the PCRF and in the access network that differ depending on PCC architecture model and UE/NW trigger of bearer procedures.

The use case with "off-path" model and UE-initiated procedures is described below (Figure 8.6). The first three steps are included in shortened form below. The full description can be found in the first use case.

1. The subscriber initiates a service, for example an IMS voice call, and performs IMS session signaling via the AF.
2. Based on service description information contained in the application signaling, the AF provides the PCRF with service-related information over the Rx interface.
3. The PCRF may request subscription-related information from the SPR.

The difference between UE-initiated and NW-initiated procedures will now become evident. In the first use case, the PCRF "pushed" the rules to the PDN GW, and the PDN GW initiated bearer procedures to ensure that the service receives the appropriate QoS treatment. In the second use case, the PCRF instead waits until the request from the UE triggers a "pull" of rules from the PCRF. This is described in the steps below:

4. The application in the UE makes an (internal) request for the access interface to request the QoS resources needed by the newly started application.
5. The UE sends a request to the network for QoS resources for this service. The UE includes the QoS class and packet filters associated with the service. The UE may also include a request for a certain GBR. The exact details regarding this request depend on which access technology the UE is using. For E-UTRAN, the UE would send a UE-requested bearer resource modification. For GERAN/UTRAN

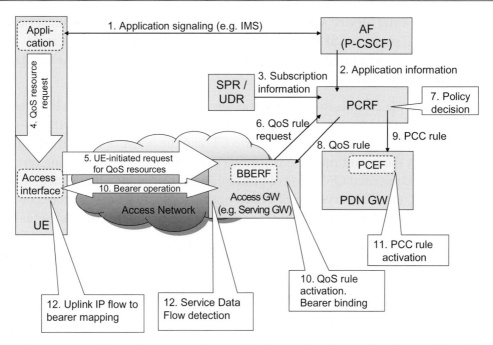

Figure 8.6: High-Level Use Case for PCC in the EPS for the "Off-Path" Model and UE-Initiated Bearer Procedures.

access, the UE would make a secondary PDP context activation or modification request. Other accesses may provide similar access-specific signaling.

6. Since off-path PCC is used, the BBERF initiates a PCRF interaction when receiving the request sent by the UE. (For comparison, in the on-path model, the request from the UE would have been forwarded from the Serving GW to the PDN GW, and the PDN GW would have sent a request for PCC rules to the PCRF.)

7. Similar to the first use case (step 4 in the first use case), the PCRF takes the session information, operator-defined service policies, subscription information, and other data into account when building policy decisions. The policy decisions are formulated as PCC rules. Since off-path PCC is used, the PCRF also derives QoS rules based on the PCC rules.

8. The PCRF sends the QoS rules to the BBERF.

9. The PCRF sends the corresponding PCC rules to the PCEF.

10. The BBERF (e.g. Serving GW) installs the QoS rules and performs bearer binding to ensure that the traffic for this service receives appropriate QoS. This may result in the establishment of a new bearer, or the modification of an existing bearer.

11. The PDN GW (PCEF) installs the PCC rules. The PCEF performs gating, bit rate enforcement, and service-level charging as defined by the PCC rule.

12. The media for the service session is now being transported across the network. The UE uses uplink packet filters to determine which bearer should carry the uplink traffic. Both the BBERF and PCEF perform SDF detection to detect the IP flow for this service. The BBERF forwards the downlink traffic over the appropriate bearer.

Note that the above two use case examples are not exhaustive in any sense. There are many other scenarios and configurations. For example, for services that do not provide an AF or Rx interface, it is still possible to use PCC. In that case step 2 of the second use case would be omitted and the PCRF could e.g. authorize PCC/QoS rules based on preconfigured policies without access to dynamic session data. As will be described further in Section 8.2.5, the PCRF may also authorize PCC/QoS rules based on application detection information provided from an application detection function.

8.2.2.3 Bearer Binding

The PCC rule needs to be mapped to a corresponding bearer in the access network to ensure that the packets receive the appropriate QoS treatment. This mapping is one of the central components of PCC. The association between a PCC/QoS rule and a bearer is referred to as bearer binding. The bearer binding is done by the Bearer Binding Function (BBF), which is located either in the PCEF (for on-path) or in the BBERF (for off-path). When the PCEF (or BBERF) receives new or modified PCC or QoS rules, the BBF evaluates whether or not it is possible to use the existing bearers. If one of the existing bearers can be used, for example if a bearer with the corresponding QCI and ARP already exists, the BBF may initiate bearer modification procedures to adjust the bit rates of that bearer. If it is not possible to use any existing bearer and NW-initiated bearer procedures are used, the BBF initiates the establishment of a suitable new bearer. In particular, if the PCC rule contains GBR parameters, the BBF must also ensure the availability of a GBR bearer that can accommodate the traffic for that PCC rule. If NW-initiated bearer procedures are used, the BBF triggers resource reservation in the access network to ensure that the authorized QoS of the PCC rule can be provided. Further details on the bearer concept can be found in Chapter 6.

For EPS and if the UE is using 3GPP accesses, the BBF uses the EPS bearer procedures when activating the PCC/QoS rules. Other accesses interworking with EPS may have other, access-specific, QoS signaling mechanisms. It is the task of the BBF to interact with the appropriate QoS procedures depending on access technology. In order to set up the right QoS resources in the access, the PCEF/BBERF not only needs to invoke the appropriate QoS procedures but may also need to map the QoS parameters. In particular, the BBF must map the QCI of the PCC/QoS rule, which is an access-independent parameter, to access specific QoS parameters. For the 3GPP family of accesses in EPS this is simple since the QCI is also used as a QoS parameter for the EPS bearers. For other accesses, the mapping may involve a "translation" from the QCI in the PCC/QoS rule to other access-specific QoS parameters that are used in that particular access.

8.2.2.4 *Service Data Flow Detection*

Once the service session is set up and the media traffic for the service is flowing, the PCEF and the BBERF use the packet filters of installed PCC and QoS rules to classify IP packets to authorized SDFs. This process is referred to as SDF detection. Each filter in the SDF filter is associated with a precedence value. The PCEF (or BBERF) matches the incoming packets against the available filters of the installed rules in order of precedence. The precedence is important if there is an overlap between filters in different PCC rules. One example of such overlap is a PCC rule that contains a wildcard filter that overlaps with more narrowly scoped filters in other PCC rules. In this case, the wildcard filter should be evaluated after the more narrowly scoped filters; otherwise the wildcard filter will cause a match before the PCEF/BBERF even tries the narrowly scoped filters. If a packet matches a filter, and the gate of the associated rule is open, then the packet may be forwarded to its destination. For the downlink part, the classification of an IP packet to an SDF also determines which bearer should be used to transfer the packet (Figure 8.7). See also Section 6.2 for more details on bearers and how packet filters are used to direct packets on to the right bearer.

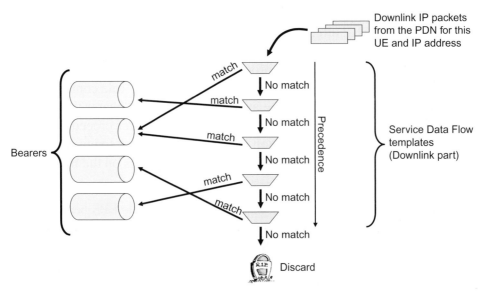

Figure 8.7: Example of SDF Detection and Mapping to Bearers for Downlink Traffic.

An additional aspect related to SDF detection occurs when DSMIPv6 is used as a mobility protocol. In that case the user-plane traffic is tunneled between UE and PDN GW, and thus also when passing through the BBERF (see Section 16.3 for more details on DSMIPv6). Since the packet filters of the PCC rule refer to untunneled packet flows, the BBERF has to "look inside" the DSMIPv6 tunnel to be able to apply the packet filters in the SDF template. This is something that could be referred to as "tunnel look-through" and is illustrated in Figure 8.8. The outer tunnel header is determined when the DSMIPv6 tunnel is established by the UE and the PDN GW. Information about the tunnel header – that is, the outer header IP addresses, etc. – is sent from the PDN GW to the BBERF via the PCRF so that the BBERF can apply the right packet filters for the tunnel.

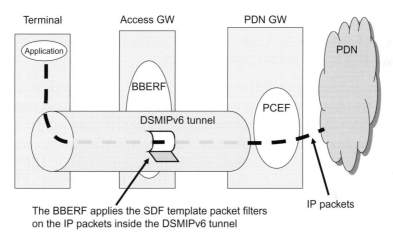

The BBERF applies the SDF template packet filters
on the IP packets inside the DSMIPv6 tunnel

Figure 8.8: BBERF "Tunnel Look-Through" If DSMIPv6 Is Used.

8.2.2.5 Events and Renewed Policy Decisions

When the PCRF makes a policy decision, information received from the access network may be used as input. For example, the PCRF may be informed about the current access technology used by the UE, or whether the user is in their home network or is roaming. During the lifetime of a session, the conditions in the access network may however change. For example, the user may move between different access technologies or different geographical areas. There may also be situations where a certain authorized GBR can no longer be maintained over the radio link. In these cases, the PCRF may want to re-evaluate its policy decisions and provide new or updated rules to the PCEF (and BBERF, if applicable). The PCRF should thus be able to keep itself up to date about events taking place in the access network. To achieve this, procedures have been defined that allow the PCRF to notify the PCEF/BBERF about which events the PCRF is interested in. In PCC terminology we say that the

PCRF subscribes to certain events, and that the PCEF/BBERF sets the corresponding event triggers. When an event occurs, and the corresponding event trigger is set, the PCEF/BBERF will report the event to the PCRF and allow the PCRF to revisit its previous policy decisions.

In the "on-path" model, information about the access network (information regarding available QoS on radio link, etc.) is available in the PDN GW and the PCEF can thus report on any status change via the Gx reference point. As mentioned above, there is no need for the BBERF in this case. In the "off-path" model, however, the PCRF will need to subscribe to events either in the PCEF or in the BBERF, depending on the nature of the event. With Mobile IP-based protocols, the access-specific bearers terminate in the BBERF instead of the PCEF. This implies that certain information about the access network is only available to the BBERF. Therefore, the BBERF detects such events and reports them over the Gxa/Gxc reference points. Other events, such as events related to multi-access mobility, are only known to the PCEF and thus reported by the PCEF also in the "off-path" model.

In the "off-path" model, the Gxa/Gxc and Gx interfaces will also be used for more generic parameter transfer. Some of the information provided by the BBERF is also needed in the PDN GW/PCEF. For example, the PDN GW may need to know which 3GPP radio technology is used (GERAN, UTRAN, or E-UTRAN) to enable proper charging and this information is not necessarily provided via the PMIP-based S5/S8 reference point. It must then be provided by the BBERF to the PCEF (PDN GW), via the PCRF, as illustrated in Figure 8.9.

Also, the AF may be interested in notifications about conditions in the access network, such as what access technology is used or the status of the connection with the UE. Therefore, the AF may subscribe to notifications via the Rx reference point. In this case it is the PCRF that reports to the AF. The notifications over Rx are not directly related to renewed policy decisions in the PCRF, but event triggers also play a role here. The reason is that if the AF subscribes to a notification over Rx, the PCRF will need to subscribe to a corresponding event via the Gx or Gxa/Gxc interface.

Figure 8.9 shows a high-level summary of the information flow in a PCC architecture.

8.2.2.6 Functional Allocations

Most functions of the PCEF are common to both "on-path" and "off-path" models. For example, service level charging, gate control, QoS enforcement, and event reporting are done in the PCEF in both cases. However, as we have also seen above, the bearer-related functions and certain event reporting need to be performed by the BBERF in the "off-path" case. Table 8.3 summarizes the allocation of different functions in the two architecture alternatives.

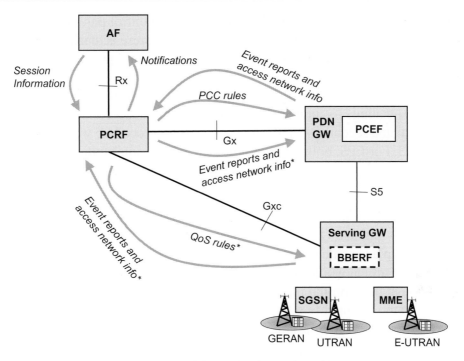

Figure 8.9: High-Level Information Flow.
Items marked with an asterisk apply to the off-path model only. Only the 3GPP
family of accesses is shown in the figure for clarity.

Table 8.3: Allocation of Functions in "On-Path" and "Off-Path" Models

	"On-Path" (BBERF Not Used)	"Off-Path" (BBERF Used)
Service level charging (flow-based charging)	PCEF	PCEF
Service level gating control	PCEF	PCEF
Service level UL and DL bit rate enforcement	PCEF	PCEF
Bearer binding	PCEF	BBERF
Event reporting	PCEF	BBERF and PCEF

8.2.3 Network vs. Terminal-Initiated QoS Control

As was already indicated in the two use cases above, there are two basic methods for initiating the QoS allocation in the access: either triggered by the UE or triggered by the network. We refer to these as the terminal-initiated and network-initiated QoS control paradigms. Below we look in more detail at a few general aspects of the two paradigms.

Originally, GPRS only supported the UE-initiated QoS control paradigm. To use UE-initiated procedures was very reasonable since there were actually no means to trigger resource reservation procedures from the network until policy control was introduced in 3GPP. However, in 3GPP Release 7, when PCC was developed, the PCC solution made it possible to trigger QoS resource reservation from the network based on application signaling. To support the network-initiated QoS control paradigm, the network-initiated secondary PDP context activation procedure was introduced in Release 7 GPRS. For EPS, both network-initiated and terminal-initiated procedures are supported for GERAN/UTRAN and E-UTRAN. CDMA2000 systems specified by 3GPP2, including eHRPD, support terminal-initiated procedures in general, while network-initiated procedures have been introduced for the eHRPD system.

With the terminal-initiated QoS control paradigm, it is the terminal that initiates the signal to set up a specific QoS towards the network. For the particular case of E-UTRAN, the terminal would send a request for bearer resources to the network. The application in the terminal must know what QoS it wants, and trigger the access interface part in the terminal (e.g. the E-UTRAN part) over a terminal–internal "interface", or Application Programming Interface (API). This API is typically not standardized and may, for example, differ between terminal vendors and access technologies; the usage of the API is illustrated in step 4 of the second use case in Section 8.2.2.2. This means that in order to specify the QoS information for the access, the client applications need to be aware of the specifics of the QoS model of the access network. With this paradigm, there is no need for a PCRF to push QoS information to the network. A PCRF may, however, still be used to authorize the QoS resources requested by the terminal, as was illustrated in the second use case above. The terminal-initiated QoS control principles are illustrated in Figure 8.10.

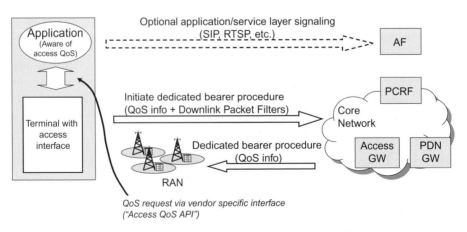

Figure 8.10: Terminal-Initiated QoS Control.

Using network-initiated QoS control, it is the network that initiates the signal to set up specific QoS towards the terminal and the radio network. For the particular case of E-UTRAN, it would be the network that initiates the dedicated bearer procedure. The trigger for this signal is received from other network nodes, typically an AF in combination with a PCRF. The signaling is sent over standardized references points such as Rx and Gx. This scenario is described by the first use case in Section 8.2.2.2 and is also illustrated in Figure 8.11.

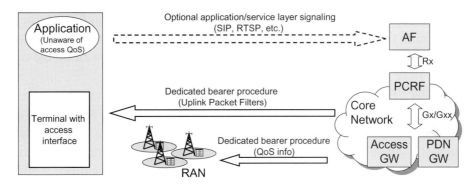

Figure 8.11: Network-Initiated QoS Control.

Using the network-initiated paradigm, the client application does not need to worry about the specifics of the QoS model of the access network. The application in the terminal can instead rely on the network to ensure that the access-specific QoS procedures are executed as needed. The application may, however, have access-agnostic knowledge of the QoS that it wants to be provided with and make a request for this QoS via the application layer. For example, the QoS to be applied to the session may be negotiated with the network by means of application-layer signaling such as SIP and Real-Time Streaming Protocol (RTSP) combined with the SDP. Note, however, that there is no access-specific information in this signaling. This property of network-initiated QoS control is very attractive since it can be used to provide QoS to access agnostic client applications, for example applications that are downloaded and installed by the user. This is not possible for terminal-initiated QoS control, which requires access-specific client applications that need to be programmed towards a vendor-specific QoS API. The possibility to be "access QoS agnostic" also enables QoS to be provided in the "split terminal" case where the client applications reside in a node (e.g. a laptop or set-top box) that is physically separated from the terminal. The signaling for a network-initiated QoS control was illustrated in the first use case in Section 8.2.2.2.

A prerequisite for the network-initiated paradigm is that the network is able to understand what QoS resources are needed for the service. However, in practice many services (e.g. Mobile-TV, IMS voice) are actually provided by the access network operator, possibly through agreements with third-party service operators, and are thus known to the operator. It is therefore reasonable that the operator also assigns the QoS level for the SDF associated with the service.

Due to the mentioned advantages with the network-initiated QoS control paradigm, we consider it to be the most beneficial in cases where the operator controls and has full knowledge of the service. For services not known to the operator it is also possible to use the terminal-initiated QoS control paradigm. An example may be that the user accesses a streaming server (not known to the operator) on the Internet and the application in the terminal wants to set up premium QoS for that service. In this case terminal-initiated QoS could be used, assuming that it is allowed by the operator.

8.2.4 PCC and Roaming

As already briefly mentioned in Chapter 2, PCC in 3GPP Release 8 supports roaming for both on-path and off-path scenarios. When a user is roaming in a visited network, we distinguish between two main roaming cases: the "Home Routed" case and the "Visited Access" case. The last alternative is often also called "Local Breakout" (LBO). In the Home Routed case, the user is connected through a PDN GW in the home network and all traffic for that IP connection is routed via the home network. In the Visited Access case, the user is connected via a PDN GW in the visited network and traffic is transported between the UE and the PDN without traversing a PDN GW in the home network. Since the PCEF is located in the PDN GW, this means that the PCEF may be located in the home or visited network. The BBERF (if applicable) is always located in the visited network when the user is roaming.

To support such roaming scenarios, two different architecture alternatives would be possible:

- One alternative would be that the PCRF in the home network directly controls the PCEF and/or BBERF in the visited network via the Gx and/or Gxa/Gxc interface.
- Another alternative would be to introduce a reference point between the PCRF in the home network and a PCRF in the visited network. The Gx/Gxa/Gxc interface would then go between the visited PCRF and the PCEF/BBERF in the visited network.

One main principle when developing the PCC architecture for these roaming scenarios has been that no policy control entity is allowed to directly control a policy enforcement entity in another operator's network. The interaction should always go via a policy control entity in the same network as the policy enforcement entity. Therefore, it was decided to go for the second alternative above, introducing a new reference point, S9, between two PCRFs, one in the home network and one in the visited network. These two PCRFs are denoted Home PCRF (H-PCRF) and Visited

PCRF (V-PCRF) respectively. The two roaming scenarios and the associated PCC architecture are illustrated in Figure 8.12. It should be noted that in the Visited Access case, the AF may be associated with either the home or the visited network.

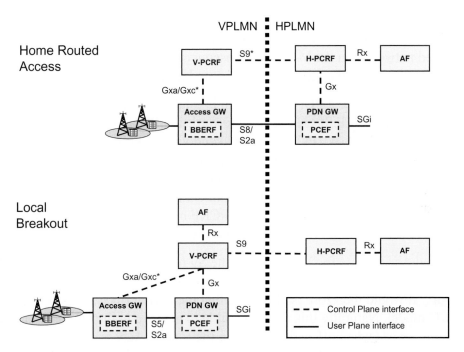

Figure 8.12: PCC Architecture for Home Routed and Visited Access (Local Breakout) Roaming Cases.
Items marked with an asterisk are only applicable to the off-path model.

Control of allowed services and the authorization of resources are always handled by the H-PCRF. Since the home operator provides this control in both roaming and non-roaming scenarios, a consistent user experience is possible. For the roaming scenarios when S9 is used, the V-PCRF may accept or reject, but not change, policy decisions coming from the home network. This allows the visited operator to control the usage of the resources in its (radio) access network.

In the Home Routed roaming scenario the PCEF is located in, and controlled by, the home operator. The PCEF connects to the H-PCRF via Gx and online charging can be performed via Gy to the OCS in a similar way as for non-roaming scenarios.

If the on-path model is used – that is, the roaming interface for home routed traffic is based on GTP – there is no need for a BBERF or Gxa/Gxc, and consequently no V-PCRF or S9 either. All QoS signaling with the visited network is taking place over

S8 using GTP. There is no need to invoke a PCRF in the visited operator's network. This roaming model is basically the same as that existing for pre-EPS GPRS.

If the off-path model is used – that is, if the roaming interface for home routed traffic is using a Mobile IP-based protocol – the S9 reference point is needed. The BBERF in the visited network is connected via a V-PCRF to the H-PCRF over the S9 reference point. For this case the H-PCRF is responsible for controlling the BBERF in the visited network, via the V-PCRF. Consequently the H-PCRF provides policy decisions (QoS rules) to the BBERF in the visited network, via the V-PCRF.

In the Visited Access case a PDN connection is established via a PDN GW in the visited operator's network. If GTP is used towards the PDN GW in the visited network, then the PDN GW (PCEF) is connected via a V-PCRF to the H-PCRF via S9. On the other hand, if a Mobile IP-based protocol is used towards the PDN GW in the visited network, then the S9 reference point and also the role of the V-PCRF become more complex. The reason is that both Gx and Gxa/Gxc procedures are handled within the same S9 session. The V-PCRF must be able to split and combine messages between S9 on one side and Gx and Gxa/Gxc on the other.

In the Visited Access case the V-PCRF and the S9 reference point are used independently of whether the visited network is using the on-path or off-path architecture. It would therefore be desirable that the S9 interface is also independent of the PCC model. Naturally this is not possible to achieve in the Home Routed case since S9 only exists when off-path is used. In the Visited Access case, however, it is to a large extent possible to "hide" the Gxa/Gxc aspects in the visited network from the S9 interface. This has been one of the goals when designing the S9 protocol. For additional details, see Chapter 15 and 3GPP TS 29.215.

For the Visited Access scenario it is possible to use AFs connected via Rx to the V-PCRF. In this situation, Rx signaling is forwarded via the V-PCRF to the H-PCRF using S9.

8.2.5 Additional PCC Features Developed Since Release 8

The previous chapters have, with a few exceptions, dealt with the basic PCC architecture and functionality as specified in 3GPP Release 8. There have, however, been significant enhancements made to PCC in 3GPP Releases 9, 10, and 11. It is the purpose of this section, and the next section (8.2.6), to describe the most significant of these enhancements in some detail.

It should be noted that these enhancements are not standalone features but, rather, functional enhancements integrated with the Release 8 baseline. It could therefore have been natural to describe these enhancements as part of the overall PCC description in the previous sections. However, in order to simplify things for a reader that is already familiar with Release 8 PCC and also to keep the description of the basic PCC functions within a reasonable scope, we have chosen to describe most of the enhancements done after Release 8 in separate sections.

In this section we will describe PCC enhancements for:

- Application Detection and Control
- Usage Monitoring Control
- Policy control based on subscriber spending limits
- PCC support for IMS emergency calls and Multimedia Priority Services
- PCC support for DSMIPv6-based IP Flow Mobility
- PCC supports for sponsored connectivity.

8.2.5.1 Application Detection and Control

When we described the basic PCC architecture earlier in this chapter it was assumed that dynamic policy and charging control can be provided for applications with explicit service session signaling. This is supported by an AF that takes part in the service signaling with the UE and also communication with the PCRF via Rx to transfer dynamic service information. An example of such application is IMS where the AF is a P-CSCF and the service session signaling is based on SIP/SDP. PCC could also provide policy and charging control where the PCC rules were preconfigured and activated already when a PDN connection was activated.

In 3GPP Release 11, PCC is enhanced to support application awareness also when there is no explicit service session signaling. A new feature, Application Detection and Control (ADC), has been introduced for this purpose. With the ADC feature it is possible to request the detection of specified application traffic and to report on the start or stop of application traffic to the PCRF. It is also possible to apply specific enforcement actions for the application traffic. The enforcement actions may include blocking of the application traffic (gating), bandwidth limitations, and redirection of traffic to another address.

There are many examples of services where there is no well-defined service session signaling protocol or where there is no AF with Rx-based control of the service. For example, the ADC feature may be used to detect the usage of a streaming video service (without AF-control) and report this usage to the PCRF. The PCRF may then, using the normal PCC procedures, instruct the PCEF to reserve appropriate resources. Another example is a point-to-point (p2p) file sharing protocol. In this case it may not be possible to report traffic filters describing the service to the PCRF. Instead, the ADC function performs enforcement to, say, limit the bit rate used by the service. Reporting the detection of start and stop of application usage to the PCRF may be useful even if it is not possible to report packet filters describing the service. The operator may use the reporting for analysis of traffic patterns, for example. It is worth noting that bit rate limitations for general applications were also possible prior to introduction of ADC using predefined PCC rules in the PCEF. However, with ADC, the bit rate limitation may be dynamically provided by the PCRF.

The application detection and control functions, including packet inspection and reporting, can either be implemented in a Traffic Detection Function (TDF) or be integrated with the PCEF. The TDF is a standalone logical entity that has been introduced with the ADC feature. The TDF resides on the SGi interface and inspects the user traffic that goes over SGi. A new reference point, Sd, is introduced between the TDF and the PCRF. The PCEF, on the other hand, is part of the Release 8 PCC architecture but will in this case become a PCEF enhanced with ADC. These two options are shown in Figure 8.13. The figure to the left shows a scenario with ADC in a standalone Traffic Detection Function. The figure to the right shows a scenario with PCEF enhanced with ADC.

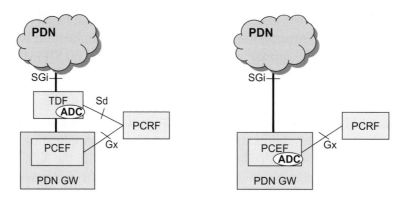

Figure 8.13: Architectures for Application Detection and Control.

It can be noted that similar functions that support application detection and reporting have also been available prior to the Release 11 time frame, but then rather as proprietary products.

The ADC feature supports two different models, solicited and unsolicited application reporting:

• In solicited application reporting the PCRF instructs the TDF (or the PCEF enhanced with ADC) about what application should be detected and reported to the PCRF. The PCRF activates so-called ADC rules in the TDF or PCEF. In addition to instructions on what applications to detect and report, the ADC rules may also contain certain enforcement actions that the TDF or PCEF should apply to the detected traffic. The PCRF takes into account subscription data, etc. when deciding on ADC rule activation (or ADC rule authorization).

- When unsolicited application reporting is used, the TDF is preconfigured with what applications to detect and report. In this case application detection by TDF is performed in the same way for all users and dynamic activation of ADC rules is not performed by PCRF. Enforcement actions are also not supported by the TDF in this case. If enforcement is needed, this is done in the PCEF using the normal policy and charging enforcement functionality as described in previous sections. Unsolicited reporting does not apply to the PCEF.

The solicited and unsolicited reporting modes have certain differences. One important aspect is the privacy issue. Typically ADC has to make use of Deep Packet Inspection (DPI) and is thus inspecting the application layers of the subscriber's traffic. There may, however, be regulations or laws that require user consent before this is performed by the operator. In the solicited model, the PCRF may therefore only activate application detection if the user profile has been configured to allow it. In the unsolicited model, the subscriber's profile is not taken into account since the TDF inspects traffic based on preconfigured rules. Therefore, to use unsolicited application reporting, it has to be assumed that a user profile configuration to indicate whether or not application detections are allowed is not required.

However, if privacy policies or other subscription-related aspects are not of concern, unsolicited reporting from TDF could be considered somewhat simpler since there is no need to set up an Sd session before there is traffic to detect for a given user. Scalability could also be an aspect if there are multiple TDFs. With unsolicited reporting there is no need for the PCRF to determine which TDF is handling a certain user's traffic.

The ADC rules are logically separate from PCC rules. They contain information that is needed to identify the rule, to detect the start and stop of certain applications, as well as information about what enforcement actions should be taken for the applications detected by the rule.

Table 8.4 lists the parameters of the ADC rules that may be provided by the PCRF. The different aspects of the ADC rules will be described in the following text.

Both predefined and dynamic ADC rules are supported. A predefined ADC rule is fixed and cannot be modified by the PCRF, only activated or deactivated. A dynamic ADC rule may be activated, modified, and deactivated by the PCRF. Note that these ADC rule operations supported on Gx and Sd (activation, modification, and deactivation) only apply to solicited reporting.

Table 8.4: ADC Rules

Type of Element	ADC Rule Element	Comment
Rule identification	ADC Rule Identifier	Uniquely identifies an ADC rule and is used between the PCRF and the TDF or PCEF for referencing ADC rules.
Application Detection	Application Identifier	References the application for which the rule applies. This parameter can not be changed by the PCRF but is fixed with the ADC rule and is reported to the PCRF when start or stop of an application is detected.
Usage Monitoring Control	Monitoring Key	This parameter is used with the Usage Monitoring Control. See Section 8.2.5.2 for further details.
Enforcement Control	Gate status	Indicates whether the detected application may pass (gate open) or will be discarded (gate closed).
	UL and DL maximum bit rates	The maximum uplink (UL) and downlink (DL) bit rates authorized for the application.
	Redirect	Has the values enabled or disabled and indicates whether the detected application should be redirected to another address.
	Redirect Destination	Defines the address to which detected application traffic should be redirected. Only used if Redirect is enabled.

The ADC rule definitions are assumed to be pre-provisioned in the TDF or PCEF enhanced with ADC. In solicited mode, the PCRF therefore does not provide the complete rule via the Sd or Gx interface but only a reference to the pre-provisioned rule. The PCRF may update some of the ADC rule elements, the gate status and the bit rate parameters, for example. This is a difference compared to dynamic PCC rules that also contain the SDF template for SDF detection. The reason for not providing traffic detection information (IP filters, etc.) over Sd or Gx reference points together with other ADC rule parameters is that the method to perform the detection of application traffic typically extends beyond the basic IP header. The traffic detection method may be complex, extending beyond simple packet filters, and it would probably be impossible to standardize the information so that it covers all applications and packet detection methods that may be of interest. Instead, the PCRF is able to reference the preconfigured ADC rule in the TDF or PCEF and also modify some of the attributes.

Once an ADC rule is active, either based on explicit ADC rule activation by the PCRF in the solicited mode, or based on ADC rules that are "always on" in the unsolicited mode, the TDF or PCEF enhanced with ADC takes certain actions:

- **Traffic detection.** Based on the information and methods implemented and configured in the TDF or PCEF enhanced with ADC, the application traffic matching the corresponding application is detected.
- **Reporting.** Once the TDF or PCEF detects events related to traffic for an application, the TDF or PCEF reports to the PCRF. The TDF or PCEF reports when it detects starting and stopping of traffic for the relevant application. The reports to the PCRF contain an Application Identifier identifying the application. The reports may also, if possible, contain service data flow filters of the detected traffic. However, it only makes sense to provide the SDF descriptions in certain cases, e.g. when there is service-specific signaling (e.g. for streaming video) and traffic can be described as one or a small number of IP 5-tuples. If the traffic filters are not deducible, e.g. for applications without service-specific signaling as with some peer-to-peer applications, this is left out of the report.
- **Enforcement actions.** Depending on the ADC rule, the TDF or PCEF enhanced with ADC may perform enforcement actions. This may imply blocking of the application traffic (gating) or limiting the application traffic to a certain uplink or downlink maximum bit rate. The enforcement actions may also include redirection of the detected application traffic to another destination, e.g. to an application server with a "top-up" or service provisioning page. If redirection is enabled, a redirection address is also contained in the ADC rule to define the destination for the redirected traffic. Redirection may not be relevant for all application types and it may, for example, only be performed on specific HTTP-based traffic. It should be noted that enforcement actions in the TDF or PCEF enhanced with ADC are an option. If the SDF descriptions cannot be provided to the PCRF, the enforcement actions would need to be performed by the TDF or PCEF enhanced with ADC. However, if the SDF descriptions are provided to the PCRF, the enforcement actions can also be performed by the PCEF as in normal PCC operation using PCC rules and SDF detection in the PCEF. The enforcement actions that are supported by the TDF or ADC function of the PCEF are, however, not the same as those supported by the PCEF. As described in previous sections, the PCEF supports bearer binding, charging, shaping, and gating. As described in this bullet, the ADC function supports shaping, gating, and redirect actions.

An example use case for using Application Detection and Control is shown in Figure 8.14 and described below:

1. The user has connected to the network. A PDN connection exists between the UE and PDN GW and an IP-CAN session corresponding to this PDN connection exists between the PDN GW and the PCRF.
2. The PCRF gets the subscription data indicating that application detection and control should be enabled for the user.

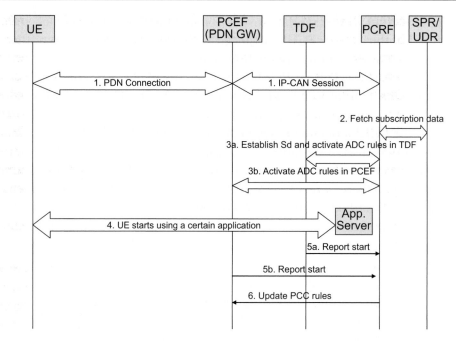

Figure 8.14: Use Case Example for Application Detection and Control.

3. If TDF is used, the PCRF initiates the Sd session with a TDF (3a). The PCRF also activates the appropriate ADC rules in either TDF (3a) or PCEF (3b).
4. The user starts an application, e.g. a video streaming application.
5. The ADC function detects that the application has started based on pre-provisioned application detection rules. The TDF (5a) or the PCEF (5b) reports to the PCRF and provides (if possible) the Service Data Flow filters for the detected application traffic.
6. The PCRF determines that this user is authorized to use premium QoS for the detected application. The PCRF therefore provides new or modified PCC rules to the PCEF corresponding to the detected SDF filters and the authorized QoS.

8.2.5.2 Usage Monitoring Control

Usage Monitoring Control is a feature added in 3GPP Release 9 that allows the operator to enforce dynamic policy decisions based on the total network usage in real time. This is supported by enhancing PCC to support monitoring of accumulated usage of network resources on a per IP session and user basis. The network resources in this case are based on traffic volume.

An example scenario where Usage Monitoring Control is useful is when the operator wants to allow a subscriber a certain high (e.g. unrestricted) bandwidth for a certain maximum volume (say 2 gigabytes) per month. If the subscriber uses more than that during the month, the bandwidth is limited to a smaller value (say 0.5 Mbit/s) for the remainder of the month. Another example is when the operator wants to set a usage cap on traffic for certain services, e.g. to allow a certain maximum volume per month for a TV or movie-on-demand service.

When usage monitoring is used for making dynamic policy decisions, it is either the PCEF or TDF that performs the counting of resources for the purpose of usage monitoring control. The PCRF sets the applicable volume thresholds and provides these to the PCEF or TDF for monitoring. The PCEF or TDF notifies the PCRF when a threshold is reached and reports the accumulated usage since the last report for usage monitoring. In order to maintain the accumulated usage also when the subscriber does not have any active IP sessions, the usage is stored in the subscriber database (SPR or UDR) when an IP session is closed. Later, when the subscriber again activates an IP session, the accumulated usage is retrieved from the subscriber database.

The usage monitoring capability can be applied to different traffic flows or groups of traffic flows. It is, for example, possible to apply for individual service data flows in the PCEF, a group of services data flows, or for all traffic of an IP session in the PCEF. The usage monitoring capability is also supported for the ADC feature. It is in this case possible to apply usage monitoring for application traffic detected by the TDF or by PCEF enhanced with ADC. In this case the usage monitoring may, for example, be performed for a particular application or a group of applications as identified by the ADC rule(s) or for all traffic belonging to a specific TDF session.

The PCRF can request usage monitoring on the complete IP Session by providing a Volume Threshold to the PCEF. The Volume Threshold indicates the overall user traffic volume after which the PCEF or TDF should report back to the PCRF. In this way the PCEF/TDF performs volume measurement of all traffic of an IP session in the PCEF or of a TDF session in the TDF. When the accumulated usage reaches the threshold, PCEF or the TDF reports to the PCRF that a usage threshold has been reached. The PCRF can then make a new policy decision and may, for example, provide updated PCC rules (or ADC rules) to the PCEF or TDF. The usage threshold report is an example of event triggers described in Section 8.2.2.

The way a more selective or granular Usage Monitoring Control (e.g. per Service Data Flow or groups of Service Data Flows) is solved is that accumulated volume usage is counted towards so-called Monitoring Keys. The operator configures the PCRF with the Monitoring Keys and associates a Volume Threshold with each Monitoring Key. The Monitoring Keys and Volume Thresholds are then provided to the PCEF or TDF. The PCEF/TDF can perform volume measurement for each

Monitoring Key separately and maintains the measurement per Monitoring Key. Since each PCC rule (or ADC rule) may include a Monitoring Key value, traffic covered by a specific PCC rule (or ADC rule) will be counted towards the accumulated usage of that specific Monitoring Key. The Monitoring Key in PCC and ADC rules can be found in Tables 8.2 and 8.4 respectively so they are not repeated here. A specific Monitoring Key may be included in one or more PCC/ADC rules. In that way the operator selects for what traffic and at what granularity to apply Usage Monitoring Control. When the accumulated usage for a specific Monitoring Key is reached, PCEF or the TDF reports to the PCRF that a usage threshold has been reached. The PCRF can then make a new policy decision and may for example provide updated PCC rules to the PCEF.

An example use case for Usage Monitoring Control is provided below and illustrated in Figure 8.15. The example shows a scenario where usage monitoring is performed by the PCEF.

1. The user has connected to the network. A PDN connection exists between the UE and PDN GW.
2. The PCEF initiates an IP-CAN session establishment towards the PCRF.
3. The PCRF contacts the subscriber database (SPR or UDR) to get the remaining usage allowance(s).
4. The PCRF sets and sends the applicable thresholds to the PCEF.
5. The PCEF counts traffic for the IP-CAN session and/or for the received monitoring keys.
6. When a usage threshold is reached, the PCEF notifies the PCRF and reports the accumulated usage since the last report.
7. If requested by the PCRF, the PCEF continues usage monitoring.
8. When the IP-CAN session is terminated, the PCRF stores the remaining usage allowance in the SPR/UDR.

8.2.5.3 Policy Decisions Based on Subscriber Spending Limits

The next PCC feature we will describe is policy decisions based on spending limits. It is a feature added in 3GPP Release 11 that allows the PCRF to take actions related to spending limits that are maintained in the OCS. A spending limit is a usage limit (e.g. monetary, volume, duration) that a subscriber is allowed to consume. When the subscriber spending limit (e.g. monetary, volume, duration) has passed a certain threshold (up or down), the system will be able to modify resources (e.g. QoS, Bandwidth, access) to services accordingly.

An example use case could be that the operator defined a daily spending limit of $2. When this subscriber spending limit has been reached, the system may, for example, trigger a QoS change. The system could also restrict access to one, several, or all

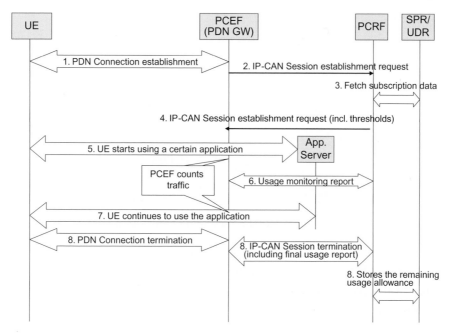

Figure 8.15: Example Call Flow Illustrating Usage Monitoring Control.

IP services based on operator predefined thresholds. In addition, the operator may redirect a user to a top-up page where the user is informed that a certain monetary limit has been reached. The user can then confirm whether continued use of the service, and continued charging, is accepted.

Since the spending limits may be related to monetary amounts, PCC needs help in order to keep track of the counters. The PCRF does not know anything about monetary amounts or how much has been spent by a user. Anything relating to actual money is handled by the charging systems. On the other hand, the charging systems do not know the QoS or charging control policies; these are handled by the PCRF. So in order to solve the use cases described above, there needs to be some coordination or interaction between the charging system and the PCRF. The solution to support policy control based on subscriber spending limits is that a new reference point, the Sy reference point, has been defined between the Online Charging System (OCS) and the PCRF. The OCS maintains counters for the subscriber spending and provides status reports to the PCRF via the Sy reference point. The PCRF uses the spending limit status as input for policy decisions related to, say, e.g. QoS control, gating, or charging conditions. The reference architecture is shown in Figure 8.16.

The solution has been defined in a way that maintains the separation between monetary aspects (OCS territory) and policy control aspects (PCC territory). In order

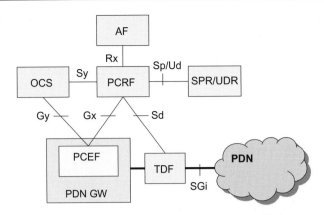

Figure 8.16: Architecture for Subscriber Spending Limits with Sy Reference Point.

to do this, a certain abstraction between the spending counters and limits themselves and the status of the counters and limits has been introduced. The OCS is enhanced to support so-called *policy counter(s)* to track spending for a subscription. A policy counter (e.g. monetary, volume, duration) in OCS may represent the spending for one or more services, one or more subscriber accounts, etc. The representation is operator dependent and is configured in the OCS. The OCS then provides the *status* of the policy counters to the PCRF via the Sy reference point. The policy counters themselves are not sent over Sy since that would mean that the PCC would need to become aware of, say, monetary values. Instead, only the *status* of the counters is sent, e.g. whether the counter lies above or below a certain policy counter threshold that is defined in OCS. The policy counter status values thus have to be configured in both OCS and PCRF, while the policy counters themselves are only known to the OCS. Based on the policy counter status, the PCRF can make a policy decision for the subscriber.

As mentioned above, the PCRF receives information relating to subscriber spending available in the OCS via the Sy interface. The PCRF requests reporting of policy counter status from the OCS and the OCS provides the corresponding status as a reply back to the PCRF. If spending limit reporting is enabled, the OCS will also notify the PCRF if the status of the policy counters change (e.g. if a daily spending limit is reached). The PCRF can then take actions (i.e. policy decisions) associated with the policy counter status received from OCS. The actions to be taken are configured in the PCRF and could be policy decisions relating to QoS control, gating, or change of charging conditions. The PCRF could, for example, downgrade the QoS for the whole IP session (i.e. APN-AMBR). The PCRF could also provide modified PCC rules to the PCEF or modified ADC rules to the TDF or the PCEF to change gating, QoS, or charging conditions for certain Service Data Flows or applications.

An example use case for spending limit reporting is provided below and illustrated in Figure 8.17. The example shows a scenario where a user reaches a certain spending limit and as a result the PCRF downgrades the APN-AMBR.

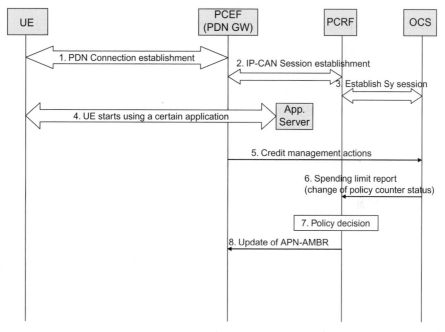

Figure 8.17: Example Call Flow Illustrating Policy Control Based on Subscriber Spending Limits.

1. The user has connected to the network. A PDN connection exists between the UE and PDN GW.
2. The PCEF initiates an IP-CAN session establishment towards the PCRF.
3. The PCRF initiates establishment of the Sy session with the OCS and activates spending limit reporting for certain policy counters.
4. The UE makes use of an application.
5. Usage is reported to the OCS via the Gy interface.
6. When the status of a policy counter changes (e.g. if a daily spending limit is reached) the OCS informs the PCRF via the Sy interface.
7. The PCRF makes a policy decision to modify the value of the APN-AMBR.
8. The PCRF provides the updated APN-AMBR to the PCEF. The PCEF enforces the new value of the APN-AMBR.

Note that spending limits may be affected not only by online charging over Gy (i.e. PS domain), but the OCS may also take into account IMS level charging, CS charging, and account top-ups.

8.2.5.4 PCC Support for IMS Emergency Calls and Multimedia Priority Services

The EPC procedure to support IMS-based emergency services for 3GPP accesses was introduced in 3GPP Release 9. An overview of the support of emergency services in EPC can be found in Chapter 11. As we will see below, emergency services also put specific requirements on PCC, since PCC needs to be enhanced to properly handle the prioritization and authorization aspects of emergency services.

Emergency services in EPS are network services provided through an Emergency APN. The PCRF determines, based on the APN, if an IP session concerns an IMS emergency session. Furthermore, an since emergency service is not a subscription-based service, the PCRF does not need to use any user subscription data when authorizing PCC rules to provide policy control for emergency calls. PCRF may instead have certain information locally configured in order to provide policy control for an emergency session without any proper credentials; Chapter 11 provides more details on this.

As described in Chapter 11, an IP session serving an IMS emergency session is dedicated to emergency services and will not serve any other service. It is one of the tasks of the PCRF to ensure that the emergency IP session is used only for IMS emergency sessions. To do this, the PCRF makes authorization and policy decisions (reflected in PCC rules) that restrict the traffic to emergency destinations, and for traffic needed for emergency services.

When an IMS Emergency call is invoked, the PCRF receives a Priority indicator over Rx from the P-CSCF (acting as AF) together with the service session information. The PCRF can use this as an indication that the call is an emergency call. When the PCRF generates PCC rules for emergency sessions it selects QoS parameters that allow proper prioritization, e.g. by setting the ARP to the specific value that is reserved for IMS Emergency calls. Furthermore, if an IMS session is being established via the emergency IP session and the AF (i.e. P-CSCF) does not provide an emergency indication to the PCRF, the PCRF will reject the session setup.

Multimedia Priority Services (MPS) is another special service that puts additional requirements on PCC. In contrast to IMS Emergency Services, MPS is a subscription based service. MPS is further described in Chapter 11. For Multimedia Priority Services, Section 8.2.2.2 describes use cases for PCC that can be reused mostly as is. An additional PCC aspect is the possible trigger of priority change by an AF providing priority service and updating the SPR data for that user. PCRF is therefore required to subscribe to the subscription event change from SPR so it is notified of the priority change for MPS users. Once the PCRF receives notification of a change in MPS EPS Priority, MPS Priority Level, and/or IMS Signaling Priority from the SPR, the PCRF must make the corresponding policy decisions (i.e. ARP and/or QCI change) and ensure that the associated IP-CAN session(s) is (are) changed accordingly. Also, dynamic invocation for MPS needs to be supported from an AF, using the Priority indicator, over Rx. PCRF is responsible for generating appropriate ARP/QCI for MPS based on the priority level of the service being authorized. On completion of a priority service, PCRF is also responsible to change the bearers from priority level to a normal level of ARP/QCI as applicable for the user.

If the MPS is an IMS-based MPS, P-CSCF interacts with the PCRF to ensure that the bearers are established according to the priority level determined on the IMS level. This may include ensuring IMS signaling is handled with a higher priority requested by P-CSCF when user MPS subscription indicates so and if an operator has configured support according to local regulations and policies.

8.2.5.5 PCC Support for Sponsored Connectivity

With the increased usage of mobile broadband, access to new IP services over mobile broadband is becoming more and more common. For example, the user may download an ebook or a game that was purchased from an online store. The user may also access various streaming services and may, for example, watch a free trailer from an online movie store to determine whether to buy the entire movie or not. In many cases, it is reasonable that a sponsor pays for the user's data usage in order to allow the user to access the Application Service Provider's services. With sponsored data connectivity, a feature introduced in 3GPP Release 10, PCC is enhanced to simplify the handling of scenarios like these. With this feature, the sponsor has a business relationship with the operator and the sponsor reimburses the operator for the user's data connectivity in order to allow the user access to associated Application Service Provider (ASP) services.

A motivation for sponsored connectivity support is to allow the operator to make money even if the mobile subscription is flat rate, by enabling additional revenue opportunities for both the ASPs and the operators. In particular, such dynamic data usage provided by the sponsor allows the operator to increase revenues from users with limited data plans. The user may have limited data plans allowing only a nominal data volume per month and the sponsor may dynamically sponsor additional volume allowance for the user to allow access to the services offered by the Application Service Providers. For example, the user may use the limited data plan to browse an online store for interesting books, but once a book is purchased, the data usage for downloading the book is not deducted from the user's data plan allowance.

The sponsor may be the same business entity as the Application Service Provider. For the example with the free movie trailer above, the online movie store (Application Service Provider) could also act as sponsor and pay for the mobile data traffic. However, the sponsor could also be a different business entity. For example, a restaurant chain (sponsor) could sponsor the mobile data traffic by handing out vouchers to their guests that gives access to content provided by an ASP. When an end-user later accesses this content using the voucher, the restaurant chain would act as sponsor. It could also be worth noting that the sponsored traffic may be granted a certain level of QoS (e.g. for video streaming). An architecture diagram describing an example network configuration is shown in Figure 8.18.

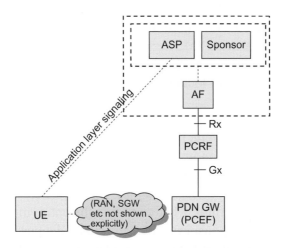

Figure 8.18: Example Architecture for Sponsored Connectivity.

To support sponsored connectivity the Rx reference point has been enhanced to allow the AF to provide the sponsor's identity, the application service provider identifier, and information identifying the application (e.g. packet filters, application identifier). The AF may also include a usage threshold to indicate the limit for how much data is sponsored. The sponsor identity and the application service provider identifier can also be included in the PCC rules and in the charging records for offline charging. Correlation of accounting records and usage data records per sponsor and/or application service provider is possible to allow proper handling and settlement by the charging systems.

In principle it was possible to support similar use cases already before the Release 10 enhancements (based on pre-Release 10 Rx and charging mechanisms), but with the support of the explicit sponsor and ASP identifier elements in Rx, Gx, and in CDRs, it is now more straightforward to configure the network to support sponsored connectivity.

8.2.5.6 PCC Support for DSMIPv6-Based IP Flow Mobility

When DSMIPv6-based IP flow mobility is supported, a single IP session in the UE and the PCRF may traverse more than one access technology. For example, some IP flows may be transported via E-UTRAN while other IP flows for the same IP session are simultaneously transported via WLAN. DSMIPv6-based IP flow mobility is described in some detail in Chapters 6 and 14.

IP flow mobility also has an impact on PCC. Without IP flow mobility, all Service Data Flows of an IP session are transported via a single access technology. This means that all PCC rules that are active for a certain IP session represent traffic that is carried over a single access. There may, now and then, be handovers between access technologies (e.g. between E-UTRAN and WLAN) but in that case all traffic belonging to an IP session are handed over simultaneously. So the assumption about a single access at a given point in time still holds. However, with IP flow mobility some Service Data Flows may be carried over one access technology while other Service Data Flows are simultaneously carried over another access technology. To the PCRF this means that the PCC rules that are active for a certain IP session may be representing traffic that is carried over different accesses.

In order to make proper policy decisions and to generate proper PCC rules, the PCRF needs to know which access technology is used to transport a certain service or IP flow. Perhaps even more importantly, if the off-path model is used the PCRF needs to know what BBERF is handling a certain service. If some services (IP flows) are carried over one access and other services over another access, these flows would also be carried over different BBERFs in the off-path model. Therefore, the PCRF must be aware of what access an IP flow is using so that the QoS rules can be provided to the appropriate BBERF. The PCRF thus needs to have some information about the IP flow mobility routing rules that are used to define over what access a certain IP flow should be routed. The PCEF provides the routing rule information to the PCRF based on the flow binding information received from the UE via DSMIPv6 signaling. The PCRF can, by analyzing the routing rules, deduce over what access traffic corresponding to a certain PCC rule is carried.

Table 8.5 lists the information contained in a routing rule on Gx.

Table 8.5: Information Contained in a Routing Rule Over Gx

Type of Element	Routing Rule Element	Comment
Rule identification	Rule Identifier	Uniquely identifies a routing rule and is used between the PCRF and the PCEF for referencing routing rules
Routing information	Precedence	Determines the order in which the routing filters are applied
	Packet filter	A list of packet filters for the detection of IP flows
	IP flow mobility Routing Address	The IP flow mobility Routing Address that the matching IP flows use

8.2.6 PCC Support for Fixed Broadband Access

As mentioned earlier in this chapter PCC is designed to be multi-access capable and to work for many different access technologies. However, to make PCC work for a specific access technology, certain adaptations may be needed, e.g. to define new information elements, event triggers, etc. that are needed to address access technology-specific aspects. One area of recent work in 3GPP has been around using PCC for fixed broadband accesses, in particular for fixed accesses defined by the Broadband Forum (BBF).

In 2008, a collaboration between 3GPP and BBF was initiated. The intent was to study solutions for interworking between 3GPP Systems and Fixed Broadband Access Networks. This work has primarily been driven inside each standardization organization but aspects affecting both organizations have been handled in joint workshops, conference calls, and liaison statements. The initial intent was to focus on interworking aspects, i.e. when network entities in the 3GPP domain interact with network entities in the BBF domain. This is also the area that has reached the most maturity and has been included in 3GPP Release 11. However, due to interest in scenarios where a single operator owns both the 3GPP domain and the fixed domain, work has also been initiated on convergence aspects. What we mean by convergence here is that a single network entity is enhanced to function both in the 3GPP domain and in the fixed domain. Different convergence aspects have been discussed, including policy convergence and user database convergence. 3GPP and BBF have started to discuss policy convergence but it is still work in progress and is not yet included in any normative specifications. In this section we will therefore focus on 3GPP–BBF interworking solutions as specified for Release 11.

Both 3GPP and BBF have been working on specifications to cover Fixed Broadband Interworking. BBF has been working on a document called Working Text 203 (WT-203) to capture the requirements and solutions. 3GPP has documented the work in TS 23.139. Details about the PCC aspects are also included in TS 23.203.

The solution addresses several scenarios for 3GPP UEs connecting via fixed broadband access and traffic is routed back to the 3GPP domain:

- 3GPP UE connecting via WLAN where traffic is routed via EPC using SWu and S2b
- 3GPP UE connecting via WLAN where traffic is routed via EPC using SWu and S2c
- 3GPP UE connecting via WLAN where traffic is routed via EPC using S2c only
- 3GPP UE connecting via Home base station (HeNB or HNB) where traffic is routed via EPC.

The solution also covers a scenario where traffic is offloaded in the fixed broadband access:

- 3GPP UE connecting via WLAN where traffic is offloaded in the fixed broadband access.

Figure 8.19 shows the reference architecture for the 3GPP–BBF interworking solution (note that not all interfaces are shown in the figure).

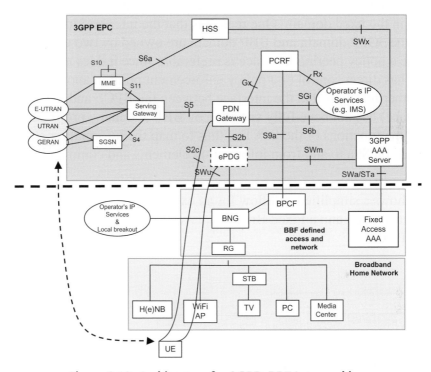

Figure 8.19: Architecture for 3GPP–BBF Interworking.

The interworking solution covers several areas, including authentication, mobility, and policy control. When it comes to authentication and mobility for WLAN scenarios, the interworking solution reuses already existing EPC solutions that have been defined for generic non-3GPP access. Access authentication in WLAN is based on EAP-AKA' and the mobility solutions reuse solutions based on S2b, SWu, and S2c reference points. In Figure 8.19, the interworking reference points STa and SWa are shown and used for authentication, authorization, and accounting (AAA) signaling. The interested reader is referred to Chapters 6 and 7 for more information about the authentication and mobility solutions, and Chapter 15 for more information on the STa and SWa reference points. For policy control, the solution uses the existing PCC architecture and reference points as a basis. However, it was not sufficient to only reuse the release 8 PCC interfaces. A new reference point, the S9a reference point, has been defined to support 3GPP–BBF interworking (see Figure 8.19).

The S9a interface is defined between the PCRF and the BBF Policy Control Function (BPCF), and is used by the PCRF to request admission control in the fixed broadband access. It is thus a policy peering interface between two policy controllers, similar to the S9 reference point. There are several reasons for defining such a policy peering interface rather than a direct interface between the PCRF and the policy enforcement point (BNG) in the fixed domain. One reason is that the interworking solution must also work if the 3GPP domain and BBF domain are owned by two different operators. In such a case a policy peering interface is preferable since it does not require that one operator has direct control of a policy enforcement point in another operator's domain. Another reason is that the interface between the BPCF and the BNG (the BBF R interface) has not been fully specified by BBF and many proprietary variants exist. One of the functions of the BPCF is thus to map the information elements and commands on S9a to and from information elements and commands on the R interface.

Figure 8.20 shows a simplified call flow for an example where a UE makes an initial attachment via WLAN and a Fixed Broadband Access and establishes a connection to EPC via an ePDG.

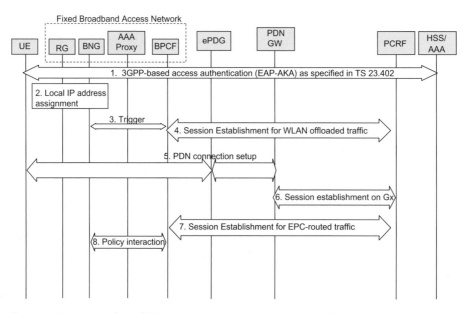

Figure 8.20: Example Call Flow for Initial Attachment via Fixed Broadband Access.

1. The UE performs access authentication in the WLAN and Fixed Broadband Access. The UE is authenticated using IEEE 802.1X and EAP-AKA' towards the home mobile operator's network.
2. The UE receives an IP address from the Fixed Broadband Access.
3. Triggered by the IP session setup in the Fixed Broadband Access, the BPCF is informed about the new IP session and the UE identity (IMSI).
4. The BPCF initiates establishment of a session over S9a. This session is a logical session (called IP-CAN session) to provide policies for the UE's traffic that is offloaded in the Fixed Broadband Access.
5. At any time later, the UE initiates a connection to EPC via en ePDG. In this example the procedures for "untrusted non-3GPP accesses" are used, establishing an IPSec tunnel towards ePDG and a GTP tunnel between ePDG and PDN GW. For more details on these procedures, see Chapters 6 and 7.
6. The PDN GW initiates a session towards the PCRF for policy control of the PDN connection in the PDN GW.
7. Triggered by the establishment of a PDN connection, the PCRF informs the BPCF that a PDN connection has been established. A new logical session (called Gateway Control Session) is established over S9a to handle policy control for traffic that is routed between the UE and the EPC over the Fixed Broadband Access.
8. The BPCF may provide policy information to the BNG for policy enforcement in the BNG.

The S9a protocol is based on Diameter. However, at the time of writing, the detailed S9a protocol definition was still work in progress.

8.3 Charging

As operators invest in new infrastructure and persuade end-users to enjoy the benefits of the newly deployed networks, the revenue-generating options become a key factor for the business cycle. How the end-users/subscribers are actually charged and how billing information is packaged towards them is very much according to the individual operator's business model and competitive environment they are operating in. From the EPS point of view, the system needs to enable collection of enough information relating to different aspects of the usage for individual users so the operator has the flexibility to determine their own variant of billing as well as packaging towards end-users. It has become increasingly important in today's competitive business environment for operators to be able to provide lucrative and competitive option packages towards their potential customers to lure business away from other operators. The process of collecting information related to charging can provide tools and means for the operators to make this possible.

For EPS, the existing charging models and mechanisms apply, except for the charging aspects of the circuit-switched domain.

The 3GPP charging infrastructure principles and mechanism did not change due to EPS, but, rather, EPS entities have been included within this infrastructure. Figure 8.21 shows the overall high-level charging system reference model.

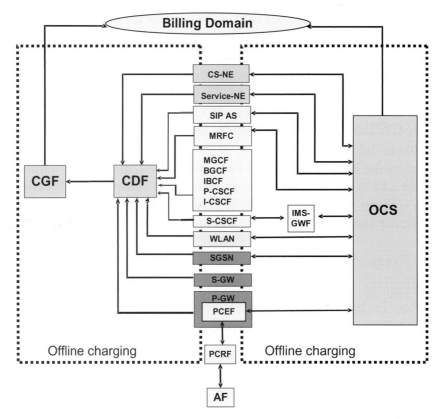

Figure 8.21: Overall Logical High-Level Reference Model for Collection of Charging Information.

The two main charging mechanisms provided by the model are offline and online charging, though the terms online and offline do not necessarily relate to how end-users are billed. These two mechanisms are the means of how the charging-related data is collected and transported to the billing system for further processing as per individual customer's billing options and for settling accounting relations between operators and between operators and subscribers.

Offline charging facilitates collection of charging-related data concurrently as the resources are being used. Offline charging data is collected in the various elements provided to support such collection. The data is collected on an individual basis and then may be sent to the billing domain according to the operator's configuration.

Online charging, on the other hand, requires the network to actually get authorization of network resource usage before such usage can occur; the OCS (Online Charging System) is the authorization entity that either grants or rejects the request made by the appropriate network element. In order to do this, the network needs to assemble the relevant charging- and resource usage-related information (known as the charging event) and send it to the OCS in real time, thus allowing the OCS to give the appropriate level of authorization. The authorization from OCS may be limited in its scope, such as the volume of data, time of usage, etc. Depending on the level of authorization, the network may need to obtain a re-authorization performed for any additional resource usage.

Note that information collected by the charging system can be used in various ways, as it may provide a statistical measure of network resource usage at certain times of day, usage behavior, application usage, etc.

In the case of offline charging, various network elements can have a distributed collection role, which would allow for more detailed availability of information. Alternatively, they may have a centralized role, with a limited collection of events capability as per the entity's role within the network. This role is specified by the Charging Trigger Function (CTF), which is an integrated component in each network entity that generates charging. See Figure 8.22 for an illustration of the logical flow of offline charging data. The CTF causes the entities to collect charging events, for example appropriate charging-related data. The CTF forwards charging events to the Charging Data Function (CDF) via the Rf interface. The CDF, in turn, receives charging events from the Charging Trigger Function and uses the information contained in the charging events to construct Charging Data Records (CDRs) with a well-defined content and format. The CDRs produced by the CDF are forwarded immediately to the Charging Gateway Function (CGF) via the Ga reference point. The CGF acts as a gateway between the 3GPP network and the Billing Domain. The CGF is responsible for sending the CDRs via the Bp reference point to the Billing Domain.

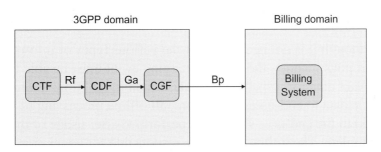

Figure 8.22: Logical Generic Offline Charging Architecture.

The CDF and CGF may be integrated in the EPC entities (e.g. PDN GW) but they may also be implemented as standalone physical elements. Figure 8.22 illustrates the interfaces between the charging functions, as well as possible mapping of the logical charging architecture to physical entities.

The heart of the offline system is the CTF, which keeps account of usage information regarding the services being delivered to the end-user based on either:

- Signaling information for sessions and service events by the users of the network or
- The handling of user traffic for these events and sessions. The offline charging system can be illustrated in a simplified manner, as shown in Figure 8.23. The EPC nodes shown in the figure may be a PDN GW, a Serving GW, or an SGSN. MME does not provide charging data.

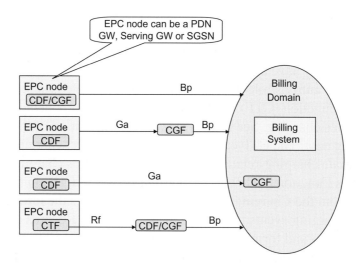

Figure 8.23: Offline Charging Entities.

The information must be made available in real time with data that uniquely associates and correlates the user's consumption of network resources and/or services.

Note that even though it is not necessary for the various types of information to be sent in a synchronous manner, the overall charging event must be able to receive and process all relevant data for a specific service/session in real time in order to provide accurate, billable data to end-users. Therefore, all offline processing of Charging Data Records (CDRs) to the end-user's billing is performed after usage of the network resources is complete. The billing domain is responsible for generation and handling of the settlement/billing process offline.

In the case of online charging, the CTF and Online Charging System (OCS) and several other network entities form the charging function. The OCS contains functions like Online Charging Function (OCF), Rating Function (RF), and Account Balance Management Function (ABMF) in order to handle the online charging process. The OCF is the entity that connects to the network elements responsible for providing the charging data – that is, it supports the CTF function. Even though the CTF performs very similar functions for both the offline and online charging mechanisms, online charging requires additional handling for authorization prior to the resource usage and thus also requires additional functions from the CTF, which are crucial for the real-time online process. Some of these functions are as follows:

- The charging events are sent to the OCF in order to retrieve authorization for the chargeable event/network resource usage requested by the user.
- The CTF must be able to delay the actual resource usage until permission has been granted by the OCS.
- The CTF must be able to track the availability of resource usage permission ("quota supervision") during network resource usage.
- The CTF must be able to enforce termination of the end-user's network resource usage when permission by the OCS is not granted or expires.

The OCF supports two methods of charging:

1. Session-Based Charging Function provides online charging for the network/user's sessions; examples of such services are the PS resource usage for IMS sessions.
2. Event-Based Charging Function (i.e. content charging) provides online charging in support of an application server or service, such as SIP AS or MMS.

The Rating Function is responsible for providing the OCF with the actual value of the network resource usage/service usage, which may be either monetary or non-monetary information. This is determined by information provided by the OCF and CTF; the actual rating or determination of the value of the usage is very much operator specific and can be quite wide ranging. Some basic examples of this rating are as follows:

- Rating of data volume (e.g. based on charging initiated by an access network entity, i.e. on the bearer level)
- Rating of session/connection time (e.g. based on charging initiated by an IMS level application)
- Rating of service events on the service level (e.g. based on charging of Web content or MMS).

The ABMF is responsible for the subscriber's account balance within OCS.

In the case of online charging, the network resource usage must be authorized and thus a subscriber must have a pre-paid account in the OCS in order for the online pre-network resource usage authorization to be performed. The two methods used to achieve this are known as Direct Debiting and Unit Reservation. As their names imply, in the case of Direct Debiting, the user is immediately debited the amount of resource usage needed for that specific service/session, whereas for Unit Reservation a predetermined unit is reserved for the usage and the user is then allowed to use that amount, or less, for that service/session. When resource usage has been completed (session terminated, or the service is completed, etc.), the actual amount of resource usage (i.e. the used units) must be returned by the network entity responsible for monitoring the usage to the OCS so that over-reserved amounts can be re-credited to the subscriber account, ensuring that the correct amount is debited.

Note that PCC makes it possible to have quite detailed charging mechanisms and allows for the possibility of operators having granular control over the subscriber's usage of the network resources. PCC also allows operators to offer various flexible charging and policy schemes towards their subscribers. More details can be found in Section 8.2 on policy control and charging. The enhancements since Release 8 to the PCC/EPC architecture, described in Section 8.2 (e.g. Application Detection and Control, Usage Monitoring Control, policy control based on subscriber spending limits, PCC supports for sponsored connectivity, Multimedia Priority Service data, CSG information), also improve capability for more complex and dynamic charging in a standardized manner. The tools available allow for customized billing to be developed by vendors catering to specific market/customer needs that gives operators the ability to distinguish themselves from each other and apply these tools in creative marketing campaigns.

For online charging, the PCEF function (as described in the PCC section) interacts with the OCS as described above and provides the online charging functions in the PS Domain. Note that the GGSN in GPRS provides the PCEF functions and related support for PCC for online charging. It also provides the necessary configuration options as specified here for PDN GW in the case of offline charging; GGSN aspects are not described within the context of EPS. Key functions relating to charging data collection triggers in the EPS can be described on a high level as follows.

Mobility management-related events relevant for charging, for example Inter-RAT handover, user's activity/inactivity during established sessions, and roaming/non-roaming status, etc., are collected at the SGSN. It may be noted that similar functions have not been specified for MME. There has not been enough interest in providing charging data from MME for mobility management. Instead, it is considered sufficient that the Serving GW generates charging data, e.g. taking into account the bearer-related events for a UE available in the Serving GW, as described further below.

EPS bearers and related functions may be collected at the SGSN and Serving GW/PDN GW. Individual service data flows within an EPS bearer according to PCC may be collected at the PDN GW but only for the GTP protocol variant. At the time of writing, for the case of PMIP, the data is collected on a per-PDN level.

The MBMS and Location Services in the EPS are supported from Release 9 and are described in Chapters 12 and 13. Charging for these functions has also been defined since Release 9.While the SGSN and Serving GW charging data collection is more related to the radio access type used, the PDN GW collects for external network-related data relevant to a subscriber. A subscriber's subscription-related data may also be relevant for charging; an example of such data would be the APN used (more on APN can be found on the session management subsection in Chapter 6). A unique charging identity is assigned; uplink and downlink data volume, date, and time are collected for individual subscribers for the purpose of charging.

One example of items that may be contained in a CDR record generated by a PDN GW, as depicted in 3GPP TS 32.251, is illustrated in Table 8.6.

Explicit identifiers are used in each of the domains: CS, PS, IMS, and applications that are involved in a specific session; this is because charging data is collected in several network entities and also is needed in order to settle inter-operator resource usage when a user is roaming. In EPS, the Charging Identity for EPC and the PDN GW Identity make up this identification.

Various levels of correlation are supported in order to achieve a complete charging information profile for the individual usage of each subscriber. Intra-level correlation aggregates the charging events belonging to the same charging session, for example over a time period, and implies the generation of interim charging records. When an end-user has accessed a service during the same session from different radio accesses or while roaming, the network entities involved in charging perform correlation of the data. Inter-level correlation combines charging events generated by the different CTFs in different 3GPP domains, and inter-network correlation for the IMS requires generation and transmission of specific identifiers across operators' networks. An example of the different levels of correlation could be an end-user in E-UTRAN access connected via EPC using MMTEL service towards an end-user connected to another operator in another network. All three types of correlation would then be generated (if supported and required) by the different operators and the specific domains.

How can all these be configured for a subscriber? Subscriber charging provides the means to configure the end-user's charging information into the network. Charging data collected by the different PLMNs involved (e.g. HPLMN, Interrogating PLMN, and VPLMN) may be used by the subscriber's home operator, depending upon the deployment and user's roaming status, to determine the network usage and the

Table 8.6: Examples of Items That May be Contained in CDRs Generated by the PDN GW

Field	Description
Served IMSI	IMSI of the served party
Served MN NAI	Mobile Node Identifier in NAI format (based on IMSI), if available
P-GW address used	The control-plane IP address of the PDN GW used
Charging ID	IP-CAN bearer identifier used to identify this IP-CAN bearer in different records created by PCNs
PDN Connection ID	This field holds the PDN connection identifier to identify different records belonging to the same PDN connection
Serving Node Address	List of SGSN/Serving GW control-plane IP addresses used during this record
Serving Node Type	List of serving node types in control plane. The serving node types listed here map to the serving node addresses listed in the field "Serving Node Address" in sequence
PGW PLMN Identifier	PLMN identifier (Mobile Country Code and Mobile Network Code) of the PDN GW
Access Point Name Network Identifier	The logical name of the connected access point to the external packet data network (network identifier part of APN)
PDP/PDN Type	PDP type or PDN type (i.e. IPv4, IPv6, or IPv4v6)
Served PDP/PDN Address	IP address allocated for the PDP context/PDN connection, i.e. IPv4 or IPv6
Dynamic Address Flag	Indicates whether served IP address is dynamic, which is allocated initial attach and UE requested PDN connectivity. This field is missing if the IP address is static
List of Service Data	Consists of a set of containers, which are added when specific trigger conditions are met. Each container identifies the configured counts (volume separated for uplink and downlink, elapsed time, or number of events) per rating group or combination of the rating group
Record Opening Time	Time stamp when IP-CAN bearer is activated in this PDN GW or record opening time on subsequent partial records
MS Time Zone	This field contains the MS Time Zone where the MS is currently located, if available
Duration	Duration of this record in the PDN GW
Cause for Record Closing	The reason for the release of record from this PDN GW
Record Sequence Number	Partial record sequence number, only present for partial records
Record Extensions	A set of network operator/manufacturer-specific extensions to the record. Conditioned upon the existence of an extension
Local Record Sequence Number	Consecutive record number created by this node. The number is allocated sequentially including all CDR types
Served MSISDN	The primary MSISDN of the subscriber
User Location Information	This field contains the user location information of the UE
Serving Node PLMN Identifier	Serving node PLMN identifier (Mobile Country Code and Mobile Network Code) used during this record
RAT Type	This field indicates the Radio Access Technology (RAT) type currently used by the Mobile Station, when available
Start Time	This field holds the time the user IP session starts, available in the CDR for the first bearer in an IP session
Stop Time	This field holds the time when the user IP session is terminated, available in the CDR for the last bearer in an IP session

services, either basic or supplemental. It may also be possible to use external Service Providers for billing.

For those subscribers handled by Service Providers, the billing information is utilized for both wholesale (Network Operator to Service Provider) and retail (Service Provider to subscriber) billing. In such cases, the charging data collected from the network entities may also be sent to the Service Provider for further processing after the Home PLMN operator has processed the information as may be desired. Figure 8.24 illustrates the different business relationships from the perspective of charging and billing. For the purposes of this book, the circuit-switched aspects, while shown in the figure, are excluded from the description.

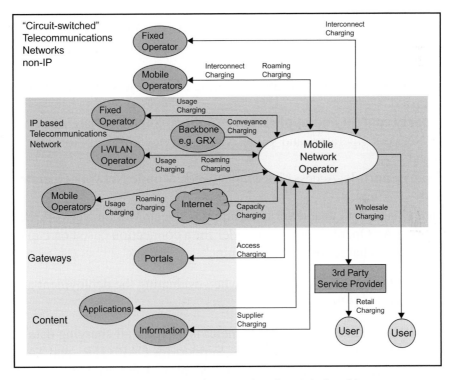

Figure 8.24: 3GPP View on Charging Relationship.
Source: TS 22.115.

The entities and their roles can be described as follows:

Users: Retail users that are charged by their Mobile Network Operator or third-party service provider. Normally users have subscriptions or similar relationship with either or both parties.

Third-Party Service Providers: Charged wholesale by the Mobile Network Operator. Responsible for providing users the billing and other charging-related customer care type services to users for services rendered.

Other telecommunications operators: Interconnect charging between Mobile Network Operator and non-IP "circuit-switched" Network Operators for call traffic carried, and usage charging between Mobile Network Operator and IP-based Network Operators for session traffic carried. For the purposes of the current section, this group is of no interest.

Other mobile operators: Roaming charging between these entities; this may require different mechanisms for IP-based types from the traditional "circuit-switched" types. Also, where mobile operators need to pass traffic to one another, there will be interconnect charging for non-IP "circuit-switched" types and usage charging for IP-based types.

I-WLAN operators: Where I-WLAN operators need to pass traffic to mobile operators or mobile operators to I-WLAN operators, there may be roaming and usage charging.

IP backbone carriers: Conveyance charging Mobile Network Operators for traffic carried.

Third-party content and application suppliers: Supplier charging between Mobile Network Operators and Value-Added Service Providers for information exchanged.

Third-party portals: Access charging between Mobile Network Operators and this entity.

Internet: Charge for capacity of connection between Mobile Network Operator and Internet. An Operator pays a provider for a connection based on capacity, for example annual charge for a 2 Mbit/s "pipe".

Some of these roles can be easily derived when evaluating the deployment and roaming scenarios for EPS.

EPS is an all-IP based network where Diameter is used for all charging data collection-related functions within the EPS. Support for CAP, GTP′, and TAP do, however, remain due to the existence of widely deployed and used billing systems, as well as for interworking and backwards compatibility with 2G/3G networks. The interfaces to the billing system rely heavily on the operators' business model and billing principles, and may also involve third-party service providers that affect end-users directly. Therefore, replacement of and/or drastic changes to these interfaces are not easily done.

As charging data can not only convey operator's business aspects, it may also reveal sensitive information regarding individual subscribers, so transfer of such information to external entities outside of an operator's secure domain must also be done in a secure manner so that information may not be disclosed to unauthorized personnel or entities. Integrity of charging information must be maintained; privacy and secrecy are integral and must be provided by the serving operator and validation of the content and receipt of charging information by the serving Network Operator must be possible. Note that by serving Network Operator, we mean both home and visited, as well as any intermediary Network Operators/third-party providers involved in the transaction and handling of the charging/billing.

CHAPTER 9

Selection Functions

9.1 Architecture Overview for Selection Functions

EPS is an all-IP-based system, and thus naturally uses DNS-based mechanisms as well as other IETF-defined node discovery mechanisms in order to find appropriate network entities within and among operators' networks. Even though it is based on IP and Internet-driven technology, EPS has specific requirements stemming from existing GPRS networks deployed today as well as the specific nature of the networks the operators manage and share resources with, and the type of services they render. In addition, the nature of the 3GPP Packet Core Network Selection Functions is also very much dictated by the fundamental principles of mobility, roaming and security, and whether the different network elements should or should not be visible from external networks (e.g. the Internet). In addition, operators also want to be able to manage their existing and Evolved Packet Core via a single selection mechanism and include the terminal's access network capabilities as part of the selection criteria.

The method for how the network entities are selected for a certain terminal also dictates how the DNS and other selection mechanisms have been developed for EPS, especially selection of entities such as MME, Serving GW, PCRF, and so on. Such core network entities will not be reachable from external networks or entities that are not governed by the roaming agreements and roaming interconnection networks known as GRX and IPX. Another important factor in the selection of key network nodes like the Serving GW and PDN GW is the role of certain information elements such as the Access Point Name (APN) identifying the target PDN, the protocol type towards the PDN GW (e.g. PMIP or GTP), and also in certain cases the terminal identity and the geographic location of some of these entities.

Even though certain adverse network conditions such as overload or even complete failure of a selected MME may also influence the selection of such entities, this is expected to be a rare situation and thus is not discussed further.

EPC and 4G Packet Networks.
DOI: http://dx.doi.org/10.1016/B978-0-12-394595-2.00009-8

The main network entities that require special means of selection are: MME, SGSN, Serving GW, and PDN GW. In addition, PCRF selection plays an important role when PCC is used.

Figure 9.1 shows a topological connection between the nodes that are involved in the selection process of a user's network attachment procedure in E-UTRAN. These nodes remain connected/active for the duration of that specific user's attachment to the network and remove/disconnect the association only when the user's session/IP connectivity is removed/closed/disconnected. Additional nodes are involved and participate in the selection process when 2G/3G and non-3GPP accesses are part of the network.

Note that for simplicity and clarity, not all interfaces are shown in Figure 9.1.

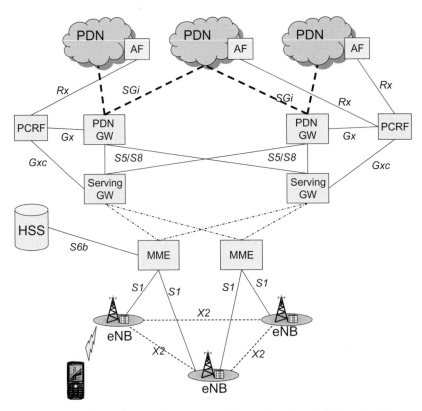

Figure 9.1: Topology Architecture and Selection Path for EPC Entities.

9.2 Selection of MME, SGSN, Serving GW, and PDN GW

9.2.1 Selection Procedure at a Glance

Let us first illustrate a simplified Initial Attachment procedure where an MME, Serving GW, and PDN GW need to be selected for the user. The actual message flows for Initial Attachment are shown in Chapter 17. In the flow chart shown in

Figure 9.2, we show only the order of events at a high level to illustrate the sequence of events for a general selection process.

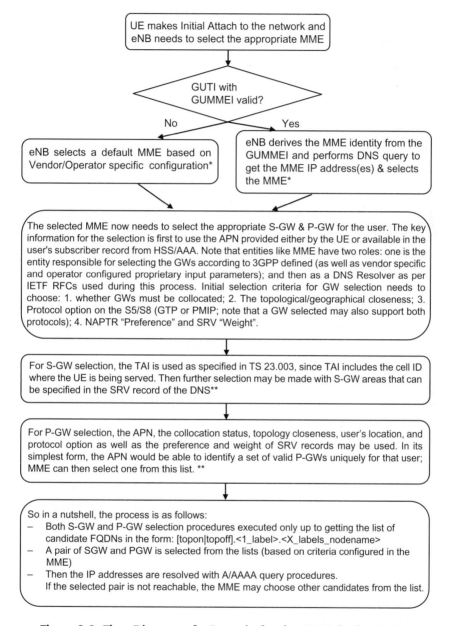

Figure 9.2: Flow Diagram of a Scenario for the GW Selection Process.

3GPP systems, including EPS, provide connectivity towards multiple PDNs with an independent and unique IP connection for each PDN (see Chapter 6 for more details). Each PDN connection established for a new APN requires an additional PDN GW selection process in the MME. In somewhat simplified terms, the DNS is used to store the mapping between APN and PDN GWs. Since for each user (i.e. UE) there can be only a single Serving GW at any given time in the network, selection of the additional PDN GW must use the current Serving GW information as part of its input data for this selection process. With an APN-FQDN created from the APN as the first query parameter to the DNS procedures, the MME will receive a list of candidate PDN GWs. Based on whether the GW collocation is preferred or not, the MME uses the Serving GW already selected as the next criterion in the selection process. When collocation is desired, the MME selects the PDN GW collocated with the already selected Serving GW, if possible. When collocation is not set, then the protocol used on the S5/S8 interface (GTP or PMIP), as well as other preference criteria set by the SRV and NAPTR records in the DNS system, may provide the next selection criterion. Otherwise MME may select a random PDN GW from the candidate list.

Until now, we have focused on the selection process for the case when a user establishes connectivity in the EPS network. In addition to that, during the mobility procedure, such as during handover between 3GPP accesses (for more details, see Section 6.4), an MME or SGSN may need to be changed (e.g. in the case of MME pool change) and in such a case, the current serving MME needs to select a new candidate MME to transfer the session to. This selection is performed by the serving node and is based on the Tracking Area Identity (TAI) Fully Qualified Domain Names (FQDNs), where the TAI is from the target Cell ID that the target eNB has provided to the serving MME. In the case of SGSN selection, either the Routing Area ID of the GERAN or UTRAN is used or, in the case of UTRAN, the RNC ID can be used in the same way as the TAI to construct the FQDN to find the right SGSN (pool) with the Release 8 SGSNs. Note that for pre-Release 8 SGSN, the existing selection mechanism must be maintained (see TS 29.303 for more details), which is simpler and does not necessarily mandate DNS usage.

In the case of Serving GW relocation, which should be an infrequent procedure that only occur if the Serving GW service area changes, the MME uses the TAI as well as the protocol supported on the S5/S8 reference point (GTPv2 or PMIPv6) in its DNS query to get the appropriate candidate Serving GW list. Use of the UE location information in the selection of a Serving GW provides flexibility of the node to be closer to the UE's current serving Radio Access network node (e.g. eNB/NB). This can be a useful criterion for features like SIPTO, where the GW selection is based on the current location of the UE.

Note that the MME is responsible for keeping track of all relevant information for the Serving GW, as well as each PDN GW, on a per-PDN basis; this includes host name, IP address, selected port (if not according to standard), and selected protocol type.

The MME needs to be configured with the variant of the protocol that is used over the S5/S8 reference point. To support roaming cases, the MME is configured on a per-HPLMN basis with the S8 protocol variant (PMIPv6 or GTP).

The HSS may also be provided with a specific PDN GW IP address and/or FQDN per APN basis and the MME will have this information available during the Initial Attachment procedure and then this input will be used to select the PDN GW for that user. In particular, to support handovers between 3GPP and non-3GPP accesses where the same PDN GW as was used in source access needs to be allocated in target access, the HSS contains information about the currently used PDN GW. This PDN GW information is provided to the node in the target access performing PDN GW selection (e.g. MME for E-UTRAN).

There are some additional aspects for the SGSN when considering selection of a GGSN or a PDN GW; these are affected by deployment and migration scenarios that are described in Chapter 3. Since an SGSN may be serving a UE that may or may not support LTE access, the MS capability (e.g. EPC support) from the UE indicates to the SGSN that a PDN GW needs to be selected for a UE instead of GGSN as the UE is capable of LTE support.

9.2.2 Use of DNS Infrastructure

Before going into the details of individual node selection, it is important to understand the 3GPP structure and requirements on the DNS and its usage. The GPRS system as well as the IP Multimedia Subsystem (IMS) in 3GPP has used DNS extensively, including DNS NAPTR and SRV records developed within the IETF. The EPS continues this and expands DNS usage to include node selection.

The DNS parameters and procedures for EPS are specified in 3GPP TS 23.003 and TS 29.303. These specifications, and DNS-related IETF RFCs such as RFC 2181, RFC 1035, RFC 3958, and RFC 2606 can be consulted for further details.

Note that the use of DNS infrastructure in order to discover the nodes is not mandatory as operators may choose to use O&M functions and other methods of provision to carry out the selection process. Use of DNS and its supporting services definitely improves the selection process as various criteria can be used an input to the process and also make it easy and flexible for inter-operator operations.

In EPS, the DNS is used to store information on mapping between APN, protocol (GTP or PMIP) and PDN GW, as well as mapping between TAI and Serving GW. The DNS can also be configured so that the DNS records provide information on collocated nodes as well as topological/geographical closeness between different nodes. In order to utilize the DNS infrastructure as well as IP networks, rules are defined for how to construct the FQDN used for DNS procedures. While GGSN selection in GPRS systems uses only address (A/AAAA) records in DNS, the EPS DNS procedures also support SRV and NAPTR records and are able to provide the more powerful features already mentioned above, e.g. to support collocated nodes and topological closeness. In particular, the Straightforward-NAPTR (S-NAPTR) procedure is used, with a suitable FQDN as input, to retrieve a candidate list of host names and corresponding service (e.g. protocol), port, and IPv4/IPv6 addresses. For more detailed information on S-NAPTR, see RFC 3958 and TS 29.303. We briefly describe some of the definitions and rules that are needed for the selection functions below.

9.2.2.1 Home Network Realm/Domain
The Home Network Realm/Domain needs to be in the form of an Internet domain name, for example operator.com, as specified in IETF RFC 1035.

The Home Network Realm/Domain needs to be derived from the IMSI as described in the following steps:

1. Take the first five or six digits of the IMSI, depending on whether a two- or three-digit MNC is used, and separate them into Mobile Country Code (MCC) and Mobile Network Code (MNC); if the MNC is two digits then a zero should be added at the beginning.
2. Use the MCC and MNC derived in step 1 to create the "mnc<MNC>. mcc<MCC>.3gppnetwork.org" domain name.
3. Add the label "epc" to the beginning of the domain name.

An example of a Home Network Realm/Domain is:

   IMSI in use: 234150999999999;

where:

   MCC is 234;
   MNC is 15;
   MSIN is 0999999999,

which gives the Home Network Realm/Domain name: epc.mnc015. mcc234.3gppnetwork.org.

9.2.2.2 DNS Subdomain for Operator Usage in EPC
The EPC node DNS subdomain (DNS zone) is derived from the MNC and the MCC by adding the label "node" to the beginning of the Home Network Realm/Domain and is constructed as:

   node.epc.mnc<MNC>.mcc<MCC>.3gppnetwork.org

This DNS subdomain is formally placed into the operator's control. 3GPP never takes this DNS subdomain back or any zone cut/subdomain within it for any purpose. As a result the operator can safely provide any DNS records under this subdomain without worrying about future 3GPP standards encroaching on the DNS names within this zone.

9.2.2.3 Access Point Name (APN)

Overall Structure

The APN is composed of two parts as follows:

1. The APN Network Identifier (APN-NI); this defines the PDN to which the UE requests connectivity and optionally a requested service by the UE. This part of the APN is mandatory.
2. The APN Operator Identifier (APN-OI); this defines in which PLMN the PDN GW (or GGSN for GPRS) is located. This part of the APN is optional.

The APN is constructed by placing the APN Operator Identifier after the APN Network Identifier.

As operators have been working towards sorting out roaming requirements on their infrastructure/network for services like VoLTE in GSMA, a few changes have been made in the APN configuration aspects. One essential change is that VPLMN roaming support for an individual APN within a VPLMN is now configured in the HSS for each user's subscription profile. Prior to this change, VPLMN APN support indication during roaming used to be indicated for all roaming partners' VPLMNs together via single parameter, thus forcing all roaming partner operators to have a roaming agreement in place for all APNs at the same time (e.g. all roaming partners VPLMN/ per APN combination basis). This is a problem since a group of operators within a wider group of roaming cooperating partners could not set up certain specific APN-based roaming agreements only among themselves. 3GPP addressed this issue by changing the HSS configuration for the subscription profile for users where now each APN has a list of supporting VPLMN allowed support indicated. This information is reflected in the profile in the following manner as described in 3GPP TS 23.401 and 3GPP TS 23.008:

> VPLMN Address Allowed   Specifies per VPLMN whether for this APN the UE is allowed to use the PDN GW in the domain of the HPLMN only, or additionally the PDN GW in the domain of the VPLMN.

This information is used during the PDN GW selection process when the user is roaming, determining if the PDN GW can be selected from the VPLMN (e.g. local breakout of the traffic) or needs to be home routed by selecting a PDN GW in the

HPLMN. General roaming agreement need to be in place between operators before proceeding with any roaming support of each other's users, even with the presence of a VPLMN Address Allowed flag being allowed being set.

APN Network Identifier

In order to guarantee the uniqueness of APN Network Identifiers within or between PLMNs, an APN Network Identifier containing more than one label should correspond to an Internet domain name. This name should only be allocated by the PLMN if that PLMN belongs to an organization that has officially reserved this name in the Internet domain. Other types of APN Network Identifiers are not guaranteed to be unique within or between PLMNs.

An APN Network Identifier may be used to access a service associated with a PDN GW. This may be achieved by defining an APN that, in addition to being usable to select a PDN GW, is also locally interpreted to be a request for a specific service by the PDN GW. An example would be a unique APN for IMS services.

APN Operator Identifier

The APN Operator Identifier is a "domain name" of the operator. It contains two labels that uniquely identify the operator (PLMN) as well as a third label that may be common for all operators. For each PLMN there is a default APN-OI that is built up using the MNC and the MCC as well as the label "gprs". The result of the APN Operator Identifier Realm/Domain will be:

mnc<MNC>.mcc<MCC>.gprs

In the roaming case, the UE may utilize the services of the VPLMN. In this case, the APN Operator Identifier needs to be constructed the same way, but replacing the home operator's MNC and MCC with that of the VPLMN.

APN-FQDN

Note that when the selection function in EPC resolves an APN in DNS, the APN is not used as is. Instead, an APN-FQDN is created by inserting the labels "apn. epc" between the APN-NI and the default APN-OI and replacing ".gprs" with "3gppnetwork.org". This results in an APN-FQDN with the following format:

<APN-NI>.apn.epc.mnc<MNC>.mcc<MCC>.3gppnetwork.org

Note that in the existing GPRS network, the suffix ".gprs" is used in the DNS instead of "3gppnetwork.org". However, with EPS the suffix "3gppnetwork.org" was chosen for the APN resolutions in DNS. One reason is that the usage of the top-level domain ".org" aligns better with the existing top-level domains of the DNS infrastructure while ".gprs" does not exist outside of the private 3GPP operator networks. Another reason is that the domain name suffix "3gppnetwork.org" is already used for IMS.

Service and Protocol Service Names for 3GPP

To perform node selection in EPS using DNS, the selection function creates an FQDN (e.g. an APN-FQDN) that is provided in the DNS query to find the host names and IP addresses of target nodes such as PDN GW. Such a target node may, however, be multi-homed (i.e. have more than one IP address) and may use different IP addresses for different protocols (e.g. for PMIP and GTP on S5/S8 interface). In this case, the node selection function needs to resolve the allowed interfaces supporting a certain service (e.g. interface and protocol type) using the DNS NAPTR procedure with certain "service parameters" describing the service. A more detailed description can be found in Section 6.5 of IETF RFC 3958 and in 3GPP TS 29.303.

Table 9.1 lists the "service parameters" to be used in the procedures specified in 3GPP TS 29.303.

Table 9.1: List of "app-service" and "app-protocol" Names

Description	IETF RFC 3958 Section 6.5 "app-service" Name	IETF RFC 3958 Section 6.5 "app-protocol" Name
PGW and interface types supported by the PGW	x-3gpp-pgw	x-s5-gtp, x-s5-pmip, x-s8-gtp, x-s8-pmip, x-s2a-pmip, x-s2a-mipv4, x-s2b-pmip, x-s2a-gtp, x-s2b-gtp
SGW and interface types supported by the SGW	x-3gpp-sgw	x-s5-gtp, x-s5-pmip, x-s8-gtp, x-s8-pmip, x-s11, x-s12, x-s4, x-s1-u, x-s2a-pmip, x-s2b-pmip
GGSN	x-3gpp-ggsn	x-gn, x-gp
SGSN	x-3gpp-sgsn	x-gn, x-gp, x-s4, x-s3
MME and interface types supported by the MME	x-3gpp-mme	x-s10, x-s11, x-s3, x-s6a, x-s1-mme

The formats follow the experimental format as specified in IETF RFC 3958. For example, to find the S8 PMIP interfaces on a PGW the service parameter "3gpp-pgw:x-s8-pmip" would be used as input in the procedures defined in IETF RFC 3958.

9.2.3 MME Selection

The MME selection function is located in the eNB only. The architecture supports multiple eNBs connected to multiple MMEs as well as Serving GWs.

When a UE attempts to attach to E-UTRAN (see Session Management section in Chapter 6 for details), it provides the eNB with parameters, i.e. GUTI (see Section 6.3 on identities for more details), which would facilitate selection of the appropriate MME based on how the GUTI is constructed. eNB is also informed of the load status of the MMEs within a pool (see Pool information in Section 6.7) via S1-AP signaling and is thus able to provide additional information to the eNBs connected to the MME pool. This allows for efficient selection of MME within a pool and also allows triggers towards the UE (using Tracking Area Update or S1 Release procedures) to reconnect to a different MME within a pool when required. The process of MME selection is designed to be efficient from the UE movement point of view and has been developed to reduce MME changes when serving within certain operating boundaries.

The eNB is responsible for the selection of an appropriate MME at UE attachment when no routing to an MME can be determined from the information provided by the UE.

It is important to understand how MME would be represented in the operators' IP network, in this case how the FQDN for the MME or MME pool would be constructed, since identifying the correct MME for an UE depends on it.

An MME within an operator's network is identified using an MME Group ID (MMEGI), and an MME Code (MMEC), which is then made available to the eNB via GUTI and/or other parameters that identify the UE in the network (see Identifiers, Section 6.3).

A subdomain name is derived from the MNC and MCC by adding the label "mme" to the beginning of the Home Network Realm/Domain.

The MME node FQDN is constructed as:

> **mmec**<MMEC>**.mmegi**<MMEGI>**.mme.epc.mnc**<MNC>
> **.mcc**<MCC>**.3gppnetwork.org**

An MME pool FQDN is constructed as:

> **Mmegi**<MMEGI>**.mme.epc.mnc**<MNC>**.mcc**<MCC>**.3gppnetwork.org**

When an MME or SGSN selects another MME, the TAI FQDN is used towards the DNS to find the appropriate MME (in a pool) and then the information is analyzed using A/AAAA queries to get the actual MME address details.

The TAI consists of a TAC, MNC, and MCC. A subdomain name is derived from the MNC and MCC by adding the label "tac" to the beginning of the Home Network Realm/Domain.

The TAI FQDN is constructed as:

> **tac-lb**<TAC-low-byte>**.tac-hb**<TAC-high-byte>**.tac.epc.mnc**<MNC>
> **.mcc**<MCC>**.3gppnetwork.org**

9.2.4 SGSN Selection Function for EPS

As described in the overview section, the SGSN selection in case of EPS is relevant for Inter-Radio Access Technology Handover (Inter-RAT HO) within 3GPP access. In such cases, the serving node is responsible for selecting the target SGSN using either RAI or RNC-ID (for UTRAN access only).

A specific SGSN within an operator's network is identified using the RAI FQDN and the Network Resource Identifier (NRI), which uniquely identifies the core network assigned to the terminal. This identifier can be used by a target MME or SGSN node to connect to the source SGSN node.

The SGSN FQDN is constructed as:

**nri-sgsn<NRI>.rac<RAC>.lac<LAC>.rac.epc.mnc<MNC>
.mcc<MCC>.3gppnetwork.org**

9.2.4.1 Routing Area Identity (RAI) – EPC

The RAI consists of a RAC, LAC, MNC, and MCC. A subdomain name for use by core network nodes based on RAI needs to be derived from the MNC and MCC by adding the label "rac" to the beginning of the Home Network Realm/Domain.

The RAI FQDN is constructed in a similar manner as the TAI and would be represented as:

Rac<RAC>.lac<LAC>.rac.epc.mnc<MNC>.mcc<MCC>.3gppnetwork.org

9.2.5 GW Selection Overview

GW selection includes the Serving GW as well as PDN GW selection process, which is performed by the MME. Compared to GPRS, selection of Serving GWs and PDN GWs has become more complex. It is very tightly coupled to the various protocol choices on certain reference points (e.g. S5/S8), as well as certain network configuration such as Local Breakout and type of operator services being provided exclusively sometimes by certain GWs (e.g. IMS); in addition to maintaining existing GPRS GGSN selection function parameters like APN.

A related aspect is the PDN GW selection by an SGSN that supports both Gn/Gp and S4 interfaces, and thus may select between GGSN and PDN GW. In this case it is possible for the SGSN to use the information about whether the UE supports E-UTRAN or not (as indicated in the UE capability information sent by the UE) as input to select either a GGSN or a Serving GW and PDN GW. UEs that do not support E-UTRAN may, for example, be allocated a GGSN while UEs that support E-UTRAN would need to be handled by a PDN GW in order to allow a potential handover to E-UTRAN.

9.2.6 PDN GW Selection Function

In the case of 3GPP access, the PDN GW selection function is located in the MME and SGSN, whereas for non-3GPP access the location of the PDN GW selection function depends on the protocol used in the appropriate reference points (i.e. S2a, S2b, and S2c).

If an APN is identified as a SIPTO APN, the PDN GW selection function takes into account the current serving location (e.g. in the form of TAI for LTE access and RAI for UTRAN access) of the UE when selecting the PDN GW for that APN. Topological naming based on topological closeness (as employed for Serving GW selection) needs to be applied for SIPTO to find the shortest path between Serving GW and PDN GW for the user-plane traffic.

1. For GTPv2 on S2a/S2b, it is the functional entity acting as GTP peer that performs the selection function. In the S2a case, where GTP-based S2a is limited to WLAN, the trusted WLAN access gateway requests the PDN GW. For S2b, the ePDG requests the PDN GW.
2. For PMIPv6 on S2a/S2b, it is the MAG functional entity that performs the selection function. For S2a, an example of such a case is the 3GPP2 HSGW entity, which is also known as the Access Gateway in a generic manner in 3GPP specifications. For S2b, the ePDG requests the PDN GW.
3. For MIPv4 FA mode on S2a, the entity requesting the PDN GW is the entity that plays the role of the FA.
4. For DSMIPv6 (S2c), the UE needs to be provided with an appropriate PDN GW address (to be used as Home Agent address for DSMIPv6). Different methods are available in EPS. When connecting over an access supporting transfer of Protocol Configurations Options (PCO), the PDN GW address may be returned to the UE in such a PCO field. When connecting towards an ePDG, the PDN GW address may be returned to the UE in the IKEv2 signaling with ePDG. If none of these methods is available, then the UE may use DHCP mechanisms or querying of a DNS server based on the PDN information (using APN) may be used to select the PDN GW.

The PDN GW selection function uses subscriber information provided by the HSS and possibly additional criteria. The PDN subscription contexts provided contain information such as the APN(s) the user has subscribed to and (for roaming cases) whether this APN(s) can be connected to using a PDN GW located in the VPLMN, or if the user has to connect via a PDN GW located in the HPLMN.

In the case of non-3GPP accesses using GTPv2, PMIPv6, or MIPv4, the PDN Gateway selection function interacts with the 3GPP AAA Server or 3GPP AAA Proxy and uses subscriber information provided by the HSS to the 3GPP AAA Server. To support separate PDN GW addresses at a PDN GW for different mobility protocols (GTP, PMIP, or MIPv4), the PDN GW selection function takes mobility protocol type into account when deriving the PDN GW address by using the Domain Name Service function, in a similar way as is done by the MME/SGSN.

For 3GPP access, the PDN GW selection function interacts with the HSS to retrieve subscriber information and uses information such as whether the Serving GW and PDN GW must be collocated, the topological/geographical closeness (described in TS 29.303), and protocol option on the S5/S8 interface (GTP or PMIP); note that a selected GW may also support both protocols.

Note that collocation of Serving GW and PDN GW is not part of the 3GPP standard specifications for non-3GPP accesses since the use of the Serving GW for non-3GPP access is only relevant for a specific case known as Chained S2a/S2b-S8. This chained scenario is not described further in this book. Interested readers should refer to TS 23.402 for more details.

If a static PDN GW is configured by the operator to be used for that user, the PDN GW is selected by either having the APN configured to map to a given PDN GW (i.e. the APN resolves to a single PDN GW only), or the PDN GW identity provided by the HSS explicitly shows the use of static PDN GW.

The APN may also be provided by the UE. In this case, this UE-provided APN is used to derive the APN-FQDN as long as the subscription allows for it.

If the user is roaming and the HSS provides a PDN subscription context that allows for allocation of a PDN GW from the visited PLMN for this APN, the PDN GW selection function selects a PDN GW from the visited PLMN. If a visited PDN GW cannot be found or if the subscription does not allow for allocation of a PDN GW from the visited PLMN, then the APN is used to select a PDN GW from the HPLMN.

If the UE requests a connection to an existing PDN via the same APN, the selection function must select the same PDN GW as it used previously to establish the PDN connection to this APN.

To support PDN GW selection using the S-NAPTR procedures, the authoritative DNS server responsible for the "apn.epc.mnc<MNC>.mcc<MCC>.3gppnetwork.org" domain must provide NAPTR records for the given APN and corresponding PDN GWs under the APN-FQDN:

<APN-NI>.apn.epc.mnc<MNC>.mcc<MCC>.3gppnetwork.org

Some example S-NAPTR procedures are included here for illustrative purposes. For detailed information on 3GPP usage of various IETF specifications and their adaptation for 3GPP network usage, TS 29.303 is an excellent source.

For 3GPP access:

- *When non-roaming and initial attachment* is performed, the S-NAPTR procedure to get the list of "candidate" PDN GWs will use the service parameters "**x-3gpp-pgw:x-s5-gtp**", "**x-3gpp-pgw:x-s5-pmip**" depending on the protocol supported.
- *When roaming*, the Serving GW is in the VPLMN and if the APN to be selected is in the HPLMN, then the selection function for the S-NAPTR procedure will use the service parameters "**x-3gpp-pgw:x-s8-gtp**", "**x-3gpp-pgw:x-s8-pmip**" depending on the protocol supported.
- *When non-roaming and an additional PDN connection* is to be established, the Serving GW is already selected (i.e. the UE has an existing PDN connection), and the selection function will use the S-NAPTR procedure service parameters "**x-3gpp-pgw:x-s5-gtp**" or "**x-3gpp-pgw:x-s5-pmip**" depending on the protocol supported.

For non-3GPP access:

- When PMIP or GTP is used for initial attachment and for both roaming and non-roaming cases, the selection function will use the S-NAPTR procedure with service parameters "**x-3gpp-pgw:x-s2a-pmip**", "**x-3gpp-pgw:x-s2b-pmip**", "**x-3gpp-pgw:x-s2b-gtp**", or "**x-3gpp-pgw:x-s2a-gtp**" depending on the protocol supported and interface used.

9.2.7 Serving GW Selection Function

The Serving GW is a mandatory node for all 3GPP accesses, while for non-3GPP accesses the Serving GW is only used in the special case of a chained scenario where an S8-S2a/S2b chained roaming is determined by 3GPP AAA Proxy in the VPLMN. This chained scenario is not further described in this book. Please refer to TS 23.402 if you are interested in learning more about this feature.

The Serving GW selection function selects an available Serving GW to serve a UE. The selection is based on network topology – that is, the selected Serving GW serves the UE's location (derived from the TAI). For overlapping Serving GW service areas, the selection criteria may prefer Serving GWs with service areas that reduce the probability of changing the Serving GW as the UE moves around the network. The S-NAPTR-based selection of the Serving GW based on the TAI using S-NAPTR ordering provides the shortest user-plane path from the UE to the Serving GW (via eNB).

Due to the possibility of using either GTP or PMIPv6 over the S5 and S8 interfaces, as well as possible multiple PDN connections involving HPLMN and VPLMN (in this case the VPLMN provides the network configuration required for Local Breakout where the PDN GW in the VPLMN needs to be selected) when roaming, Serving GWs may need to support both protocols for a single UE connected to different

PDNs. This may, for example, be needed if a UE has two PDN connections, one with PDN GW in Visited PLMN and one with PDN GW allocated in the Home PLMN. In this case PMIPv6 may be used on S5 between the Serving GW and the PDN GW in the Visited PLMN, but GTP can be used for the other PDN connection with PDN GW in the Home PLMN. In case of handover (with or without MME relocation), an MME may determine that a Serving GW needs to be relocated to a new one due to, e.g. the connectivity requirements with the target eNB serving the UE or other reasons. The MME in such scenario follows the same Serving GW selection function procedure to determine the new GW to be used.

Again, we show a few examples of S-NAPTR procedures. The detailed information can be found in TS 29.303.

For 3GPP accesses:

- *When non-roaming* and TAU procedure requires Serving GW change, the Serving GW selection function shall use the S-NAPTR procedure with the TAI FQDN to get the list of "candidate" Serving GW service parameters of **"x-3gpp-sgw:x-s5-gtp"** or **"x-3gpp-sgw:x-s5-pmip"**.
- *When roaming*, the Serving GW selection is needed due to TAU procedure in the VPLMN, and then the selection function for the S-NAPTR procedure shall use the TAI FQDN and service parameters of **"x-3gpp-sgw:x-s8-gtp"** or **"x-3gpp-sgw:x-s8-pmip"** and make use of the operator's preference for the roaming protocol (GTP/PMIP).
- *When non-roaming and with additional PDN connection*, the Serving GW is already selected (i.e. the UE has an existing PDN connection), and the selection function shall use the S-NAPTR procedure with the APN-FQDN and service parameters of **"x-3gpp-pgw:x-s5-gtp"** or **"x-3gpp-pgw:x-s5-pmip"**.

For non-3GPP access when the Serving GW acts as a local mobility anchor for non-3GPP access S8-S2a/b chained roaming, Serving GW selection is not specified and thus may be left to the vendor-specific implementation choice of an operator's networks.

9.2.8 Handover (Non-3GPP Access) and PDN GW Selection

Once the selection of PDN GW has occurred, the *PDN Gateway identity* is registered in the HSS so that it can be provided to the target access in the case of an inter-access handover. The *PDN GW identity* registered in HSS can be either an IP address or an FQDN. Registering the PDN GW IP address is appropriate if the PDN GW only has a single IP address or can use the same IP address for all the mobility protocols it supports (or if it only supports one mobility protocol). Registering the FQDN is more flexible in the sense that it allows the PDN GW to use multiple IP addresses depending on mobility protocol. So if a UE hands over from one access where PMIPv6 is used to another access where GTP is used, the PDN GW selection function could use the FQDN to derive different PDN GW IP addresses depending on the mobility protocol.

If the terminal activated the PDN connection in a non-3GPP access, it is the PDN GW that registers the PDN GW identity and its association with a UE and the APN in the HSS. For 3GPP access types, the MME/SGSN updates the HSS with the selected PDN GW identity. Once registered in the HSS, the HSS (possibly via 3GPP AAA Server or Proxy) can then provide the association of the PDN Gateway identity and the related APN for the UE later on. It may be noted that the PDN GW information is provided to target access only if mobility is performed using a network-based mobility protocol in the target access (PMIP and GTP). For handover to a non-3GPP access using DSMIPv6, it is instead the UE that knows the address of the PDN GW (Home Agent).

When the UE is moving between 3GPP and non-3GPP accesses, PDN Gateway selection information for subscribed PDNs the UE has not yet connected to is returned to the target access system like during initial attachment. For the PDNs the UE has already connected to, the PDN GW information is transferred as below:

1. If a UE hands over to a non-3GPP access using PMIPv6 in target access and it already has assigned PDN Gateway(s) due to a previous attach in a 3GPP access, the HSS provides the PDN GW identity for each of the already allocated PDN Gateway(s) with the corresponding PDN information to the 3GPP AAA server. The AAA server forwards the information to the PDN GW selection function in the target access.
2. If a UE hands over to a 3GPP access and it already has an assigned PDN Gateway(s) due to a previous attachment in a non-3GPP access, the HSS provides the PDN GW identity for each of the already allocated PDN Gateway(s) with the corresponding PDN information to the MME.

9.3 PCRF Selection

The PCRF and its role in the EPS are described in detail in Section 8.2. For the purpose of PCRF discovery, one or more PCRF nodes in the HPLMN may serve a PDN GW and associated AF. When roaming and needing to support local breakout scenarios, one or more PCRF nodes in the VPLMN may serve the PDN GW and the associated AF. There may be either one PCRF per PLMN allocated for all PDN connections of a UE, or the different PDN connections may be handled by different PCRFs. The choice of which option to use is based on the operator's network configuration. When DSMIPv6 is used, however, the only option is to use one PCRF for all PDN connections of a UE.

The Rx, Gx, Gxa/Gxc, and S9 sessions for the same IP-CAN sessions must be handled by the same PCRF. The PCRF must also be able to link the different sessions (Rx session, Gx session, S9 session, etc.) for the same IP-CAN session. This also applies to roaming cases where one V-PCRF handles the PCC sessions (Gx, Rx, Gxa/Gxc, etc.) belonging to one PDN connection of the UE. Therefore, the selection of PCRF and an individual UE's different PCC sessions over the multiple PCRF interfaces for a UE IP-CAN session are very much related. This means that the information carried over these different interfaces must be able to be correlated for the purpose of providing the overall PCC functions to individual users.

In order to ensure that the same PCRF is selected for all sessions (Gx, Rx, etc.) that make up an IP-CAN session, a logical functional entity called the Diameter Routing Agent (DRA) is used during the selection/discovery of the PCRF. 3GPP defines the DRA in TS 29.213 as:

> *A functional element that ensures that all Diameter sessions established over the Gx, S9, Gxx, and Rx reference points for a certain IP-CAN session reach the same PCRF when multiple and separately addressable PCRFs have been deployed in a Diameter realm. The DRA is not required in a network that deploys a single PCRF per Diameter realm.*

DRA complies with standard Diameter functionality as specified by IETF where the Diameter protocol and its applications are specified. DRA also needs to (both Redirect and Proxy DRA) advertise the specific applications supported by it, which includ es information like support of Gx, Gxx, Rx, and S9, as well as necessary and relevant authentication and vendor-related information for 3GPP.

When operating in the Redirect mode, the DRA client will use the value within the Redirect Host AVP of the redirect response in order to obtain the PCRF identity (see Diameter protocol details in Chapter 16).

When in Proxy mode, the DRA will only select a PCRF if one has not already been selected for the UE and if the request is an IP-CAN session establishment of gateway control. After successful establishment of the PCRF for that UE, that PCRF must be used until all IP-CAN sessions for that UE have been terminated and removed from the DRA.

When DRA is deployed in a network for the PCRF realm, the DRA is to be the first contact point from the clients for these sessions. If DRA is deployed in a roaming scenario, for either home routed or local breakout, the V-PCRF is selected by the

DRA located in the visited PLMN, and the H-PCRF is selected by the DRA located in the home PLMN.

In the case of roaming, depending on which PLMN the PCRF has been selected and the network configuration (i.e. Local Breakout or not, home routed or both), the Serving GW, the non-3GPP Access GW, and PDN GW may be the selection function trigger towards DRA. The DRA keeps the status of the assigned PCRF for a certain UE and IP-CAN session. It is assumed that there is a single logical DRA serving a Diameter realm (Figure 9.3).

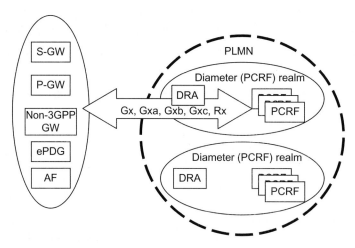

Figure 9.3: Logical DRA and Diameter Realm for 3GPP EPC.

The parameters available to enable the DRA to determine the already allocated PCRF depend on the reference point over which the DRA is contacted. Different information is available on Rx compared to, for example, Gx and Gxa/Gxc. Concluding this chapter, we have illustrated in a simplified manner various network node selection criteria, as well as methods used in 3GPP.

CHAPTER 10

Subscriber Data Management

An important area for mobile operators, but sometimes not very visible, is Subscriber Data Management (SDM). SDM is a term that may have different meanings in different contexts, but it is used here to describe the subscription handling for all processes related to, for example, privacy, authentication, authorization, policy control, and mobility management of end-users. Sometimes this is also referred to as User Data Management (UDM).

In mobile networks there are many functions and processes that require subscription-related information. The most obvious example of user subscription data that is used in LTE/EPC networks may be the user identity and security credentials that are required when an end-user device connects to an LTE/EPC network and performs authentication. The user identity (IMSI) and the security keys are stored in the USIM card in the device and the same information is also stored for each user in the operator's core network, in the Home Subscriber Server (HSS).

In mobile networks, however, a subscriber has many other parameters associated with the subscription. For example, subscriptions may differ in terms of what services they can access, what Quality of Service they will get, and what access technologies they can use, if they are charged in real time (pre-paid) or after usage (post paid), the charging model for the data consumed, etc.

In this chapter we will look at different areas of Subscriber Data Management and look at the functions and entities in the Evolved Packet Core that handle subscription data. First, we will give an overview of the different logical entities defined in EPC that maintain permanent subscriber data. These entities include the Home Subscriber Server (HSS) and the Subscriber Profile Repository (SPR). Some of this has been handled as part of previous chapters, e.g. on PCC and mobility, but in this chapter we will focus on the SDM aspects.

We will then also describe the User Data Convergence (UDC) framework defined in 3GPP. With UDC, databases for different subscriber data functions are consolidated while still maintaining compatibility towards other network entities that request subscription data, such as MME or SGSN. By consolidation of user data for different

EPC and 4G Packet Networks.
DOI: http://dx.doi.org/10.1016/B978-0-12-394595-2.00010-4

network functions and access types it is possible to provide a more efficient SDM infrastructure and to find new customer propositions.

In addition to entities permanently holding subscription data, there are different entities in EPC, e.g. the MME and the 3GPP AAA Server, that maintain subscriber data while there is an active PDN connection for a UE or may cache the subscriber data when the user is detached. However, these entities do not hold permanent subscriber data.

10.1 Home Subscriber Server (HSS)

The HSS can be described as the master database for a given user. It is the entity containing the subscription-related information to support the network entities handling mobility and user IP sessions. HSS also supports the entities handling circuit-switched calls.

Prior to 3GPP Release 5, the Home Location Register (HLR) was the main subscriber database for GPRS and circuit-switched (CS) services. The Authentication Center (AuC) was the database holding the security data for authentication of the subscriber and ciphering communication with the subscriber. From Release 5 onwards, however, the HLR and AuC functions in 3GPP specifications are considered a subset of HSS.

Figure 10.1 shows the interfaces towards the HSS. The HSS is accessible from an MME via the S6a interface, from an S4-SGSN via the S6d interface, and from a Gn/Gp SGSN via the Gr interface. The HSS is furthermore available from the 3GPP AAA

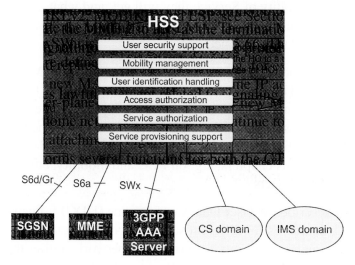

Figure 10.1: Interfaces to the HSS.
Only interfaces to EPC are shown explicitly.

Server via the SWx interface. The other interfaces to the HSS, e.g. to support the CS domain and the IMS domain, are not further described in this book, but they are the C or D interface (MAP) towards the CS domain and the Cx, Sh interfaces (Diameter) towards the IMS domain.

The S6a/S6d and SWx interfaces use the Diameter protocol and are described in more detail in Chapter 15.

The Gr interface is based on the MAP protocol and is inherited from the pre-EPC GPRS core network. Even though the EPC architecture uses S6d between S4-SGSN and HSS, the use of Gr from S4-SGSN is not precluded, e.g. for the transition from Gn/Gp SGSNs to using SGSNs supporting S4 without simultaneously needing to migrate from MAP-based HLR to Diameter-based HSS. Figure 10.2 shows a scenario where an operator has deployed S4-SGSN but kept the MAP-based Gr interface with a legacy HLR.

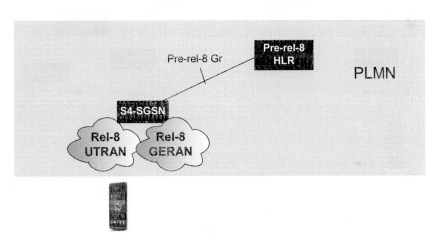

Figure 10.2: Migration Scenario with S4-SGSN and Legacy Gr/HLR.

In order to facilitate migration from HLR to HSS and to support roaming between operators with different deployments of HLR and/or HSS, 3GPP has also specified an Interworking Function (IWF) that provides protocol translation between Gr and S6a/S6d. Figure 10.3(a) shows an interworking scenario where a user from an operator with pre-Release 8 HLR is roaming in a VPLMN with EPC deployed. The IWF maps between pre-Release 8 Gr messages and S6a/Sd messages. Note that in order for this scenario to support E-UTRAN and mapping between S6a and Gr, the pre-Release 8 HLR and Gr interface must at least be enhanced to support EPS security (i.e. to deliver EPS Authentication Vectors for E-UTRAN) towards the VPLMN. Another scenario is where both operators support EPS interfaces S6a/S6d but wants to reuse the MAP/SS7 roaming infrastructure. Figure 10.3(b) shows an example scenario of such use of the IWF.

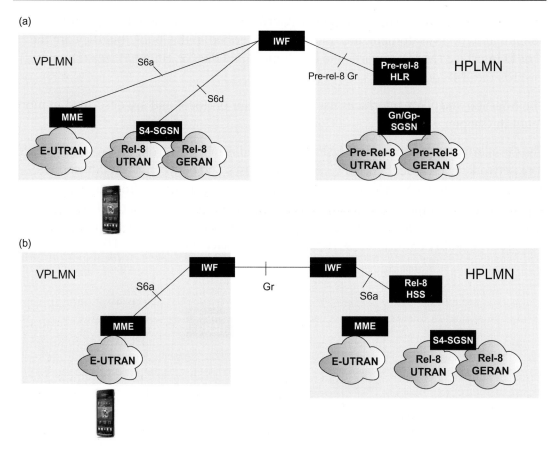

Figure 10.3: Example Interworking Scenarios Using IWF.
(a) Interworking scenario between EPC operator and operator with pre-Release 8 HLR. (b) Scenario with EPC operators roaming using MAP/SS7 roaming infrastructure.

One additional aspect when using legacy Gr and HLR with S4-SGSN and/or MME is that pre Release 8 Gr supports only delivery of GPRS subscription data. The receiving S4-SGSN and MME must therefore map the GPRS subscription data into EPS subscription data for use in EPS. How this mapping is done is not specified by 3GPP but is left to implementation and operator configuration. There are, however, a few limitations due to using pre-Release 8 Gr (either via IWF or direct) instead of using S6a/S6d. The legacy Gr interface, for example, does not support features available in the EPS subscription profile such as subscribed AMBR, dual-stack IPv4v6 bearers, handover to/from non-3GPP accesses, etc. In Release 8, Gr has, however, been enhanced to also carry the EPS subscription profile and, with such support in Gr/HLR, mapping of subscription data in MME/SGSN would not be needed.

The interested reader can consult 3GPP TS 29.305 for additional interworking scenarios using one IWF between operators with different levels of support for Gr and S6a/S6d or using two IWFs.

Even though not directly related to the topic of this section, it can also be mentioned that the IWF also supports mapping between the S13/S13′ and Gf interfaces, for equipment identity verification. See Chapter 15 for more information on the S13/S13′ interfaces.

The subscriber data and functionality of the HSS are used for a large number of functions in 3GPP networks. In the list below, we describe some of these functionalities at a high level. Many of these HSS functions are naturally reflected in other chapters of this book, where the system level functionality is described (e.g. security and mobility). Also, the functionalities of the S6a/S6d and SWx interfaces are described in Chapter 15. However, in the list below we focus on the functionality of the HSS entity as such. The functionality of HSS includes (also illustrated in Figure 10.1):

- User security support. The HSS supports authentication and security procedures for network access by providing credentials and keys towards network entities such as SGSN, MME, and 3GPP AAA Server. This aspect has been described in Chapter 7.
- Mobility management. The HSS supports user mobility by, for example, storing information about what SGSN/MME is currently serving the user. In similar ways, the HSS also supports the circuit-switched domain and IMS domain with mobility management functionality.
- User identification handling. The HSS provides the appropriate relations among all the identifiers uniquely determining the user in the system: CS domain, PS domain, and IMS (e.g. IMSI and MSISDNs for CS domain; IMSI and MSISDNs for PS domain; private identity and public identities for IMS).
- Access authorization. The HSS authorizes the user for mobile access when requested by the MSC/VLR (for CS access) and by the SGSN, MME, or 3GPP AAA Server (for PS access), by checking that the user is allowed to roam to a particular visited network.
- Service authorization support. The HSS provides basic authorization for mobile terminated call/session establishment and service invocation.
- Service provision support. The HSS provides access to the service profile data for use within the CS domain, PS domain, and/or IMS. For the PS domain, the HSS provides the APN profiles that include what APNs the user is authorized to use. The HSS also communicates with IMS entities to support Application Services.

An operator may need to have more than one HSS if the number of subscribers is too large to be handled by a single HSS. In order to support user identity to HSS resolution in such a case, Diameter agents can be deployed. The Diameter agent will relay, proxy, or redirect the request to the appropriate HSS handling the specific user (see the Diameter protocol description in Chapter 16 for more information on the different types of Diameter agents). The Diameter agent used for user identity to HSS resolution has many similarities with the Diameter Routing Agent (DRA) defined for Diameter interfaces to the PCRF (see Chapter 9 for more details on the DRA).

Table 10.1 shows a subset of the subscriber data related to EPC access that is contained in the HSS for each given subscriber (note that the list is not exhaustive). The HSS also contains other types of subscriber data, e.g. for IMS. Readers interested in subscriber data management for IMS may consult a book dedicated to IMS, e.g. Camarillo and Garcia-Martin (2008).

10.2 Subscriber Profile Repository (SPR)

The Subscriber Profile Repository (SPR) is the database that was originally defined to hold subscription data for the PCC framework. (Later, as we will see below, an option to use User Data Convergence for PCC was also introduced.) Compared to the HSS, the SPR stores the more dynamic business rules that are needed for PCC, while the HSS contains the more static subscription data needed for network access. The reference point between the PCRF and the SPR is called Sp (see Figure 10.4). The SPR may be a standalone database but is also in many cases integrated with the PCRF.

The SPR and the Sp reference point have not been standardized in detail by 3GPP. A reason for this is that the subscription-related information that is needed by the PCRF to perform policy control is very much dependent on the services that the operator provides to its end-users. Since the type of services provided and related policies are tightly coupled to the operator business model and business offering, it is quite difficult to standardize this information. Therefore, the descriptions of the SPR and the Sp interface have been purposely left rather vague by 3GPP. Note also that there is a key difference between the S6a/S6d interfaces with HSS and the Sp interface. The interfaces S6a/S6d between the MME/SGSN and the HSS may be a roaming interface between two operators where standardization is very important, while the interface between the PCRF and the SPR is always internal to an operator.

Nevertheless, it is possible to describe at a high level what type of information may be supported by SPR. For example, the SPR may provide the following subscription profile information:

- Subscriber's allowed services
- Information on subscriber's allowed QoS
- Subscriber's charging-related information (e.g. location information relevant for charging)

Table 10.1: Subset of the Subscription Data Contained in HSS for EPC Access

Field	Description
IMSI	IMSI is the main reference key
MSISDN	The basic MSISDN (i.e. telephone number) of the UE
IMEI/IMEISV	International Mobile Equipment Identity – Software Version Number (IMEI-SV) is the identity of the actual used terminal. This is provided to HSS when the UE attaches to the NW
MME Identity	The identity of the MME currently serving this MS
Access Restriction	Indicates the access restriction subscription information, i.e. what access (e.g. GERAN, UTRAN, E-UTRAN) is not allowed
EPS Subscribed Charging Characteristics	The charging characteristics for the MS, e.g. normal, pre-paid, flat rate, and/or hot billing subscription
Subscribed-UE-AMBR	The Maximum Aggregated uplink and downlink MBRs to be shared across all non-GBR bearers according to the subscription of the user
Each subscription profile contains one or more APN profiles:	
APN Profile	
PDN Address	Indicates subscribed IP address(es)
PDN Type	Indicates the subscribed PDN Type (IPv4, IPv6, IPv4v6)
Access Point Name (APN)	A label according to DNS naming conventions describing the access point to the packet data network (or a wildcard)
SIPTO Permissions	Indicates whether the traffic associated with this APN is allowed or prohibited for SIPTO
LIPA Permissions	Indicates whether the PDN can be accessed via Local IP Access Possible values are: LIPA-prohibited, LIPA-only, and LIPA-conditional
EPS Subscribed QoS Profile	The bearer level QoS parameter values for that APN's default bearer (QCI and ARP)
Subscribed-APN-AMBR	The maximum aggregated uplink and downlink MBRs to be shared across all non-GBR bearers, which are established for this APN
VPLMN Address Allowed	Specifies whether for this APN the UE is allowed to use the PDN GW in the domain of the HPLMN only, or additionally the PDN GW in the domain of the VPLMN
PDN GW Identity	The identity of the PDN GW used for this APN. The PDN GW identity may be either an FQDN or an IP address. The PDN GW identity refers to a specific PDN GW
PDN GW Allocation Type	Indicates whether the PDN GW is statically allocated or dynamically selected by other nodes. A statically allocated PDN GW is not changed during PDN GW selection

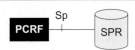

Figure 10.4: Sp Interface.

- Subscriber category
- Subscriber's usage monitoring-related information
- Multimedia Priority Service (MPS) EPS Priority and MPS Priority Level
- Subscriber's profile configuration indicating whether application detection and control should be enabled
- Spending limits profile
- A list of Application Service Providers and their applications per sponsor identity.

For more information on PCC concepts such as usage monitoring, application detection and control, and spending limits, see Chapter 8.

10.3 User Data Convergence (UDC)

In a 3GPP network, data relating to the subscriber may be managed by different network entities such as HLR/HSS and SPR. Furthermore, as we saw in the description about HSS above, there may be more than one HSS where each HSS stores subscription data for a subset of the subscribers. In that case there is a need for a Diameter agent to find the HSS that was handling a particular user.

Managing this set of databases for different kinds of subscriber data is complex and results in operational and management challenges. For example, introduction of a new user or modification of an existing subscription requires updates to multiple databases. There may also be issues with data duplication and synchronization between different locations. To resolve some of these issues, 3GPP introduced in Release 9 a new solution for User Data Convergence (UDC).

UDC aims to provide convergence of user data in order to enable smoother management and deployment of new services and networks. As we will see in more detail below, the UDC concept supports a layered architecture, keeping the actual data separate from the application logic in the 3GPP system. It does so by storing user data in a logically unique user data repository and allowing access to this data from EPC and service layer entities.

UDC aims to provide a number of benefits:

- Simplification of overall network topology and interfaces
- A single point of provision of subscriber data
- Overcoming the data capacity bottleneck of a single entry point
- Separate scaling of processing resources and data storage
- Avoidance of data duplication and inconsistency
- Avoidance of data fragmentation
- Reduction of CAPEX and OPEX for the operator.

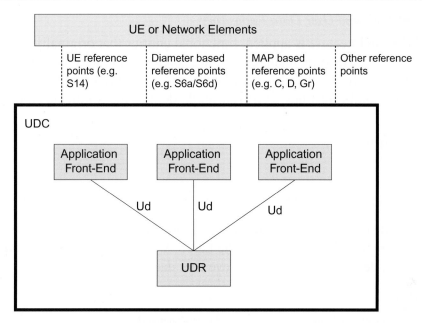

Figure 10.5: Schematic UDC Architecture.

10.3.1 UDC Overall Description

Figure 10.5 shows the logical representation of the layered architecture that separates the user data from the application logic. The user data is stored in a logically unique repository called the User Data Repository (UDR). Entities that do not store user data but need to access user data are called Application Front-Ends (FE). These Front-Ends implement the application logic for handling and operating on the user data, but they do not permanently store any user data. Examples of Front-Ends are the HSS, ANDSF, and PCRF. Access to the user data in UDR is enabled through the Ud interface.

Figure 10.6 compares a network with UDC deployed with a network where UDC is not deployed. In the non-UDC case, the network elements may have their own database storing persistent user data or they may access an external database. In the case where UDC is applied, the persistent user data is moved to the UDR. The network elements that previously stored subscription data or accessed dedicated external databases now become Application Front-Ends.

One very important aspect of UDC is that it does not affect the existing network interfaces between network entities. This can also be seen in Figure 10.6. The difference in the UDC architecture is only that a network element, which in its original form had both application logic and persistent data storage (e.g. HSS), will become an Application FE maintaining existing interfaces to other network entities, while the persistent data storage is moved to the UDR.

10.3.2 Front-Ends and User Data Repository

When the UDC architecture is applied, functional entities that originally maintained subscriber data (e.g. HSS) maintain the application logic, but they do not locally store user data permanently. As already mentioned above, these data-less functional entities become Application Front-Ends. 3GPP currently defines the following Application Front-Ends:

- HSS (and HLR/AuC)
- ANDSF
- PCRF
- IMS Application Server (AS).

The UDC also defines so-called Provisioning Front-Ends (see also Figure 10.6). These entities are used for provision of the UDR. A Provisioning Front-End provides means to create, delete, modify, and retrieve user data.

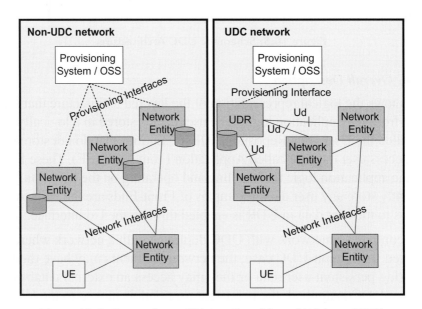

Figure 10.6: Comparison of Networks with and Without UDC.

The User Data Repository (UDR) is a functional entity that acts as a single logical repository that stores user data. The user-related data that in a non-UDC network is stored into different logical databases such as the HSS, SPR, etc. is now stored in the UDR. UDR facilitates the share and the provision of user-related data throughout services of the 3GPP system.

The UDR provides a unique reference point to one or more Applications Front-Ends. The UDR stores both permanent and more dynamic subscriber data. Permanent subscriber data relates, for example, to the necessary information the system ought to know to perform the service. User identities (e.g. MSISDN, IMSI), service data (e.g. service profile in IMS), and authentication data are examples of subscription data. This kind of user data has a lifetime as long as the user is permitted to use the service and may be modified by administrative means. The UDR also stores temporary subscriber data. This is data that may be changed as a result of normal operation of the system or traffic conditions. Examples of temporary data are the SGSN address, user status, etc.

CHAPTER 11

Voice and Emergency Services

In Chapter 5 we gave a brief overview of the various ways an operator can provide voice services in EPS. In this chapter we go a bit further into the realization of voice services in EPS, including MMTel/VoLTE, SRVCC, and CS Fallback. Finally, we touch briefly on the migration options from circuit-switched technology to VoLTE.

11.1 Voice Services Based on Circuit-Switched Technology

Circuit switching is the traditional technology used in a telephony network, where a continuous link is established between two end-users in a phone call. One vision for EPS is that IP technology will be used for all services, including voice, and effectively replacing circuit-switched services. In order to understand how voice services will be delivered using IP technology, it is necessary to have a basic understanding of the technology that it is meant to replace. This section therefore briefly describes circuit-switched technology while subsequent sections cover the implementation of voice on mobile networks using IP technology. In the EPC, carrier-grade multimedia services are provided with a technology called IMS, which is covered in the next section.

A central part of the circuit-switched network architecture is the Mobile services Switching Center (MSC). This is the core network function supporting voice calls, handling both the signaling related to the calls and switching the actual voice calls. Modern deployments of circuit-switched core networks are normally designed with a separation of the signaling functions (handled by the MSC Server) from functions handling the media plane (handled by the Media Gateway). Figure 11.1 shows a simplified architecture.

Here, the MSC Server includes call control and mobility control functions, while the media, i.e. the actual data frames making up the voice calls, flow through a Media Gateway that can convert between different media and transport formats, as well as invoke specific functions into the voice calls, such as echo cancellation or conferencing functions. The MSC Server controls the actions taken by the Media

EPC and 4G Packet Networks.
DOI: http://dx.doi.org/10.1016/B978-0-12-394595-2.00011-6

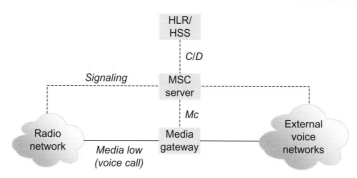

Figure 11.1: A Simplified Architecture for CS Voice.

Gateway on a specific call and interacts with the Home Location Register/Home Subscriber Server (HLR), which handles subscription data for users of circuit-switched services.

While voice calls in mobile networks have been converted into streams of digital data since the early 1990s, the data frames themselves are not sent between mobile devices and the networks using shared channels or IP technology.

This means that unique resources in the network need to be dedicated to each voice call throughout the duration of the call. The connection is established at call setup, and is maintained until call termination, when the network resources are released. Circuit-switched connections therefore consume network resources with a fixed bandwidth and a fixed delay for the duration of the call. This is also valid if no actual communication takes place, i.e. if neither party has anything to say. As long as the call is ongoing, the allocated network resources are not available to other users. There is no obvious way to optimize these resources across multiple users.

It should be noted, however, that this is something of a simplification. In order to improve the resource usage for circuit-switched services, some mechanisms have been designed to enable a somewhat more efficient usage of the available bandwidth, for example through taking advantage of silent periods in the voice calls and enabling multiplexing of several users onto a common channel. Also, in a wireless system, the available bandwidth varies to some extent, due to the characteristics of the radio channel changing during the call. This may result in variations of the voice quality as the voice coder adapts to the changing radio environment.

Since the voice data for circuit-switched services are not transported using IP packets between the devices and the network, there is also no way to multiplex several services onto the same service stream, nor to provide a standard Application Programming Interface (API) to other services or applications in the device.

The packet data services in GSM, WCDMA, and LTE, however, offer IP connectivity between the mobile device and a gateway node. This IP connectivity can be used for any IP-based application and may be utilized by multiple applications simultaneously. One of these applications is naturally voice. Furthermore, the call as such can be more than a voice call and consist of several media components in addition to the voice medium itself.

We now turn to the implementation of voice using IP technology, which within EPS using 3GPP specifications, is achieved with the IP Multimedia Subsystem – the IMS.

11.2 Voice Services with IMS Technology

IMS (IP Multimedia Subsystem) was originally designed by 3GPP in order to enable IP-based multimedia services over GSM and WCDMA GPRS systems, but was later expanded to support other access networks. It should be noted that, after the merger of the work between TISPAN, Packet Cable, 3GPP2, and 3GPP into the "Common IMS", all IMS specifications are now handled by 3GPP. Both fixed and mobile networks therefore use the same IMS specifications. It is also important to note that the voice services described in this chapter can be used with non-3GPP accesses, as well as the ones specified by 3GPP. As stated before, one of the key design points for the EPC was ensuring that both 3GPP and non-3GPP accesses would be able to connect to and utilize such services.

Where appropriate, 3GPP reuses protocol from other standards bodies within their specifications and this is what happened with IMS. The IMS concept is built around the Session Initiation Protocol (SIP), defined by the IETF in RFC 3261. This protocol was designed by the IETF as a signaling protocol for establishing and managing media sessions, for example voice and multimedia calls, over IP networks.

IMS itself is defined as a subsystem within the mobile network architecture. It consists of a number of logical entities interconnected via standardized interfaces. Note that these are logical entities, and that vendors of IMS infrastructure equipment may combine some of these entities on the same physical product or products (Figure 11.2).

At the core of the IMS subsystem is the **Call Session Control Function (CSCF)**. This is the node handling the SIP signaling, invoking applications and controlling the media path. The CSCF is logically separated into three different entities:

- The Proxy CSCF (P-CSCF)
- The Serving CSCF (S-CSCF)
- The Interrogating CSCF (I-CSCF).

These three entities may very well reside as different software features on the same physical product.

The primary role of the P-CSCF is as an SIP proxy function. It is in the signaling path between the terminal and the S-CSCF, and can inspect every SIP message that is flowing between the two endpoints. The P-CSCF manages quality of service and authorizes the usage of specific bearer services in relation to IMS-based services. The P-CSCF also maintains a security association with the terminal and may also optionally support SIP message compression/decompression.

The S-CSCF is the central node of the IMS architecture. It manages the SIP sessions and interacts with the HSS (Home Subscriber Server) for subscriber data management. The S-CSCF also interacts with the Application Servers (AS).

The primary role of the I-CSCF is to be the contact point for SIP requests from external networks. It interacts with the HSS to assign an S-CSCF that handles the SIP sessions.

The HSS manages IMS-related subscriber data and contains the master database with all subscriber profiles. It includes functionality to support access and service authorization, mobility management, and user authentication. It also assists the I-CSCF in finding the appropriate S-CSCF.

The **Media Resource Function Processor (MRFP)** is a media plane node that can (but does not need to) be invoked to process media streams. Examples of use cases where the media data are routed via the MRFP are conference calls (where mixing of multiple media streams are needed) and transcoding between different IP media formats.

The **Media Resource Function Controller (MRFC)** interacts with the CSCF and controls the actions taken by the MRFP.

The **Breakout Gateway Control Function (BGCF)** handles routing decisions for outgoing calls to circuit-switched networks. It normally routes the sessions to a Media Gateway Control Function (MGCF).

The **Media Gateway (MGW)** provides interworking including conversion and transcoding between different media formats used for IMS/IP and circuit-switched networks.

The **MGCF** provides the logic for IMS interworking with external circuit-switched networks. It controls the media sessions through ISUP signaling towards the external network and SIP signaling towards the S-CSCF, and through controlling the actions of the MGW.

The **Session Border Controller (SBC)** is an IP gateway between the IMS domain and an external IP network. It manages IMS sessions and provides support for controlling security and quality of the session. It also supports functions for firewall and NAT traversal, i.e. when the remote IMS terminal resides behind a device (for instance, a home or corporate router) that provides IP address conversion.

The **AS (Application Server)** implements a specific service and interacts with the CSCF in order to deliver it to end-users. Services may be defined within 3GPP, but because of the use of IP technology, it is not necessary for all services to be standardized. One example of a service that is defined by 3GPP is Multimedia Telephony (MMTel). MMTel has been designed to support voice calls using IMS, but MMTel can provide more than telephony in that other media can be added to the voice call, turning this into a complete multimedia session.

It is beyond the scope of this book to provide a more detailed description of IMS. Interested readers are referred to Camarillo (2008) for more information.

11.3 MMTel

As discussed in Chapter 5, MMTel is the standardized service offering for voice calls. It is built on the IMS and offers more possibilities than traditional circuit-switched voice calls in order to enhance the communication experience for end-users. For example, video or text may be added to the voice component. Since EPS is designed to efficiently carry IP flows between two IP hosts, MMTel is a natural choice for offering voice services when in LTE coverage. The IMS has been built to combine telecommunications quality service with modern IP technology and is an open and extendable standard on which to build services. In addition, MMTel complies with regulatory requirements for the delivery of voice services in most countries.

11.3.1 MMTel Architecture

The MMTel standard is described in 3GPP TS 24.173 and provides telecommunications grade voice services over IP technology. It allows for interoperable services between operators and towards legacy networks, e.g. PSTN. The standardized interfaces mean that operators can use multiple vendors within a network and integrate with Internet services. MMTel also complies with regulatory requirements associated with voice services, in contrast to most OTT VoIP services. The architecture of MMTel within a converged network scenario is illustrated in Figure 11.2.

MMTel is the 3GPP standard for voice. MMTel is a very broad and complex standard. In order to ensure take-off of MMTel, therefore, both network operators and network vendors worked within GSMA to produce a "profile" of the MMTel specifications in something called "voice over LTE" (VoLTE). VoLTE ensures basic compatibility between operator networks for multimedia services. The next section provides an overview of VoLTE.

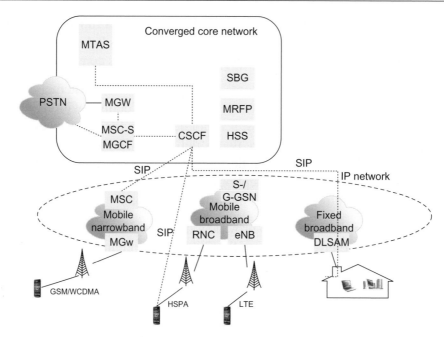

Figure 11.2: MMTel Architecture.

It is beyond the scope of this book to provide a detailed description of MMTel; instead, interested readers may find more information in both Camarillo (2008) and Noldus et al. (2011).

11.4 VoLTE

The MMTel standard is rather complex and provides many choices and options for the realization of voice services on packet radio. The functions that form the basis of the MMTel standard are also spread across multiple documents with no description of how these functions fit together in a live network implementation. This would lead to difficulties if an operator tried to implement them in a live network – how could they guarantee interoperability with other mobile network operators? Without some means to ensure that services could work across several operator networks, fragmentation of the market would result.

The response to this potential fragmentation was VoLTE, a GSMA profile (IR.92) that outlines a subset of 3GPP MMTel and SMS over IP functions that simulate the legacy 2G/3G CS services that end-users now expect. GSMA IR.92 is a UNI specification that spans every layer of the network, including telephony service features, IMS features, media features, bearer management features, LTE radio features, and common functions such as the IP version. The profile is primarily based on 3GPP Release 8 specifications with some specific additions that are based on 3GPP Release 9.

The telephony service requirements put on bearer management and charging mandate that an LTE/Evolved Packet Core (EPC) system used for MMTel according to the VoLTE profile must be equipped with PCC, which is described in Chapter 8. Two standardized classes are used for VoLTE flows; QCI 徊 1 (a guaranteed bit-rate bearer for VoIP media) and QCI 徊 5 (a high-priority non-guaranteed bit-rate bearer for IMS SIP signaling and XCAP).

IR.92 essentially outlines the *minimum* set of functions and features that are mandatory for LTE devices and networks to implement. This means that these IMS-based telephony and SMS services are guaranteed to be interoperable across operator networks implementing IR.92. There are two modes of operation:

1. All-IP solution, which is fully based on VoLTE voice services, including emergency calls delivered using the IMS.
2. CS in addition to IP solution. In this mode of operation, emergency services are provided by CS access and SRVCC complements the VoLTE voice service.

The telephony service features – commonly referred to as supplementary services – that are included in this profile are listed in Table 11.1.

Table 11.1: Telephony Service Features Included in IR.92

Name	Specification
Originating identification presentation	3GPP TS 24.607
Terminating identification presentation	3GPP TS 24.608
Originating identification restriction	3GPP TS 24.607
Terminating identification restriction	3GPP TS 24.608
Communication forwarding unconditional	3GPP TS 24.604
Communication forwarding on not logged in	3GPP TS 24.604
Communication forwarding on busy	3GPP TS 24.604
Communication forwarding on not reachable	3GPP TS 24.604
Communication forwarding on no reply	3GPP TS 24.604
Barring of all incoming calls	3GPP TS 24.611
Barring of all outgoing calls	3GPP TS 24.611
Barring of outgoing international calls	3GPP TS 24.611
Barring of outgoing international calls – ex home country	3GPP TS 24.611
Barring of incoming calls – when roaming	3GPP TS 24.611
Communication hold	3GPP TS 24.610
Message waiting indication	3GPP TS 24.606
Communication waiting	3GPP TS 24.615
Ad-hoc multi-party conference	3GPP TS 24.605

In order to ensure functionality between IMS networks, the following are mandated in IR.92:

* Support for IMS authentication and key agreement (IMS-AKA), IMS subscriber identity module (ISIM). A universal SIM (USIM) may be used if the network service provider has not deployed ISIM.
* IPSec protection of signaling over the UNI.
* Both Tel URI and SIP URI addressing schemes are supported.
* SIM-based authentication of supplementary service self-management operations using XCAP over Ut according to the GBA architecture is recommended.

IR.92 is illustrated in Figure 11.3.

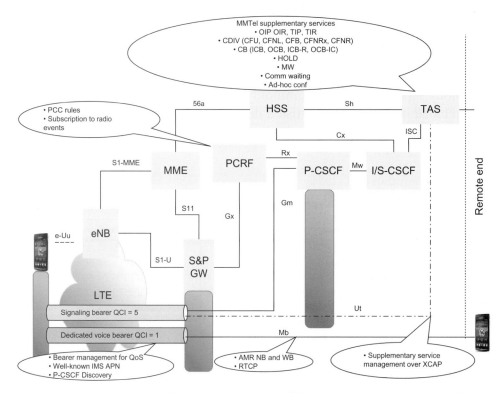

Figure 11.3: VoLTE Architecture.

In addition to IR.92 GSMA has also defined IR.94, which provides video extensions to the VoLTE profile.

11.5 T-ADS

Terminating-Access Domain Selection, or T-ADS (3GPP TS 23.292), is a function that determines where a call will be terminated for a VoLTE user and ensures that the IMS routes the call to the UE when it is in 2G/3G coverage. If LTE coverage is lost, the UE should still be able to use CS voice services, even if SR-VCC is not available. In order to route MO and MT calls appropriately, the IMS must know whether the UE is in LTE or UTRAN/GERAN CS coverage. T-ADS is the function that provides this support.

T-ADS is implemented within the SCC-AS. Using the Sh interface, the HSS provides T-ADS information (most recent IMS over PS support and RAT type as indicated by the access node) to the SCC-AS. The result of this query will be that the UE is either VoLTE and PS registered, or PS and CS registered.

On a terminating call setup received by SCC-AS, the T-ADS function takes the following knowledge/data into consideration when deciding where to terminate the call:

- Ongoing or recently terminated VoLTE session access domain is PS or CS access network.
- Registered contacts information:
  - Access domain (CS or PS)
  - UE terminal type
  - IMS voice over PS support for the most recently used PS access network retrieved from the HSS.

The T-ADS may then result in the call:

- Being terminated over LTE PS
- Being terminated over CS
- Being terminated over LTE PS and CS (leaving it to CSCF forking)
- Being rejected.

11.5.1 Ensuring Service Coverage

In order to ensure that an end-user is able to have voice service everywhere, however, it is not possible to rely on the fact that LTE coverage is present in all places. As discussed in Chapter 5, until there is universal LTE coverage, therefore, ensuring that an end-user has full service coverage for voice calls requires that other access networks complement LTE coverage and that the device used to establish the voice call also supports those access technologies. In addition, it is necessary that intersystem handovers between these access technologies is possible. The following sections cover the use cases described in Chapter 5: Single Radio Voice Call Continuity (SRVCC) and Circuit-Switched Fallback (CS Fallback).

11.6 Single Radio Voice Call Continuity (SRVCC)

As described in Chapter 5, SRVCC is a solution that addresses the problem that there may not be full coverage for MMTel VoIP services used on E-UTRAN. SRVCC solves this problem by offering a mechanism where the UE performs a coordinated radio level handover in combination with a change from IMS VoIP to circuit-switched voice using IMS procedures for service continuity.

The initial version of SRVCC was developed in Release 8 and has been enhanced over the next releases to include support for video and finally, in Release 11, support for return SRVCC was added, which allows for handover from a CS call on 2G/3G to MMTel/VoLTE on LTE networks.

11.6.1 Entities with Additional Functions to Support SRVCC

In order to support SRVCC, additional functionality is required on several network entities, as shown in Figure 11.4. This section briefly describes the extra functionality required on each node.

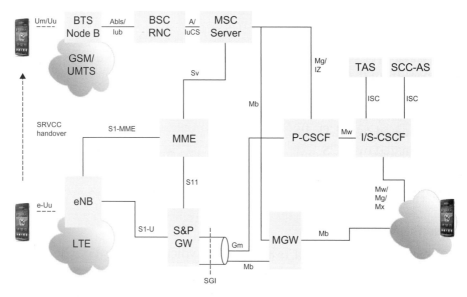

Figure 11.4: SRVCC Handover Overview Diagram.

11.6.1.1 MSC

In addition to the standard MSC procedures, an MSC supporting SRVCC must also support the relocation preparation procedure, which is requested by the MME/SGSN for voice components of a call. The MSC must also coordinate the CS handover session transfer procedures and handle the MAP_Update_Location procedure. In SRVCC, Map_Update_Location is part of the handover procedure; as such, the MSC needs to handle this procedure with the UE triggering it.

The MSC must also negotiate with the SCC_AS to determine whether it should perform an SRVCC or vSRVCC procedure. In the case of vSRVCC, in addition to the requirements above, the MSC must also initiate the handover to the target system for BS30 when it receives an Sv request from the MME that indicates a vSRVCC handover.

In order to support rSRVCC, the MSC informs the RAN about the possibility of performing CS to PS SRVCC by sending a "CS to PS SRVCC operation possible" to the RNC/BSC. The possibility of performing such a handover is based on the capability of the UE and the registration of the UE within the IMS.

11.6.1.2 MME
An MME (operator) that supports interworking to 3GPP CS must comply with the rules and procedures prescribed in 3GPP TS 23.401. Briefly, the MME performs the PS bearer splitting by separating the voice PS bearer from the non-voice PS bearers. For vSRVCC sessions, the MME identifies the vSRVCC marked video PS bearer in addition to the voice PS bearer and handles the non-voice PS bearer handover with the target cell. The MME initiates and coordinates the handover for the voice and PS and (v)SRVCC procedures.

11.6.1.3 SGSN
In order to support the SRVCC, the SGSN must interwork towards the MSC Server, as described by 3GPP TS 23.060. In support of SRVCC, the SGSN performs the PS bearer splitting function by separating the voice PS bearer from the non-voice PS bearers. The SGSN also initiates the SRVCC procedure for handover of the voice component to the target cell via the Sv interface, and coordinates the PS handover and SRVCC handover procedures when both are performed.

In order to support rSRVCC, the SGSN needs to support the interworking to the 3GPP CS as outlined in 3GPP TS 23.060. The SGSN also handles the non-voice PS bearers and allocates the resources with the target RAN upon receiving CS to PS HO request from MSC via the Sv interface.

11.6.1.4 HSS
During the E-UTRAN attach procedure, the MME downloads, among other things, the STN-SR and vSRVCC flag from the HSS. The ICS flag is used by the MSC Server enhanced for (v)SRVCC to behave also as MSC Server enhanced for ICS in 3GPP TS 23.292 if supported by the network. For rSRVCC sessions, the MSC Server downloads the "CS to PS SRVCC allowed" and optional ICS flag from the HSS during the attach procedure.

11.6.1.5 UE

Naturally, the UE must also be able to handle SRVCC, which is described in 3GPP TS 36.300 for interaction between UE and LTE and in 3GPP TS 25.331 for the interaction between UE and UTRAN (HSPA). When a UE is being configured for IMS speech service support by the home operator, it indicates that it is SRVCC capable to the network.

For vSRVCC, the UE will initiate the multimedia codec negotiation with the CS domain after vSRVCC. The UE and the network may support the MONA codec negotiation mechanisms described in ITU-T recommendation H.325, Annex K.

11.6.1.6 E-UTRAN/eNodeB

For SRVCC handover, the E-UTRAN selects a target cell and sends an indication to the MME that SRVCC is required for this handover.

The E-UTRAN may be capable of determining the neighbor cell list based on simultaneous presence of established QCI = 1 (for voice part) and vSRVCC marked bearers for a specific UE.

11.6.1.7 PCC

When an IMS session is anchored in the SCC AS, the PCRF enforces the architecture principle to use QC = 1, traffic-class = "conversational", and source statistics descriptor = "speech". There are two possible methods for this to be achieved – either through deploying the S9 reference point or through configuration and roaming agreements between operators.

11.6.1.8 Sv Interface between MSC and MME

The Sv interface connects the MSC and MME and enables handover signaling between the MME and the MSC for SRVCC.

11.7 IMS Centralized Services (ICS)

ICS provides mechanisms to support the use of CS media bearer for IMS sessions. With ICS, IMS sessions using CS media are treated as standard IMS sessions for the purpose of service control and service continuity.

In short, ICS is about centralization of voice and SMS services in a "homogeneous" network, where VoLTE will have less coverage than a CS network (2G/3G).

There are two underlying drivers for IMS Centralized Services:

- Homogeneous service experience in a heterogeneous network, where VoIP will have less coverage than CS access.
- Service interaction and service enhancement are easier to do in a single service engine, providing a consistent service experience to end-users.

See Figure 11.5 for an overview of ICS principles. IMS is the service engine; specifically, the Telephony Application Server (TAS) is where the VoIP services are executed and delivered.

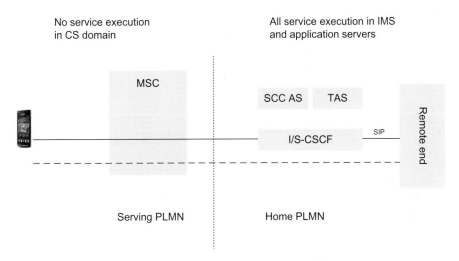

Figure 11.5: Overview of ICS Principles.

11.7.1 Service Centralization and Continuity Application Server (SCC-AS)

The IMS Service Centralization and Continuity Application Server (SCC-AS) is a key component in the IMS Centralized Services (ICS) solution. This is a logical node and co-located inside the TAS but it can also be provided as a standalone AS.

ICS enables the support of the VoLTE service engine (TAS) over legacy CS access from a VoLTE device.

One important function of the SCC-AS is terminating access domain selection (T-ADS), described in Section 11.5.

The details of SRVCC and related procedures reach beyond the EPS into the IMS and the CS core network. This chapter will outline the SRVCC procedure and the impact on the EPC network at a high level without going too far into the details of other parts of the system. The 3GPP TS 23.216 elaborates more on the SRVCC procedure and impacts the EPC while the 3GPP TS 23.237 details how the IMS handles service continuity. It is beyond the scope of this book to describe the IMS in detail; refer to Camarillo (2008) for more information.

The solutions for SRVCC towards GERAN/UTRAN and SRVCC to CDMA are not exactly the same due to the differences in the interworking of EPS with CDMA and GERAN/UTRAN respectively.

11.7.2 SRVCC from E-UTRAN to GERAN or UTRAN

Figure 11.6 Shows High Level Flows Regarding how SRVCC is Executed from E-UTRAN Towards UTRAN/GERAN.

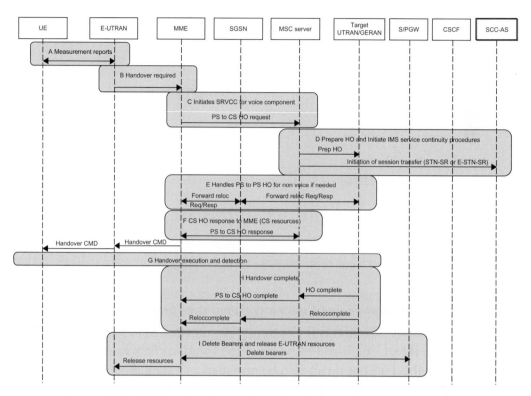

Figure 11.6: Call Flow for SRVCC from E-UTRAN to GERAN or UTRAN.

- **A/B:** Based on measurement reports received from the UE, the source E-UTRAN decides to trigger an SRVCC handover to UTRAN/GERAN.
- **C/D:** The source E-UTRAN sends a Handover Required message to the source MME indicating that it is a CS + PS handover. The source MME initiates the PS–CS handover procedure for the voice bearer by sending an SRVCC PS to CS Request message to the MSC Server. In turn, the MSC Server interworks the PS–CS handover request with a CS inter-MSC handover request by sending a Prepare Handover Request message to the target MSC. The target MSC requests resource allocation for the CS relocation by sending the Relocation Request/ Handover request to the target RNS or BSS. For non-emergency sessions, the MSC Server initiates the Session Transfer by using the STN-SR, e.g. by sending an ISUP IAM (STN-SR) message to the IMS. During the execution of the Session Transfer procedure, the SCC-AS updates the remote end with the SDP of the CS

access leg. This causes the flow of VoIP packets to be switched to the CS access leg. The original IMS access leg is released by the SCC-AS.

- **E/F:** In parallel to C/D, the MME splits the voice bearer from all other PS bearers and initiates their relocation towards the MSC Server and the SGSN. The source MME sends a Forward Relocation Request message to the target SGSN. The target SGSN requests resource allocation for the PS relocation by sending the Relocation Request/Handover request message to the target RNS/BSS. Once the target RNS/BSS receives both the CS relocation/handover request with the PS relocation/handover request, it assigns the appropriate CS and PS resources. The target RNS/BSS acknowledges the prepared PS relocation/handover by sending the Relocation Request Acknowledge/Handover Request message to the target SGSN, which then sends a Forward Relocation Response to the source MME. In parallel to E, the target RNS/BSS acknowledges the prepared CS relocation/handover by sending the Relocation Request Acknowledge/Handover Request Acknowledge message to the target MSC. The target MSC then sends a Prepare Handover Response message to the MSC Server. A circuit between the target MSC and MGW associated with the MSC Server is established.

- **F:** The MSC Server sends a SRVCC PS to CS Response (Target to Source Transparent Container) message to the source MME. The source MME synchronizes the two prepared relocations and sends a Handover Command message to the source E-UTRAN. E-UTRAN sends a Handover from E-UTRAN Command message to the UE. The UE tunes to the target UTRAN/GERAN cell.

- **G:** Handover detection at the target RNS/BSS occurs. The UE sends a Handover Complete message via the target RNS/BSS to the target MSC. At this stage, the UE re-establishes the connection with the network and can send/receive voice data.

- **H:** CS relocation/handover is complete. The Target RNS/BSS sends a Relocation Complete/Handover Complete message to the target MSC. The target MSC sends an SES (Handover Complete) message to the MSC Server. The speech circuit is through connected in the MSC Server/MGW. The MSC Server sends an SRVCC PS to CS Complete Notification message to the source MME. The source MME acknowledges the information by sending an SRVCC PS to CS Complete Acknowledge message to the MSC Server. The target RNS/BSS sends a Relocation Complete/Handover Complete message to target SGSN. The Target SGSN sends a Forward Relocation Complete message to the source MME. The source MME acknowledges the information by sending a Forward Relocation Complete Acknowledge message to the target SGSN.

- **I:** The source MME deactivates the voice bearer towards S-GW/P-GW and sets the PS-to-CS handover indicator to Delete Bearer Command message. The MME sends a Delete Session Request to the SGW. The source MME sends a Release

Resources message to the Source eNodeB. The Source eNodeB releases its resources related to the UE and responds back to the MME.

11.8 SRVCC from E-UTRAN to CDMA 1xRTT

The high level procedure for SRVCC between IMS Voice Telephony on E-UTRAN to CS telephony on CDMA is shown in Figure 11.7.

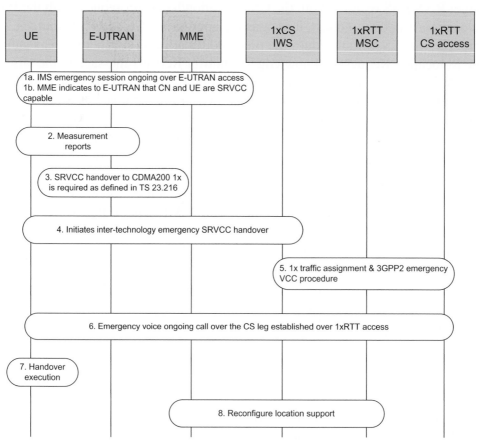

Figure 11.7: SRVCC to CDMA High Level Flows.

The architecture for SRVCC between IMS Voice Telephony on E-UTRAN to CS telephony on CDMA looks similar to the one for SRVCC to GERAN and UTRAN, but there are a few key differences. Instead of an Sv interface between the MME and the interface to trigger handovers there is a different type of mechanism employed. The UE communicates directly with the CDMA MSC using NAS message tunneling. The MME is, in principle, just a signaling relay and the interworking function tunnels the signaling messages and interworks them towards the CDMA MSC. This allows the UE to communicate with the CDMA MSC to trigger the service continuity procedure in IMS and to prepare the access before performing the handover on the radio layer. For details on the SRVCC solution for CDMA, refer to 3GPP TS 23.216 and 3GPP2 TS X.S0042-0.

11.9 Circuit-Switched Fallback

Circuit-Switched Fallback (CSFB) provides the operators with an option to support LTE with voice and SMS services before IMS and VoLTE services are deployed. SMS over SGs (SMSoSGs) is a subset of the CSFB solution that was developed to support devices that only use SMS via CS Domain and not CS voice, e.g. simple modems in meters or tablets such as the iPad.

The main idea behind CS Fallback is to allow UEs to camp on LTE and utilize the LTE for data services but reuse the GSM, WCDMA, or CDMA network for circuit-switched voice services.

There are special additions to the attach and tracking area update procedures that activate an interface, called SGs, between the MME and the MSC. This interface is used by the MSC to send paging messages for CS calls to the UE camping on LTE (see Figure 11.8).

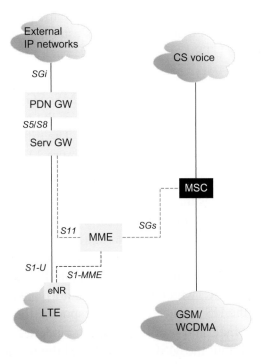

Figure 11.8: CSFB: Architecture Additions to Support CSFB.

The SGs reference point is used for the mobility management and paging procedures between EPS and CS domains. The SGs reference point is also used for the delivery of both mobile originating and mobile terminating SMS.

For CSFB the MME is enhanced to support combined EPS/IMSI attach and combined tracking area update. These are the procedures that establish and maintain the SGs association between the MME and the MSC. The MME also supports paging and transfer of messages between MSC and UE.

The MSC maintain the SGs association towards MME for EPS/IMSI attached UEs. The MSC can request the MME to page the UE and for SMSoSGs it can use the MME as an NAS signaling relay.

E-UTRAN needs to differentiate between normal PS-to-PS mobility procedures and CSFB mobility in terms of the target cells it selects. The operator may also have specific preferences on CSFB, e.g. a preference to fall back to 3G cells. Mobility procedures from E-UTRAN have been enhanced and it is now possible for E-UTRAN to use a redirect procedure that includes multiple target cells. This enhances the CSFB performance as the time-consuming measurements on target cells can be eliminated; instead, the UE uses the faster process to tune to the strongest cells among those included in the redirect procedure. For SMS over SGs, no specific E-UTRAN functionality is required.

From 3GPP Release 10 the specifications also include the possibility for the MSC to inform the UTRAN or GERAN that a CS call was established as a result of CSFB. This allows the GERAN or UTRAN to send the UE back to LTE as soon as the CS call is terminated.

To illustrate the principles of the CS Fallback and SMSoSGs we will explore some of the procedures below. The first procedure to explore is the attachment procedure for CSFB and SMSoSGs, called EPS/IMSI attachment. The differences compared to a normal EPS attachment procedure are outlined in Figure 11.9.

A. The UE starts the attachment procedure by the sending an Attach request to the MME. The UE indicates that it requests a combined EPS/IMSI attachment and includes its capabilities and preferences. If the UE only needs SMS service (SMoSGs) it indicates "SMS-only" in the request.
B. On receipt of the attachment request the MME triggers the normal EPS attach procedure, which is described in Chapter 12. In parallel, the MME sends a location update to the MSC, which performs the normal CS domain location update procedure to the HSS. Once the HSS responds, the MSC creates the SGs association with the MME and sends a Location Update Response to the MME.
C. The MME waits until both the MSC responds and the normal EPS attachment procedure are ready until it sends an Attach Accept back to the UE. The Accept

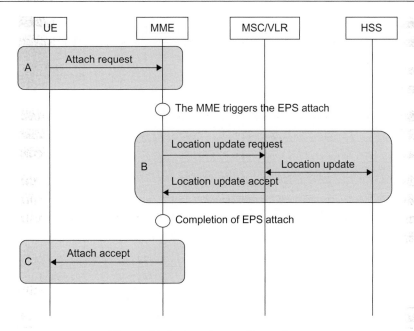

Figure 11.9: Attachment Procedure.

message includes information that tells the UE if the IMSI attachment was successful and if all sevices are available or only SMSoSGs.

After the attachment procedure is performed the network is aware of the UE location and can route terminating calls and SMS and the UE can initiate calls or SMS.

The next procedure to explore is the mobile terminating CS call procedure (see Figure 11.10). This procedure is supported when the Network and UE supports CSFB and the UE is EPS/IMSI attached. If the UE is only attached for SMS (SMSoSGs) the MSC will block terminating CS calls.

A. The MSC receives an incoming voice call and sends a CS Page to the MME over an SG interface. The MME uses the TMSI (or IMSI) received from the MSC to find the S-TMSI (which is used as the paging address on the LTE radio interface). The MME forwards the paging to the eNodeB in the tracking areas where the UE is registered. The eNodeBs perform the paging procedures in all of the cells in the indicated tracking areas. The paging message includes a special CS indicator that informs the UE that the incoming paging is for a terminating CS call.
B. On receipt of the Paging message, the UE performs a service request procedure, which establishes the RRC connection and sends the service request to the MME. The Service Request Message includes a special CS Fallback indicator

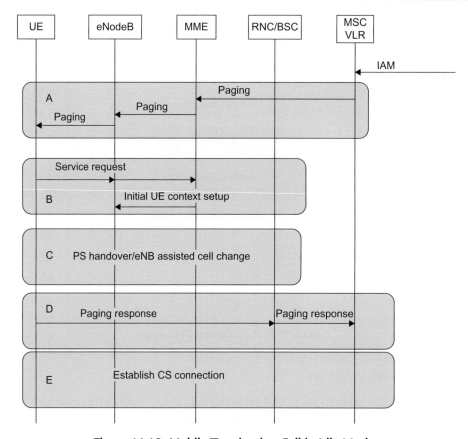

Figure 11.10: Mobile Terminating Call in Idle Mode.

that informs the MME that the CS Fallback is needed. This triggers the MME to activate the bearer context in the eNodeB with an indication to perform fallback to GERAN or UTRAN.

C. The eNodeB selects a suitable target cell, possibly by triggering the UE to send measurements on neighboring cells, and initiates a handover or cell change procedure. The selection between handover or cell change procedure is based in the target cell capabilities and is configured in the eNodeB.

D. After handover/cell change the UE detects the new cell, establishes a radio connection, and sends a paging response to the MSC, via the target RAN.

E. When the page response arrives at the MSC a normal mobile terminated call setup continues and the CS call is activated towards the UE.

On completion of the CS call in 2G/3G, the MSC informs the GERAN/UTRAN that the call was triggered by CSFB. This information may be used by GERAN/UTRAN to immediately redirect the UE back to LTE.

SMS over SGs is, as the name implies, a mechanism to send SMS from MSC via the MME to the UE and vice versa. This means that the UE can remain on LTE while sending and receiving SMS. A mobile terminated SMS procedure is outlined in Figure 11.11.

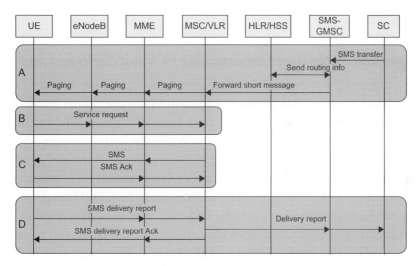

Figure 11.11: Mobile Terminating SMS in Idle Mode.

A prerequisite for this procedure is that the UE EPS/IMSI is for CSFB or for SMS only (SMSoSGs):

A. The SMS Service Center (SC) sends the SMS to an SMS-GWMSC function that asks the HLR a routing number for SMS services and the SMS message is forwarded to the VLR where the UE is CS attached. The MSC/VLR sends a paging request to the MME, which initiates the paging messages to each eNodeB with cells belonging to the tracking area(s) in which the UE is registered. The eNodeBs page the UE.

B. When the UE receives the paging message it triggers a service request procedure that establishes a signaling connection between the UE and the MSC via the MME. The service request procedure is described in more detail in Chapter 12.

C. The MSC forwards the SMS message to the MME in a Downlink Unitdata message. The MME encapsulates the SMS message in an NAS message and sends the message to the UE. The UE acknowledges receipt of the SMS message to the MSC.

D. The UE returns a delivery report. The delivery report is encapsulated in an NAS message and sent to the MME. The MME forwards the delivery report to the MSC in an Uplink Unitdata message. The delivery report is forwarded to the SC and the MSC/VLR acknowledges receipt of the delivery report to the MS/UE.

The MME should not use the SGs Release Request message as a trigger for the release of S1 resources.

The CS Fallback specifications cover all necessary procedures to support fallback to GSM, WCDMA, and CDMA (1xRTT). For mobile originated and mobile terminating calls in both idle and active modes, the procedures for CSFB and SMSoSGs are specified in 3GPP TS 23.272.

11.10 Migration Paths and Coexistence of Circuit-Switched and VoLTE

As we have shown in the previous sections, there are several methods to handle the lack of full service coverage for IP voice on mobile networks. This provides operators with some options about how they will migrate from circuit-switched to VoLTE. Figure 11.12 illustrates the migration options. While these may appear to be distinct, it is quite possible that operators could combine all of these steps at the same time – there is no reason why CSFB, VoLTE, and SRVCC cannot all be implemented on the same operator network. This would allow a decision to be taken based on coverage and capacity of the part of the network the end-user was in at that point in time whether to use CSFB or SRVCC to provide full-service coverage for voice. There are numerous aspects that may affect the selection of which migration path to take, including terminal support, operator strategies, and availability of emergency call support.

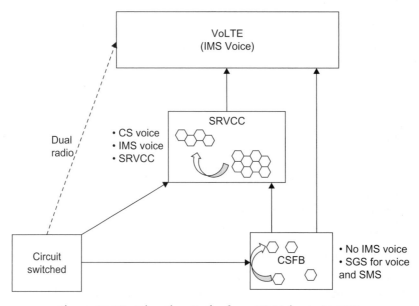

Figure 11.12: Migration Paths from CS Voice to VoLTE.

As we have discussed, there are a few different paths that can be followed in order to implement GSMA VoLTE on an operator network. For example, an operator could reasonably start with CSFB, migrate to SRVCC, and finally step to full VoLTE. Alternatively, an operator could skip CSFB altogether and implement SRVCC directly. Finally, if an operator implements a dual radio solution, circuit-switched and VoLTE can be run in parallel. Figure 11.12 illustrates potential migration paths from CS Voice to VoLTE.

For example, if an operator has implemented LTE, but with limited or "spotty" coverage, it may be most useful to use CSFB at this stage. Operators can reuse existing CS networks to deliver continuous voice services to LTE subscribers. In this scenario, SMS continues to be used for messaging and device configuration as in existing networks. The service architecture is maintained by using SMS over SGs to permit sending SMS to LTE clients.

Alternatively, once LTE is implemented in more areas, a VoLTE subscriber may require handover mechanisms towards CS telephony. SRVCC and ICS address this requirement through anchoring voice calls in the IMS domain and delivering voice services via VoLTE.

Once an operator has greater broadband coverage through LTE and HSPA, however, all telephony can be provided via the IMS. This means that IMS will provide all voice services and SMS may be provided using SMS over IP.

11.11 EPS Emergency Bearer Service for IMS Emergency

In 3GPP, support for emergency calls via IMS has been available for some time now. Support for emergency sessions in the Packet Core over 3GPP Radio Accesses, however, was not introduced until Release 9 of the 3GPP specifications. As CS domain use of emergency call support is very well established, the need for support of emergency services via EPC only became relevant and critically important once LTE deployment started ramping up and interests for deploying VoLTE grew. For more on this, see Chapter 5 and Section 11.10, which give details on voice service realization using E-UTRAN and IMS. Requirements for support of emergency PS bearer services remain the same as the requirements for CS emergency calls, which is based on local and legal jurisdictions within each country or set of cooperating regions. Emergency services can also be enabled for UEs without a UICC, for UEs that fail authentication, and for scenarios where authentication cannot or may not be performed.

Emergency sessions in the Packet Core over 3GPP Radio Accesses are supported using the emergency PS bearer service. The basic architecture as specified for 3GPP EPC applies for the emergency bearer service as well. However, there are certain architectural aspects that must be respected due to the fact that the emergency services, in the form of a mandatory regulatory service, must be provided by the local operator in the local country where the user is attempting to make the emergency access. Emergency services are thus not a subscription-based service. A default emergency profile in an MME that supports emergency bearer services assures that a consistent user experience can be achieved according to local regulations and operator policy to deliver emergency support from the system. Limited Service State includes support for emergency sessions in situations where, for example, a UE may not have UICC or the UE is in an area where it is normally restricted to make access, or it is not a member of a CSG cell it is in, or may have failed other system-level verification after successful authentication. Depending on the local regulations and operator policy configured in the local network, emergency services may be delivered to the following UE types:

1. Normal users with valid subscription only (Limited Service State access is not allowed).
2. UEs that have been authenticated successfully and have valid IMSI and certain Limited Service State access is allowed (e.g. UE without UICC is not allowed, UE in a restricted location).
3. UEs with valid IMSI but authentication is optional and certain Limited Service State access is allowed (e.g. UE without UICC not allowed).
4. All UEs, with or without UICC allowed, IMEI used if not valid IMSI (e.g. no UICC), and IMEI is retained by the network for operational use purposes and certain applicable Limited Service State access is allowed (e.g. UE without UICC allowed).

The nature of emergency services requires that in the case of roaming, the packet-switched domain must utilize local breakout and select a PDN GW in the VPLMN. Figure 11.13 depicts a complete but simplified architecture for IMS-based emergency architecture using EPC for LTE access. The details of the IMS architecture components are not discussed further in this section, but readers should be familiar with the IMS aspects in Section 11.2 and those interested in further details may consult 3GPP specifications such as TS 23.167 and TS 23.228.

Since the actual emergency session/service is provided by the IMS, the same principle also applies to the IMS domain as required and specified in 3GPP TS 23.167. Since Release 11, 3GPP emergency session support has included additional media support other than voice (i.e. real-time video (simplex, full duplex), synchronized with speech if present; session mode text-based instant messaging; file transfer; video clip sharing, picture sharing, audio clip sharing). The most efficient way to achieve prioritized and media-specific support is to use the PCC architecture and control the

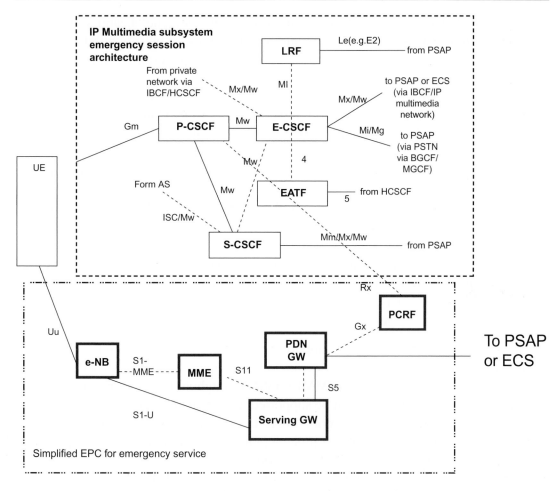

Figure 11.13: Simplified 3GPP E-UTRAN/IMS Architecture.

EPS bearers via PCC and IMS. 3GPP has adopted this principle and thus, for IMS emergency session support using EPC emergency bearers, PCC is mandatory in the network. Other aspects that are different compared to a non-emergency bearer service are as follows:

- During the mobility management procedure (e.g. Attach, Tracking Area Update) the MME needs to indicate to the UE whether emergency bearer services are supported by the network.
- eNB broadcasts support of UICC-less access for emergency services.
- A default configuration profile in MME supporting the emergency bearer profile may contain a dedicated PDN GW for emergency services or an emergency APN, which leads to a specific PDN GW.

- Additionally, the MME may be configured when an emergency bearer service can be supported or when authentication/authorization fails, or be supported in areas where the UEs are not allowed to make normal calls (e.g. certain Limited Service States).
- When a request for a PDN connection is for emergency access, the UE indicates this to the network via an indication in the Attach or PDN Connection request. If any APN is provided by the UE in this procedure the network ignores it.
- A special establishment cause code at the Access level (e.g. RRC signaling in 3GPP accesses) indicates an emergency session, which allows (where applicable) higher priority access for the UE.
- It is the network and specifically the PCRF and the PDN GW that ensures that no UE requested bearer modification or non-emergency bearers are allowed on an emergency PDN connection.
- The MME/SGSN ensures that various mobility-related restrictions such as CSG verification, handover restrictions, etc. do not affect any ongoing emergency service.
- At any given time, a UE is only allowed to have one active emergency PDN connection to an emergency APN.

To enable consistent and ubiquitous support of emergency services throughout a PLMN, it is expected that an operator deploying emergency support in the EPC should upgrade the core network nodes like MME/SGSN throughout the network so that handover and session continuity can be supported during emergency calls. In order to support access from UEs that are not allowed to normally attach to the network, support for an Emergency Attachment procedure has been introduced where the network assumes that the UE does not have proper credentials to access the network and, in this case, is able to only access emergency services via IMS. PCC and MME/SGSN are responsible, ensuring that no fraudulent use of this emergency access can occur by the UE. Indicating in the Attach that the request is for emergency access performs the Emergency Attachment procedure. This information instructs the MME to bypass procedures for subscription handling, authorization and authentication, and any other verification that may prevent a user under normal conditions to be denied access to the system. Note that in order for a UE to initiate Emergency Attachment, the broadcast message destined for the UE from that PLMN must indicate support for this service. The MME and PCC function then uses the default profile information to set up appropriate emergency bearer(s) for the user. The UE needs to detach from the network before it can initiate a normal attachment procedure in the network again to retain services. The network (e.g. PDN GW) also maintains a timer to disconnect any bearers that may have remained from an emergency bearer service after the timer's expiration.

Table 11.2 lists the emergency-related data configured in an MME that support emergency bearer service.

Table 11.2: MME Data Configured for Emergency Bearer Service

Field	Description
Emergency Access Point Name (em APN)	A label according to DNS naming conventions describing the access point used for Emergency PDN connection (wild card not allowed).
Emergency QoS Profile	The bearer level QoS parameter values for Emergency APN's default bearer (QCI and ARP). The ARP is an ARP value reserved for emergency bearers.
Emergency APN-AMBR	The Maximum Aggregated uplink and downlink MBR values to be shared across all Non-GBR bearers, which are established for the Emergency APN, as decided by the PDN GW.
Emergency PDN GW Identity	The statically configured identity of the PDN GW used for emergency APN. The PDN GW identity may be either an FQDN or an IP address.
Non-3GPP HO Emergency PDN GW Identity	The statically configured identity of the PDN GW used for emergency APN when a PLMN supports handover to non-3GPP access. The PDN GW identity may be either an FQDN or an IP address. The FQDN always resolves to one PDN GW.

One of the key aspects for the terminals would be to deal with the choice of whether to select CS domain or EPC/IMS-based emergency services since we have to deal with the complexity of features like CS Fallback and SRVCC and coexistence with the pure CS domain in UTRAN access. In this regard, a number of factors are involved in making the final decision. The decision is based, for example, on terminal capabilities, initial current status/conditions of the UE in the serving network, network support for IMS and PS emergency services, and type of emergency that would be best suited to use, as well as operators' policy as dictated by the local regulations. Table 11.3 provides an overview of the domain selection rules for emergency session attempts for UTRAN and E-UTRAN.

IMS emergency sessions are supported from E-UTRAN and UTRAN accesses. To perform SRVCC of the IMS emergency session in E-UTRAN or HSPA to the CS domain (e.g. GERAN or UTRAN), the IMS emergency session needs to be anchored in the serving IMS (i.e. in the visited PLMN when roaming), as detailed in 3GPP TS 23.167.

The E-UTRAN/UTRAN initiates the SRVCC procedure for regular voice over IMS session. The MME/SGSN is aware of the emergency session status and sends an indication to the MSC Server enhanced for SRVCC, informing of the nature of the call. The MSC Server then initiates the IMS service continuity procedure with the locally configured E-STN-SR to the serving IMS. Since we have not delved into the details of aspects of IMS in this book, we recommend interested readers look up the appropriate 3GPP specifications given in the References section of this book.

Table 11.3: Domain Selection Rules for Emergency Session Attempts for UTRAN and E-UTRAN as Specified in 3GPP TS 23.167

	CS Attached	PS Attached	VoIMS	EMS	First EMC Attempt	Second EMC Attempt
A	N	Y	Y	Y	PS	CS if available and supported
B	N	Y	N	Y	PS or CS if the emergency session includes at least voice. PS if the emergency session contains only media other than voice.	PS if first attempt in CS CS if first attempt in PS
C	N	Y	Y or N	N	CS if available and supported and if the emergency session includes at least voice.	
D	Y	N	Y or N	Y or N	CS if the emergency session includes at least voice. PS if available and EMS is "Y" and emergency session contains only media other than voice.	PS if available and EMS is "Y"
E	Y	Y	Y	Y	If the emergency session includes at least voice, follow rules in TS 22.101, which say to use the same domain as for a non-EMC. PS if the emergency session contains only media other than voice.	PS if first attempt in CS CS if first attempt in PS
F	Y	Y	Y or N	N	CS if the emergency session includes at least voice.	
G	Y	Y	N	Y	CS if the emergency session includes at least voice. PS if the emergency session contains only media other than voice.	PS

EMC = Emergency Session.
VoIMS = Voice over IMS over PS sessions supported as indicated by IMS Voice over PS session as defined in TS 23.401 and TS 23.060.
EMS = IMS Emergency Services supported as indicated by Emergency Service Support indicator as defined in TS 23.401 and TS 23.060.

The example session flows in Figures 11.14 and 11.15 show some of the IMS/CS interactions. When handover of the emergency session has been completed, the MME/SGSN or the MSC Server may initiate location continuity procedures for the UE as described in Chapter 13.

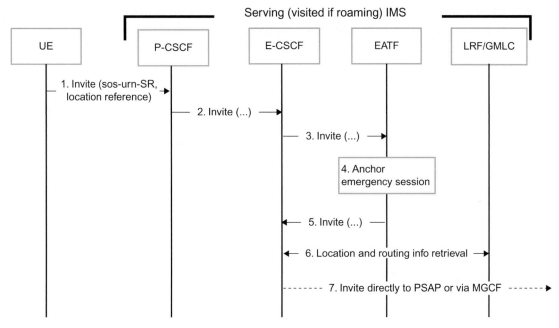

Figure 11.14: UE Initiated IMS Emergency Session Anchored in the Serving IMS.

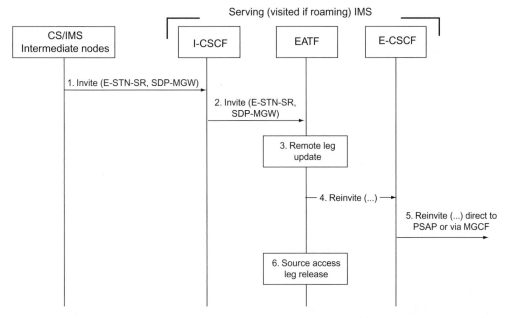

Figure 11.15: SRVCC for IMS Emergency Session with E-STN-SR.

Inter-PLMN mobility during an ongoing emergency session is not supported in the specifications today due to the possible complexity of multiple operators and/or multiple regulatory boundaries that may be encountered in such scenarios.

11.12 Multimedia Priority Service (MPS)

Multimedia Priority Service has been developed to provide a special high-priority access with a very selective group of users in mind and applies to the end-to-end session. In the context of this book, we focus on the E-UTRAN/EPC and IMS-based priority services. Although priority services are available in 2G and 3G networks for the CS domain, we do not go into the details of the CS Priority functions. In the case of CS Fallback of priority services, a user's MPS subscription information allows for transfer of a Priority CS call to GERAN or UTRAN CS access according to the requirements of the target system itself. The target users are usually authorized government, security, and special service users who need to be able to maintain communications via public mobile networks during special situations where public safety and security may be at risk. Such assignment is only allowed by regional/national authorities and is done through the mobile operator subscription mechanism with support for a special subscription profile for Multimedia Priority Service (MPS). In reality the benefit of an MPS user is seen during adverse situations like natural disasters, security threats, etc. coinciding with possible failure of any possible private networks that may be otherwise used for priority/emergency communication by authorities. Since reserving a certain network capacity for this type of usage takes away possible service availability from regular users for that PLMN operator, there may be a limit depending on the situation and regional/local regulations on how much of the traffic volume can be reserved for MPS usage.

So, how does an MPS service user benefit from a 3GPP system compared to a normal user? MPS is intended for Voice, Video, and Data bearer services in the packet-switched (PS) domain and the IP Multimedia Subsystem (IMS) and it needs to be invoked on a per-session basis. A system that supports MPS is able to provide end-to-end priority access to that MPS service user according to the level of priority assigned by the appropriate authority in situations like congestion or special disaster events. This is indicated by the MPS priority level in the user's MPS subscription profile in the HSS and downloaded during mobility management procedures in the MME. This then provides preferential treatment for access and core network resources associated with the session (i.e. signaling- and media bearer-related resources within an operator domain like in 3GPP) when possible and is also supported across multiple operator domains when the session is destined towards, for example, PSTN (e.g. end-to-end priority). MPS is based on the ability to invoke, modify, maintain, and release sessions with priority even at the expense of other normal users if needed, and deliver the priority media packets under network congestion conditions. Support of MPS in a roaming situation requires necessary regulations in place and the networks' ability to provide MPS.

The ability to support the priority service in the EPC and in the IMS relies on the MPS setting in the user's subscription profile in the HSS, as well as the priority level setting, which may also be configured via SPR and applied through PCRF. During the development of the MPS feature in 3GPP, UDR work was still in progress. As such, no specific or explicit mention of UDR instead of SPR in relation to MPS has been made in the specification. But according to the authors' understanding, UDR should be applicable for MPS where UDR provides SPR function.

In order to support priority access over the radio access (e.g. to control originating attempts or paging), the system requires a specific Access Class to be assigned to the user via the USIM configuration (i.e. usually, of 15 access classes, 10 are reserved for normal users and five are reserved for high-priority users). Operators can, under special conditions, prevent even emergency access attempts and allow only specific MPS users to access the network. 3GPP specifications provide Access Class information guidelines as shown in Table 11.4 (for more detailed information, see 3GPP TS 22.011). Note, though, that this table is a simplification of a very complex selection mechanism for 3GPP radio accesses and many additional variables and settings need to be taken into account for the specific PLMN in question to make a final assessment.

An MPS subscription includes an indication in the HSS for support of EPS bearer priority service, IMS priority service, and CS Fallback priority service for the user. The priority level in the subscription also indicates the appropriate level that is applied for the EPS bearer and IMS.

In the case of SRVCC from E-UTRAN to GERAN/UTRAN (described in Section 11.7.2), for MPS services using IMS, it is provided according to the priority-level mapping between IMS and target CS domain during the establishment of the CS call in the 2G/3G system. In the source network, MME indicates the priority level accordingly for the MSC Server providing SRVCC during the session setup. MME detection is based on the ARP associated with the EPS bearer used for IMS signaling. The priority indication corresponds to the ARP information element. If SRVCC with priority is supported, and the MSC Server receives the priority indication in the SRVCC PS to CS Request, the MSC Server/MGW sends this priority indication to the Target MSC. The MSC Server maps the ARP to the priority level, pre-emption capability/vulnerability for CS services based on local regulation or operator settings. If the MSC Server indicated priority, the target Radio Access (e.g. BSS or

Table 11.4: Access Class Setting

Class	Reserved Usage	Applicable To
0–9	Normal users	Home and Visited PLMNs Depending on the Class in question
10	In addition to other Class, this indicates whether or not network access for Emergency Calls is allowed for UEs with access classes 0–9 or without an IMSI. For UEs with Access Classes 11–15, Emergency Calls are not allowed if both "Access Class 10" and the relevant Access Class (11–15) are barred. Otherwise, Emergency Calls are allowed.	
11	For PLMN Use	Home PLMN only if the Equivalent HPLMN(EHPLMN) list is not present or any EHPLMN
12	Security Services	Home PLMN and visited PLMNs of home country only. For this purpose the home country is defined as the country of the MCC part of the IMSI
13	Public Utilities	Home PLMN and visited PLMNs of home country only. For this purpose the home country is defined as the country of the MCC part of the IMSI
14	Emergency Services	Home PLMN and visited PLMNs of home country only. For this purpose the home country is defined as the country of the MCC part of the IMSI
15	PLMN Staff	Home PLMN only if the EHPLMN list is not present or any EHPLMN

RNC) allocates the radio resource based on the existing CS procedures with priority indication. The MSC Server includes priority indication (received from the MME) to the IMS and the IMS entity handles the session transfer procedure with priority. The mapping of the priority level is based on operator policy and/or local configuration, and the IMS priority indicator should be the same as for the original IMS created over PS. Note that successful continuation of the MPS services during the SRVCC procedure relies on the fact that the target system supports CS priority.

MPS can be on demand or not, based on regional/national regulatory requirements and operator settings. On-demand service is based on Service User invocation/revocation explicitly and is applied to the PDN connections for an APN. When an MPS user is not on demand, MPS service does not require invocation and provides priority treatment for all EPS bearers for a given Service User after attachment to the EPS network. Priority levels are not determined based on whether the setting is on demand or not, but if many MPS service users are defined as not on demand, then this may cause possible issues with the number of high-priority users in a network.

In the case of an originating MPS session, the priority is set according to the originating user's setting in the originating network. For the terminating session, the priority is set in the terminating network based on the originating user's priority setting when available in the terminating network. MPS is requested by including an MPS code/identifier in the session origination request, or optionally by using an MPS input string (e.g. for an IMS session request it can be an MPS public user identity). The charging system is populated with certain information regarding MPS user service invocation/reception and level of priority treatment.

Like emergency services, the MPS also requires the support of the PCC infrastructure, and details of PCC impacts as well as interactions, and setting of priority-related data for the bearers are further explained here, but readers should be familiar with the overall PCC infrastructure and QoS aspects for 3GPP E-UTRAN access to fully understand the relationship. MPS reuses extensively the basic PCC principles with some additional triggers to facilitate the Application Function and SPR database usage.

Figure 11.16 shows a simplified flow of an MPS IMS session setup sequence, illustrating the main aspects where the priority service impacts the procedures. It is evident that the sequence is consistent with the basic PCC procedures and follows a setup mechanism as described in Chapter 8, with the possible change of EPS bearers according to MPS priority and profile.

An originating MPS session requires setting of high-priority access for the radio resource establishment request, which is then validated by the MME by checking the user's MPS subscription profile setting against the request. The session setup procedure then continues with high priority over the necessary interfaces. Then specific ARP and QCI setting according to operator policy are established for the non-GBR bearers and the priority is set as follows:

- EPS bearers (including default bearer) are assigned ARP value settings corresponding to the priority level of the Service User.
- Setting ARP pre-emption capability and vulnerability for MPS bearers, subject to operator policies and depending on national/regional regulatory requirements.
- Pre-emption of non-Service Users over Service Users during network congestion situation, subject to operator policy and national/regional regulations.

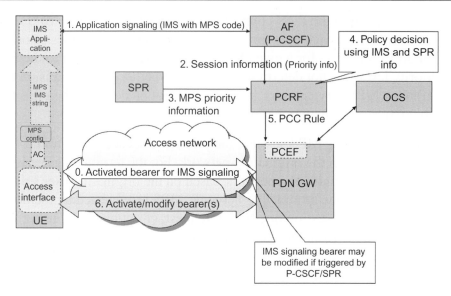

Figure 11.16: IMS MPS Originating Session.

The specifics for IMS-based MPS service support are not described further here, though it should be noted that in order to provide priority treatment end to end, EPS priority support is also required. There may be MPS-specific applications that can invoke MPS service via the PCC infrastructure by modifying the user's priority level accordingly.

The specific PCC aspects that dictate how an MPS EPS bearer service is provided by a 3GPP network depend on MPS subscription, IMS signaling priority (specific to IMS-based priority services; where an IMS signaling bearer exists, it gets priority treatment via high ARP), and Priority Indication from Application Functions providing MPS-specific service. The AF priority information represents session/application priority and is separate from the MPS EPS priority. MPS EPS priority and MPS priority level may also be configured in the SPR and then it must be ensured that the subscription profiles in HSS and SPR contain consistent data. The priority level determines the MPS user priority level received in the system.

The MPS subscription data are made available to PCRF via the Sp reference point from SPR, or it may also be received via the Ud reference point from the UDR. PCRF must subscribe to an MPS subscription information update event from SPR for Priority EPS service. This procedure is not specific to MPS but it is a mandatory requirement for the operation of MPS EPS. Dynamic invocation for MPS is provided from an AF, using the Priority indicator, over Rx. Figure 11.17 is a simplified diagram of such an invocation, though note that any AF-specific details are not subject to

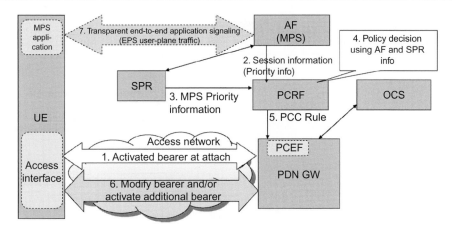

Figure 11.17: AF Triggered MPS Session Trigger and Establishment of EPS Priority Bearer Service.

3GPP standardization. Bearer modifications based on information received from the AF have always been possible in PCC, but the priority indication provided by the AF will then be used to generate appropriate QCI/ARP values. Due to the presence of MPS data, the PCRF must generate appropriate QCI/ARP values for the associated EPS bearers according to the MPS priority. If an AF triggers an MPS EPS service, the PCRF needs to then trigger appropriate PCC rule modification procedures suitable to the priority level for the service. The PDN GW will then, in turn, trigger the appropriate bearer updated procedures, if needed.

On-demand service is based on Service User invocation/revocation explicitly and applied to the PDN connections for a specific APN. This in turn requires explicit invocation/revocation via an SPR MPS user profile update by the AF (any third party AF itself and such AF-SPR interaction is not part of 3GPP standardization), which communicates with the PCC to upgrade/downgrade. When MPS EPS bearer service is not due to on-demand invocation, then the system provides priority treatment for all EPS bearers for a given Service User after attachment to the EPS network. When priority treatment is applied to certain EPS bearers, the associated EPS bearer priority needs to be adjusted to a level not associated with MPS at the end of the service.

One of the key aspects, as can be understood from the information presented here, is that PCC support for MPS includes the management of change of MPS subscription and other relevant information. When a PCRF receives notification of a change in MPS EPS priority, MPS Priority Level, and/or IMS Signaling Priority from the SPR, the PCRF makes the corresponding policy decisions (i.e. ARP and/or QCI change) and initiates the necessary IP-CAN session modification procedure(s) on Gx to apply the change. The flow shown in Figure 11.18 provides an example interaction

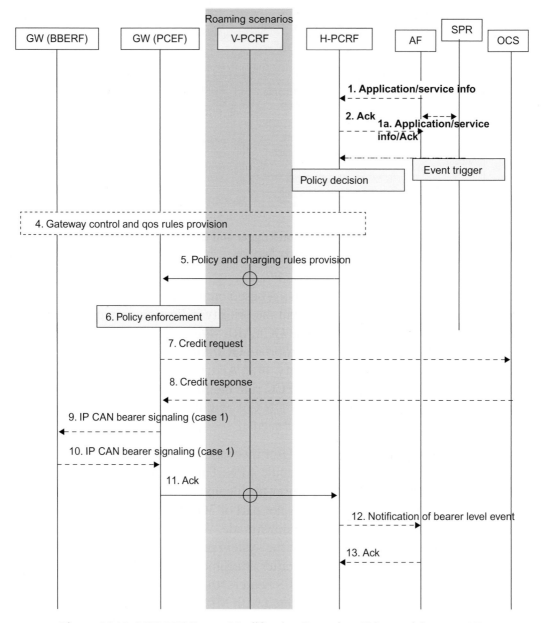

Figure 11.18: MSP EPS Bearer Modification Procedure Triggered from an AF.

for an MPS EPS bearer service. In this example, the AF may directly trigger change in PCRF or it may interact via SPR to change the Priority level according to the Priority Indicator. Specifications allow any of these mechanisms to trigger EPS bearer modifications to suit the need for the application. But 3GPP specification does not define any interface or protocol between the AF and the SPR.

An AF-based MPS may be more relevant for an MPS EPS Priority Service, especially for the on-demand case. The role of the AF and the type of signaling/interaction that may be applied is not subject to standardization. However, the PCC infrastructure provides triggers for the AF to get information/events related to the bearer associated with the application and thus makes it possible for the AF to interact with the EPC and manage the application via the Rx interface. The P-CSCF from IMS is one possible AF using the mechanism from PCC to manage EPS bearers associated with IMS applications.

CHAPTER 12

LTE Broadcasting

12.1 Background and Main Concepts

The rapid growth of mobile data usage creates a potential challenge for operators in terms of providing enough capacity and service quality for every user. It is therefore natural that various ways of optimizing the overall network resource usage are being proposed and investigated.

In those cases where the same content is to reach a large group of users, a natural proposal is to utilize broadcasting mechanisms instead of allocating dedicated network resources to every single user, hence achieving a point-to-multipoint distribution mechanism. Examples of content suitable for such a distribution scheme include broadcasting of TV channels, mass distribution of MMS, and push of software to be used for over-the-air upgrades of the terminal fleet.

IP multicast is a well-proven technology for optimization of transport network usage in fixed IP networks. Instead of sending a unique copy of the same content to every destination host, content is instead addressed to specific multicast addresses, propagated through the IP network (which needs to be multicast-enabled), and then delivered to every host that has joined a specific multicast group associated with this specific content.

Standard IP multicast, however, would not help scaling the distribution of content to end-users. There are two reasons for this. First, LTE/EPC (like the packet data solutions for GSM and WCDMA) relies on point-to-point tunneling of IP packets between the PDN GW and the end-user device. These IP packets are carried as payload inside other IP packets (and separated by a GTP header), and hence multicasting on end-user IP addresses is not possible between the end-user device and the PDN GW. Secondly, the air interface is normally the most resource-constrained interface in modern networks, and here addressing is not done based on IP addresses.

In order to create an efficient distribution mechanism all the way to end-user devices, including the air interface, 3GPP defined a broadcasting architecture for mobile systems – Multimedia Broadcast Multicast Service, or MBMS.

EPC and 4G Packet Networks.
DOI: http://dx.doi.org/10.1016/B978-0-12-394595-2.00012-8

The solution relies on broadcasting support on the air interface, usage of IP multicast for the transport of IP packets between the GW and the base stations, and specific procedures for establishing and maintaining MBMS sessions. Note that these sessions, and the content transported in the associated IP packets, are not tied to specific users as far as the Packet Core or Radio network is concerned. There is no user awareness in this part of the infrastructure. The solution for handling subscriber access to the broadcasted content is supported by other parts of the network solution but is, as said, transparent to Radio and Packet Core networks.

Figure 12.1 shows the unicast case, where an identical copy of the content is sent individually to each device, while Figure 12.2 shows the MBMS solution based on content broadcasting.

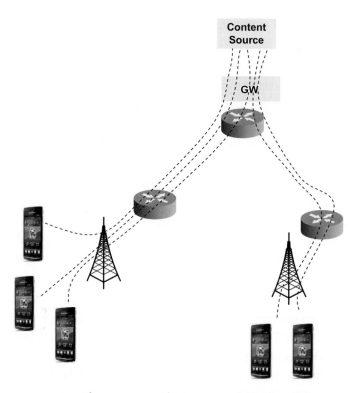

Figure 12.1: Unicast Transmission.

The benefit of MBMS is that it offers a way to optimize the network resource usage when distributing very popular content. However, MBMS does not come without penalties. For LTE, there is no dynamic control over when a cell shall transmit specific content using broadcast or unicast. This is instead semi-static and configured

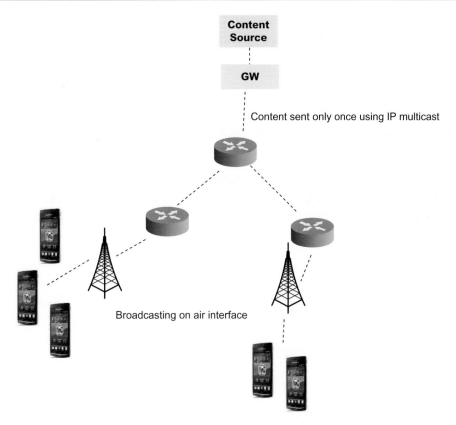

Figure 12.2: MBMS Transmission.

by the operator. This means that expensive radio spectrum might be wasted in some cells for transmission of content that no one is receiving. Operators considering deploying MBMS must therefore carefully consider if the most efficient way of using their available radio spectrum is to use unicasting or broadcasting. The conclusion will depend on factors such as spectrum availability and cost, as well as the uptake rate of services for which broadcasting can be an option.

12.2 MBMS Solution Overview

MBMS is specified in 3GPP technical specification 23.246. The two key components of the MBMS solution are the MBMS User Service and the MBMS Bearer Service.

The MBMS Bearer Service is a point-to-multipoint content distribution mechanism, realized through specific functionality in the mobile network infrastructure and the terminals. This enables an efficient distribution of content to multiple users.

When initially specified for WCDMA and GSM/GPRS access in 3GPP Release 6, two modes of MBMS Bearer Service operation were defined:

- Broadcast mode
- Multicast mode.

When MBMS was enhanced in Release 9 under the name of eMBMS "Evolved MBMS", only the Broadcast mode was included. In addition to WCDMA support LTE support was added, but eMBMS is not supported over GSM/GPRS access.

Table 12.1 gives an overview of the different variants of MBMS, including WCDMA IMB, short for "Integrated Mobile Broadcast", which defines WCDMA-based MBMS broadcasting in the TDD spectrum. The focus of this text is, however, to describe MBMS for LTE.

Table 12.1: Overview of MBMS Variants

3GPP Release	GSM/GPRS	WCDMA	LTE
Release 6	Broadcast + Multicast	Broadcast + Multicast	Not included
Release 8	As Release 6	Enhanced with IMB for TDD	Not included
Release 9+	Not included	Broadcast only	Broadcast only

In Broadcast mode, the MBMS Bearer Service has no relation to specific users – it is simply a broadcasting mechanism that is "user and device unaware". There are no user- or device-related sessions in the MBMS Bearer Service, only sessions related to different media content. This is different from the more complex Multicast mode, where MBMS UE contexts are dynamically created, and the Packet Core network tracks all devices utilizing MBMS services.

The MBMS User Service, on the other hand, is a service offered to users, or, alternatively formulated, a service used by user devices when specific end-user applications are activated. MBMS User Services are enabled by the so-called BM-SC (Broadcast Multicast Service Center), which is a logical function in the network that may be realized either as a network node itself, or as a function residing in another network node.

The BM-SC is the node that controls MBMS sessions. It initiates notifications that MBMS will be used to broadcast specific content, and it manages functionality associated with terminals that want to join the sessions, such as authentication, authorization, data encryption, and charging for the service. All interaction between the BM-SC and the terminals is handled over normal unicast (since MBMS has no uplink capabilities). This then means that the BM-SC also has a connection to the PDN GW.

Figure 12.3 shows a simplified overview of the BM-SC interaction with the mobile network infrastructure. We will come back to the details hidden inside the Mobile Network cloud later in this chapter.

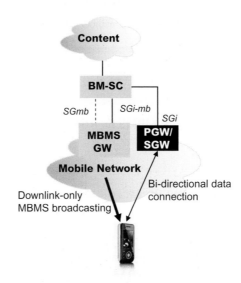

Figure 12.3: BM-SC Connection to the Mobile Network Infrastructure.

MBMS User Services rely on the transport mechanisms offered by the MBMS Bearer Services, but a major difference is that the MBMS User Service is user aware, i.e. sessions are controlled on a per-user basis.

Figure 12.4 illustrates the dependency between MBMS User Services and MBMS Bearer Services.

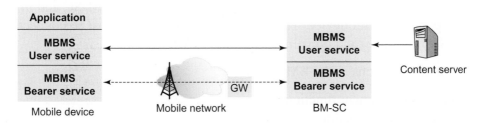

Figure 12.4: MBMS User Services and MBMS Bearer Services.

In fact, 3GPP specifies three different functional layers (Figure 12.5) associated with data transport:

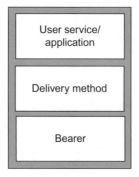

Figure 12.5: MBMS Functional Layers.

- Services and Applications
- Delivery methods (downloading or streaming)
- Bearers (unicast or MBMS).

In practice, different applications will use different combinations. As an example, downloading of a specific file to multiple devices might rely on using the downloading delivery method over an MBMS bearer, while some other services might instead utilize the streaming delivery method over a unicast bearer.

Since the topic of this chapter is MBMS, it is important to make clear that the MBMS Bearer Services can support both the streaming and downloading delivery methods.

12.3 MBMS User Services

There are multiple mechanisms involved to provide MBMS User Services. While a detailed description of the MBMS User Service mechanisms is beyond the scope of this book, we present an overview of the functionality specified to offer end-user service support in this section.

The mechanisms relating to MBMS User Services reside in the BM-SC and in the MBMS-capable end-user device. They interact via a normal unicast IP connection (over the SGi interface) and downlink over the MBMS bearers.

The three main functions in the BM-SC are:

- Service Announcement function
- Key Management function
- Session and Transmission function.

Service Announcement is a procedure carried out by the BM-SC to notify terminals that a session utilizing MBMS is about to take place. The typical use case involves announcing controlling metadata to the terminals in advance of a broadcast; however, there are also use cases where it is necessary to update the metadata during a broadcast file download delivery and, sometimes, after it has completed. 3GPP does not specify exactly how the Service Announcement takes place, but instead suggests a number of possible technologies such as using SMS, MMS, Wap Push, or HTTP-based notifications. This means that the operator needs to make sure that the terminals intended to be used for MBMS reception in the network support Service Announcements using the method implemented in the mobile network.

The Key Management function handles user service registration, including authentication and user service deregistration. It also generates encryption keys to be used to secure the MBMS transmission, and distributes these keys to the terminals and to the Session and Transmission function within the BM-SC. Details on security for MBMS, such as key generation schemes and algorithms, are beyond the scope of this book but can be found in the 3GPP technical specification 33.246.

The Session and Transmission function does exactly what its name indicates – it transfers MBMS data to terminals, either using point-to-point unicast connections (over the Gi interface), or by utilizing the MBMS Bearer Services for content broadcasting. In the case of MBMS Bearer Services being used, the BM-SC interacts with the MBMS GW over the SGmb interface.

The Session and Transmission function in the BM-SC may optionally apply encryption (using the keys mentioned above) and header compression to the MBMS data stream. It may optionally also add forward-error correction (FEC) information to the data frames to allow the receiving UEs to correct data frames that have been corrupted during transmission. Finally, the BM-SC may add time synchronization information to the data frames to ensure correct radio network operation in multi-cell environments.

Figure 12.6 outlines the functionality related to MBMS User Services inside the BM-SC and the mobile terminal. Figure 12.7 shows an example of a simplified call flow for an MBMS User Service.

1. The first step is that the MBMS Service metadata used to control the service on the terminal is announced by the BM-SC. As described above, this announcement can utilize various mechanisms. In Figure 12.6 either an existing unicast data connection can be used with http- or WAP-based notifications (option a in the figure), or SMS using control channel signaling (option c). An additional option is that existing MBMS Bearers can be used to transfer notification of new MBMS User Services (option b).

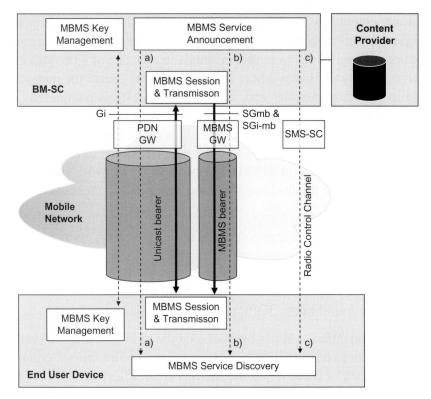

Figure 12.6: Main Functionality Related to MBMS User Services.

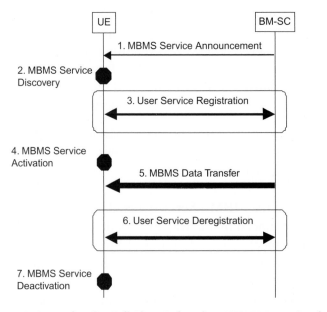

Figure 12.7: Example of a Call Flow Related to MBMS User Services.

2. The next step is the terminal itself discovering the Service Announcement and taking action based on that. No signaling to the network takes place in this step.

3. If the MBMS User Service is to be protected, a User Service Registration then takes place via an interaction over the normal data connection between the terminal and the Key Management function in BM-SC. The user is authenticated and, when authorized, encryption keys to be used for decrypting MBMS content are sent to the terminal.

4. MBMS reception is activated in the terminal.

5. Data is broadcast using the MBMS Bearer Services in the mobile network, and terminals which are within MBMS coverage and having correct decryption keys can receive and decode the content.

6. If User Service Registration was required for the specific MBMS User Service, User Service Deregistration is also required. This takes place by interaction over the normal data connection between the terminal and the Key Management function in the BM-SC.

7. MBMS reception is deactivated in the terminal.

12.3.1 Associated Delivery Procedures

Two types of associated delivery procedures are specified in 3GPP technical specification 26.346:

- File repair
- Reception reporting.

Both types are controlled by the Service Announcement metadata.

12.3.1.1 File Repair

3GPP specifies both a point-to-point and point-to-multipoint file repair service.

A terminal running a file download MBMS User Service may request the repair of missing data from the BM-SC. The file repair parameters, such as the File Repair Server Address, are provided with the metadata description in the Service Announcement. It is possible for the terminal to request either individual UDP packets or entire files to be repaired.

In order to protect the uplink and the BM-SC against overload conditions, a wait-time window is defined in the metadata controlling the service. Terminals that want to use the file repair service randomly distribute unicast file retrieval requests to the BM-SC using a wait-time window.

12.3.1.2 Reception Reporting

The reception reporting procedure allows an operator to instruct the terminal to gather statistics about the reception conditions and the Quality of Experience (QoE) of the MBMS session. This allows an operator to monitor the reception quality, the receiver group size, and possibly even coverage weaknesses. The gathered statistics are uploaded as a reception report message from the terminal.

12.4 Mobile Network Architecture for MBMS

12.4.1 Architecture Overview

If all nodes and interfaces that are not involved in the MBMS services are stripped away from a complete mobile network architecture, the remaining parts for the case of broadcasting over LTE are those shown in Figure 12.8.

As mentioned above, the BM-SC interfaces with the PDN GW for interaction with the terminals, but since this is no different from any other data connection over LTE, the PGW/SGW is shown as black in Figure 12.8 and is not described further below.

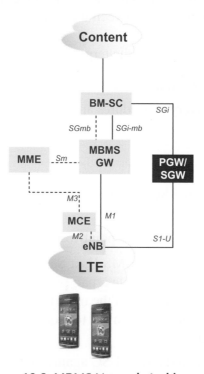

Figure 12.8: MBMS Network Architecture.

The MBMS GW interfaces the BM-SC for both signaling and data transfer, and is responsible for transmitting session data downstream to the base stations, as well as for invoking the MME in MBMS signaling to control the broadcast sessions. The MBMS GW can optionally also support charging for bearer service usage; this is in order to be able to charge content providers for network usage. (Charging of end-users for MBMS usage is not possible in the MBMS GW, since it is user unaware.)

The MME communicates with the LTE RAN in the selected service areas, relaying session control information received from the MBMS GW.

The MBMS architecture within LTE RAN consists of two logical entities – the eNodeB (the base station) and the MCE (Multi-cell/multicast Coordination Entity) interconnected using the RAN-internal M2 interface.

The reason for adding the MCE into the network is the usage of a technique called MBMS Single-Frequency Network (MBSFN), which allows for the best possible signal-to-interference ratio when the mobile devices receive MBMS broadcasting content. This is achieved by synchronization of all MBMS transmissions from a number of cells within an MBSFN Area. This is then perceived by the mobile devices as a single transmission of the same content. There may be multiple MBSFN Areas in a network, and one cell may belong to more than one MBSFN Area. One MCE coordinates and controls MBMS broadcasting resources within one MBSFN Area.

It can be noted that 3GPP specifies one architecture option where an MCE is integrated into each eNodeB. More information on this option can be found in 3GPP technical specification 36.300. This option simplifies the solution but limits the size of an MBSFN Area to only the cells that are controlled by one eNodeB. This is an RAN-internal implementation decision, and does not affect the EPC part of the MBMS solution as such.

Since the focus of this book is on EPC and not on radio technology, the Radio network architecture in the call flows of this chapter has in fact been assumed to be simplified with MCEs integrated into each eNodeB. Hence, there is no visible M2 interface within LTE RAN and we describe the M3 interface as being between the MME and the eNodeB.

The eNB receives control signaling from the MME (via the MCE) and data from the MBMS GW, and uses dedicated MBMS radio channels for broadcasting of control information and data respectively. Once MBMS has been activated in a specific cell and data is being received, this broadcasting happens regardless of whether there are any terminals in that cell that will receive content sent over MBMS or not. The behavior of the terminals with respect to MBMS is invisible to both LTE RAN and EPC. Access to MBMS content is instead controlled by the BM-SC as part of the MBMS User Services.

12.4.2 Interfaces

12.4.2.1 Control Plane
SGmb Interface

SGmb carries signaling between the BM-SC and the MBMS GW. A specific MBMS application on top of the Diameter base protocol is used together with standard Diameter AVPs (see Section 16.5 for more information on the Diameter protocol).

In normal operation, the BM-SC controls start, stop, and mid-session modifications of MBMS sessions using this interface. Figure 12.9 shows the protocol stack for the SGmb interface.

Diameter
TCP/SCTP
IP
L2/L1

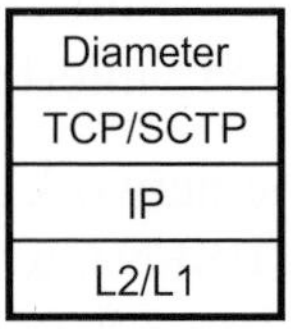

Figure 12.9: Protocol Stack for the SGmb Interface.

Sm Interface

Sm carries MBMS-related signaling between the MBMS GW and the MME. MBMS-specific signaling is realized through specific GTPV2-C messages, carried over UDP as any other GTP interface, to start, stop, and modify MBMS sessions. Figure 12.10 shows the protocol stack for the Sm interface.

GTPv2-C
UDP
IP
L2/L1

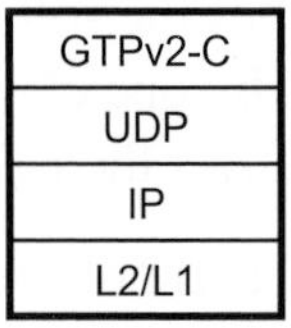

Figure 12.10: Protocol Stack for the Sm Interface.

M3 and M2 Interfaces

M3 carries MBMS-related signaling between the MME and the affected MCE and M2 between MCE and eNB. If the MCE is integrated into each eNB, only the M3 interface exists in the network solution. The 3GPP-specific M3-AP (and M2-AP) protocols are used, carried over SCTP to provide a reliable connection between the MME and the eNB. M3-AP (and M2-AP) messages are used to start, stop, and modify MBMS sessions. Figures 12.11 and 12.12 show the protocol stacks for the M2 and M3 interfaces respectively.

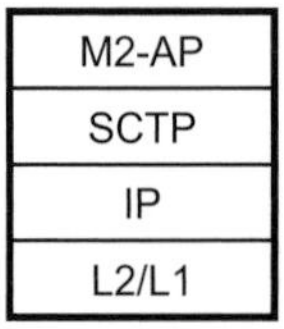
M2-AP
SCTP
IP
L2/L1

Figure 12.11: Protocol Stack for the M2 Interface.

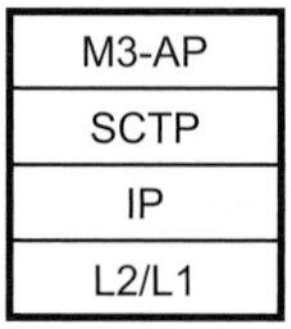
M3-AP
SCTP
IP
L2/L1

Figure 12.12: Protocol Stack for the M3 Interface.

12.4.2.2 User Plane

SGi-mb Interface

The SGi-mb interface carries user data from the BM-SC to the MBMS GW. The data frames carrying the content are carried inside the 3GPP-specific SYNC protocol (basically a set of additional headers), which enables the BM-SC to set time stamps to each data frame to ensure that MBMS data broadcasted over multiple cells is synchronized in time. Figure 12.13 shows the protocol stack for the SGi-mb interface.

Content data
SYNC
L4
IP
L2/L1

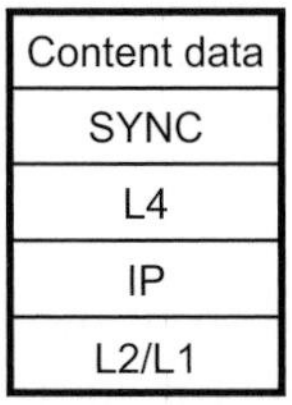

Figure 12.13: Protocol Stack for the SGi-mb Interface.

M1 Interface

The M1 interface is used for carrying MBMS data from the MBMS GW to the affected eNBs using IP multicast as the transport mechanism. In order to preserve the timing information, the MBMS GW not only leaves the data frames (received from the BM-SC over SGi-mb) intact, but also the headers assigned as part of the SYNC protocol. The MBMS GW is fully transparent to these upper protocol layers. Figure 12.14 shows the protocol stack for the M1 interface.

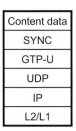

Content data
SYNC
GTP-U
UDP
IP
L2/L1

Figure 12.14: Protocol Stack for the M1 Interface.

12.5 MBMS Bearer Services

MBMS Bearer Services are functions provided by the mobile network to the BM-SC, and allow for controlling sessions and transfer of data to the terminals. For the Broadcast mode of MBMS, the MBMS Bearer Services are unrelated to and unaware of specific users or terminals.

In normal operation the BM-SC uses three different procedures towards the mobile network for control of broadcast sessions:

- Session Start
- Session Update
- Session Stop.

In addition, the connection between the BM-SC and the mobile network is used to transfer the data itself. This data is then broadcast over a specified area by utilization of mechanisms in the mobile network, which is totally unaware of whether there are users receiving the data or not.

12.5.1 Session Start

The Session Start procedure is initiated from the BM-SC, triggering all involved nodes to take necessary actions to allow for starting MBMS transmission. The call flow is shown in Figure 12.15.

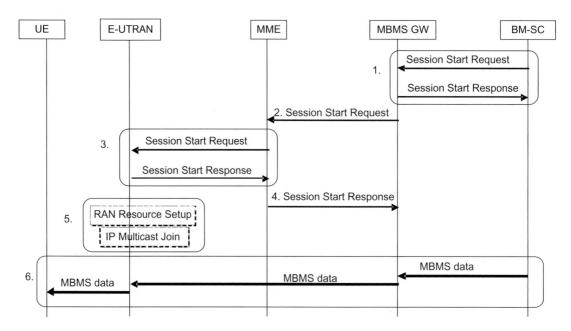

Figure 12.15: MBMS Session Start Procedure.

1. The procedure starts with the BM-SC sending a Session Start Request message to the MBMS GW that is configured as the default GW in the BM-SC. This message may optionally be sent to multiple MBMS GWs in parallel depending on the configuration in the BM-SC. The Session Start message contains information

uniquely identifying and specifying the upcoming MBMS session, including which area that will be affected, what QoS to use, etc. It also contains information for the MBMS GW to identify which MMEs to include in the next step of the procedure. The MBMS GW then creates an MBMS Bearer context, allocates IP multicast transport resources for the upcoming session, and responds back to the BM-SC with a Session Start Response to confirm that it is ready to receive MBMS data over the SGi-mb interface.

2. The MBMS GW then sends Session Start Request messages to all MMEs listed by the BM-SC in step 1. The contents of these messages again uniquely identify and specify the upcoming MBMS session. It now also includes the IP multicast address for the base stations to use for reception of MBMS data.

3. Each MME creates an MBMS Bearer context and in turn sends a Session Start Request message either to all connected LTE base stations (eNB) or to the base stations matching a Service Area Identity filter (an addition specified by 3GPP in Release 11). Each affected eNB creates an MBMS Bearer context and responds to the MME with a Session Start Response message. It can be noted that 3GPP specifications also allow for the Session Start Response message to be sent by the MME directly after receiving a Session Start Request from the MBMS GW.

4. Each MME then stores a list of eNBs associated with the specific MBMS Bearer Context, and sends a Session Start Response message to the MBMS GW.

5. MBMS radio resources are established by each affected eNB. Details are beyond the scope of this book. Furthermore, each affected eNB joins the applicable IP multicast group utilizing information on the IP multicast address received from the MBMS GW via the MME. The base stations are now ready to receive data.

6. The MBMS content is sent by the BM-SC to the MBMS GW and then uses IP multicast transmission to each eNB that has joined the applicable IP multicast group. The content is broadcast in the applicable cells using dedicated MBMS radio channels, and is received and decoded only by the user terminals that are authorized to join the specific MBMS service.

12.5.2 Session Stop

The Session Stop procedure is naturally used to terminate an ongoing MBMS session. The call flow is shown in Figure 12.16.

1. The BM-SC starts the procedure by sending a Session Stop Request message to the MBMS GW (identifying the session to be stopped). The MBMS GW responds to the BM-SC with a Session Stop Response message and releases the bearer resources associated with the specific MBMS session.

2. The MBMS GW sends a Session Stop Request message to each affected MME, which in turn respond to the MBMS GW with a Session Stop Response message.

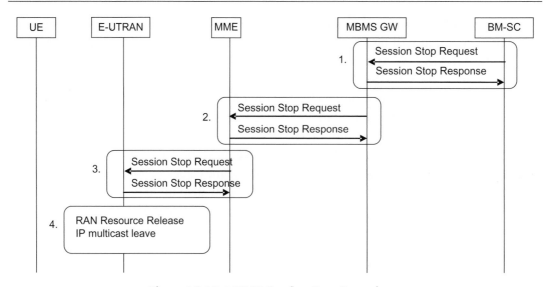

Figure 12.16: MBMS Session Stop Procedure.

3. The MME sends a Session Stop Request message to all affected eNBs. Each eNB responds back with a Session Stop Response message and the MME releases the applicable MBMS bearer context.
4. The eNBs release the MBMS radio resources associated with the stopped session and leave the IP multicast group using standard IETF mechanisms.

12.5.3 Session Update

When there is a need to modify some of the attributes related to an ongoing MBMS Session, the Session Update procedure is used. There are three use cases defined by 3GPP for which this procedure can be used:

- Changing the MBMS service area
- Changing the set of MMEs
- Removing or adding a new radio access (as MBMS is also specified for WCDMA).

Typically this procedure is used when the MBMS service area is to be modified, through either adding or removing cells, and potentially also impacting the set of affected MMEs. The other use case on adding or removing MBMS over WCDMA is not applicable in an LTE-only solution as described here. The call flow for the main case (changing the MBMS Service Area) is shown in Figure 12.17.

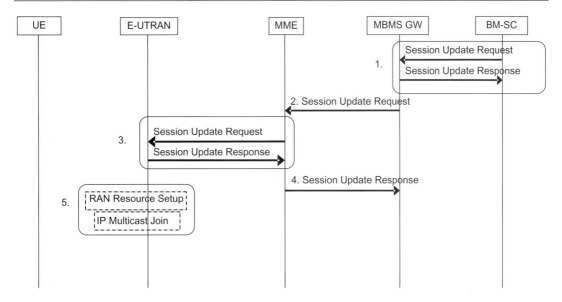

Figure 12.17: MBMS Session Update Procedure.

1. The procedure starts with the BM-SC sending a Session Update Request message to the MBMS GW. This message contains an updated MBMS Service Area and a list of MMEs for the MBMS GW to use, a list which potentially may be updated. The MBMS GW responds back to the BM-SC with a Session Update Response message.

2. The MBMS GW then takes one of three actions based on a comparison of the old and new lists of affected MMEs:
    - If there are any MMEs added to the new list, a Session Start Request message is sent to each of these as described in Section 12.5.1.
    - If there are MMEs removed in the new list, a Session Stop Request message is sent to each of these, as described in Section 12.5.2.
    - For all other MMEs, a Session Update Request message is sent (this is what is shown in the remainder of this call flow).

3. Each MME then sends a Session Update Request message to all connected eNBs. Each eNB affected by the session updates the MBMS context as necessary to reflect the updated Service Area, and responds back to the MME with a Session Update Response message.

4. The MME updates the session context and responds back to the MBMS GW using a Session Update Response message.

5. LTE RAN radio resources are added/modified/removed to reflect the updated Service Area and, if needed, new base stations join the IP multicast group (as described for the Session Start procedure) while removed base stations will leave the IP multicast group (as described for the Session Stop procedure).

CHAPTER 13

Positioning

Positioning is the process of determining the geographical location of a device – such as a mobile phone, laptop or tablet computer, or navigation or tracking equipment. Once the coordinates of a device have been established, they can be mapped to a location – such as a road, a building, a park, or an object – and then delivered back to the requesting service. The mapping function and the delivery of location information are part of location services (LCS) – which, for example, Emergency Services depend on. Customer services that offer added value by being location aware are known as location-based services (LBSs). In addition to supporting customer services for users, location services can also be used to optimize network performance and to enhance automated services such as network self-learning and self-optimization.

Some common examples of LBSs include Emergency Services, localized weather forecasts, targeted advertising, and applications that can position the nearest bus stop, or find the location of an object – such as a subscriber's car keys.

Positioning in wireless networks is dependent on the mobility of users and the dynamic nature of both the environment and radio signals. Positioning QoS is typically defined in terms of accuracy, confidence level, and the time it takes to obtain a positioning result.

Users naturally presume that applications will work regardless of where they are and whether they are in a fixed location or on the move. They expect the same level of performance whether they are indoors at home or at work, outdoors in a rural or urban environment, or traveling.

Different applications may also require varied levels of positioning accuracy. LTE supports a high level of application-adaptive requirements created by more advanced user needs and application development. To meet the requested positioning QoS, the best mix of positioning technologies can be selected for each case.

Operators are responsible for compliance with regulatory standards established to ensure reliable positioning in emergency situations (for example, E911 in North America and E112 in Europe).

EPC and 4G Packet Networks.
DOI: http://dx.doi.org/10.1016/B978-0-12-394595-2.00013-X
Copyright © 2013, 2009 Elsevier Ltd. All rights reserved.

The current Wireless E911 Location Accuracy Requirements specify that carriers must, over time, satisfy these standards at either a county level or at a Public Safety Answering Point (PSAP) geographical level, as well as being able to provide confidence and uncertainty data for all E911/E112 calls. The main challenge here is to achieve the required levels of accuracy for indoor locations.

This chapter provides an overview of the EPS positioning solution and is partly based on Ericsson White Paper on LTE Positioning (2011). For additional details beyond the scope of this book, see Ericsson White Paper on LTE Positioning (2011), 3GPP TS 23.272, and 3GPP TS 36.305.

13.1 Positioning Solutions

GPS-capable devices are increasing user expectations and often meet stringent positioning QoS requirements. While many new mobile devices are likely to be equipped with GPS receivers, there are devices that are not. In addition, no single positioning method, including GPS, works well in all environments. GPS, for example, fails to provide a reasonable level of positioning accuracy in indoor and urban canyon environments. In today's world, where more than 50% of mobile phone calls are made from indoor locations, there is a need for positioning methods that can provide the required level of accuracy in all environments.

Rural deployment of base stations is costly; as a result, the distances between sites in rural networks tend to be long, cells tend to be larger, and there are fewer detectable neighboring cells. Accurate positioning in rural areas is subsequently more difficult owing to the longer distances involved and larger coverage areas. Due to the maximum-power limitation of terminals, network-based positioning is both more coverage-limited and less efficient from a battery perspective than terminal-assisted positioning. To enhance positioning accuracy for all types of environments, LTE uses complementary positioning methods. The main location technologies used are Observed Time Difference of Arrival (OTDOA) and Assisted Global Navigation Satellite System (A-GNSS), due to the high level of accuracy these methods can achieve with no requirement for additional radio network equipment (where OTDOA is used for indoor locations and A-GNSS for outdoor environments) as further described in Section 13.3. To improve positioning results, these methods can be complemented with additional technologies such as self-learning fingerprinting or proximity location. The use of a combination of technologies can also enhance positioning performance.

Individual positioning techniques do not perform with the same success rates in all environments, so they should be used to complement each other rather than used as standalone technologies. Integrated positioning solutions that effectively combine different positioning techniques can meet a wide range of accuracy and performance requirements, while allowing efficient use of network and device resources. Such solutions must operate well in synchronous and asynchronous networks, and with FDD and TDD.

The approach used in integrated positioning solutions should also be applied to terminal-assisted, terminal-based, and network-based positioning. Terminal-assisted positioning can make use of terminal measurements together with the available knowledge about the radio environment accumulated in the network. Terminal-assisted positioning also has advantages over standalone network-based positioning, which relies on network measurements and network knowledge, is constrained by the maximum terminal power, and cannot benefit from measurements at the actual user location.

Table 13.1 illustrates a set of positioning methods available in LTE. The set of methods operates as a unit, responding to network capability and architecture, meeting positioning QoS demands, and taking into account the radio propagation environment. The methods discussed differ in their typical accuracy ranges, and all of them may be used within a hybrid technique. Each method calculates positions using different measurements and signals from different sources. For example, satellite-based measurements enable the best performance for terminals with GNSS-capable receivers in suburban and rural areas. A method based on Time Difference of Arrival (TDOA), such as OTDOA, may be a better choice for indoor locations and urban canyon environments, while Adaptive Enhanced Cell Identity (AECID) is a good fit in all environments and is especially suitable for terminals that are not equipped with GNSS receivers. More information on the individual positioning methods is presented in chapter 13.3 below.

Table 13.1: Positioning Methods in LTE

Positioning Method	Environment Dependency	Response Time in LTE	Horizontal Uncertainty	Vertical Uncertainty
CID proximity location	No	Very low	High	N/A
E-CID	No	Low	Medium	N/A
E-CID/AoA	Rich multipath	Low	Medium	N/A
RF fingerprinting	Rural (audibility)	Low-medium	Low-medium	Medium
AECID	No	Low	Low-medium	Medium
UTDOA	Suburban/rural (audibility)	Medium	<100 m	Medium
OTDOA	Rural (audibility)	Medium	<100 m	Medium
A-GNSS	Rural (audibility)	Medium-high	<5 m	<20 m

The ideal positioning system should be self-learning and environmentally adaptive, capable of building up information databases that store actual observations, and employ smart data analysis mechanisms. By using more measurements, new ways of collecting them, and more advanced algorithms, LTE has the capability to support flexible self-learning and network-adaptive positioning systems.

13.2 Positioning Architecture and Protocols

LTE positioning functionality is distributed across LTE radio nodes, eNodeBs, and the positioning node. The eNodeBs, for example, ensure proper configuration of positioning reference signals, provide information to the Enhanced Serving Mobile Location Center (E-SMLC), enable UE inter-frequency measurements if necessary, and provide network-based measurements on request from the E-SMLC.

The positioning node determines which positioning method to use, builds up and provides assistance data to facilitate calculating measurements, collects the necessary measurements, works out the position, and communicates the result to the requesting client.

Positioning is supported over both the control and user planes. In the control plane, a positioning request is always sent via the Mobility Management Entity (MME) to the E-SMLC, and in the delivery of a response, including positioning data, the Gateway Mobile Location Center (GMLC) controls user authorization and charging information. In the user plane, positioning information is exchanged over data channels using the Secure User Plane Location (SUPL) protocol in the application layer.

LTE positioning architecture contains three key functions: the LCS client, LCS target, and LCS server. The LCS server is a physical or logical entity that manages positioning for an LCS target device. It collects measurements and other location information, assists the UE in calculating measurements when necessary, and estimates the LCS target location. An LCS client is a software and/or hardware entity that interacts with an LCS server to obtain location information for LCS targets and may reside in the LCS target. An LCS client sends a request to the LCS server to obtain location information; the LCS server processes the request and sends the positioning result and, optionally, a velocity estimate back to the LCS client. A positioning request can originate from an LCS client in either the UE or the network.

LTE operates two positioning protocols via the radio network: LTE Positioning Protocol (LPP) and LPP Annex (LPPa). LPP is a point-to-point protocol for communication between an LCS server and an LCS target device, and is used to position the device. LPP can be used both in the user plane and control plane, and multiple LPP procedures are allowed in series and/or in parallel (reducing latency). LPPa is a communication protocol between an eNodeB and an LCS server for control-plane positioning – although it can assist user-plane positioning by querying eNodeBs for information and measurements. The SUPL protocol is used as transport for LPP in the user plane.

Figure 13.1 illustrates LTE's high-level positioning architecture, where the LCS target is a terminal and the LCS server is an E-SMLC or an SLP.

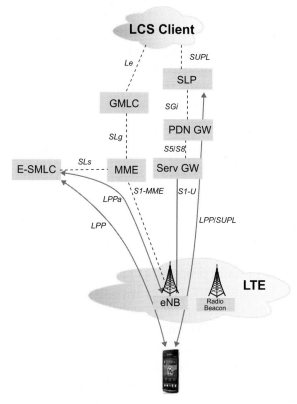

Figure 13.1: Overview of the EPS Positioning Architecture.

Deploying additional positioning architecture elements, such as radio beacons, can enhance the performance of individual positioning methods. Deploying extra radio beacons and, for example, using proximity location techniques is a cost-efficient solution that can significantly improve positioning performance both indoors and outdoors.

13.3 Positioning Methods

LTE networks support a range of complementary positioning methods. The basic method – Cell ID (CID) – utilizes cellular system knowledge about the serving cell of a specific user; the user location area is thus associated with the serving CID. Support for this method has been mandatory since Release 8, and the following methods became available with Release 9:

- Enhanced Cell ID (E-CID) – UE-assisted and network-based methods that utilize CIDs, RF measurements from multiple cells, timing advance, and Angle of Arrival (AoA) measurements.
- OTDOA – UE-assisted method based on reference signal time difference (RSTD) measurements conducted on downlink positioning reference signals received from multiple locations, where the user location is calculated by multilateration.
- A-GNSS – UE-based and UE-assisted methods that use satellite signal measurements retrieved by systems such as Galileo (Europe) and GPS (US). LTE supports positioning with existing satellite systems and will develop as new satellite systems become available.

The following commonly known methods do not require additional standardization and are also included in LTE Release 9:

- RF fingerprinting, a method of finding a user's position by mapping RF measurements obtained from the UE onto an RF map, where the map is typically based on detailed RF predictions or site surveying results.
- AECID, a method that enhances the performance of RF fingerprinting by extending the number of radio properties that are used, where at least CIDs, timing advance, RSTD, and AoA may be used in addition to received signal strengths, and where the corresponding databases are automatically built up by collecting high-precision OTDOA and A-GNSS positions, tagged with measured radio properties.
- Hybrid positioning, a technique that combines measurements used by different positioning methods and/or results delivered by different methods.

Uplink TDOA (UTDOA), an uplink alternative method to OTDOA, is standardized in Release 11. UTDOA utilizes uplink time of arrival (ToA) or TDOA measurements performed at multiple receiving points. The measurements will be based on Sounding Reference Signals (SRSs).

For some environments, positioning based on measurements of radio signals can be challenging. Alternative methods, such as enhanced proximity location, can be applied as complements to CID-based methods to improve positioning results. A proximity method may, for example, utilize knowledge about the set of detected networks or radio devices. As civic address information associated with a cell or network node is both comprehensible by a person and the native format for PSAPs, a proximity method may use this information instead of geographical coordinates.

CID is the fastest available measurement-free positioning method that relies on the cell ID of the serving cell – typically available information – and the location associated with that cell, but its accuracy depends on the size of the serving cell. A-GNSS, including A-GPS, is the most accurate positioning method in satellite-friendly environments. The most accurate terrestrial method is OTDOA, which is based on downlink measurements of positioning reference signals transmitted by radio nodes such as eNodeBs or beacon devices. OTDOA and A-GNSS provide highly accurate positioning in most parts of a cellular network and for most typical environments. UTDOA performance may approach that of OTDOA in some deployment scenarios that are not UL-coverage-limited, assuming the use of enhanced UL receivers. To improve positioning in challenging radio environments, these methods can be complemented, for example with hybrid positioning, proximity location, and new positioning methods in the middle accuracy range, including AoA, RF fingerprinting, and AECID. Note that the AECID method utilizes a wider set of measurements than the RF fingerprinting method – including, for example, timing measurements – meaning that AECID is significantly less subject to environment limitations. In the future, as networks become denser, the role of proximity methods will become important.

13.4 Position-Reporting Formats

Seven position-reporting formats, each associated with a Geographical Area Description (GAD) shape, are supported in 3GPP for LTE, UMTS, and GSM. All seven formats can be used for positioning, although certain formats may be more typically associated with particular positioning methods.

13.5 EPS Positioning Entities and Interfaces

This section provides some further details on the EPS positioning related nodes and interfaces shown in Figure 13.1.

13.5.1 GMLC

The Gateway Mobile Location Center (GMLC) is the first node an external LCS client accesses in a Mobile Network. The GMLC may request routing information from the HSS. It supports routing of positioning requests and responses. The GMLC also performs authorization and checks the subscriber's privacy profile.

- The "Requesting GMLC" is the GMLC, which receives the request from the LCS client.
- The "Visited GMLC" is the GMLC, which is associated with the serving node of the target mobile.

- The "Home GMLC" is the GMLC residing in the target mobile's home PLMN, which is responsible for the control of privacy checking of the target mobile.

13.5.2 E-SMLC

The Enhanced Serving Mobile Location Center (E-SMLC) supports the LCS services function and coordinates positioning of a UE that is attached to LTE. It calculates the final location and velocity estimate and estimates the achieved accuracy.

13.5.3 SLP

The SUPL Location Platform (SLP) supports user-plane positioning and is defined by Open Mobile Alliance (OMA) in OMA AD SUPL: "Secure User Plane Location Architecture" (see References).

13.5.4 Le Interface

The Le interface is used by an external application (LSC client) to send location requests to a GMLC.

13.5.5 SLg Interface

SLg is the interface used by the GMLC to convey a location request to the MME. The interface is also used by the MME to return location results to the GMLC.

13.5.6 SLs Interface

SLs is the interface between the MME and the E-SMLC. The SLs interface is used to convey location requests and reports between the MME to the E-SMLC. It is also used for tunneling measurement requests from the E-SMLC to the eNodeB.

13.5.7 LTE Positioning Protocol (LPP)

The LTE Positioning Protocol (LPP) is a protocol between a UE and a positioning server (the E-SMLC in the control-plane case or SLP in the user-plane case). LPP supports both control-plane and user-plane protocols as underlying transport. 3GPP specifies the control-plane use of LPP in 3GPP TS 36.305, for example. User-plane support of LPP is defined in the OMA SUPL 2.0 specifications. LPP supports positioning- and location-related services (e.g. transfer of assistance data).

LPP messages are carried as transparent PDUs across intermediate network interfaces using the appropriate protocols. The LPP protocol is intended to enable positioning for LTE using a multiplicity of different positioning methods, while isolating the details of any particular positioning method from the specifics of the underlying transport.

LPP supports hybrid positioning, in which two or more position methods are used concurrently to provide measurements and/or a location estimate or estimates to the server.

LPP is specified in 3GPP TS 36.355.

13.5.8 LTE Positioning Protocol Annex (LPPa)

The LTE Positioning Protocol Annex (LPPa) carries information between the eNodeB and the E-SMLC. The LPPa protocol is transparent to the MME. The MME acts as a router of LPPa messages without further knowledge of the LPPa transaction.

LPP is specified in 3GPP TS 36.455.

13.6 Positioning Procedure

Figure 13.2 outlines a control-plane positioning procedure where an application outside the mobile network requests the position of a target UE.

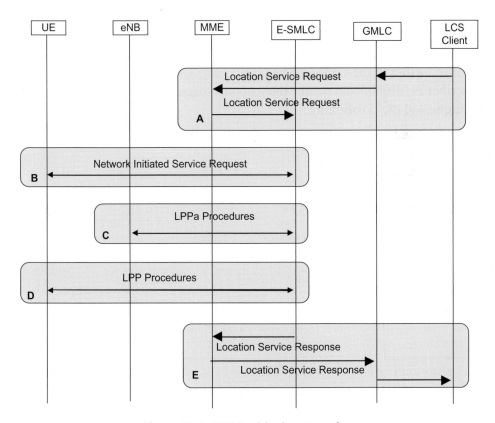

Figure 13.2: EPS Positioning Procedure.

A. An external LCS client requests the current location and optionally the velocity of a target UE. The LCS client may also request a specific position QoS. The GMLC authorizes the LCS client. The GMLC queries the HSS to provide MME address of the target UE and forwards the message to the MME.

B. If the target UE is in Idle mode the MME performs the Network Initiated Service Request procedure to establish a connection with the UE. This means that the MME pages the UE and, when the UE responds, indicating that a signaling connection between the UE and MME is established. In addition, the user may be informed that the UE positioning has been requested and the user may need to confirm that positioning is allowed.

C. Depending on the requested positioning QoS and supported capabilities in the eNodeB, the E-SMLC may initiate an LPPa location procedure towards the eNodeB. The LPPa procedures support mechanisms to obtain positioning measurements for the E-CID positioning method or assistance data for the OTDOA method.

D. In addition to, or instead of step C, the E-SMLC may initiate LPP procedures towards the UE. The LPP procedure supports mechanisms to obtain a location estimate or positioning measurements and to transfer location assistance data to the UE.

E. The E-SMLC provides a location service response to the MME which passes the response further to the GMLC. The GMLC checks that the Privacy profile of the subscriber is fulfilled and passes the response back to the Application/LCS client that requested the positioning.

CHAPTER 14

Offload Functions and Simultaneous Multi-Access

14.1 Introduction

One discussion that started to take definitive shape in the standards after 3GPP completed the architecture work on LTE and EPS is "offloading traffic" (e.g. user-plane data) from the operator's core network or from the 3GPP radio network via different means. Though not all operators share the same view when it comes to the idea of "offloading" and the need for standardizing solutions, certain companies were very enthusiastic on embarking upon another round of architecture redesign exercises. But the reality is that GPRS was already designed to accommodate certain traffic to be offloaded from the core network via some smart and simple configuration and design ideas. In addition, the explosion of smart phones with integrated WiFi also provided additional means for offloading from the Radio Access networks that suddenly were faced with huge data traffic increases as well as a high increase in the number of active users always on due to social networking user habits.

Before describing solution details, we first analyze the various requirements required for offloading. There are roughly two main types of offloading, which may be characterized as "radio access offloading" and "core network offloading". Radio access offloading allows operators to avoid certain congestion situations in locations where radio access network capacity does not support demand for various reasons and it is not possible to rectify the situation quickly enough. Core network offloading is for certain types of traffic closer to the user's current location away from their central service centers or hubs, so the traffic does not need to traverse up to the central location. Additionally, core network offload traffic cannot be part of any kind of specific filter/restricted type such as "child access block" or "secure delivery" or "time restrictions". Figure 14.1 illustrates different scenarios for offloading that will be described in this chapter.

EPC and 4G Packet Networks.
DOI: http://dx.doi.org/10.1016/B978-0-12-394595-2.00014-1

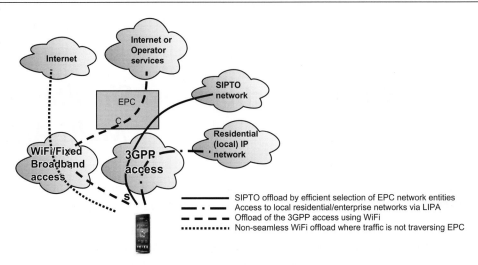

Figure 14.1: Offload Scenarios.

14.2 Offloading the 3GPP RAN – Simultaneous Multi-Access

When an operator is looking to offload users from a radio access network (e.g. 2G, 3G, and LTE), Wireless LAN offloading is an often used alternative. Interworking and mobility with WiFi has been available in EPC since Release 8, as described in earlier chapters of this book. The assumption in Release 8, however, is that the terminal is only active on one access at a time, i.e. either a 3GPP access or WiFi access, if mobility and session continuity is to be supported. Another assumption in the Release 8 architecture is that when the terminal is connected via WiFi, traffic is always routed via a PDN GW in EPC due to the assumption on the functional support as mentioned before. Though nothing prevented a terminal providing support for accessing WiFi without connecting via EPC, in that case the scenario and requirements were not specified in 3GPP. From Release 10, however, 3GPP also defines mechanisms for simultaneous connection on multiple accesses in the form of simultaneous dual-radio (one 3GPP and one non-3GPP access) support. 3GPP also included in its Release 10 specifications explicit support for connecting via WiFi where traffic is routed directly from the WLAN access without traversing a PDN GW in EPC. The following combinations of simultaneous connectivity in multiple accesses are provided:

- Multi-access PDN connectivity (MAPCON): The support for having one (or more) PDN connection(s) in 3GPP (2G/3G/LTE) access and one (or more) PDN connection(s) in a non-3GPP access. Mobility of each PDN connection between 3GPP and non-3GPP access is also supported.
- IP flow mobility (IFOM): The support for having one PDN connection over both 3GPP access and WLAN access simultaneously and choosing over which access

to route traffic on a per-IP-flow basis. Movement of IP flows seamlessly between 3GPP and WLAN access is supported.

- Non-seamless WLAN offloading (NSWO): The support for routing traffic over WLAN without traversing the Evolved Packet Core. Mobility (IP session continuity) with 3GPP access is not supported.

If an operator is looking to offload traffic from its 3G/4G access networks, then the MAPCON and IFOM options are attractive as the operator retains control of the user and is able to provide its own services through its network and service domain.

As discussed further below, however, IFOM has not been such an attractive option so far due to its tight coupling to the use of the terminal-based mobility protocol DSMIPv6.

The NSWO mechanisms may be useful in the case of general Internet access. However, once the traffic is directed towards NSWO, that traffic is no longer under any direct control from the 3GPP operator.

As described in Chapter 6, the Access Network Discovery and Selection Function (ANDSF) can be used to control the steering of traffic towards the different accesses and in turn steer towards a possible offload mechanism if supported.

14.2.1 Multi-Access PDN Connectivity (MAPCON)

In the case of MAPCON, a UE with simultaneous dual-radio connection (e.g. one in a 3GPP-defined radio access like HSPA or LTE and the other being WLAN) may connect to different PDN connections over each access network via the operator's core (i.e. EPC) network. In this case, the user has a subscription to the same operator for both access technologies and the UE is configured to enable MAPCON (e.g. via ANDSF ISRP policies configured).

These two different PDN connections towards different APNs are independent of each other and allow the operator as well as the user to offload traffic, for example downloading large amounts of data, over one access network without affecting the user's service over the other. This is illustrated in Figure 14.2, where the UE is connected to two different PDN GWs in EPC via two different access networks simultaneously (3GPP and non-3GPP). The APNs used over different accesses must be distinct in MAPCON. If the user has more than one PDN connection to a single APN, those PDN connections must be over the same access. If the user hands over one of the PDN connections to another access, the other PDN connections for that APN also have to be handed over.

14.2.2 IP Flow Mobility (IFOM)

In the case of IFOM, one can also consider part of the IP flows being offloaded for similar reasons, such as heavy data-driven apps like a video download from the

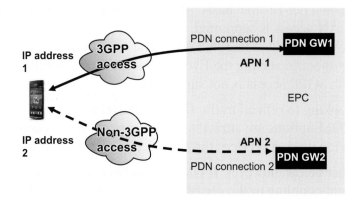

Figure 14.2: Example Scenario for Multi-Access PDN Connectivity.

Internet. For IFOM, when supported and configured in the UE (e.g. via ANDSF ISRP policies), a UE is able to use a single PDN connection (i.e. one APN) over multiple accesses where certain IP flows of that PDN connection are directed over one access (e.g. 3GPP-defined radio access technologies like HSPA, LTE) and the other IP flows are directed over WLAN. Figures 14.3 and 14.4 illustrate examples of such a connection for various different applications being used in the UE making use of multiple access networks simultaneously. These IP flows may belong to a single application or different applications; there is no technical limitation in realizing the flow mobility/connectivity. Mobility support for moving the individual IP flows between 3GPP access and WLAN access is also included in the IFOM solution.

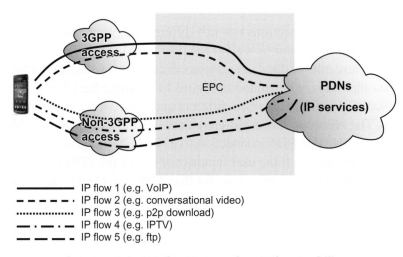

IP flow 1 (e.g. VoIP)
IP flow 2 (e.g. conversational video)
IP flow 3 (e.g. p2p download)
IP flow 4 (e.g. IPTV)
IP flow 5 (e.g. ftp)

Figure 14.3: Example Use Case for IP Flow Mobility.

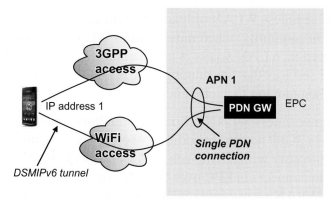

Figure 14.4: IFOM Solution with Single PDN Connection Over Multiple Accesses.

The IFOM solution requires certain architectural considerations to ensure that all IP flows traverse through a single PDN GW regardless of the different access technologies used to deliver the data. In this case the IP address the UE has for that PDN connection is the same and any possible handling of the end-user experience from possible application impacts (e.g. possible time difference and quality) due to the use of two different access networks is left to the application level to resolve. IFOM requires that a DSMIPv6-based solution (i.e. S2c as per architecture) is used. The DSMIPv6 protocol is used to signal the IP filters between UE and PDN GW that determine what IP flows goes over what access. More details on how this works can be found in Section 16.3.6. The support for functions like IFOM in network-based mobility protocols like GTP and PMIP is under investigation in 3GPP.

14.2.3 Non-Seamless WLAN Offloading (NSWO)

In the case of non-seamless WLAN offloading, the UE connects via WLAN and traffic is routed directly from the WLAN access network to the target network, typically the Internet. The traffic does not pass via a PDN GW in EPC. The way this is specified in TS 23.402 is that the UE gets a local IP address from the WLAN network. This IP address is thus not associated with a PDN connection and is not allocated by a PDN GW. Figure 14.5 illustrates the use of NSWO.

Figure 14.5: Non-Seamless WLAN Offloading.

As the name "non-seamless WLAN offloading" indicates, there is no support for mobility (IP session continuity) when moving from WLAN to a 3GPP access. The local IP address used in WLAN access is not maintained by an anchor point such as the PDN GW. The services that are running over NSWO would thus experience an IP address change when moving to another access.

A UE may use non-seamless WLAN offloading simultaneously with having active PDN connections in 3GPP access. In addition, the UE may use non-seamless WLAN offloading simultaneously with having active PDN connections over WLAN. When using S2b or S2c to connect via EPC over WLAN, the UE has both a local IP address received from the WLAN access network and IP address(es) received from the PDN GW(s) associated with PDN connections, which was described in Chapter 6 (Session Management and Mobility).

14.3 Offloading the Core and Transport Network – Selected IP Traffic Offload (SIPTO)

In the event that offloading the operator's core and transport network (including the interconnection infrastructure towards a centralized Service Network not necessarily close to a user's current location). Selected IP Traffic Offload (SIPTO) is an attractive and quite simple alternative method. For 3GPP accesses, this is achieved by selecting a PDN GW (and Serving GW) for the specific APN that is physically located near a specific user based on the location of the user, as illustrated in Figure 14.6. The intention of selecting a physically close PDN GW is to have a short path between the UE's base station and the PDN GW, for efficient routing of the user data. When/if the user moves far enough away from the location served by the selected PDN GW, the PDN GW may no longer be optimally located. In that case, the MME can request that the UE disconnects from the current PDN and reactivates the PDN connection again, giving the network an opportunity to connect to another PDN GW closer to the user's latest location.

Initially focused on the Macro network, SIPTO utilizes existing 3GPP GW selection and a UE redirect/reconnect mechanism with some enhancement to achieve this specific option. One can easily envision SIPTO as a special type of local breakout based on the user's subscription and the type of service he/she intends to access. The main principle behind the mechanism involves enabling the PDN GW (as well as S-GW where necessary) selection for a given APN based on the user's current location information (e.g. TA, RA), in addition to the criteria already specified in Chapter 9.

The APN may be provided by the UE. If not provided by the UE, the MME/SGSN selects the default APN retrieved as part of the user's profile downloaded from the HSS. The APN indicates the type of PDN (e.g. service network, Internet, or specific service domain like IMS) the user wants to connect to. With the APN profiles provided

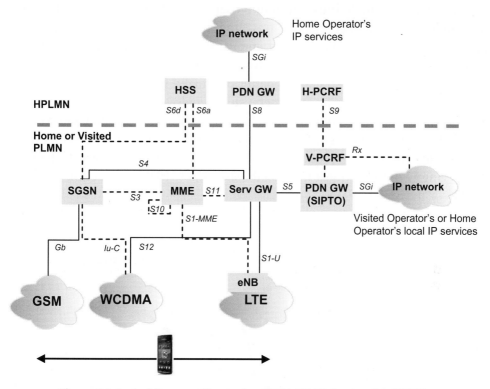

Figure 14.6: Architecture Illustrating PDN GW Selection for SIPTO.

as part of the subscription data from the HSS, including an indication per APN whether SIPTO is allowed or prohibited, operators can control SIPTO-based offloading being offered or not. In addition, this simple profile-based control mechanism also makes it possible to avoid having any misrouting of the data or misconnection to, say, the Internet. This is especially important for scenarios where restrictions such as parental control or dedicated filtering for content control/site control need to be applied.

For roaming scenarios, SIPTO can pose some concerns if the VPLMN operator does not want to provide connection to its local PDN GWs or the roaming agreement requires home routing to be performed. The MME/SGSN in the local network is thus configured specifically to allow or disallow SIPTO selection even prior to checking the user's subscription data.

The specifications 3GPP TS 23.401 and 3GPP TS 29.303 (DNS procedures for GW selection) provide details on how SIPTO is realized with the enhanced DNS mechanism and how subscription data and additional configuration on an APN level allow operators to control how and to whom SIPTO is delivered. SIPTO function has no impact on whether GTP- or PMIP-based mobility is applied in the network.

Figure 14.7 shows a simple illustration of a user with SIPTO PDN and other PDN (i.e. not subject to SIPTO) connection traveling through two LTE access networks while retaining the connections (steps 1 and 2). When the UE moves into a UTRAN access network, the SIPTO PDN connection may be better served by a local PDN GW and the Serving GW also may be better optimized with a Serving GW closer to the UTRAN access network. The intersystem handover ensures that both PDN connections have been moved to the new Serving GW via S1-based handover with MME and Serving GW change. After the handover is completed, the SGSN triggers a PDN disconnection with a request to reactivate the connection towards the UE. The UE complying with this request then establishes a new PDN connection via a new PDN GW for the SIPTO PDN.

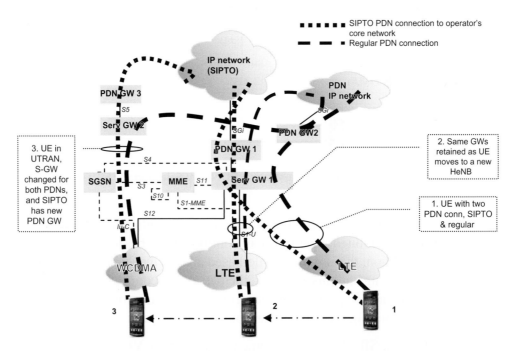

Figure 14.7: Disconnection with Reconnection for SIPTO.

Reconnection of the UE and reselection of PDN GW should not be performed when the user is actively connected to the PDN GW and possibly transmitting/receiving data, since that will definitely be a disruptive experience for the end-user for the applications in use. This possibility to deactivate an active PDN connection to an APN and, in the disconnection request to the UE, adding the request to reactivate to the same APN is an existing GPRS functionality and the feature was also added to the LTE/EPC system. One of the major advantages of the SIPTO feature as it is specified is that operators can activate/deploy this without having to worry about terminal support, since the existing terminal base should have the necessary function support already available. SIPTO as such does not require any new functions in the devices.

14.4 Access to Local Networks – Local IP Access (LIPA)

Development of LIPA (Local IP Access) support in 3GPP has taken many twists and turns due to the occasional lack of strong support from the operator community. Within 3GPP Release 10, the initial focus for the Local IP Access function was to facilitate access to devices like printers and other services within a user's home premises served by a H(e)NB. Since LIPA is very much an end-user service (in contrast to SIPTO, which is an operator-driven feature), where essentially the H(e)NB is most likely owned by the same person who is also getting access to other local devices, it requires a specific APN identifying an explicit user request to connect to the services connected by local IP access. Similarly to SIPTO, the use of LIPA can be controlled by the operator using a per-APN LIPA authorization configuration in the user's subscription profile in HSS. A typical scenario for LIPA is shown in Figure 14.8.

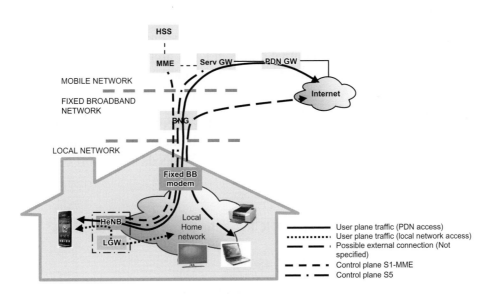

Figure 14.8: Scenarios for LIPA.

The connectivity (user-plane traffic or data) towards these local services does not traverse the mobile operator's core network, though it may traverse part of the Radio Access network such as the H(e)NB subsystem. SIPTO and LIPA are considered mutually exclusive and, as such, it is important to set up the user's profile in the HSS correctly for LIPA and SIPTO. One example is a user with SIPTO status set to "allowed" for a given APN having, for the same APN, LIPA status set to "prohibited". In the HSS, LIPA APN support is indicated per CSG subscription data for a given APN. LIPA can be supported as "conditional" for APN(s) that can be authorized to LIPA service when the UE is using a specific CSG. In the absence of LIPA support, connection using a non-LIPA PDN connection may be enabled for APNs that are configured as "conditional" LIPA APN. A LIPA APN may explicitly be prohibited access for certain users by setting the HSS profile for that LIPA APN for a specific CSG.

Due to the local nature of services that seem useful (local devices that are not connected to the Internet, those that do not provide any extensive services other than connectivity towards local printers, etc.) to provide services outside of an operator's network, the choice of the architecture can be simplified somewhat. So the focus became a simplified mechanism to deliver connectivity and it was possible to provide only limited network functions and eliminate more advanced ones such as QoS, multiple bearer support, mobility to/from other H(e)NB or macro access networks, charging, legal intercept, and any specific security. Also, it was decided that when a user is connected to a LIPA PDN, there was no need to connect to the operator's network from that PDN connection. So it should be clear to the reader by now that the LIPA feature is geared towards small cell deployment scenarios like an H(e)NB. The deterministic characteristic for the final compact LIPA architecture was then selected to be a Local GW (which has a subset of PDN-GW and Serving GW) co-located with the H(e)NB node. The LIPA reference architecture is shown in Figure 14.9.

The functions that need to be supported in the L-GW depend on the functions that the LIPA PDN connection would support. As already mentioned, a number of key functions were not needed for LIPA PDN. The L-GW needs to support only the following key functions:

- UE IP address allocation
- DHCPv4 (server and client) and DHCPv6 (client and server) functions
- Packet screening
- Neighbor discovery for IPv6 (RFC 4861).

Additionally, the Local GW needs to support additional functions due to the "nature of local connection/co-location" with the H(e)NB entity:

- ECM-Idle mode downlink packet buffering: This makes it possible to trigger a page towards UE that has gone Idle and data arrives for that UE. L-GW buffers the data until it can either be delivered or discarded.
- ECM-Connected mode direct tunneling towards the H(e)NB: This enables connection to a local network without traversing the operator's network.

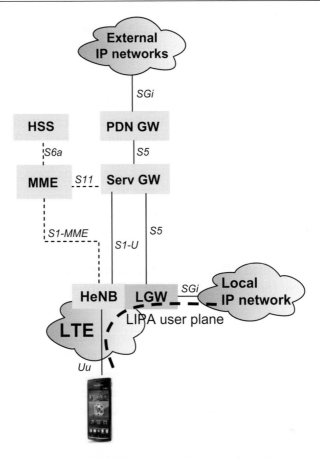

Figure 14.9: Reference Architecture for LIPA.

Some of the key differences between establishing a normal PDN connection compared to a LIPA PDN connection are the following:

- The user must have LIPA status as allowed or conditional in his/her subscription profile for a specific APN that the user intends to connect to for LIPA services and allowed access according to the CSG subscription.
- LIPA support is restricted to certain APN(s) and each LIPA enabled APN is valid only for a specific CSG.
- The UE must explicitly ask for the LIPA APN during the Attach/PDN connection request.
- During the S1-AP setup procedure, the LIPA-capable H(e)NB indicates to the MME/SGSN about the support of LIPA and the IP address of the L-GW and TEID (for GTP-based S5)/GRE Key (for PMIP-based S5) to be used towards the L-GW to support user-plane connection for that PDN.

- The Correlation ID enabling direct user-plane connection between H(e)NB and L-GW for the LIPA PDN connection uses the TEID/GRE Key. When Correlation ID is present, H(e)NB establishes a direct user-plane path towards the L-GW.
- When MME/SGSN detects the L-GW information from the H(e)NB and the UE makes a request to the LIPA APN and is allowed to do so, then MME/SGSN bypasses the PDN-GW selection procedure and uses the L-GW IP address as its "PDN-GW" address.
- There can only be a single bearer (e.g. default) for LIPA PDN connection (i.e. no QoS support).
- There is no support for Policy Control function (i.e. no support for PCRF).
- If LIPA status is conditional in the user's profile and the UE requests connection to a LIPA APN and the H(e)NB does not provide the L-GW information during S1 connection setup, then MME/SGSN uses a standard DNS mechanism to select an alternative (non-LIPA) PDN-GW to establish non-LIPA PDN connection.

One of the consequences of collocating the L-GW with H(e)NB has been that seamless mobility and handover to/from an H(e)NB to other (Radio) access networks (regardless of its being 3GPP defined or not) cannot be supported easily without significant impact on the network itself while maintaining the LIPA architecture. As such, the current system ensures that the LIPA PDN connection(s) is smoothly disconnected by the H(e)NB when any form of mobility/handover/system change is detected.

Since some operators consider the nature of the LIPA/HeNB deployment "unsecure" from the perspective of operator control and responsibility, whenever possible the operator's network entity (e.g. MME/SGSN) responsible for the user's mobility ensures that any LIPA PDN connection is disabled before any change of the UE location on the currently serving HeNB. Note that in the case of HNB, since the SGSN is not informed of the UE change of location (e.g. Intra-RNC), it is not possible to ensure that the LIPA PDN connection has been disconnected. However, for HeNB architecture, that is not the case since the MME is aware of the UE change of location, though at different parts of the event (e.g. Path Switch Message); this is described in more detail in Chapter 17 and confirms that no LIPA PDN connection remains connected.

Figures 14.10 and 14.11 illustrate two scenarios where the UE moves out of the combined HeNB and L-GW coverage. The first diagram illustrates the case where the PDN connection is retained in the EPC and the second diagram illustrates a SIPTO PDN connection that is disconnected after completion of handover and SGSN has detected that a local SGW/PDN GW would serve the UE better.

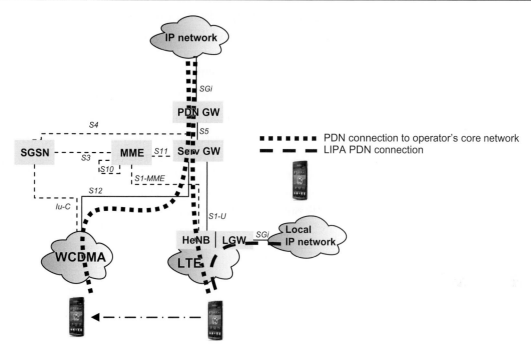

Figure 14.10: After Handover to UTRAN, When Only the PDN Connection Via the Operator's Core Network Remains, the LIPA PDN Connection Is Terminated.

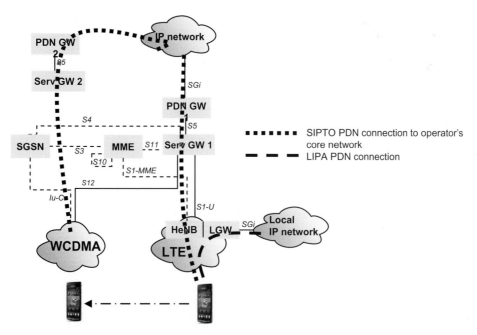

Figure 14.11: After Handover to UTRAN, When Only the (Re-established) SIPTO PDN Connection Via the Operator's Core Network Remains, the LIPA PDN Connection Is Terminated.

An additional restriction for CS Fallback and LIPA PDN connections is that if LIPA PDN is the only PDN connection for that UE, then CS Fallback should not use the PS Handover procedure as that will cause the UE to be detached from the system. So, in such cases, other alternatives such as the "Release with Redirect" mechanism should be used to direct the UE towards a CS domain appropriate for CS Fallback.

During the writing of this book, 3GPP has postponed further work on expanding the existing LIPA architecture that would provide support for mobility/handover, QoS, and more flexibility of support of GWs that are not preconfigured (thus enabling functions like load balancing). However, a technical study conducted has shown that it would not be too complicated to enhance the existing 3GPP architecture to support an architecture not just tied to HeNB but any small cells in a campus/enterprise/office environment.

PART IV

The Nuts and Bolts of EPC

CHAPTER 15

EPS Network Entities and Interfaces

In this chapter we describe in more detail the different network entities, interfaces, protocols, and procedures used in EPS. This chapter can be used as a reference for readers wanting more details than were given in preceding chapters. Before jumping into the actual descriptions, however, it is useful to take a look at what we actually mean by a network entity, an interface, a protocol, and a procedure.

A network entity in EPS is sometimes called a "logical entity" this means in the 3GPP standard it is a logically separate entity with a well-defined functionality. There are also well-defined interfaces between different network entities. This does not, however, imply that actual physical "boxes" implemented by vendors and deployed in real networks have to correspond one-to-one with the network entities in the standard. Vendors may implement a network entity as a standalone product, or may choose to combine different network entities in the same product. It may, for example, be beneficial to combine a Serving GW and a PDN GW in the same node in order to reduce the number of physical nodes that the user plane has to traverse in non-roaming cases.

Although 3GPP uses the term *reference point* to denote an association between two logical network entities, we have chosen the more commonly used term *interface* in this book. There are some differences in the formal definition of a reference point and an interface, but for the purpose of this book the difference has no practical implications. Interfaces in 3GPP typically have a prefix letter and one or two additional letters. In GPRS, most interfaces start with the letter "G" while in EPS most interfaces start with the letter "S".

A protocol is defined in TS 29.905 as "A formal statement of the procedures that are adopted to ensure communication between two or more functions within the same layer of a hierarchy of functions"; this definition is taken from ITU-T document I.112. This could possibly be described simply as a well-defined set of rules for sending information between two network entities. These rules typically cover transmission, message and data formatting, error handling, etc. EPS uses protocols defined by

EPC and 4G Packet Networks.
DOI: http://dx.doi.org/10.1016/B978-0-12-394595-2.00015-3

the Internet Engineering Task Force (IETF) as well as protocols defined by 3GPP. When IETF protocols are used, such as Diameter or IPSec, 3GPP specifies how these protocols are applied in the 3GPP architecture and what protocol options and amendments should or should not be used. In some cases new IETF RFCs are created to specify protocol amendments as required by 3GPP.

It is important to note that there is not a one-to-one mapping between protocols and interfaces. The same protocol may be used on multiple interfaces and multiple protocols may be used on a single interface. One obvious example of the latter is, of course, that different protocols are used on different layers in the protocol stack. Another example is the S5/S8 interface that supports two protocol alternatives, Proxy Mobile IP (PMIP) and GPRS Tunneling Protocol (GTP), to implement similar functions. It is also possible to use different protocols to implement different functions on an interface. For example, on the S2c interface IKEv2 is used to establish a security association (SA), while DSMIPv6 is used for mobility purposes.

EPS is an "all-IP" system where all protocols are transported over IP networks. This is different from the original GPRS standard, where some interfaces supported protocols based on, for example, ATM and Signaling System No. 7 (SS7).

Even though there are numerous interfaces in the EPS architecture, the protocols supported over those interfaces can be grouped into a relatively small number of groups. Figure 15.1 illustrates a few of the most significant protocols and interfaces

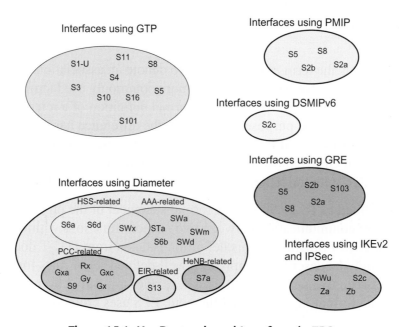

Figure 15.1: Key Protocols and Interfaces in EPS.

in EPS, grouped by protocol type. It should be noted that in some cases multiple protocols are supported over given interfaces.

The procedures, i.e. the message flows, define how commands and messages are transferred between network entities in order to implement a function (e.g. a handover between two base stations). It may be worth noting that the messages and information content shown in the procedures described in Chapter 17 do not necessarily correspond one-to-one with the actual protocol messages. Messages with different names in the message flows may, for example, be implemented using the same protocol message (or vice versa). Another example is that different logical information elements may be combined into a single parameter when defining the actual protocol fields.

This chapter begins with a brief overview of the EPS network entities. The EPS interfaces are then described, with an overview of the functions supported and the protocols used over each interface. Chapter 16 describes the protocols, aiming to give a basic overview of what protocols are used in EPS and their basic properties. Chapter 17 provides a brief introduction to some of the procedures used in EPS. It should be noted that it is not possible to give a complete description of all procedures that exist in EPS. Instead, we have chosen a few procedures that we think will give a good overview for the intended readership without omitting any conceptual information of the key functions supported by EPS. Interested readers may also consult the 3GPP technical specifications 23.060, 23.401, and 23.402 for complete descriptions.

15.1 Network Entities

The network architecture of SAE consists of a few different network entities; each network entity has a distinct role in the architecture. This section covers the roles of the different nodes: the eNodeB, the Mobility Management Entity (MME), the Serving GW, the PDN GW, and the PCRF.

15.1.1 eNodeB

The eNodeB provides the radio interface and performs radio resource management for Long-Term Evolution (LTE), including radio bearer control, radio admission control, and scheduling of uplink and downlink radio resources for individual UEs. The eNodeB also supports IP header compression and encryption of the user-plane data. eNodeBs are interconnected to one another via an interface called X2; this interface has several uses, e.g. handover. eNodeBs are also connected to the EPC via the S1 interface, which is split up into the user plane and the control plane. The control-plane interface is referred to as S1-MME and terminates in the MME. The S1-U interface, meanwhile, terminates at the Serving GW and handles user-plane traffic. The S1 interface supports pooling, i.e. a many-to-many relation between the eNodeBs and the MMEs, and also between the eNodeBs and the Serving GW. The S1 interface also supports network sharing. This allows operators to share the radio network, i.e. the eNodeBs, while maintaining their own EPC networks.

15.1.2 Mobility Management Entity

From a Core Network perspective, the MME is the main node for control of the LTE access network. It selects the Serving GW for a UE during the initial attachment and also during handover, if necessary, between LTE networks. It is responsible for the tracking and paging procedures for UEs in Idle mode and also the activation and deactivation of bearers on behalf of a UE. The MME, via interaction with the HSS, is responsible for authenticating the end-user. For UEs that are roaming, the MME terminates the S6a interface towards the UE's home HSS. The MME also ensures that the UE has authorization to use (camp on) an operator's PLMN and also enforces any roaming restrictions that the UE may have.

In addition, the MME provides control-plane functionality for mobility between LTE and 2G/3G access networks. The S3 interface terminates at the MME from the SGSN.

An MME is selected by the MME selection function. Selection is based on network topology, dependent on which MME serves the particular location that a UE is in. If several MMEs serve a particular area, the selection procedure is based on a few different criteria, for example selecting an MME that reduces the need to change it later or alternatively based on load balancing needs. A full description of the MME selection function is given in Section 9.2.3.

The MME is also responsible for Non-Access Stratum (NAS) signaling, which terminates at the MME; the MME also acts as the termination point in the network for the security of NAS signaling, handling the ciphering protection and management of security keys.

The MME also handles lawful intercept related to signaling.

15.1.3 Serving GW

The Serving GW performs several functions for both the GTP-based and PMIP-based network architectures. The Serving GW terminates the interface towards E-UTRAN; every UE that attaches to an EPS is associated with a single Serving GW. In the same way as the MME, the Serving GW is selected for the UE based on network topology and UE location. The Domain Name Service (DNS) may be used to resolve a DNS string of possible Serving GW addresses that serve the UE's location. The selection of Serving GW may be affected by a few criteria. First, a Serving GW may be selected based on the fact that its service area may reduce the necessity to change the Serving

GW at a later time. Secondly, Serving GW selection may be based on the need for load balancing between different Serving GWs. A full description of the selection procedure is given in Chapter 9.

Once a UE is associated with a Serving GW, it handles the forwarding of end-user data packets and also acts as a local anchor point when required for inter-eNodeB handover. During handover from LTE to other 3GPP access technologies (inter-RAT handover for other 3GPP access technologies), the Serving GW terminates the S4 interface and provides a connection for the transfer of user traffic from 2G/3G network systems and the PDN GW. During both the inter-NodeB and inter-RAT handovers, the Serving GW sends one or more "end-markers" to the source eNodeB, SGSN, or RNC in order to assist the reordering function in the eNodeB.

When a UE is in Idle state, the Serving GW will terminate the downlink (DL) path for data. If new packets arrive, the Serving GW triggers paging towards the UE. As part of this, the Serving GW manages and stores information relevant to the UE, for example parameters of the IP bearer service or internal network routing information.

The Serving GW is also responsible for the reproduction of user traffic in the case of lawful intercept.

15.1.4 PDN GW

The PDN GW provides connectivity to external PDNs for the UE, functioning as the entry and exit point for the UE data traffic. A UE may be connected to more than one PDN GW if it needs to access more than one PDN. The PDN GW also allocates an IP address to the UE. These PDN GW functions apply to both the GTP-based and the PMIP-based versions of the SAE architecture. This is covered in more detail in Section 6.1.

In its role as a gateway, the PDN GW may perform deep packet inspection or packet filtering on a per-user basis. The PDN GW also performs service-level gating control and rate enforcement through rate policing and shaping. From a QoS perspective, the PDN GW also marks the uplink and downlink packets with, for example, the DiffServ Code Point. This was covered in more detail in Chapter 8.

Another key role of the PDN GW is to act as the anchor for mobility between 3GPP and non-3GPP technologies such as WiFi and 3GPP2 (CDMA/HRPD).

15.1.5 Policy and Charging Rules Function

The Policy and Charging Rules Function (PCRF) is the policy and charging control element of the SAE architecture and encompasses policy control decision and flow-based charging control functionalities. This means that it provides network-based control related to service data flow detection, gating, QoS, and flow-based charging towards the Policy and Charging Enforcement Function (PCEF). It should be noted, however, that the PCRF is not responsible for credit management.

The PCRF receives service information from the Application Function (AF) and decides how the data flow for a particular service will be handled by the PCEF. The PCRF also ensures that the user-plane traffic mapping and treatment is in accordance with the subscription profile associated with an end-user. The PCRF functions are described in more detail in Section 8.2.

15.1.6 Home eNodeB Subsystem and Related Entities

The Home eNodeB Subsystem (HeNS) consists of a Home eNodeB (HeNB), optionally a Home eNodeB Gateway (HeNB-GW), and optionally a Local GW (L-GW). The Home eNodeB Subsystem is connected via the S1 interface to the MME and the Serving GW.

A Home eNodeB is customer-premises equipment offering E-UTRAN coverage. The HeNB supports the same functionality as an eNodeB and it supports the same procedures towards the MME and Serving GW as en eNodeB. X2-based HO between HeNBs is allowed between closed/hybrid access HeNBs having the same CSG ID or when the target HeNB is an open access HeNB.

A Home eNodeB Gateway is an optional gateway through which the Home eNodeB accesses the core network. The HeNB GW serves as a concentrator for the control plane, specifically the S1-MME interface. The S1-U interface from the HeNB may be terminated at the HeNB GW, or a direct logical U-plane connection between HeNB and S-GW may be used. The HeNB GW appears to the MME as an eNB. The HeNB GW appears to the HeNB as an MME. The S1 interface between the HeNB and the EPC is the same, regardless of whether the HeNB is connected to the EPC via a HeNB GW or not.

A Local GW is a gateway towards local IP networks (e.g. residential/enterprise networks) associated with the HeNodeB. The Local GW is co-located with the Home eNodeB.

15.1.6.1 CSG List Server

The CSG List Server provides the Allowed CSG list and the Operator CSG list on the UE using management procedures such as Over The Air (OTA) procedures or OMA DM procedures. The CSG List Server is located in the subscriber's home network.

15.1.6.2 CSG Subscriber Server (CSS)

The CSS is an entity in the Visited network that stores and manages CSG subscription-related information for roaming UEs. The CSS is used to enable autonomous CSG roaming in visited networks. The CSS supports download of CSG subscription information upon request from the serving MME, via the S7a interface. The CSS also supports service provision, including updating the MME with modifications of the CSG membership granted to the subscriber.

15.2 Control Plane Between UE, eNodeB, and MME

15.2.1 S1-MME

15.2.1.1 General

The E-UTRAN-Uu interface is defined between the UE and the eNodeB and the S1-MME interface is defined between the eNodeB and the MME, as shown in Figure 15.2.

Figure 15.2: S1-MME Interface.

In the case of HeNS the S1-MME interface is defined between HeNB and MME, and between HeNB, HeNB-GW, and MME, as shown in Figure 15.3.

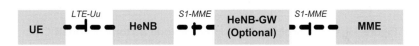

Figure 15.3: S1-MME Interface for the HeNB and HeNB-GW.

15.2.1.2 Interface Functionality

The S1-MME interface provides support for functionality such as paging, handover, UE context management, E-RAB management, and transparent transport of messages between the MME and UE.

15.2.1.3 Protocol

The protocol stack for E-UTRAN-Uu and S1-MME is depicted in Figures 15.4 and 15.5.

As can be seen in Figures 15.4 and 15.5, the NAS protocols run directly between the UE and the MME, while the eNodeB and HeNS just act as a transparent relay. The protocol layers below NAS on E-UTRAN-Uu and S1-MME are called the Access Stratum (AS).

The AS protocols on E-UTRAN-Uu (RRC, PDCP, RLC, MAC, and the physical LTE layer) implement Radio Resource Management and support the NAS protocols by transporting the NAS messages across the E-UTRAN-Uu interface. Similarly, the

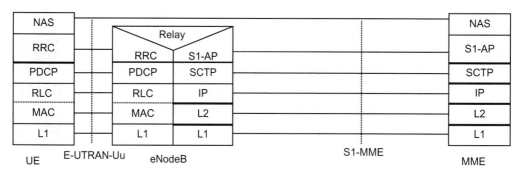

Figure 15.4: Protocol Stack for E-UTRAN-Uu and S1-MME.

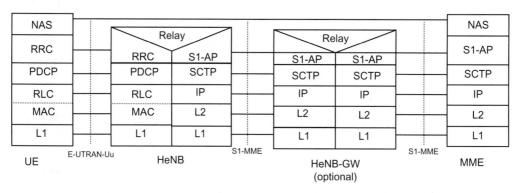

Figure 15.5: Protocol Stack for Home eNode B Home eNodeB GW.

AS protocols on S1-MME (S1-AP, SCTP, IP, etc.) implement functionality such as paging, handover, UE context management, E-RAB management, and transparent transport of messages between MME and eNodeB and HeNS. The S1-MME interface is defined in 3GPP TS 36.410.

The NAS layer consists of an EPS Mobility Management (EMM) protocol and an EPS Session Management (ESM) protocol. The EMM protocol provides procedures for the control of mobility and security for the NAS protocols. The ESM protocol provides procedures for the handling of EPS bearer contexts. Together with the bearer control provided by the Access Stratum, this protocol is used for the control of user-plane bearers. The NAS protocols are defined in 3GPP TS 24.301.

15.3 GTP-Based Interfaces

15.3.1 Control Plane

More detailed information about the procedures and functions of GPTv2-C interfaces can be found in 3GPP technical specifications 23.401 and 23.060.

The detailed messages and parameters for control-plane GTPv2-C protocol are defined in 3GPP TS 29.274. At the time of writing, this specification is still under development, and so may not have a complete set of functions. Note that both IPv4 and IPv6 are supported on the transport layer.

Table 15.1 shows some messages supported over various interfaces using GTPv2-C.

Table 15.1: Messages Supported Over Various Interfaces Using GTPv2-C

Message Name	Entity Involved	Interfaces
Create Session Request/Response	SGSN/MME to Serving GW to PDN GW (response in opposite path)	S4/S11, S5/S8
	Trusted WiFi access/ePDG to PDN GW (response in opposite path)	S2a/S2b
Modify Bearer Request/Response	SGSN/MME to Serving GW to PDN GW (response in opposite path)	S4/S11, S5/S8
Create Bearer Request/Response	PDN GW to Serving GW to SGSN/MME (response in opposite path)	S5/S8, S4/S11
	PDN GW to trusted WiFi access/ePDG (response in opposite path)	S2a/S2b
Update Bearer Request/Response	PDN GW to Serving GW to SGSN/MME (response in opposite path)	S5/S8, S4/S11
	PDN GW to trusted WiFi access/ePDG (response in opposite path)	S2a/S2b
Context Request/Response	MME to MME	S10
	SGSN to/from MME	S3
	SGSN to/from SGSN	S16
Create Forwarding Tunnel Request	MME to Serving GW	S11
Downlink Data Notification	Serving GW to SGSN/MME	S4/S11
Detach Notification	SGSN to MME, MME to SGSN	S3
Delete Session Request	SGSN/MME to Serving GW to PDN GW (response in opposite path)	S4/S11, S5/S8
	ePDG/trusted WiFi access to PDN GW (response in opposite path)	S2a/S2b

15.3.2 MME ↔ MME (S10)

The S10 interface is defined between two MMEs; this interface uses GTPv2-C exclusively and is for LTE access only. The main function over this protocol is to transfer the contexts for individual terminals attached to the EPC network and thus sent on a per-UE basis. This interface is used primarily when MME is relocated. Figure 15.6 shows the protocol stack for S10 interface.

15.3.3 MME ↔ Serving GW (S11)

The S11 interface is defined between MME and the Serving GW; this interface exclusively uses GTPv2-C and is for LTE access only. Due to the separation of the

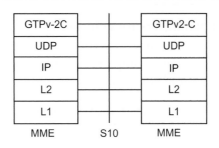

Figure 15.6: S10 Interface.

control- and user-plane functions between MME and Serving GW, the S11 interface is used to create a new session (i.e. to establish the necessary resources for the session) and then manage these sessions (i.e. modify, delete, and change) for any sessions of a terminal (for each PDN connection) that has established connection within EPS.

The S11 interface is always triggered by some events either directly via NAS level signaling from the terminal, such as a device attaching to the EPS network, adding new bearers to an existing session, or creating a connection or when creating a connection towards a new PDN, or it may be triggered during network-initiated procedures such as PDN GW-initiated bearer modification procedures. As such, the S11 interface keeps the control- and user-plane procedures in sync for a terminal during the period that the terminal is active/attached in the EPS. In the case of handover, the S11 interface is used to relocate the Serving GW when appropriate, establish a direct or indirect forwarding tunnel for user-plane traffic, and manage the user data traffic flow.

Figure 15.7 shows the protocol stack across the S11 interface.

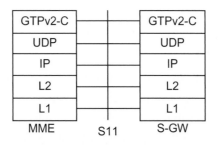

Figure 15.7: S11 Interface.

Note that some of the interactions (e.g. signaling between MME and Serving GW) also need to be performed between the Serving GW and PDN GW over the S5/S8 interface as described below, though in such cases, depending on the protocol choice over S5/S8, the messages are either continued over GTP or transferred over PMIP.

The S11 interface shares common functions with interface S4 as well, which can be seen later in this section.

15.3.4 Serving GW ↔ PDN GW (S5/S8)

The S5/S8 interface is defined between the Serving GW and the PDN GW. The S5 interface is used in non-roaming scenarios where the Serving GW is located in the home network, or in roaming scenarios with both the Serving GW and the PDN GW are located in the visited network. The latter scenario is also referred to as Local Breakout. The S8 interface is the roaming variant of S5 and is used in roaming scenarios with the Serving GW in the visited network and the PDN GW in the home network.

When the GTP variant of the protocol is used, the S5/S8 interface provides the functionality associated with creation/deletion/modification/change of bearers for individual users connected to EPS. These functions are performed on a per-PDN connection for each terminal. The Serving GW provides the local anchor for all bearers for a single terminal and manages them towards the PDN GW. Figure 15.8 shows the protocol stack over the S5/S8 interface. Section 15.4 contains information on the PMIP variant of S5/S8.

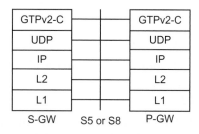

Figure 15.8: S5 or S8 Interface (GTP Variant).

15.3.5 SGSN ↔ MME (S3)

The S3 interface is between S4-based SGSN (this SGSN is enhanced to support EPS mobility compared to the Gn/Gp-based SGSN, as described in Chapter 2) and MME in order to support handover to/from 2G/3G radio access network for 3GPP accesses. The functions supported include transfer of information relating to the terminal that is being handed over and handover/relocation messages; thus, the messages are on an individual terminal basis. This interface exclusively supports GTPv2-C and the protocol stack is shown in Figure 15.9.

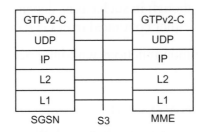

Figure 15.9: S3 Interface.

15.3.6 SGSN ↔ Serving GW (S4)

The S4 interface is between SGSN supporting 2G/3G radio access and Serving GW; it has functions equivalent to those of the S11 interface, but only for 2G/3G radio access networks, and supports related procedures to support legacy terminals connecting via EPS as well. This interface supports exclusively GTPv2-C and provides procedures to enable user-plane tunneling between SGSN and the Serving GW if the 3G network has not enabled direct tunneling for user-plane traffic from RNC to/from the Serving GW. Figure 15.10 shows the protocol stack for the S4 interface.

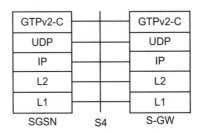

Figure 15.10: S4 Interface.

15.3.7 SGSN ↔ SGSN (S16)

The S16 interface is defined between two SGSNs and this interface exclusively uses GTPv2-C and for 2G/3G accesses only when on an EPS network. The main function of this protocol is to transfer the contexts for individual terminals attached to the EPC network; these are thus sent on a per-UE basis as in the case of the S10 interface. Figure 15.11 shows the protocol stack for the S16 interface.

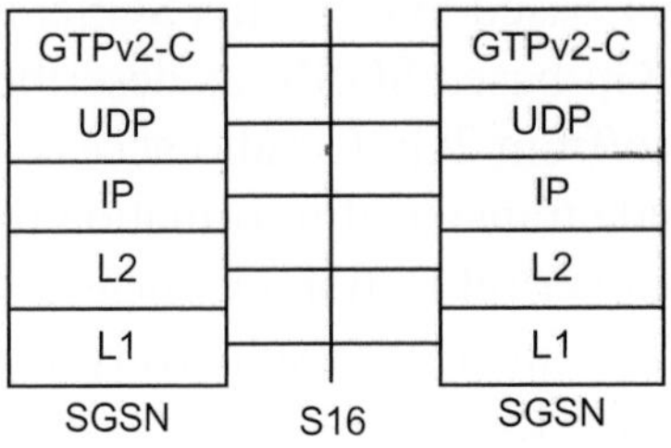

Figure 15.11: S16 Interface.

15.3.8 Trusted WLAN Access Network ↔ PDN GW (S2a)

The S2a interface is defined between a trusted WLAN access network (TWAN) and the PDN GW. The S2a interface may be used in non-roaming scenarios where the TWAN is directly connected to the home network or in roaming scenarios with TWAN connected to a home network via a visited PLMN. In the roaming scenario, the PDN GW may be located in the home network or the visited network. The latter scenario is also referred to as Local Breakout.

When the GTP variant of the protocol is used, the S2a interface provides the functionality associated with creation/deletion/modification/change of bearers for individual users connected to EPS via a trusted WLAN access network. These functions are performed on a per-PDN connection for each terminal. Figure 15.12 shows the protocol stack over the S2a interface. Section 15.4 contains information on the PMIP variant of S5/S8.

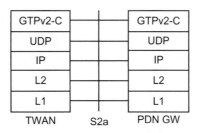

Figure 15.12: S2a Interface.

15.3.9 ePDG ↔ PDN GW (S2b)

The S2b interface is defined between an ePDG and the PDN GW. The S2b interface may be used in non-roaming scenarios where the ePDG and PDN GW is in the home network. The S2b interface may also be used in roaming scenarios with ePDG located in the visited network. In the roaming scenario, the PDN GW may be located in the home network or the visited network. The latter scenario is also referred to as Local Breakout.

When the GTP variant of the protocol is used, the S2b interface provides the functionality associated with creation/deletion/modification/change of bearers for individual users connected to EPS via a trusted WLAN access network. These functions are performed on a per-PDN connection for each terminal. Figure 15.13 shows the protocol stack over the S2b interface. Section 15.4 contains information on the PMIP variant of S5/S8.

15.3.10 User Plane

The user-plane protocols use GTPv1-U as defined in 3GPP TS29.281. This runs over X2-U, S1-U, S4 user-plane, S5/S8 user-plane, and S12 user-plane interfaces for the

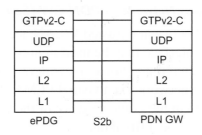

Figure 15.13: S2b Interface.

GTP variant of the EPS architecture. GTPv1-U is also used for 2G/3G Packet Core (also known as GPRS) over the Gn, Gp, and Iu-U interfaces. Note that both IPv4 and IPv6 are supported on the transport layer of IP.

15.3.11 eNodeB ↔ Serving GW (S1-U)

The S1-U is the user-plane interface carrying user data traffic between the eNodeB and Serving GW received from the terminal. The protocol stack for S1-U is shown in Figure 15.14. The HeNB supports the same protocol stack as the eNodeB.

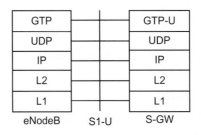

Figure 15.14: S1-U Interface.

15.3.12 UE ↔ eNodeB ↔ Serving GW ↔ PDN GW (GTP-U)

Figure 15.15 shows end-to-end user-plane traffic in the case of LTE access using the GTPv1-U protocol stack (over the S1-U-S5/S8 GTP variant). As can be seen in Figure 15.15, the protocol stack on SGi is the Application protocol over IP. For the HeNB case the HeNB and the (optional) HeNB-GW support the same protocol stack as E-UTRAN.

15.3.13 UE ↔ BSS ↔ SGSN ↔ Serving GW ↔ PDN GW (GTP-U)

Figure 15.16 shows end-to-end user-plane traffic for the case of 2G access over EPC using the GTPv1-U protocol stack (over the S4 and S5/S8 GTP variant). As can be seen in Figure 15.16, the protocol stack on SGi is the Application protocol over IP.

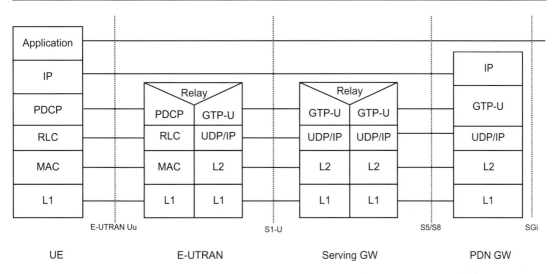

Figure 15.15: User Plane for LTE Access (GTP-Based S5/S8).

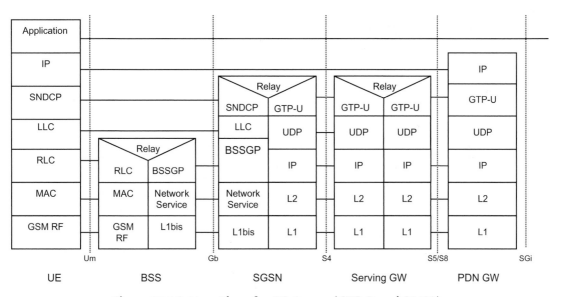

Figure 15.16: User Plane for 2G Access (GTP-Based S5/S8).

15.3.14 UE ↔ UTRAN ↔ Serving GW ↔ PDN GW (GTP-U)

Figure 15.17 shows end-to-end user-plane traffic in the case of 3G access over EPC using the GTPv1-U protocol stack (over the Iu-U-S5/S8 GTP variant) and where the SGSN is no longer in the user-plane path (direct tunnel established between RNC and Serving GW). As can be seen in Figure 15.17, the protocol stack on SGi is the Application protocol over IP.

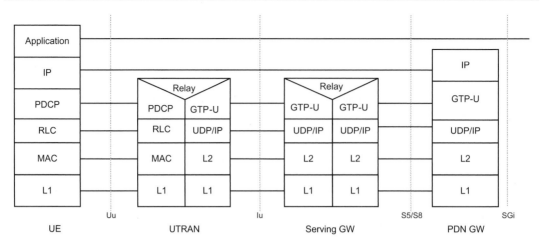

Figure 15.17: User Plane for 3G Access with "Direct Tunnel" (GTP-Based S5/S8).

15.3.15 UE ↔ UTRAN ↔ SGSN ↔ Serving GW ↔ PDN GW (GTP-U)

Figure 15.18 shows end-to-end user-plane traffic in the case of 3G access over EPC using the GTPv1-U protocol stack (over the Iu-U-S5/S8 GTP variant) and where the SGSN is in the user-plane path (direct tunnel not used). As can be seen in Figure 15.18, the protocol stack on SGi is the Application protocol over IP.

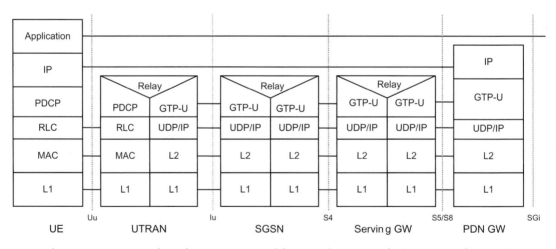

Figure 15.18: User Plane for 3G Access Without "Direct tunnel" (GTP-Based S5/S8).

15.3.16 UE ↔ Trusted WLAN Access Network ↔ PDN GW (GTP-U)

Figure 15.19 shows end-to-end user-plane traffic in the case of trusted WLAN access network (TWAN) access over EPC using the GTPv1-U protocol stack (over the S2a GTP variant). As can be seen in Figure 15.19, the protocol stack on SGi is the Application protocol over IP.

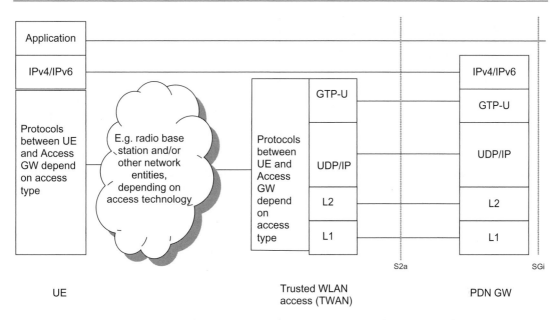

Figure 15.19: User Plane for Trusted WLAN Access with GTP-Based S2a.

15.3.17 UE ↔ ePDG ↔ PDN GW (GTP-U)

Figure 15.20 shows end-to-end user-plane traffic in the case of access to EPC using the GTPv1-U protocol stack (over the S2b GTP variant). As can be seen in Figure 15.20, the protocol stack on SGi is the Application protocol over IP.

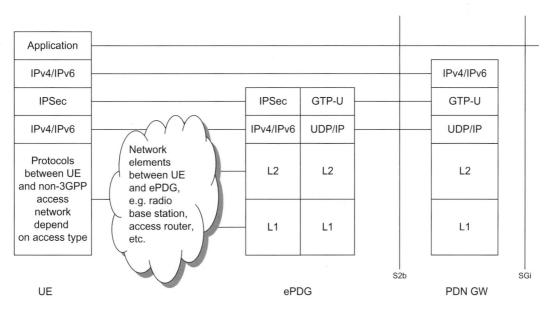

Figure 15.20: User Plane for 3G Access Without "Direct Tunnel" (GTP-Based S5/S8).

15.4 PMIP-Based Interfaces

15.4.1 Serving GW–PDN GW (S5/S8)

15.4.1.1 General

The S5/S8 interface is defined between the Serving GW and the PDN GW. The S5 interface is used in non-roaming scenarios where the Serving GW is located in the home network, or in roaming scenarios with both Serving GW and PDN GW located in the visited network (see Figure 15.21). The latter scenario is also referred to as Local Breakout. The S8 interface is the roaming variant of S5 and is used in roaming scenarios with the Serving GW in the visited network and the PDN GW in the home network.

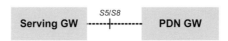

Figure 15.21: S5/S8 Interface.

15.4.1.2 Interface Functionality

The S5/S8 interface provides user-plane tunneling and tunnel management between the Serving GW and the PDN GW. It is used for Serving GW relocation due to UE mobility and if the Serving GW needs to connect to a non-collocated PDN GW for the required PDN connectivity.

15.4.1.3 Protocol

Two protocol alternatives have been specified for the S5/S8 interface: a PMIP-based variant and a GTP-based variant. The protocol stacks for the GTP-based alternative are described in Section 15.3. The protocol stacks for the PMIP-based variant of S5/S8 are shown in Figure 15.22 for PMIP-S5-CP (control plane) and Figure 15.23 for PMIP-S5-UP (user plane). The Serving GW acts as a Mobile Access Gateway (MAG) for PMIPv6 and the PDN GW acts as an LMA. The user plane is tunneled using Generic Routing Encapsulation (GRE). The key field extensions of GRE are used (IETF RFC 5845), where the key field value is used to identify a PDN connection.

The definition of the PMIP-based S5/S8 interface and its functionality are given in 3GPP TS 23.402. The PMIPv6-based protocol for S5/S8 is defined in TS 29.275.

15.4.2 Trusted Non-3GPP IP Access–PDN GW (S2a)

15.4.2.1 General

The S2a interface is defined between an access GW in the trusted non-3GPP access network and the PDN GW. The S2a interface is defined for both roaming and non-roaming scenarios (see Figure 15.24).

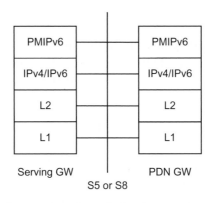

Figure 15.22: Control-Plane Protocol Stack for the S5/S8 Interface (PMIPv6 Variant).

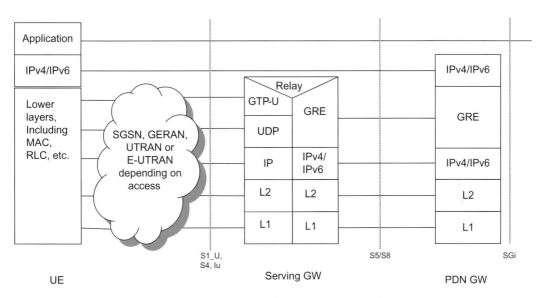

Figure 15.23: User-Plane Protocol Stack for the S5/S8 Interface (PMIPv6 Variant).

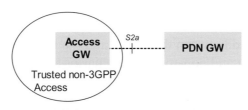

Figure 15.24: S2a Interface.

15.4.2.2 Interface Functionality

The S2a interface provides user-plane tunneling and tunnel management between the trusted non-3GPP access and the PDN GW. It provides mobility support for mobility within the trusted non-3GPP access and between different accesses.

15.4.2.3 Protocol

Two protocol alternatives have been specified for the S2a interface: a PMIP-based variant and a GTP-based variant. The GTP variant is limited to the case when the trusted non-3GPP access is a trusted WLAN access. The protocol stacks for the GTP-based alternative are described in Section 15.3. The protocol stacks for the PMIP-based variant are shown in Figures 15.25 (control plane) and 15.26 (user plane). An Access GW in the trusted non-3GPP access acts as MAG for PMIPv6 and the PDN GW acts as LMA. The user plane is tunneled using GRE. The key field extensions of GRE are used (IETF RFC 5845), where the key field value is used to identify a PDN connection.

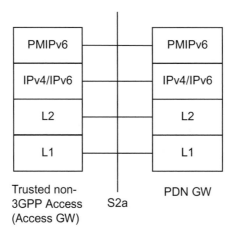

Figure 15.25: Control-Plane Protocol Stack for the S2a Interface.

The definition of the S2a interface and its functionality are given in 3GPP TS 23.402. The PMIPv6-based protocol for S2a is defined in TS 29.275.

15.4.3 ePDG–PDN GW (S2b)

15.4.3.1 General

The S2b interface is defined between the ePDG and the PDN GW. The S2b interface is defined for both roaming and non-roaming scenarios (see Figure 15.27).

15.4.3.2 Interface Functionality

The S2b interface provides user-plane tunneling and tunnel management between the ePDG and the PDN GW.

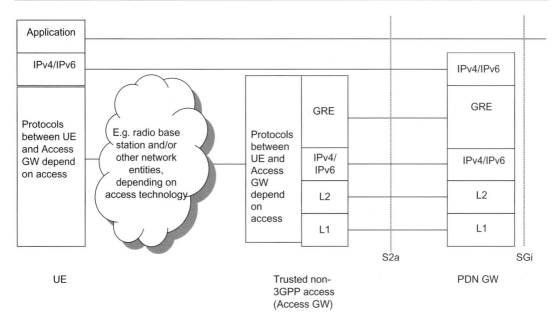

Figure 15.26: User-Plane Protocol Stack for the S2a Interface.

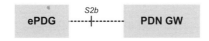

Figure 15.27: S2b Interface.

15.4.3.3 Protocol

Two protocol alternatives have been specified for the S2b interface: a PMIP-based variant and a GTP-based variant. The protocol stacks for the GTP-based alternative are described in Section 15.3. The protocol stacks for the PMIP-based variant are shown in Figures 15.28 (control plane) and 15.29 (user plane). The ePDG acts as an MAG for PMIPv6 and the PDN GW acts as an LMA. The user plane is tunneled using GRE. The key field extensions of GRE are used (IETF RFC 5845), where the key field value is used to identify a PDN connection.

The definition of the S2b interface and its functionality are given in 3GPP TS 23.402. The PMIPv6-based protocol for S2b is defined in TS 29.275.

15.5 DSMIPv6-Based Interfaces

15.5.1 UE–PDN GW (S2c)

15.5.1.1 General

The S2c interface is defined between the UE and the PDN GW. The S2c interface is defined for both roaming and non-roaming scenarios (see Figure 15.30).

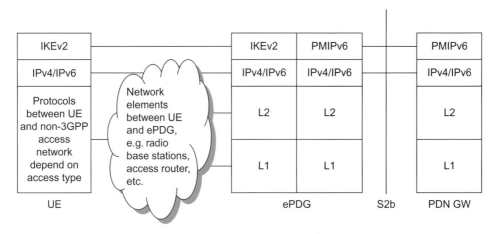

Figure 15.28: Control-Plane Protocol Stack for the S2b Interface.

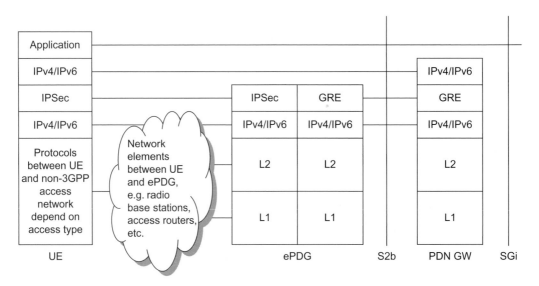

Figure 15.29: User-Plane Protocol Stack for the S2b Interface.

Figure 15.30: S2c Interface.

15.5.1.2 Interface Functionality

The S2c interface provides the user plane with related control and mobility support between UE and the PDN GW. This interface is implemented over trusted and/or untrusted non-3GPP access and/or 3GPP access. Special considerations apply for S2c when the UE is attached over a 3GPP access (see below).

15.5.1.3 Protocol

The protocol over the S2c interface is based on DSMIPv6. The protocol stacks when using S2c over a trusted non-3GPP access are shown in Figure 15.31 while the protocol stack for using S2c over an untrusted non-3GPP access is shown in Figure 15.32. The user plane is tunneled using IP-in-IPv4/IPv6 encapsulation or using IP-in-UDP-in-IPv4 encapsulation to support NAT traversal.

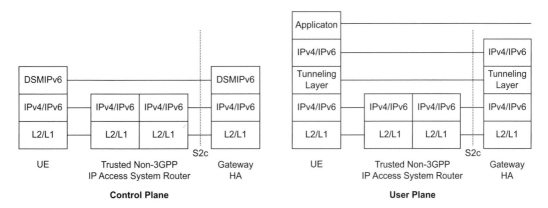

Figure 15.31: Protocol Stacks for the Control Plane and User Plane When Using S2c Over a Trusted Non-3GPP Access.

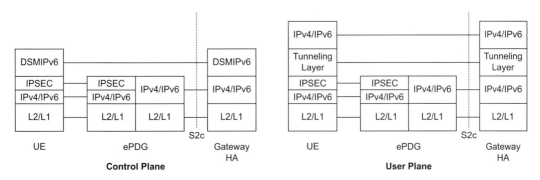

Figure 15.32: Protocol Stacks for the Control Plane and User Plane When Using S2c Over an Untrusted Non-3GPP Access.

When the UE is connected via a 3GPP access, the UE is considered to be on its home link (in a DSMIPv6 sense). There is thus no S2c user-plane tunneling over 3GPP accesses. S2c is used only for DSMIPv6 bootstrapping and DSMIPv6 de-registration (Binding Update with Lifetime equals zero) when the UE is connected via 3GPP access. More information on these and other DSMIPv6 aspects can be found in Section 16.3.

The definition of the S2c interface and its functionality are given in 3GPP TS 23.402. The DSMIPv6-based protocol for S2c is defined in TS 24.303.

15.6 HSS-Related Interfaces and Protocols

15.6.1 General

In this section we describe the interface S6a between the HSS and the MME, as well as interface S6d between HSS and SGSN. The interface between the 3GPP AAA server and the HSS is described in Section 15.7.

15.6.2 MME–HSS (S6a) and SGSN–HSS (S6d)

15.6.2.1 General

The interface S6a is defined between the HSS and the MME, and the interface S6d is defined between HSS and SGSN. The S6d interface is used in EPS and thus applies to the S4-based SGSN only. For a Gn/Gp-based SGSN, the Gr interface between SGSN and HSS/HLR applies (see Figure 15.33).

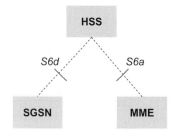

Figure 15.33: S6a and S6d Interfaces.

15.6.2.2 Interface Functionality

The interfaces between HSS and SGSN as well as between HSS and MME are used for multiple purposes. They allow the MME/SGSN and HSS to:

• Exchange location information. As described in Sections 6.4 and 6.5, the MME/ SGSN currently serving the UE notifies the HSS about the MME/SGSN identity. In some cases, for example if the UE attaches to a new MME/SGSN, the MME/ SGSN downloads information from the HSS about the MME or SGSN that previously served the UE.
• Authorize a user to access the EPS. The HSS holds the subscription data, including for example the allowed Access Point Names (APNs) and other information relating to the user's authorized services. The subscription profile is downloaded to the MME/SGSN and used when granting a user access to the EPS.

- Exchange authentication information. As described in Chapter 7, the HSS provides authentication data (the EPS Authentication vector) to the MME/SGSN when the user is being authenticated.
- Download and handle changes in the subscriber data stored in the server. When the subscriber data in the HSS is modified, for example if the subscription is withdrawn, or access to certain APNs is withdrawn, the updated subscription data is downloaded to the MME/SGSN currently serving the UE. Based on the updated subscription data, the MME/SGSN can modify the ongoing session, or even detach the UE completely.
- Upload the PDN GW identity and the APN being used for a specific PDN connection. The information about currently active PDN connections (PDN GW identity and APN) is stored in the HSS to support mobility with non-3GPP accesses.
- Download the PDN GW identity and APN pairs being stored in HSS for an already ongoing PDN connection. This occurs, for example, during the handover from a non-3GPP access to a 3GPP access.

15.6.2.3 Protocol

The same protocol is used on both S6a and S6d. This S6a/S6d interface protocol is based on Diameter and is defined as a vendor-specific Diameter application, where the vendor is 3GPP. The S6a/S6d Diameter application is based on the base Diameter protocol but defines new Diameter commands and attribute–value pairs (AVPs) to implement the functions described in the previous section. Diameter messages over the S6a/S6d interfaces use the Stream Control Transmission Protocol (SCTP; see IETF RFC 2960) as a transport protocol. The protocol stack is illustrated in Figure 15.34.

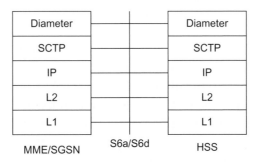

Figure 15.34: Protocol Stack for S6a/S6d.

The protocol over the S6a/S6d interface, including the S6a/S6d Diameter application, is defined in 3GPP TS 29.272. Table 15.2 lists the Diameter commands used in the S6a/S6d Diameter application.

Table 15.2: Diameter Commands Used by the S6a/S6d Diameter Application

Command Name	Abbreviation	Comment
Update-Location-Request	ULR	Sent from MME or SGSN to HSS
Update-Location-Answer	ULA	Sent by HSS in response to the ULR command
Cancel-Location-Request	CLR	Sent from HSS to MME or SGSN
Cancel-Location-Answer	CLA	Sent by MMR/SGSN in response to the AIR command
Authentication-Information-Request	AIR	Sent from MME or SGSN to HSS
Authentication-Information-Answer	AIA	Sent by HSS in response to the AIR command
Insert-Subscriber-Data-Request	IDR	Sent from HSS to MME or SGSN
Insert-Subscriber-Data-Answer	IDA	Sent by MMR/SGSN in response to the IDR command
Delete-Subscriber-Data-Request	DSR	Sent from HSS to MME or SGSN
Delete-Subscriber-Data-Answer	DSA	Sent by MMR/SGSN in response to the DSR command
Purge-UE-Request	PUR	Sent from MME or SGSN to HSS
Purge-UE-Answer	PUA	Sent by HSS in response to the PUR command
Reset-Request	RSR	Sent from HSS to MME or SGSN
Reset-Answer	RSA	Sent by MMR/SGSN in response to the DSR command
Notify-Request	NOR	Sent from MME or SGSN to HSS
Notify-Answer	NOA	Sent by HSS in response to the NOR command

15.7 AAA-Related Interfaces

15.7.1 General

The network nodes specific to 3GPP accesses, such as the MME and SGSN, connect directly to the HSS. Network entities related to other accesses, for example as described by 3GPP2, however typically interface an AAA server instead. Therefore, a 3GPP AAA server is used in the EPC architecture to interface entities related to, for example, CDMA accesses and other accesses outside the 3GPP family. In order to access, for example, subscription data and other data available in the HSS, the 3GPP AAA server interfaces the HSS via the SWx interface.

The different AAA-related interfaces covered in this section are illustrated in Figure 15.35. S6b, STa, SWa, and SWm may connect to either the 3GPP AAA server or the 3GPP AAA proxy depending on whether it is a roaming or non-roaming scenario. For the roaming scenario, S6b may connect to the AAA server or AAA proxy depending on whether the PDN GW is located in the visited network or in the home network (see Chapter 2 for a more complete description of the architecture alternatives).

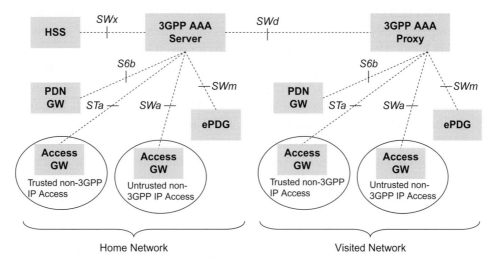

Figure 15.35: AAA-Related Interfaces.

15.7.2 AAA Server–HSS (SWx)

15.7.2.1 General

The SWx interface is defined between the HSS and the 3GPP AAA server (see Figure 15.36).

Figure 15.36: SWx Interface.

15.7.2.2 Interface Functionality

The SWx interface is used for subscriber profile management as well as for non-3GPP access-related location management.

Non-3GPP access location management procedures on SWx include the following functionalities:

- AAA server registration. The 3GPP AAA server registers the current 3GPP AAA server address in the HSS for a given user when a new subscriber has been authenticated by the 3GPP AAA server.
- Upload PDN GW identity and APN. The 3GPP AAA server informs the HSS about the current PDN GW identity and APN being used for a given UE, or that a certain PDN GW and APN pair is no longer used. This occurs, for example, when a PDN connection is established or closed. This corresponds to a similar functionality as is available on S6a/S6d when the UE is in 3GPP access.
- PDN GW identity and APN download. The 3GPP AAA server downloads the PDN GW identity and APN information being stored in HSS for already ongoing

PDN connections for a given UE. This is for the case when the UE has already been assigned PDN GW(s) due to a previous attachment in a 3GPP access (when the UE is handed over from a 3GPP access to a non-3GPP access).

- AAA-initiated de-registration. The 3GPP AAA server may de-register the currently registered 3GPP AAA server in the HSS for a given user and purge any related non-3GPP user status data in the HSS. This occurs if the UE has, for some reason, been disconnected from the non-3GPP access.
- HSS-initiated de-registration. The HSS may initiate a de-registration procedure to purge the UE from the 3GPP AAA server: this happens when the user's subscription has been canceled or for other operator-determined reasons. As a result, the 3GPP AAA server should deactivate any UE tunnel in the PDN GW and/or detach the UE from the access network.

The subscriber profile management procedures over SWx include the following functionalities:

- Subscriber profile push. The HSS may decide to send the subscriber profile to a registered 3GPP AAA server. This occurs, for example, when the subscriber profile has been modified in the HSS and the 3GPP AAA server needs to be updated.
- Subscriber profile request. The 3GPP AAA server may also request the user profile data from the HSS. This procedure is invoked when, for some reason, the subscription profile of a subscriber is lost or needs to be updated.

15.7.2.3 Protocol

The SWx interface protocol is based on Diameter and is defined as a vendor-specific Diameter application, where the vendor is 3GPP. The SWx Diameter application has its own Diameter application identifier but reuses Diameter commands from the Diameter Cx/Dx application, which is a 3GPP vendor-specific application defined for IMS. New AVPs are, however, defined for the SWx application to implement the functions described above. The protocol stack is illustrated in Figure 15.37.

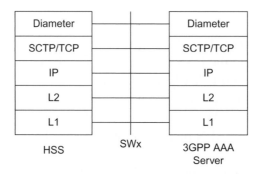

Figure 15.37: Protocol Stack for SWx.

The definition of the interface and its functionality are given in 3GPP TS 23.402. The specification of the SWx Diameter application can be found in 3GPP TS 29.273. Table 15.3 lists the Diameter commands used in the SWx Diameter application.

Table 15.3: Diameter Commands Used by the SWx Diameter Application

Command Name	Abbreviation	Comment
Multimedia-Authentication-Request	MAR	Sent by the 3GPP AAA Server to the HSS in order to request security information
Multimedia-Authentication-Answer	MAA	Sent by a server in response to the MAR command
Push-Profile-Request	PPR	Sent by the HSS to the 3GPP AAA Server in order to update the subscription data
Push-Profile-Answer	PPA	Sent by the HSS in response to the PPR command
Server-Assignment-Request	SAR	Sent by the 3GPP AAA Server to the HSS to register, de-register a user, and/or download the user profile
Server-Assignment-Answer	SAA	Sent by the HSS to the 3GPP AAA Server to confirm the registration, de-registration, or user profile download procedure
Registration-Termination-Request	RTR	Sent by the HSS to the 3GPP AAA Server to request the de-registration of a user
Registration-Termination-Answer	RTA	Sent by the 3GPP AAA Server in response to the RTR command

15.7.3 Trusted Non-3GPP Access–3GPP AAA Server/Proxy (STa)

15.7.3.1 General

The STa interface is defined between the trusted non-3GPP IP access and the 3GPP AAA server in the non-roaming case. In the roaming case it is defined between the trusted non-3GPP IP access and the 3GPP AAA proxy (see Figure 15.38).

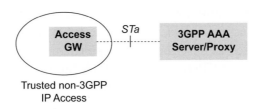

Figure 15.38: STa Interface.

15.7.3.2 Interface Functionality

The STa interface includes the following functionality to:

- Authenticate and authorize a user when the user attaches to a trusted non-3GPP IP access.

- Transport subscription data such as the APN-AMBR and default QoS profile from the 3GPP AAA server to the trusted non-3GPP IP access. The 3GPP AAA server has in turn received the subscription data from the HSS.
- Transport mobility parameters that are needed for the S2a interface, i.e. when PMIPv6 or Mobile IPv4 is used to connect the UE to the EPC. In particular, this information may include the PDN GW identity(s) and APN(s) currently allocated to a UE during a previous attachment in a 3GPP access.
- Transport mobility parameters related to the S2c interface, i.e. when the UE attaches to EPC using DSMIPv6. In particular, the Home Agent IP address or Fully Qualified Domain Name (FQDN) may be sent from the 3GPP AAA server to the gateway of the trusted non-3GPP access for Home Agent discovery based on DHCPv6.
- To transport information about IP Mobility Mode Selection. This includes information from the trusted non-3GPP IP access to the 3GPP AAA server/proxy about the mobility features supported by the non-3GPP access (e.g. if the Access GW in the non-3GPP IP access supports MAG functionality for PMIP) as well as information from the 3GPP AAA server/proxy to the Access GW regarding the selected mobility mechanism.

15.7.3.3 Protocol

This STa interface protocol is based on Diameter and is defined as a vendor-specific Diameter application, where the vendor is 3GPP. The STa Diameter application is based on the Diameter base protocol and also includes commands from the following specifications:

- The Diameter Network Access Server (NAS) application, which is a Diameter application used for AAA services in the Network Access Server (NAS) environment (IETF RFC 4005).
- The Diameter EAP application, which is a Diameter application to support EAP transport over Diameter (IETF RFC 4072). The EAP methods EAP-AKA and EAP-AKA' may be used as described in Chapter 7.
- Extensions relevant for PMIPv6 defined in IETF Internet Draft, Diameter Proxy Mobile IPv6 and extensions relevant for DSMIPv6 are defined in IETF RFC 5447.

The protocol stack for STa is illustrated in Figure 15.39.

The definition of the interface and its functionality are given in 3GPP TS 23.402. The STa Diameter application is defined in TS 29.273. In Table 15.4 are listed the Diameter commands used in the STa Diameter application.

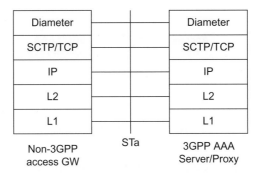

Figure 15.39: Protocol Stack for STa.

Table 15.4: Diameter Commands Used by the STa Diameter Application

Command Name	Abbreviation	Comment
Diameter-EAP-Request	DER	Sent from a non-3GPP access network to a 3GPP AAA Server
Diameter-EAP-Answer	DEA	Sent by a server in response to the DER command
Abort-Session-Request	ASR	Sent from a 3GPP AAA Server/Proxy to a non-3GPP access network
Abort-Session-Answer	ASA	Sent by a non-3GPP access network in response to the ASR command
Session-Termination-Request	STR	Sent from a trusted non-3GPP GW to a 3GPP AAA Server/Proxy
Session-Termination-Answer	STA	Sent by a server in response to the STR command
Re-Auth-Request	RAR	Sent from a 3GPP AAA Server/Proxy to a non-3GPP access network to request reauthorization
Re-Auth-Answer	RAA	Sent by a non-3GPP access network in response to the RAR command
AA-Request	AAR	Sent from a trusted non-3GPP GW to a 3GPP AAA Server/Proxy
AA-Answer	AAA	Sent by a server in response to the AAR command

15.7.4 Untrusted Non-3GPP IP Access–3GPP AAA Server/Proxy (SWa)

15.7.4.1 General
The SWa interface is defined between the untrusted non-3GPP IP access and the 3GPP AAA server (non-roaming case) or 3GPP AAA proxy (roaming case) (see Figure 15.40).

15.7.4.2 Interface Functionality
The SWa interface is used for 3GPP-based access authentication and authorization with an untrusted non-3GPP access. It also supports reporting of accounting information generated by the access network.

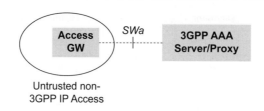

Figure 15.40: SWa Interface.

As described in Section 7.3, access authentication in untrusted non-3GPP IP access is optional. This is because when accessing an untrusted non-3GPP access, the UE will anyway be authenticated and authorized for EPS access using the tunnel procedures with an ePDG (using SWu and SWm interfaces). Use of SWa is therefore optional when a UE accesses an untrusted non-3GPP IP access.

15.7.4.3 Protocol

This SWa interface protocol is based on Diameter and is defined as a vendor-specific Diameter application, where the vendor is 3GPP. It uses the base Diameter protocol and includes commands from the following two applications:

- The Diameter EAP application, which is a Diameter application to support EAP transport over Diameter (IETF RFC 4072). The EAP methods EAP-AKA and EAP-AKA′ may be used as described in Chapter 7.
- The Diameter Network Access Server (NAS) application, which is a Diameter application used for AAA services in the Network Access Server (NAS) environment (IETF RFC 4005) (see Figure 15.41).

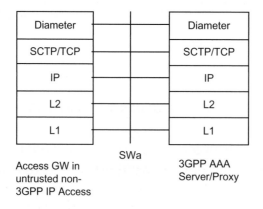

Figure 15.41: Protocol Stack for SWa.

The definition of the interface and its functionality are given in 3GPP TS 23.402. The SWa protocol is defined in TS 29.273. In Table 15.5 are listed the Diameter commands used in the SWa Diameter application.

Table 15.5: Diameter Commands Used by the SWa Diameter Application

Command Name	Abbreviation	Comment
Diameter-EAP-Request	DER	Sent from a non-3GPP access network to a 3GPP AAA Server
Diameter-EAP-Answer	DEA	Sent by a server in response to the DER command
Abort-Session-Request	ASR	Sent from a 3GPP AAA Server/Proxy to a non-3GPP access network
Abort-Session-Answer	ASA	Sent by a non-3GPP access network in response to the ASR command
Session-Termination-Request	STR	Sent from a trusted non-3GPP GW to a 3GPP AAA Server/Proxy
Session-Termination-Answer	STA	Sent by a server in response to the STR command
Re-Auth-Request	RAR	Sent from a 3GPP AAA Server/Proxy to a non-3GPP access network to request reauthorization
Re-Auth-Answer	RAA	Sent by a non-3GPP access network in response to the RAR command

15.7.5 ePDG–3GPP AAA Server/Proxy (SWm)

15.7.5.1 General

The SWm interface is defined between the ePDG and the 3GPP AAA server or between the ePDG and the 3GPP AAA proxy (see Figure 15.42).

Figure 15.42: SWm Interface.

15.7.5.2 Interface Functionality

The SWm interface includes functionality to:

- Authenticate and authorize a user at tunnel setup on the SWu interface (i.e. between UE and ePDG).
- Transport subscription profile data from the 3GPP AAA server to the ePDG. The 3GPP AAA server has in turn received the subscription profile data from the HSS.
- Transport mobility parameters that are needed for the S2b interface, i.e. when PMIPv6 is used to connect the UE to the EPC. In particular, this information may include the PDN GW identity(s) and APN(s) currently allocated to a UE during a previous attachment in a 3GPP access.
- Transport mobility parameters related to the S2c interface, i.e. when the UE attaches to EPC using DSMIPv6. In particular, the Home Agent IP address or FQDN may be sent from the 3GPP AAA server to the gateway of the trusted non-3GPP access for Home Agent discovery based on IKEv2.

- To transport information about IP Mobility Mode Selection. This includes both information from the ePDG to the 3GPP AAA server/proxy about the mobility features supported by the ePDG (e.g. if the ePDG supports MAG functionality for PMIP), as well as information from the 3GPP AAA server/proxy to the ePDG regarding the selected mobility mechanism.
- To transport session termination indications and requests. This includes both a session termination indication from the ePDG to the 3GPP AAA server/proxy if the session with the UE has been terminated, as well as session termination requests from the 3GPP AAA server/proxy to request the ePDG to terminate a given session.

15.7.5.3 Protocol

This SWm interface protocol is based on Diameter and is defined as a vendor-specific Diameter application, where the vendor is 3GPP. The SWm Diameter application is based on the base Diameter protocol and includes commands from the following specifications:

- The Diameter Network Access Server (NAS) application, which is a Diameter application used for AAA services in the Network Access Server (NAS) environment (IETF RFC 4005).
- The Diameter EAP application, which is a Diameter application to support EAP transport over Diameter (IETF RFC 4072). The EAP methods EAP-AKA and EAP-AKA′ may be used, as described in Chapter 7.
- Extensions relevant for PMIPv6 defined in IETF Internet Draft, Diameter Proxy Mobile IPv6 and extensions relevant for DSMIPv6 defined in IETF RFC 5447.

The protocol stack for SWm is illustrated in Figure 15.43.

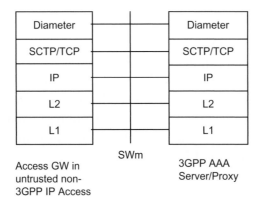

Figure 15.43: Protocol Stack for SWm.

The definition of the interface and its functionality are given in 3GPP TS 23.402. The SWm Diameter application is defined in TS 29.273. In Table 15.6 are listed the Diameter commands used in the SWm Diameter application.

Table 15.6: Diameter Commands Used by the SWm Diameter Application

Command Name	Abbreviation	Comment
Diameter-EAP-Request	DER	Sent from an ePDG to a 3GPP AAA Server
Diameter-EAP-Answer	DEA	Sent by a server in response to the DER command
Abort-Session-Request	ASR	Sent from a 3GPP AAA Server/Proxy to an ePDG
Abort-Session-Answer	ASA	Sent by an ePDG in response to the ASR command
Session-Termination-Request	STR	Sent from an ePDG to a 3GPP AAA Server/Proxy
Session-Termination-Answer	STA	Sent by a server in response to the STR command
Re-Auth-Request	RAR	Sent from a 3GPP AAA Server/Proxy to an ePDG to request reauthorization
Re-Auth-Answer	RAA	Sent by an ePDG in response to the RAR command
AA-Request	AAR	Sent from an ePDG to a 3GPP AAA Server/Proxy
AA-Answer	AAA	Sent by a server in response to the AAR command

15.7.6 PDN GW–3GPP AAA Server/Proxy (S6b)

15.7.6.1 General

The S6b interface is defined between the PDN GW and the 3GPP AAA server (for the non-roaming case, or roaming with home routed traffic to PDN GW in the home network) and between the PDN GW and the 3GPP AAA proxy (for the roaming case with PDN GW in the visited network, i.e. local breakout) (see Figure 15.44).

Figure 15.44: S6b Interface.

15.7.6.2 Interface Functionality

The S6b interface is not utilized when the UE is attached in GERAN, UTRAN, or E-UTRAN. In these cases it is instead the S6a/S6d interfaces that provide the necessary functionality, as described in Section 15.6. When a UE attaches over another access not in the 3GPP family of accesses, S6b provides the functionality described below.

The S6b interface is used to inform the 3GPP AAA server/proxy about current PDN GW identity and the APN being used for a given UE, or that a certain PDN GW and APN pair is no longer used. This occurs, for example, when a PDN connection is established or closed. (The information is then forwarded to the HSS via the SWx interface.) S6b may be used to retrieve specific subscription-related parameters such as a subscribed QoS profile for non-3GPP accesses.

The above functionality of S6b is common for all mobility protocols that can be used when a UE attaches over a non-3GPP access, i.e. PMIPv6-based S2a or S2b, Mobile IPv4-based S2a, or DSMIPv6-based S2c.

Other main functions of the S6b interface depend on what mobility protocol is used when connecting a UE to the EPC.

When the UE attaches to the EPC using the DSMIPv6-based S2c interface, the S6b interface is also used to authenticate and authorize the UE. It is also used to indicate to the PDN GW that a PDN GW reallocation should be performed (see TS 23.402 for more details on the PDN GW reallocation procedure and the scenarios when it is used). When S2c is used the S6b interface can also be used to transport a session termination indication from the 3GPP AAA server/proxy to the PDN GW, to trigger a termination of a PDN connection (the 3GPP AAA server/proxy may receive a trigger for sending the termination indication from the HSS via the SWx interface).

When the UE attaches using the Mobile IPv4-based S2a interface, S6b is also used to authenticate and authorize the Mobile IPv4 Registration Request message that was sent by the UE.

15.7.6.3 Protocol
This S6b interface protocol is based on Diameter and is defined as a vendor-specific Diameter application, where the vendor is 3GPP. The S6b Diameter application is based on the Diameter base protocol with the following additions:

- The Diameter Network Access Server (NAS) application, which is a Diameter application used for AAA services in the Network Access Server (NAS) environment (IETF RFC 4005).
- Extensions relevant for PMIPv6 defined in IETF Internet Draft, Diameter Proxy Mobile IPv6 and extensions relevant for DSMIPv6 defined in IETF RFC 5447.
- The Diameter EAP application, which is a Diameter application to support EAP transport over Diameter (IETF RFC 4072). The EAP method EAP-AKA is used as described in Chapter 7.

The protocol stack for S6b is illustrated in Figure 15.45.

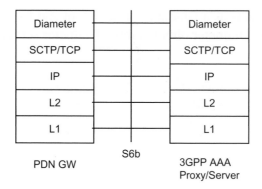

Figure 15.45: Protocol Stack for S6b.

The definition of the interface and its functionality are given in 3GPP TS 23.402. The S6b Diameter application is defined in TS 29.273. In Table 15.7 are listed the Diameter commands used in the S6b Diameter application.

Table 15.7: Diameter Commands Used by the S6b Diameter Application

Command Name	Abbreviation	Comment
Diameter-EAP-Request	DER	Sent from a PDN GW to a 3GPP AAA server
Diameter-EAP-Answer	DEA	Sent by a server in response to the DER command
Abort-Session-Request	ASR	Sent from a 3GPP AAA Server/Proxy to a PDN GW
Abort-Session-Answer	ASA	Sent by a PDN GW in response to the ASR command
Session-Termination-Request	STR	Sent from a PDN GW to a 3GPP AAA Server/Proxy
Session-Termination-Answer	STA	Sent by a server in response to the STR command
Re-Auth-Request	RAR	Sent from a 3GPP AAA Server/Proxy to a PDN GW to request reauthorization
Re-Auth-Answer	RAA	Sent by a PDN GW in response to the RAR command
AA-Request	AAR	Sent from a PDN GW to a 3GPP AAA Server/Proxy
AA-Answer	AAA	Sent by a server in response to the AAR command

15.7.7 3GPP AAA Proxy–3GPP AAA Server/Proxy (SWd)

15.7.7.1 General

The SWd interface is defined between the 3GPP AAA proxy and the 3GPP AAA server. The SWd interface is used in roaming scenarios where the 3GPP AAA proxy is located in the visited network and the 3GPP AAA server is located in the home network.

The 3GPP AAA proxy acts as a Diameter proxy agent and forwards Diameter commands between the Diameter client and the Diameter server (see Figure 15.46).

Figure 15.46: SWd Interface.

15.7.7.2 Interface Functionality

The prime purpose of the protocols crossing this interface is to transport AAA signaling between home and visited networks in a secure manner. The SWd interface may be used in connection with any of the interfaces SWa, STa, SWm, and S6b depending on the particular roaming scenario. The functionality of these interfaces applies to SWd as well.

15.7.7.3 Protocol

The SWd interface uses the Diameter applications and extensions that are used on the SWa, STa, SWm, and S6b interfaces. There is thus no separate Diameter application defined for the SWd interface; instead, the SWa, STa, SWm, and S6b applications are proxied onto SWd. The same Diameter commands as defined for the SWa, STa, SWm, and S6b interfaces are used on SWd as well, depending on the specific roaming scenario.

The protocol stack for SWd is illustrated in Figure 15.47.

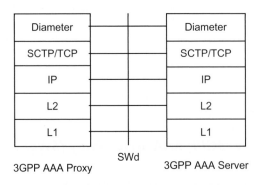

Figure 15.47: Protocol Stack for SWd.

The definition of the interface and its functionality are given in 3GPP TS 23.402. The SWd protocol is defined in TS 29.273.

15.8 PCC-Related Interfaces

15.8.1 General

The PCRF-related interfaces include Gx, Gxa, Gxc, Rx, S9, and Sp (see Section 8.2 for more details on PCC and where these interfaces are located in the architecture). Below we will go through the different interfaces related to PCC and briefly describe each of them.

15.8.2 PCEF–PCRF (Gx)

15.8.2.1 General

The Gx interface is defined between the PCEF (PDN GW) and the PCRF (see Figure 15.48).

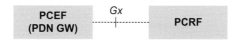

Figure 15.48: Gx Interface.

15.8.2.2 Interface Functionality

The main purpose of the Gx interface is to support PCC rule handling and event handling.

PCC rule handling over the Gx interface includes the installation, modification, and removal of PCC rules. All these operations can be made upon any request coming from the PCEF or due to some internal decision in the PCRF.

The event handling procedures allows the PCRF to subscribe to those events it is interested in. The PCEF then reports the occurrence of an event to the PCRF.

For more details on PCC rule handling and event reporting, see Section 8.2.

15.8.2.3 Protocol

This Gx protocol is based on Diameter and is defined as a vendor-specific Diameter application, where the vendor is 3GPP. 3GPP Release 8 reuses the Gx Diameter application that was defined for Gx in 3GPP Release 7. Only minor updates have been done to Gx in Release 8. The Gx Diameter application is based on the base Diameter protocol and also incorporates commands and AVPs from the Diameter Credit Control Application (DCCA) defined in IETF RFC 4006.

The protocol stack for Gx is illustrated in Figure 15.49.

The definition of the Gx interface and its functionality are given in 3GPP TS 23.203. The Gx Diameter application is defined in TS 29.212. Table 15.8 lists the Diameter commands used in the Gx Diameter application.

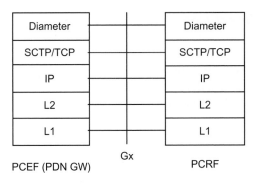

Figure 15.49: Protocol Stack for Gx.

Table 15.8: Diameter Commands Used by the Gx Diameter Application

Command Name	Abbreviation	Comment
CC-Request	CCR	Sent from a PCEF (PDN GW) to a PCRF, e.g. in order to request PCC rules
CC-Answer	CCA	Sent by a PCRF in response to the CCR command
Re-Auth-Request	RAR	Sent by the PCRF to the PCEF in order to provide PCC rules using the push procedure
Re-Auth-Answer	RAA	Sent by a PCEF in response to the RAR command

15.8.3 BBERF–PCRF (Gxa/Gxc)

15.8.3.1 General

The Gxa and Gxc interfaces are located between the PCRF and the BBERF. Gxc applies when the BBERF is located in the Serving GW and Gxa applies when the BBERF is located in an Access GW in a trusted non-3GPP IP access (see Figure 15.50).

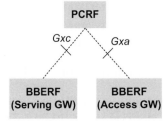

Figure 15.50: Gxa/Gxc Interfaces.

15.8.3.2 Interface Functionality

The main purpose of the Gxa and Gxc interfaces is to support QoS rule and event handling for PCC. This is similar to the Gx interface, with the difference that the Gx interface handles PCC rules instead, while Gxa and Gxc handle QoS rules.

For more details on QoS rule handling and event reporting, see Section 8.2.

15.8.3.3 Protocol

The protocol over the Gxa and Gxc interfaces is based on Diameter. A new Diameter vendor-specific application, the Gxx Diameter application, has been defined and is used on both Gxa and Gxc. The Gxx Diameter application is, similarly to the Gx application, based on the base Diameter protocol and incorporates commands and AVPs from the DCCA defined in IETF RFC 4006.

The protocol stack for Gxa and Gxc is illustrated in Figure 15.51.

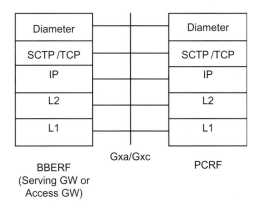

Figure 15.51: Protocol Stack for Gxa/Gxc.

The definition of the Gxa/Gxc interface and its functionality are given in 3GPP TS 23.203. The Gxx Diameter application is defined in TS 29.212. Table 15.9 lists the Diameter commands used in the Gxx Diameter application.

15.8.4 PCRF–AF (Rx)

15.8.4.1 General

The Rx interface is defined between the PCRF and the AF (see Figure 15.52).

15.8.4.2 Interface Functionality

The main purpose of the Rx interface is to transfer session information from the AF to the PCRF. Rx is also used by the AF to subscribe to notifications about traffic plane events, for example that an IP session has been closed or that the UE has handed over

Table 15.9: Diameter Commands Used by the Gxx Diameter Application

Command Name	Abbreviation	Comment
CC-Request	CCR	Sent from a PCEF (PDN GW) to a PCRF, e.g. in order to request QoS rules
CC-Answer	CCA	Sent by a PCRF in response to the CCR command
Re-Auth-Request	RAR	Sent by the PCRF to the PCEF in order to provision QoS rules using the push procedure
Re-Auth-Answer	RAA	Sent by a PCEF in response to the RAR command

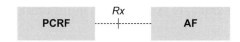

Figure 15.52: Rx Interface.

to a different access technology. The PCRF will notify the AF of the occurrence of a subscribed traffic plane event.

For more details on Rx procedures, see Section 8.2.

15.8.4.3 Protocol
The protocol over the Rx interface is based on Diameter. 3GPP Release 8 reuses the Rx Diameter application that was defined for Rx in 3GPP Release 7. The Rx Diameter application is based on the Diameter base protocol and also incorporates commands from the Diameter Network Access Server (NAS) application defined in IETF RFC 4005. It can be noted, however, that the concept of an NAS (Network Access Server) is not used with Rx; it is merely that Diameter NAS application commands are reused for the Rx protocol, not its functional framework.

The protocol stack for Rx is illustrated in Figure 15.53.

The definition of the Rx interface and its functionality are given in 3GPP TS 23.203. The Rx Diameter application is defined in TS 29.214. Table 15.10 lists the Diameter commands used in the Rx Diameter application.

15.8.5 TDF–PCRF (Sd)

15.8.5.1 General
The Sd interface is defined between the TDF and the PCRF, as illustrated in Figure 15.54.

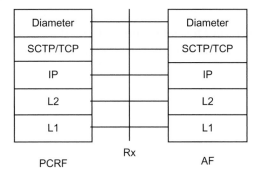

Figure 15.53: Protocol Stack for Rx.

Table 15.10: Diameter Commands Used by the Rx Diameter Application

Command Name	Abbreviation	Comment
AA-Request	AAR	Sent by an AF to the PCRF in order to provide it with the Session Information
AA-Answer	AAA	Sent by the PCRF to the AF in response to the AAR command
Re-Auth-Request	RAR	Sent by the PCRF to the AF in order to indicate an Rx-specific action
Re-Auth-Answer	RAA	Sent by the AF to the PCRF in response to the RAR command
Session-Termination-Request	STR	Sent by the AF to inform the PCRF that an established session shall be terminated
Session-Termination-Answer	STA	Sent by the PCRF to the AF in response to the STR command
Abort-Session-Request	ASR	Sent by the PCRF to inform the AF that bearer for the established session is no longer available
Abort-Session-Answer	ASA	Sent by the AF to the PCRF in response to the ASR command

Figure 15.54: Sd Interface.

15.8.5.2 Interface Functionality

The main purpose of the Sd interface is to support ADC rule handling, usage monitoring control of TDF sessions and of detected applications, and reporting of the start and the stop of a detected application's traffic and transfer of service data flow descriptions for detected applications, if available, from the TDF to the PCRF.

For more details on ADC rule handling and application reporting, see Section 8.2.

15.8.5.3 Protocol

This Sd protocol is based on Diameter and is defined as a vendor-specific Diameter application, where the vendor is 3GPP. The Sd Diameter application is defined from Release 10. The Sd Diameter application is based on the Diameter base protocol and also incorporates commands and AVPs from the Diameter Credit Control Application (DCCA) defined in IETF RFC 4006.

The protocol stack for Sd is illustrated in Figure 15.55.

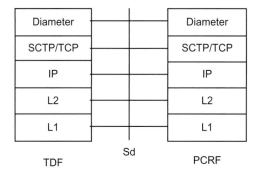

Figure 15.55: Protocol Stack for Sd.

The definition of the Sd interface and its functionality are given in 3GPP TS 23.203. The Sd Diameter application is defined in TS 29.212. Table 15.11 lists the Diameter commands used in the Sd Diameter application.

Table 15.11: Diameter Commands Used by the Sd Diameter Application

Command Name	Abbreviation	Comment
TDF-Session-Request	TSR	Sent by the PCRF to the TDF in order to establish the TDF session and to provision the ADC rules
TDF-Session-Answer	TSA	Sent by a TDF in response to the TSR command
CC-Request	CCR	Sent from a TDF to a PCRF, e.g. in order to request ADC rules and to inform PCRF about the requested application detection
CC-Answer	CCA	Sent by a PCRF in response to the CCR command
Re-Auth-Request	RAR	Sent by the PCRF to the TDF in order to provision ADC rules using the push procedure
Re-Auth-Answer	RAA	Sent by a TDF in response to the RAR command

15.8.6 OCS–PCRF (Sy)

15.8.6.1 General

The Sy interface is defined between the OCS and the PCRF (see Figure 15.56).

Figure 15.56: Sy Interface.

15.8.6.2 Interface Functionality

The main purpose of the Sy interface is to support transfer of information relating to subscriber spending from OCS to PCRF. The Sy interface also allows PCRF to request policy counter status reporting from OCS, and the OCS to notify the PCRF about policy counter status changes.

For more details on subscriber spending handling by PCC, see Section 8.2.

15.8.6.3 Protocol

This Sy protocol is based on Diameter and is defined as a vendor-specific Diameter application, where the vendor is 3GPP. The Sy Diameter application is defined from Release 11. The Sy Diameter application is based on the Diameter base protocol.

The protocol stack for Sy is illustrated in Figure 15.57.

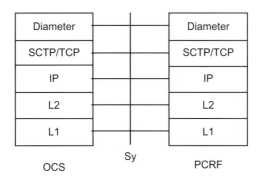

Figure 15.57: Protocol Stack for Sy.

The definition of the Sy interface and its functionality are given in 3GPP TS 23.203. The Sy Diameter application is defined in TS 29.219. Table 15.12 lists the Diameter commands used in the Sy Diameter application.

Table 15.12: Diameter Commands Used by the Sy Diameter Application

Command Name	Abbreviation	Comment
Spending-Limit-Request	SLR	Sent by the PCRF to the OCS as part of the Initial or Intermediate Spending Limit Report Request procedure
Spending-Limit-Answer	SLA	Sent by an OCS in response to the SLR command
Spending-Status-Notification-Request	SNR	Sent by an OCS to the PCRF as part of the Spending Limit Report procedure
Spending-Status-Notification-Answer	SNA	Sent by the PCRF in response to the SNR command
Session-Termination-Request	STR	Sent by the PCRF to the OCS
Session-Termination-Answer	STA	Sent by an OCS in response to the STR command

Note: At the time of writing, the Sy protocol was still a work in progress as part of 3GPP Release 11.

15.8.7 PCRF–PCRF (S9)

15.8.7.1 General

The S9 interface is defined between the PCRF in the home network (H-PCRF) and a PCRF in the visited network (V-PCRF). S9 is an inter-operator interface and is only used in roaming scenarios (see Figure 15.58).

Figure 15.58: S9 Interface.

15.8.7.2 Interface Functionality

The main purpose of the S9 interface is to transfer policy decisions (i.e. PCC rules or QoS rules) generated in the home network into the visited network and transport the events that may occur in the visited network to the home network.

The S9 interface can also be used to transfer session information in specific roaming scenarios. The two main roaming scenarios are the home routed case (with PDN GW/PCEF in the home network) and the Local Breakout (with PDN GW/PCEF in the visited network). In the Local Breakout case, it is also possible to use an AF either in the home network or in the visited network. When Local Breakout is used and the AF is in the visited network, the S9 interface carries service session information from the V-PCRF to the H-PCRF (see Section 8.2 for further details regarding PCC usage in roaming scenarios).

15.8.7.3 Protocol

The protocol over the S9 interfaces is based on Diameter. Two 3GPP vendor-specific Diameter applications are used over the S9 interface: the S9 application and the Rx application.

The S9 Diameter application is a new vendor-specific application defined in 3GPP Release 8. It is based on the base Diameter protocol and incorporates commands and AVPs from the DCCA defined in IETF RFC 4006.

The Rx Diameter application is described above for the Rx interface. It is used over the S9 interface in the case of Local Breakout with the AF in the visited network, as described above.

The protocol stack for S9 is illustrated in Figure 15.59.

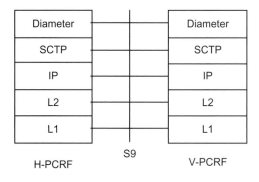

Figure 15.59: Protocol Stack for S9.

The definition of the S9 interface and its functionality are given in 3GPP TS 23.203. The protocols over the S9 interface, including the S9 Diameter application, are defined in TS 29.215. The Rx Diameter application is defined in TS 29.214. Table 15.13 lists the Diameter commands used in the S9 Diameter application. The Rx Diameter application commands, used on the S9 interface, are listed in the description of the Rx interface above.

15.8.8 BPCF–PCRF (S9a)

15.8.8.1 General

The S9a interface is defined between the BPCF in a Fixed Broadband Access network and a PCRF. The PCRF may be either in the home network (PCRF in non-roaming

Table 15.13: Diameter Commands Used by the S9 Diameter Application

Command Name	Abbreviation	Comment
CC-Request	CCR	Sent by the V-PCRF to the H-PCRF in order to request PCC/QoS rules or to indicate bearer or PCC/QoS rule-related events
CC-Answer	CCA	Sent by the H-PCRF to the V-PCRF in response to the CCR command. It is used to provide PCC/QoS rules and event triggers
Re-Auth-Request	RAR	Sent by the H-PCRF to the V-PCRF in order to provide QoS/PCC rules and event triggers
Re-Auth-Answer	RAA	Sent by the V-PCRF to the H-PCRF in response to the RAR command

case) or in a visited network (V-PCRF roaming case). S9a is an inter-operator interface and is only used in roaming scenarios (see Figure 15.60).

Figure 15.60: S9a Interface.

15.8.8.2 Interface Functionality

The main purpose of the S9a interface is to transfer policy decisions (i.e. PCC rules or QoS rules) generated in the 3GPP network into the Fixed Broadband Access network.

The S9 interface can also be used to transfer session information in specific traffic scenarios. The two main roaming scenarios are the EPC routed case (with traffic passing a PDN GW in the 3GPP network) and the non-seamless WiFi offloading case (with traffic offloaded in the Fixed Broadband Network without passing EPC). In the WiFi offloading case, it is also possible to use an AF either in the 3GPP network or in the Fixed Broadband Access network. When an AF in Fixed Broadband Access is used, the S9a interface may also carry service session information from the BPCF to the PCRF (see Section 8.2 for further details regarding PCC usage in WiFi offloading scenarios).

15.8.8.3 Protocol

The protocol over the S9a interfaces is based on Diameter. Two 3GPP vendor-specific Diameter applications are used over the S9a interface: the S9a application and the Rx application.

The S9a Diameter application is a new vendor-specific application defined in 3GPP Release 11. It is based on the Diameter base protocol and incorporates commands and AVPs from the DCCA defined in IETF RFC 4006.

The Rx Diameter application is described above for the Rx interface. It is used over the S9a interface in the case of WiFi offloading with the AF in the Fixed Broadband Network, as described above.

The protocol stack for S9a is illustrated in Figure 15.61.

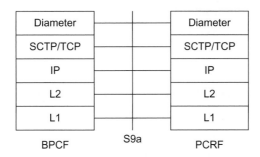

Figure 15.61: Protocol Stack for S9a.

The definition of the S9a interface and its functionality are given in 3GPP TS 23.203. The protocols over the S9a interface, including the S9a Diameter application, are defined in TS 29.215. The Rx Diameter application is defined in TS 29.214. At the time of writing, the S9a protocol was still a work in progress as part of 3GPP Release 11.

15.8.9 SPR–PCRF (Sp)

15.8.9.1 General
The Sp interface is defined between the PCRF and the Subscriber Profile Repository (SPR) (see Figure 15.62).

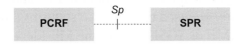

Figure 15.62: Sp Interface.

15.8.9.2 Interface Functionality
The Sp interface is used to transport subscription data from the SPR to the PCRF. The PCRF may request subscription data for a given user. The SPR may also notify the PCRF if the subscription data has been modified.

The SPR may contain subscription data relating to the transport level policies for the access network. The details regarding the subscription data contained in the SPR have not been further specified. A reason for not specifying the detailed subscription data is that the policies may depend significantly on the operator's business models and the types of subscriptions and services offered. It is thus reasonable to avoid detailed specification in order not to put unnecessary restrictions on the type of policies that can be kept in the SPR.

15.8.9.3 Protocol
The Sp interface and its functionality are specified in 3GPP TS 23.203. The protocol over the Sp interface is, however, not specified.

15.9 EIR-Related Interfaces

15.9.1 MME-EIR and SGSN-EIR Interfaces (S13 and S13′)

15.9.1.1 General
The interface S13 is defined between the Equipment Identity Register (EIR) and the MME, and the interface S13′ is defined between EIR and SGSN. The S13′ interface applies to the S4-based SGSN only (see Figure 15.63).

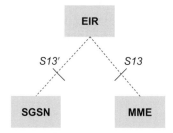

Figure 15.63: S13 and S13′ Interfaces.

15.9.1.2 Interface Functionality
The S13 and S13′ interfaces between the MME and the EIR and between the SGSN and the EIR respectively are used to check the status of the UE (e.g. if it has been reported stolen). The MME or SGSN checks the ME Identity by sending the Equipment Identity to an EIR and analyzing the response.

15.9.1.3 Protocol
The same protocol is used on both S13 and S13′. This protocol is based on Diameter and is defined as a vendor-specific Diameter application, where the vendor is 3GPP. The S13/S13′ Diameter application is based on the Diameter base protocol but defines new Diameter commands and AVPs to implement the functions described above.

Diameter messages over the S13 and S13′ interfaces use the SCTP (IETF RFC 2960) as a transport protocol. The protocol stack is illustrated in Figure 15.64.

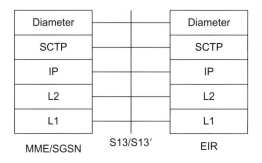

Figure 15.64: Protocol Stack for S13/S13′.

The protocol over the S13/S13′ interface, including the S13/S13′ Diameter application, is defined in 3GPP TS 29.272. Table 15.14 lists the Diameter commands used in the S13/S13′ Diameter application.

Table 15.14: Diameter Commands Used by the S13/S13′ Diameter Application

Command Name	Abbreviation	Comment
ME-Identity-Check-Request	ECR	Sent from MME or SGSN to EIR
ME-Identity-Check-Answer	ECA	Sent from EIR to MME or SGSN as reply to the ECR

15.10 I-WLAN-Related Interfaces

15.10.1 UE–ePDG (SWu)

15.10.1.1 General

The SWu interface is defined between the UE and the ePDG. The interface runs over an untrusted non-3GPP IP access (see Figure 15.65).

Figure 15.65: SWu Interface.

15.10.1.2 Interface Functionality

The SWu interface supports procedures for establishment or disconnection of an end-to-end tunnel between the UE and the ePDG. The tunnel establishment is always initiated by the UE, whereas the tunnel disconnection can be initiated by the UE or the ePDG. The SWu interface also supports tunnel modification procedures in order to update the tunnel if the UE has acquired a new IP address from the untrusted non-3GPP IP access, for example if the UE has moved to another untrusted non-3GPP IP access. For further details on tunnel management over untrusted non-3GPP IP access, see Section 7.3.

15.10.1.3 Protocol

The tunnel between UE and ePDG is an IPsec tunnel. The UE and ePDG use IKEv2 to establish the IPSec security association (SA) for the tunnel.

The UE uses standard DNS mechanisms in order to select a suitable ePDG. As input to the DNS query, the UE creates an FQDN based on the operator identity. As a reply from the DNS system, the UE receives one or more IP addresses of suitable ePDG(s). Once the ePDG has been selected, the UE initiates the IPsec tunnel establishment procedure using the IKEv2 protocol. Public key signature-based authentication with certificates, as specified for IKEv2, is used to authenticate the ePDG. EAP-AKA within IKEv2 is used to authenticate the UE. As part of the IKEv2 procedure, an IPSec SA is established. IPSec Encapsulated Security Payload (ESP) in tunnel mode is used for the IPSec tunnel between the UE and the ePDG.

SWu also supports the mobility extensions for IKEv2 defined by MOBIKE (IETF RFC 4555). This allows the IPSec SA to be updated if the UE acquires a new IP address in the untrusted non-3GPP IP access.

For further details on IKEv2, MOBIKE, and ESP, see Section 16.9.

The protocol stack for SWu is illustrated in Figure 15.66. The tunnel management procedures for SWu are defined in 3GPP TS 33.402.

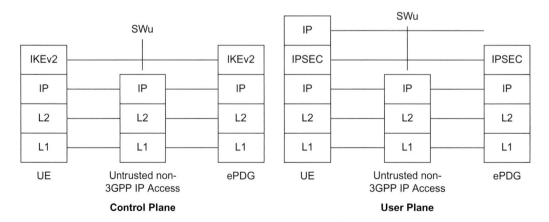

Figure 15.66: Protocol Stack for SWu.

15.11 ANDSF-Related Interfaces

The Access Network Discovery and Selection Function (ANDSF) is a mechanism that allows the UE to be provided with relevant parameters for intersystem mobility policy and access network discovery. This is done using the S14 interface, which utilizes Open Mobile Alliance (OMA) Device Management (DM). A brief outline of OMA DM is provided below in order to place the discussion regarding the ANDSF-UE interface in the correct context. A detailed description of OMA DM is beyond the scope of this book and the interested reader is referred to Brenner and Unmehopa (2008). A general overview of ANDSF can be found in Section 6.4.

The OMA DM v1.2 specifications are based on the OMA DM v1.1.2 specifications and make use of the OMA SyncML Common v1.2 specifications as given in the OMA SyncML Common specifications Enabler Release Definition for SyncML.

DM is the technology that allows the ANDSF to configure the UE on behalf of the operator and the end-user. Using DM, the ANDSF remotely sets parameters via the use of a Managed Object (MO). The MO is organized in nodes: interior nodes and leaf nodes. The leaf nodes contain the actual parameter values. The ANDSF MO contains nodes related to Inter System Mobility Policy (ISMP), Discovery Information, UE Location, and Inter System Routing Policy (ISRP). There is also an additional node defined, Ext, for vendor-specific requirements. The overall ANDSF MO is outlined in Figure 15.67.

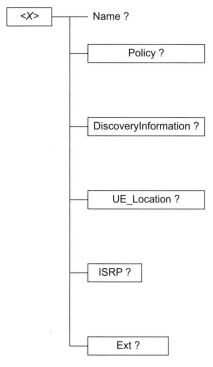

Figure 15.67: ANDSF MO.

The relation between ISMP, ISRP, and Discovery Information is that ISMP prioritizes the access network when the UE is not capable of connecting to the EPC through multiple accesses. ISRP indicates how to distribute traffic among available accesses when the UE is capable of connecting to the EPC through multiple accesses (i.e. the UE is configured for IFOM, MAPCON, non-seamless WLAN offloading, or any combination of these capabilities), while Discovery Information provides further information for the UE to access the access network defined in the ISMP or in the ISRP.

15.11.1 ISMP Policy Node

The policy node acts as a placeholder for policies relating to intersystem mobility, including the rules and also the priority in which the policies should be applied. It also allows the ANDSF to set a particular access technology as the prioritized access, meaning that this is the one that should be searched for first, including the access network ID. Also included within the Policy node is the Validity Area for a particular rule; this is used in the case where a particular rule is applicable to a particular location that the UE may find itself in. The IMSP Policy node is illustrated in Figure 15.68.

With regard to Validity Areas, these can be different for different access networks; as a result, there are leaves that describe the validity areas and rules associated with them for each radio access technology type: 3GPP, 3GPP2, WiMAX, and WiFi.

In the case where a UE is roaming, the roaming leaf contains roaming conditions that should be applied in this case. The UE will only apply such rules, however, if the UE's roaming state matches the one specified in the roaming value.

Particular rules may be applicable at a certain time of day and as a result there is also a leaf named TimeOfDay that handles this scenario.

If the UE finds itself in a situation where it feels that a rule is no longer valid, it uses the UpdatePolicy leaf to determine whether it needs to request an update of its intersystem mobility policy or not.

15.11.2 Discovery Information Node

Using the Discovery Information node, an operator may provide information on the access networks that are available for a UE to connect to. The UE, meanwhile, may use this information in order to decide which access network to connect to.

The Discovery Information node, therefore, provides information regarding the type of access network, the area that the access network covers. The leaf describing the access network areas again covers all of the different network access types: 3GPP,

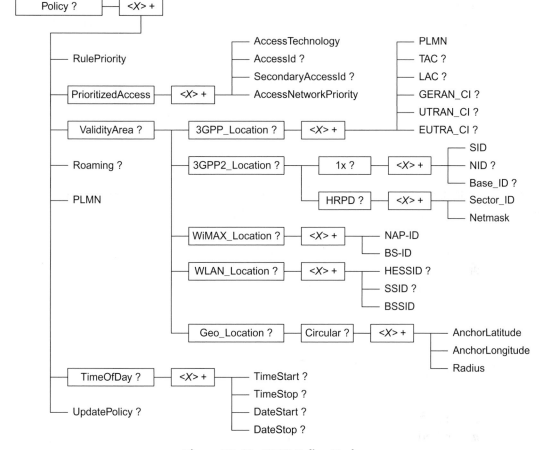

Figure 15.68: ISMP Policy Node.

3GPP2, WiMAX, WiFi. It also covers the use of Geo_Location, which acts as a placeholder for the geographic location of one or more access networks (see Figure 15.69).

15.11.3 UE Location Node

The UE Location node (see Figure 15.70) acts as a placeholder for location descriptions; a UE therefore inserts information regarding all of the access networks that it can discover into this node. For 3GPP networks, this includes PLMN, Tracking Area Code, Location Area Code, and Cell Global Identity. For 3GPP2 networks, this includes SID, NID, and Base ID. For WiMAX networks, UE Location includes NAP-ID and BS-ID. For WiFi networks, SSID and BSSID are captured.

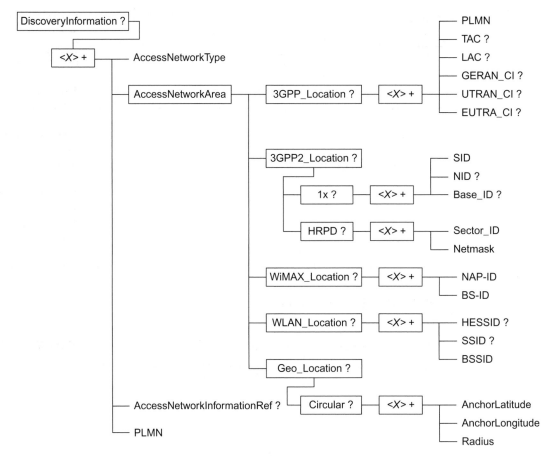

Figure 15.69: Discovery Information Node.

15.11.4 ISRP Node

The ISRP information consists of a set of one or more ISRP rules. Each ISRP rule contains indication of traffic distribution for UEs that are configured for IFOM, MAPCON, or non-seamless WiFi offloading (see Figure 15.71).

Each ISRP rule has a PLMN leaf and an optional Roaming leaf. Roaming and PLMN leaves are used by the UE to determine if an ISRP rule is applied. At any point in time there will be at most one ISRP rule applied.

An ISRP rule can contain one or more flow distribution containers for flow distribution rules. These are: ForFlowBased for IFOM service (Figure 15.72), ForServiceBased for MAPCON (Figure 15.73), and ForNonSeamlessOffload for non-seamless WiFi offloading (Figure 15.74). A UE evaluates only the supported flow distribution containers of the "active" ISRP rule.

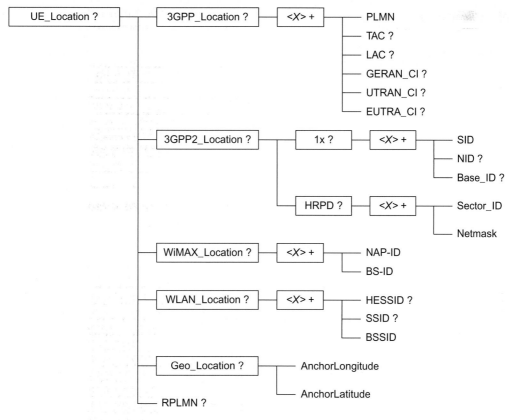

Figure 15.70: UE Location Node.

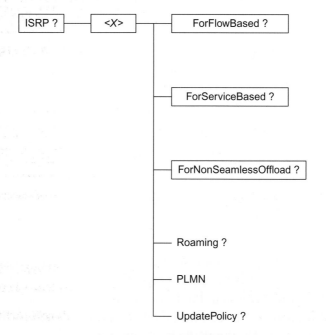

Figure 15.71: ISRP Node.

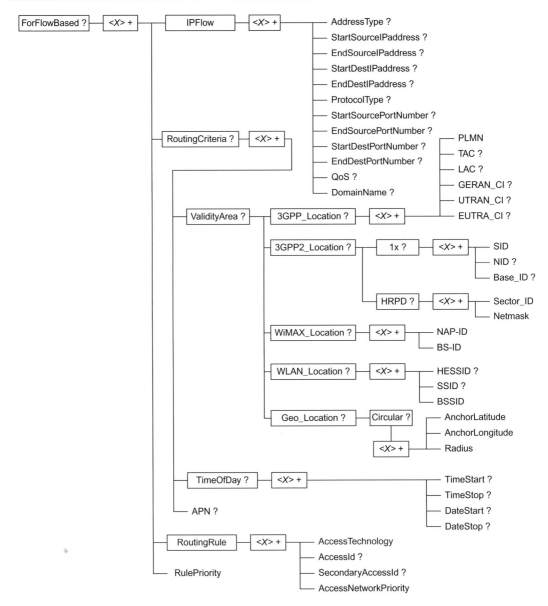

Figure 15.72: ForFlowBased Distribution Container.

A flow distribution rule has a number of results (e.g. preferred access technology and restricted access technology) defined in the RoutingRule node to be used whenever the flow distribution rule is applied. Each flow distribution rule also has a mandatory node identifying the data traffic (e.g. based on APN or IP flow description) to which the results contained in the RoutingRule node apply.

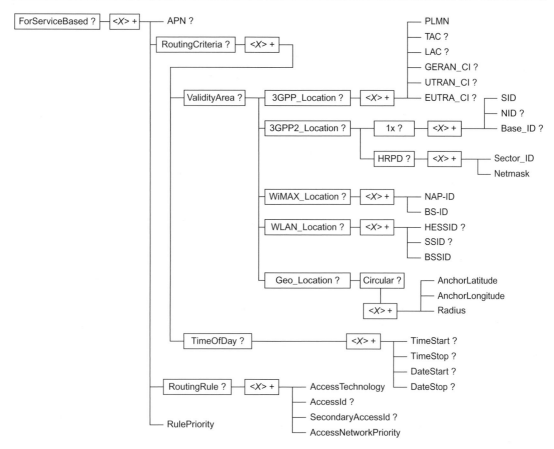

Figure 15.73: ForServicesBased Distribution Container.

15.11.5 Ext Node

Ext is the node where vendor-specific information about the ANDSF MO is placed. For the purposes of this node, vendor means application vendor, device vendor, etc. This is generally indicated by a vendor-specific name under the Ext node. As can be expected, since this is a vendor-specific node, the leaves under the Ext node are left undefined. If a vendor wishes to utilize extensions, they define the interior nodes and leaves themselves. These are therefore naturally not within the scope of the standardization.

For further details on the ANDSF MO, see 3GPP TS 24.312.

15.12 HRPD IW-Related Interfaces

15.12.1 Optimized Handover and Related Interfaces (S101 and S103)

In order to support optimized handover between LTE and eHRPD networks, as explained in the preceding chapter, one control-plane (S101) and one user-plane

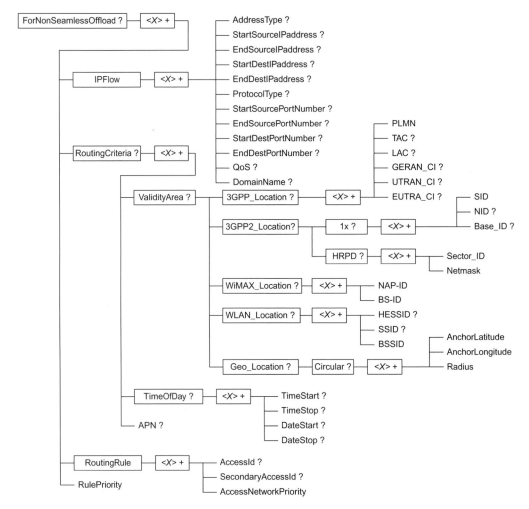

Figure 15.74: ForNonSeamlessOffload Distribution Container.

(S103) interface have been added to the architecture. Below we describe the functions and protocol supported over these two interfaces.

15.12.2 MME ↔ eHRPD Access Network (S101)

S101 is a tunnel between the MME and eHRPD access network where messages are carried over the serving access network towards a target access network (where the handover may occur) in order for preparation for handover via pre-registration; it then maintains the resources via session maintenance and then the actual handover. These are messages tunneled over S101-AP, where the GTPv2-C protocol functions are used with explicit utilization for the S101 interface. The GTPv2-C message type used for S101 is as follows:

0	Reserved
1	Echo Request
2	Echo Response
3	Version Not Supported Indication
4	Direct Transfer Request message
5	Direct Transfer Response message
6	Notification Request message
7	Notification Response message
8–24	For future S101 interface use
25–31	Reserved for Sv interface
32–255	Reserved for GTPv2-C spec

The protocol itself is segmented to provide the preconfigured tunnel via Path management general messages and then for specific messages used for information transfer over the control plane (see Figure 15.75).

Figure 15.75: S101 Interface.

3GPP TS 29.276 and TS 29.274 describe the details for the messages. Procedures are covered in TS 23.402.

The messages carried over this interface are not modified by MME but, rather, forwarded to/from the source and the target access network (in this case between eNodeB and HRPD AN). Each message must have a unique identifier (also known as Session Identifier) to be able to identify uniquely in a global network the individual terminal the message is destined to or coming from.

The preconfigured tunnel carries messages from the MME or eHRPD access network towards its peer in the target network using Direct Transfer Message and Response. Depending on where the message originates, the content is according to that specific access network (e.g. if an HRPD message is to be tunneled then this message shall be carried over a transparent container) (see Figure 15.76).

In case of a Notification Request Message, the information mainly carries events such as completion of handover process.

The GTPv2-C Path management and Reliability procedures are used to manage the preconfigured tunnel S101.

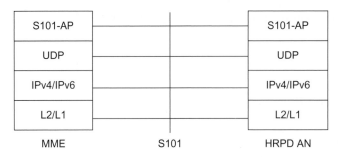

Figure 15.76: Protocol Stack for S101.

The S101 message header takes the following form (bits marked with an asterisk are spare and set to zero):

Examples of Direct Transfer Message information elements are described below. The parameter usage during the optimized handover procedure is described in Chapter 17.

Information Element	Presence requirement
IMSI	Mandatory
HRPD Sector ID	Conditional
S101 Transparent Container	Mandatory
PDN GW PMIP GRE Tunnel Info	Conditional
S103 GRE Tunnel Info	Conditional
S103 HSGW IP Address	Conditional
Handover Indicator	Conditional
Tracking Area Identity	Conditional
Recovery	Conditional
Private Extension	Optional

Note that, for interested readers, 3GPP TS 29.276 is the specification for S101.

15.12.3 Serving GW ↔ HSGW (S103)

The S103 interface is from the Serving GW to the HSGW in the CDMA HRPD network (see Figure 15.77).

Figure 15.77: S103 Interface.

This interface provides support for forwarding of downlink data during handover from LTE to HRPD. The purpose of this interface is to reduce loss of user data during the handover procedure. Signaling procedures and parameters available on the S101 interface are used to set up a GRE tunnel on the S103 interface. For details on GRE tunnels, see Section 16.6 for the overview and usage of GRE. The protocol stack for S103 is described in Figure 15.78.

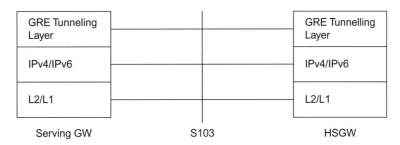

Figure 15.78: Protocol Stack for S103.

The S103 interface must be able to identify the user data traffic on a per-terminal and per-PDN connection basis.

15.13 Interface to External Networks

15.13.1 General

The SGi interface is defined between the PDN GW and external IP networks (also called "PDNs"), as shown in Figure 15.79. The external networks may be Internet and/or Intranets and IPv4 and/or IPv6 may be used.

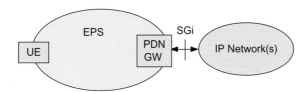

Figure 15.79: IP Network Interworking.

The PDN GW is the access point of the EPS and the EPS will look like any other IP network or subnetwork. From the external IP network's point of view, the PDN GW is seen as a normal IP router.

15.13.2 Functions

Access to Internet, Intranet, or an ISP involves functions such as IPv4 address allocation and IPv6 address autoconfiguration, and may also involve specific functions such as authentication, authorization, and secure tunneling to Intranet/ISP.

An operator may offer direct transparent access to the Internet and operator services, and will in that case offer at least basic ISP functions. An operator may also offer so-called non-transparent access to an Intranet or ISP.

In both the transparent and non-transparent cases, the UE is given an IPv4 address and/or an IPv6 prefix. The difference is that in the transparent case the IP addresses belong to the operator while in the non-transparent case the addresses belong to the Intranet/ISP addressing space.

IPv4 address and/or IPv6 prefix are assigned either statically tied to the subscription or dynamically allocated at PDN connectivity establishment. This IPv4 address and/or IPv6 prefix is used for packet forwarding between the Internet and the PDN GW and within the packet domain. With IPv6, Stateless Address Autoconfiguration will be used to assign an IPv6 address to the UE. The PDN GW may use a local IP address pool, or use DHCP or AAA protocols to retrieve UE IP addresses from the external IP network. For more information on IP address allocation for different cases, see 3GPP TS 29.061 and 23.401.

The PDN GW prevents IP spoofing by verifying the source IP address of the IP packets issued by the UE and comparing it with the allocated address.

To support IMS the PDN GW will also need to provide a list of P-CSCF addresses to the UE on request. The UE needs this information in order to register for IMS services. In addition, IMS requires a bearer for the IMS signaling and bearers for media when IMS sessions are established. This means that PDN GW needs to be configured to allow the IMS signaling and it needs to support the Gx interface to the PCRF in order to allocate bearers for the IMS media flows.

15.14 CSS Interface

15.14.1 MME–CSS Interface (S7a)

15.14.1.1 General
The interface S7a is defined between the CSG Subscriber Server (CSS) and the MME (see Figure 15.80).

15.14.1.2 Interface Functionality
The S7a interface is used to transfer CSG subscription data for roaming subscribers between the MME and the CSS. The CSS is an optional element that stores VPLMN-

Figure 15.80: S7a Interface.

specific CSG subscription information and provides it to the MME. The CSS is always in the same PLMN as the current MME. If the same CSG ID exists in both CSS subscription data and HSS subscription data, the CSG subscription data from the HSS shall take precedence over the data from CSS.

15.14.1.3 *Protocol*

The protocol on S7a is based on Diameter and is defined as a vendor-specific Diameter application, where the vendor is 3GPP. The S7a Diameter application is based on the Diameter base protocol but defines new Diameter commands and AVPs to implement the functions described above. Diameter messages over S7a use SCTP (IETF RFC 2960) as a transport protocol. The protocol stack is illustrated in Figure 15.81.

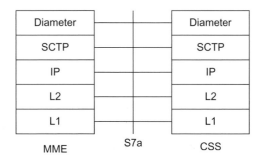

Figure 15.81: Protocol Stack for S7a.

The protocol over the S7a interface, including the S7a Diameter application, is defined in 3GPP TS 29.272. At the time of writing, the stage 3 work for S7a had just started and the Diameter commands were not specified.

CHAPTER 16

Protocols

16.1 Introduction

This chapter considers the main protocols used in the EPS, with the aim of giving a basic overview of these protocols and their basic properties.

16.2 GPRS Tunneling Protocol Overview

The original version of the GTP protocol was the result of GSM standards development to cater for specific needs such as mobility and bearer management and tunneling of user data traffic for GPRS. Then 3GPP further enhanced GTP for use in 3G. During the development of EPS, the GTP track of the architecture was enhanced considerably to improve the bearer handling and thus the GTP control-plane protocol was upgraded to GTPv2-C.

The two main components of GTP are the control-plane part of GTP (GTP-C) and the user-plane part of GTP (GTP-U). GTP-C is used to control and manage tunnels for individual terminals connecting to the network in order to establish user data path. The GTP-U uses a tunnel mechanism to carry the user data traffic. There also exists GTP′, which is defined under the GTP protocol umbrella for the purpose of charging, but in this book we will not discuss this legacy protocol usage of GTP. There exist three versions of GTP-C: GTPv0, GTPv1, and GTPv2. There are two versions of GTP-U: GTPv0 and GTPv1. In this book we will provide some background on GTPv1-C for better understanding of GTPv2-C, which is used exclusively for EPS. We will also discuss some details on GTPv1-U. Since Release 8, 3GPP no longer supports changes to GTPv0 protocol or interworking with GTPv1 protocol.

In order to understand the functions of the GTP protocol, it is useful to have a look at how GTP has been used in GPRS for 2G and 3G. Figure 16.1 illustrates the interfaces that use GTP.

In the case of GPRS, the Gn interface between SGSNs and between SGSN and GGSN (when the entities are within an operator's PLMN), and the Gp interface

EPC and 4G Packet Networks.
DOI: http://dx.doi.org/10.1016/B978-0-12-394595-2.00016-5

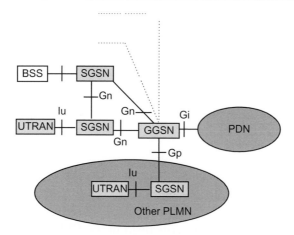

Figure 16.1: GTP Interfaces for GPRS.

between SGSN and GGSN (inter-PLMN or inter-operator may be more common term) support GTPv1-C and GTPv1-U protocols. For 3G packet core using WCDMA/ HSPA radio accesses, Iu supports the GTPv1 user-plane protocol.

In the case of EPS, the interfaces between SGSN and MME, between MMEs, between SGSNs, between Serving GW and PDN GW, and between Serving GWs use the GTPv2-C, and the interface between the HRPD Access Network and MME uses GTPv2-C tunnels to carry the tunneled messages. GTPv1-U is used between eNodeB and Serving GW, between RNC and Serving GW, and between SGSN and Serving GW, as well as between Serving GWs. Thus, the GTPv2-C is used on S3, S4, S5, S8, S10, S11, and S16 interfaces, and GTPv1-U is used on S1-U, Iu-U, S4, S5, S8, S12, and X2-U interfaces. In non-3GPP access support via EPC, GTPv2-C has been introduced on the reference points S2a and S2b, further expanding the network-based mobility protocol usage outside of 3GPP accesses. Figure 16.2 illustrates all the EPS interfaces that support GTPv2-C and GTP-U. For more details on the use of GTPv2-C on S2a/S2b, see Section 6.5.

So what is this protocol suite and how does it work? As can easily be seen from the 3GPP architecture, the entities supporting the GTP protocol need to support one-to-many and many-to-many relationships with each other. A single SGSN must be able to connect to multiple RNCs, SGSNs, as well as many GGSNs within and between different operators' networks. Similarly, a GGSN must be able to connect to multiple SGSNs from different operators' networks spanning significant geographical areas in order to support its own subscribers who may be in their home network or in a roaming partner's network. Development of GTP protocol caters to such diverse

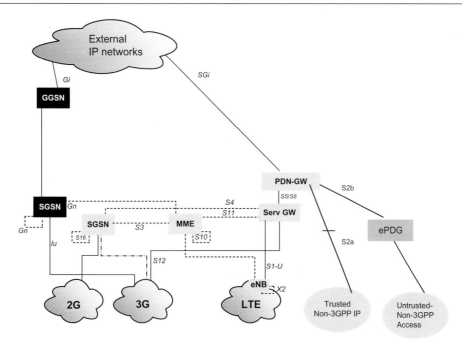

Figure 16.2: GTP Interfaces for 3GPP and Non-3GPP Access Using EPC.

deployment requirements that it is a cornerstone of the success of mobile systems worldwide.

If we look at the original GTP message structure (see Table 16.1), it becomes quite obvious that it serves to manage a cellular network by developing and grouping messages according to the functional needs of a cellular system. Then when we look at the GTPv2 messages developed for EPS in the next section (see Table 16.3) we clearly see the evolution of the protocol in a more generic manner catering to more flexible bearer management as well as simplified/unified network element interactions and support for mobility/common core functions for non-3GPP access networks, and better error/failure/network management as well as restoration and recovery handling for network elements such as MME and Serving GW/PDN GW.

The key functions of GTP protocol (GTPv1 C) were originally built upon for GPRS and additional extensions made to support EPS in GTPv2-C are described next.

1. **Mobility Management.** The set of messages created as part of this function includes managing a mobile device's identification and maintaining presence/ status among various network elements in a coordinated manner, handling data transfer between entities during/at handover/relocation of the mobile terminal.

Table 16.1: GTPv1 Control-Plane Messages (GPRS)

Message Type Value (Decimal)	Message	GTP-C
0	For future use. Shall not be sent. If received, shall be treated as an Unknown message	
1	Echo Request	X
2	Echo Response	X
3	Version Not Supported	X
4	Node Alive Request	
5	Node Alive Response	
6	Redirection Request	
7	Redirection Response	
8–15	For future use. Shall not be sent. If received, shall be treated as an Unknown message	
16	Create PDP Context Request	X
17	Create PDP Context Response	X
18	Update PDP Context Request	X
19	Update PDP Context Response	X
20	Delete PDP Context Request	X
21	Delete PDP Context Response	X
22	Initiate PDP Context Activation Request	X
23	Initiate PDP Context Activation Response	X
24–25	For future use. Shall not be sent. If received, shall be treated as an Unknown message	
26	Error Indication	
27	PDU Notification Request	X
28	PDU Notification Response	X
29	PDU Notification Reject Request	X
30	PDU Notification Reject Response	X
31	Supported Extension Headers Notification	X
32	Send Routing Information for GPRS Request	X
33	Send Routing Information for GPRS Response	X
34	Failure Report Request	X
35	Failure Report Response	X
36	Note MS GPRS Present Request	X
37	Note MS GPRS Present Response	X
38–47	For future use. Shall not be sent. If received, shall be treated as an Unknown message	
48	Identification Request	X
49	Identification Response	X
50	SGSN Context Request	X
51	SGSN Context Response	X
52	SGSN Context Acknowledge	X
53	Forward Relocation Request	X
54	Forward Relocation Response	X
55	Forward Relocation Complete	X
56	Relocation Cancel Request	X
57	Relocation Cancel Response	X

(Continued)

Table 16.1: GTPv1 Control-Plane Messages (GPRS) (Continued)

Message Type Value (Decimal)	Message	GTP-C
58	Forward SRNS Context	X
59	Forward Relocation Complete Acknowledge	X
60	Forward SRNS Context Acknowledge	X
61–69	For future use. Shall not be sent. If received, shall be treated as an Unknown message	
70	RAN Information Relay	X
71–95	For future use. Shall not be sent. If received, shall be treated as an Unknown message	
96–105	MBMS	X
106–111	For future use. Shall not be sent. If received, shall be treated as an Unknown message	
112–121	MBMS	X
122–127	For future use. Shall not be sent. If received, shall be treated as an Unknown message	
128	MS Info Change Notification Request	X
129	MS Info Change Notification Response	X
130–239	For future use. Shall not be sent. If received, shall be treated as an Unknown message	
240	Data Record Transfer Request	
241	Data Record Transfer Response	
242–254	For future use. Shall not be sent. If received, shall be treated as an Unknown message	
255	G-PDU	

Messages in this category include Forward Relocation, Context Request, Identity Request, and Detach Handling.

2. **Tunnel Management.** This involves creation and deletion of the end-user's session, and creation, modification, and deletion of bearers established during the period the user is connected and actively involved in services by the network. Downlink traffic triggering Downlink Data Notification messages to page the UE and maintaining GTP tunnels for the UE all fall into this category. Simply stated, these messages keep the user's different service requirements maintained in the network as the user moves around within and between PLMNs.

3. **Service-Specific Functions.** For GTPv1 these include mainly support of MBMS-related functions. For GTPv2, MBMS service has been developed and protocol effects on GTPv2 include addition of messages supporting procedures like Session Start/Stop/Update and the corresponding responses. GTPv2 provides messages in order to support CS Fallback/SRVCC procedure like Resume/Suspend, optimized handover with 3GPP2, and non-3GPP mobility such as Create Forwarding Tunnel Request/Response messages.

4. **Mobile Terminal Information Transfer.** For GTPv2 this is incorporated within Mobility Management and is only supported for GERAN/UTRAN accesses.

5. **System Maintenance (path management/error handling/restoration and recovery/trace).** This supports network level functions in order to handle overall robustness of the tunnels and recovery from failure in network entities. These messages (such as Echo Request/Response) have been supported in GTPv1 and now in GTPv2 but, wherever possible, for GTPv2 improvements have been made in error handling and recovery procedures. PDN GW restart notification message, indicating PDN GW failure status to MME/SGSN, is supported in GTPv2.

Some messages have been removed on going from GTPv1 to GTPv2 since their associated functions are no longer supported in the system; one example is the messages relating to the function Network Initiated PDP Context Setup.

16.2.1 *Protocol Structure*

Let us now first take a look at the GTPv1 protocol structure. It can be seen in Figure 16.1, where the GTP-C protocol provides the messages to carry out functions such as mobility management, bearer management (also referred to as tunnel management), location management, and mobile terminal status reporting. GTPv2 follows a similar structure, but some groups of messages are not required for systems operation and are thus not supported, as discussed in Section 16.2 above. It should be quite clear that the GTP-C and GTP-U tunnels are associated with each other for any single specific user, since their role is to establish connections throughout the network so that the terminal can send/receive data. Table 16.1 illustrates the key GTPv1 control-plane messages for GPRS.

For GTPv1-C, some example messages that carry out the functions mentioned above are provided here before we delve into the protocol details.

For GTPv1-C, some example message flows between SGSN and GGSN are shown in Table 16.2.

Table 16.2: Functional Messages in GTPv1 in GPRS

Functions	Message Name	Entities	Interface
Mobility management	SGSN Context Request	SGSN–SGSN	Gn
	Forward Relocation Request	SGSN–SGSN	Gn
Tunnel management	Create PDP Context	SGSN–GGSN	Gn/Gp
	Update PDP Context	SGSN–GGSN	Gn/Gp
Path management	Echo Request	SGSN–GGSN	Gn/Gp

Similar messages for the EPS network are shown in detail along with the interface details later on.

Examples of GTPv1-U messages can be described similarly, though note that the main purpose of these control messages is to ensure "smooth" user data traffic handling in the uplink and downlink directions for the end-user. These messages include Echo request/response for path management purposes and Error Indication messages for exception handling. A GTP entity may use the Echo request to find out if the other GTP entity is live. The Error Indication messages can be used to inform the other GTP entity that there is no EPS bearer (or PDP context in the case of GPRS) corresponding to a received user-plane packet. The actual control signaling for GTP-U is performed over S1-AP (for MME and eNodeB) and GTPv2-C (for the core network entities), and over RANAP and GTPv1/v2-C for RNC and core network entities (Figure 16.1).

Let us now go into more detail on the GTP tunnels and their basic structure. For those readers interested in the details of the protocols, such as all the messages, the coding of the parameters, and the interworking of the formats themselves, we recommend the specifications where GTP-C protocols are defined, i.e. 3GPP TS 29.060 (GTPv1) and TS 29.274 (GTPv2-C); the GTP-U protocol is defined in 3GPP TS 29.060 and TS 29.281.

Figures 16.3–16.5 illustrate the format of GTP-C messages.

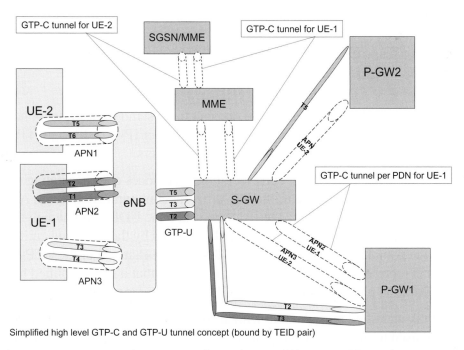

Simplified high level GTP-C and GTP-U tunnel concept (bound by TEID pair)

Figure 16.3: GTP Tunnel Representation Between Different Entities and UE Bearers.

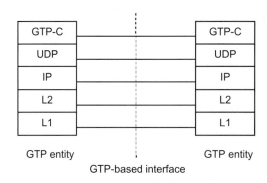

Figure 16.4: GTP Control Plane Protocol Stack.

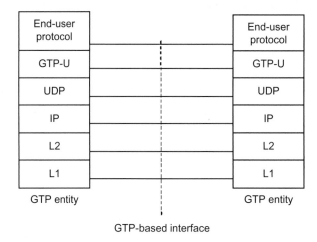

Figure 16.5: GTP Control User Protocol Stack.

A few concepts must be described before one can understand GTP protocol handling. GTP is a tunneling protocol over UDP/IP (it can be either IPv4 or IPv6). GTP is a tunnel with specific tunnel definition and tunnel identifiers.

GTP tunnels are used between two corresponding GTP nodes communicating over a GTP-based interface to separate traffic into different communication flows. A local Tunnel Endpoint Identifier (TEID), the IP address, and the UDP port uniquely identify a tunnel endpoint in each node, where the TEID assigned by the receiving entity must be used for the communication. Figure 16.3 illustrates an example of the GTP-C and GTP-U tunnel representation in EPS for terminals. Note that it is a simplified high-level view for illustration purposes on how the GTP tunnels are represented in the system.

A GTP path is identified in each node with an IP address and a UDP port number. A path may be used to multiplex GTP tunnels and there may be multiple paths between two entities supporting GTP.

Another important feature of the GTP protocol is its usage of Cause values in response messages. Cause values represent the actual status of the action requested (e.g. Accept/Reject) as well as additional useful information that would facilitate the receiving entity to make a more informed decision on the possible course of action. For EPS, a list of these Cause values can be found in the specification TS 29.274.

16.2.2 Control Plane (GTPv2-C)

Tunnels are established, used, managed, and released via GTP-C messages. A path may be maintained by keep-alive echo messages. The GTPv2-C protocol stack is shown in Figure 16.4.

For the control plane, for each endpoint of a GTP-C tunnel there is a control-plane TEID (TEID-C). The scope of the GTP tunnel and the TEID-C depends on the interface and its functions (such as whether the interface is used on a per-terminal connection basis such as the S3 interface, or per-PDN connection basis like the S5/S8 interface):

- The TEID-C is a unique per-PDN connection on GTP-based S5 and S8. The same tunnel is shared for the control messages relating to all bearers associated with the PDN connection. A TEID-C on the S5/S8 interface is released after all its associated EPS bearers are deleted.
- There is only one pair of TEID-Cs per UE on each of the S3 and the S10 interfaces. The same tunnel is shared for control messages relating to the same UE operation. A TEID-C on the S3/S10 interface is released after its associated UE context is removed or the UE is detached.
- There is only one pair of TEID-Cs per UE over the S11 and the S4 interfaces. The same tunnel is shared for control messages relating to the same UE operation. A TEID-C on the S11/S4 interface is released after all its associated EPS bearers are deleted.

GTP defines a set of messages between two associated EPC entities. The messages are defined in 3GPP TS 29.274 and are shown here for illustration purposes. For the most detailed and up-to-date information, readers should consult the latest version of the specification.

Message types for GTPv2 for EPS are listed in Table 16.3.

16.2.3 User Plane (GTPv1-U)

GTP-U tunnels are used to carry encapsulated payload (original Packet Data Unit to be tunneled) and signaling messages between a given pair of GTP-U Tunnel Endpoints. The TEID-U that is present in the GTP header indicates which tunnel a

Table 16.3: GTPv2 Control-Plane Messages (EPS According to 3GPP TS 29.274)

Message Type Value (Decimal)	Message	Reference	Initial	Triggered
0	Reserved			
1	Echo Request		X	
2	Echo Response			X
3	Version Not Supported Indication			X
4–24	Reserved for S101 Interface	TS 29.276		
25–31	Reserved for Sv Interface	TS 29.280		
SGSN/MME/TWAN/ePDG to PGW (S4/S11, S5/S8, S2a, S2b)				
32	Create Session Request		X	
33	Create Session Response			X
36	Delete Session Request		X	
37	Delete Session Response			X
SGSN/MME to PGW (S4/S11, S5/S8)				
34	Modify Bearer Request		X	
35	Modify Bearer Response			X
38	Change Notification Request		X	
39	Change Notification Response			X
40–63	For future use			
164	Resume Notification		X	
165	Resume Acknowledge			X
Messages Without Explicit Response				
64	Modify Bearer Command (MME/ SGSN/ TWAN/ePDG to PGW – S11/ S4, S5/S8, S2a, S2b)		X	
65	Modify Bearer Failure Indication (PGW to MME/SGSN/ TWAN/ePDG – S5/S8, S11/S4, S2a, S2b)			X
66	Delete Bearer Command (MME/ SGSN to PGW – S11/S4, S5/S8)		X	
67	Delete Bearer Failure Indication (PGW to MME/SGSN – S5/S8, S11/ S4)			X
68	Bearer Resource Command (MME/ SGSN to PGW – S11/S4, S5/S8)		X	
69	Bearer Resource Failure Indication (PGW to MME/SGSN – S5/S8, S11/ S4)			X
70	Downlink Data Notification Failure Indication (SGSN/MME to SGW – S4/S11)		X	
71	Trace Session Activation (MME/ SGSN/ TWAN/ePDG to PGW – S11/ S4, S5/S8, S2a, S2b)		X	

(Continued)

Table 16.3: GTPv2 Control-Plane Messages (EPS According to 3GPP TS 29.274) (Continued)

Message Type Value (Decimal)	Message	Reference	Initial	Triggered
72	Trace Session Deactivation (MME/ SGSN/ TWAN/ePDG to PGW – S11/ S4, S5/S8, S2a, S2b)		X	
73	Stop Paging Indication (SGW to MME/SGSN – S11/S4)		X	
74–94	For future use			
PGW to SGSN/MME/TWAN/ePDG (S5/S8, S4/S11, S2a, S2b)				
95	Create Bearer Request		X	X
96	Create Bearer Response			X
97	Update Bearer Request		X	X
98	Update Bearer Response			X
99	Delete Bearer Request		X	X
100	Delete Bearer Response			X
PGW to MME, MME to PGW, SGW to PGW, SGW to MME, PGW to TWAN/ePDG, TWAN/ePDG to PGW (S5/S8, S11, S2a, S2b)				
101	Delete PDN Connection Set Request		X	
102	Delete PDN Connection Set Response			X
103–127	**For future use**			
MME to MME, SGSN to MME, MME to SGSN, SGSN to SGSN (S3/S10/S16)				
128	Identification Request		X	
129	Identification Response			X
130	Context Request		X	
131	Context Response			X
132	Context Acknowledge			X
133	Forward Relocation Request		X	
134	Forward Relocation Response			X
135	Forward Relocation Complete Notification		X	
136	Forward Relocation Complete Acknowledge			X
137	Forward Access Context Notification		X	
138	Forward Access Context Acknowledge			X
139	Relocation Cancel Request		X	
140	Relocation Cancel Response			X
141	Configuration Transfer Tunnel		X	
142–148	For future use			
152	RAN Information Relay		X	
SGSN to MME, MME to SGSN (S3)				
149	Detach Notification		X	
150	Detach Acknowledge			X

(Continued)

Table 16.3: GTPv2 Control-Plane Messages (EPS According to 3GPP TS 29.274) (Continued)

Message Type Value (Decimal)	Message	Reference	Initial	Triggered
151	CS Paging Indication		X	
153	Alert MME Notification		X	
154	Alert MME Acknowledge			X
155	UE Activity Notification		X	
156	UE Activity Acknowledge			X
157–159	For future use			
SGSN/MME to SGW, SGSN to MME (S4/S11/S3)				
SGSN to SGSN (S16), SGW to PGW (S5/S8)				
162	Suspend Notification		X	
163	Suspend Acknowledge			X
SGSN/MME to SGW (S4/S11)				
160	Create Forwarding Tunnel Request		X	
161	Create Forwarding Tunnel Response			X
166	Create Indirect Data Forwarding Tunnel Request		X	
167	Create Indirect Data Forwarding Tunnel Response			X
168	Delete Indirect Data Forwarding Tunnel Request		X	
169	Delete Indirect Data Forwarding Tunnel Response			X
170	Release Access Bearers Request		X	
171	Release Access Bearers Response			X
172–175	For future use			
SGW to SGSN/MME (S4/S11)				
176	Downlink Data Notification		X	
177	Downlink Data Notification Acknowledge			X
179	PGW Restart Notification		X	
180	PGW Restart Notification Acknowledge			X
SGW to SGSN (S4)				
178	Reserved. Allocated in earlier version of the specification			
181–199	For future use			
SGW to PGW, PGW to SGW (S5/S8)				
200	Update PDN Connection Set Request		X	
201	Update PDN Connection Set Response			X
202–210	For future use			

(Continued)

Table 16.3: GTPv2 Control-Plane Messages (EPS According to 3GPP TS 29.274) (Continued)

Message Type Value (Decimal)	Message	Reference	Initial	Triggered
MME to SGW (S11)				
211	Modify Access Bearers Request		X	
212	Modify Access Bearers Response			X
213–230	For future use			
MBMS GW to MME/SGSN (Sm/Sn)				
231	MBMS Session Start Request		X	
232	MBMS Session Start Response			X
233	MBMS Session Update Request		X	
234	MBMS Session Update Response			X
235	MBMS Session Stop Request		X	
236	MBMS Session Stop Response			X
237–239	For future use			
Other				
240–255	For future use			

particular payload belongs to. Thus, packets are multiplexed and demultiplexed by GTP-U between a given pair of Tunnel Endpoints.

In the case of LTE/EPC, GTP-U tunnels are established using S1-MME or GTP-C (e.g. EPS bearer establishment process) and in the case of the 3G packet core, it is established as mentioned before using RANAP and GTP-C (e.g. PDP context activation process). The protocol stack for GTP-U is shown in Figure 16.5.

Since there are different protocol versions, the version-not-supported indicator is used to determine which version the peer GTP endpoint supports.

16.2.4 Protocol Format

The control-plane GTP uses a variable length header. Control-plane GTP header length is a multiple of four octets, as shown in Figure 16.6, according to 3GPP TS 29.274.

The GTP-C header may be followed by subsequent information elements depending on the type of control-plane message. The format of a GTPv2-C message is illustrated in Figure 16.7.

In GTPv2-C, the information elements are added for new parameters if needed in future instead of using extension headers that were formerly in use for GTPv1-C.

					Bits			
Octets	8	7	6	5	4	3	2	1
1	Version			P	T	Spare	Spare	Spare
2	Message Type							
3	Message Length (1st Octet)							
4	Message Length (2nd Octet)							
m to k(m+3)	If T flag is set to 1, then TEID shall be placed into octets 5–8. Otherwise, TEID field is not present at all.							
n to (n+2)	Sequence Number							
(n+3)	Spare							

Figure 16.6: General Format Header for GTP-v2 Control Plane.

					Bits			
Octets	8	7	6	5	4	3	2	1
1	Version			P	T=1	Spare	Spare	Spare
2	Message Type							
3	Message Length (1st Octet)							
4	Message Length (2nd Octet)							
5	Tunnel Endpoint Identifier (1st Octet)							
6	Tunnel Endpoint Identifier (2nd Octet)							
7	Tunnel Endpoint Identifier (3rd Octet)							
8	Tunnel Endpoint Identifier (4th Octet)							
9	Sequence Number (1st Octet)							
10	Sequence Number (2nd Octet)							
11	Sequence Number (3rd Octet)							
12	Spare							

Figure 16.7: GTP-C EPC Specific Message Header Format.

					Bits			
Octets	8	7	6	5	4	3	2	1
1 to m	GTP-C header							
m+1 to n	Zero or more Information Element(s)							

Figure 16.8: GTP-C Header Followed by Subsequent Information Element.

For EPS, the GTPv2-C header takes the form shown in Figure 16.8 (EPC functional message specific header format that does not include messages such as Echo type, etc.).

A user-plane GTP header would, for example, have the format shown in Figure 16.9, as specified in 3GPP TS 29.281.

Octets	Bits							
	8	7	6	5	4	3	2	1
1	Version			PT	(*)	E		S	PN
2	Message Type							
3	Length (1st Octet)							
4	Length (2nd Octet)							
5	Tunnel Endpoint Identifier (1st Octet)							
6	Tunnel Endpoint Identifier (2nd Octet)							
7	Tunnel Endpoint Identifier (3rd Octet)							
8	Tunnel Endpoint Identifier (4th Octet)							
9	Sequence Number (1st Octet) (1)(4)							
10	Sequence Number (2nd Octet) (1)(4							
11	N-PDU Number (2)(4)							
12	Next Extension Header Type (3)(4)							

NOTE 0: (*) This bit is a spare bit. It shall be sent as '0'. The receiver shall not evaluate this bit.

NOTE 1: (1) This field shall only be evaluated when indicated by the S flag set to 1.

NOTE 2: (2) This field shall only be evaluated when indicated by the PN flag set to 1.

NOTE 3: (3) This field shall only be evaluated when indicated by the E flag set to 1.

NOTE 4: (4) This field shall be present if and only if any one or more of the S, PN, and E flags are set.

Figure 16.9: GTP-U Header Format.

16.3 Mobile IP

16.3.1 General

The basic IP stack does not provide support for mobility. If a UE has been allocated an IP address, this IP address is used not only to identify the UE in the sense that packets sent to this IP address are really destined for that UE. The IP address is also used to identify the network where the UE has attached. Each global IP address belongs to a certain IP subnetwork. Routers connecting different subnetworks will, with the help of routing protocols, make sure that packets destined for this IP address will reach the subnetwork to which this IP address "belongs". If the UE connects to another IP subnetwork the IP packets destined for the old IP address will still be routed to the old subnetwork. The UE will thus no longer be reachable using the old IP address. Furthermore, even packets sent by the UE in its new subnetwork may be discarded. The reason is that routers or firewalls may perform egress filtering of traffic leaving the subnetwork and discard packets sent with IP addresses not belonging to the network. This is illustrated in Figure 16.10. The change of subnetwork may, for example, occur if the UE moves and connects to another network using the same interface (e.g. a UE that moves between WLAN hotspots) or connects to another network using another access technology (e.g. goes from using a 3GPP access to using WLAN).

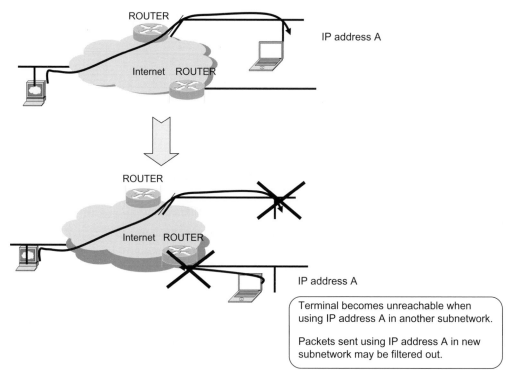

Figure 16.10: A Node Becomes Unreachable with Its Original IP Address When Moving to Another Subnetwork.

Packets destined for the UE's old IP address will continue to end up on the old subnetwork corresponding to that IP address. The UE thus needs to change IP address and get an IP address from the IP address range corresponding to the new point of attachment. In this way the UE will be reachable at its new point of attachment using its new IP address. However, if the UE replaces its old IP address with a new one, ongoing IP sessions need to be terminated and then restarted with the new IP addresses.

Mobile IP (MIP) is intended to solve these problems by providing mobility support on the IP layer. Mobile IP allows the UE to change its point of attachment (i.e. subnetwork) while continuing to use the same IP address and maintaining ongoing IP sessions. Since Mobile IP operates on the IP layer, it can provide mobility support for different kinds of lower layers. Mobile IP is thus suitable for providing mobility not just when moving across different IP networks for the same access technology, but also across heterogeneous access technologies. How this works is explained below.

EPS makes use of Mobile IP to provide IP-level mobility when the UE moves between different access technologies, for example from an access in the 3GPP family of accesses to a WLAN access.

Mobile IP is specified by the IETF. In fact, the IETF has specified different variants of Mobile IP applicable to IPv4, IPv6, or both IPv4 and IPv6. The different variants are more or less related to one another. Mobile IPv4 (IETF RFC 3344) is applicable to IPv4 and was specified first. The Mobile IP version for IPv6, Mobile IPv6 (MIPv6) (IETF RFC 3775), reuses many of the basic concepts developed for Mobile IPv4, but is still a distinct protocol. Dual-stack MIPv6 (IETF RFC 5555) is based on MIPv6 and contains the necessary enhancements for dual-stack IPv4/IPv6 operation. There is also a network-based version of MIPv6 called Proxy Mobile IPv6 (PMIPv6) (IETF RFC 5213). Figure 16.11 illustrates the different variants and also indicates their relationships. In addition, there are numerous RFCs containing amendments, optimizations, and enhancements, for example to improve handover performance (not illustrated in the figure). There have also been proposals for Proxy Mobile IPv4 and dual-stack Mobile IPv4 variants, but these are not covered here.

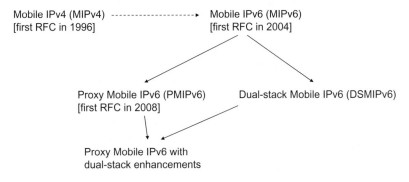

Figure 16.11: Mobile IP Family Tree.

It is not feasible in a book like this to describe all the different variants of Mobile IP or even all aspects and details of a single Mobile IP variant. Instead, we provide a high-level overview of primarily MIPv6 and dual-stack MIPv6. In Section 16.4, PMIPv6 is covered. These are the main Mobile IP-based protocols used in EPS. Also, Mobile IPv4 is supported by EPS to some extent. However, we regard dual-stack MIPv6 and PMIPv6 as the most general and future-proof Mobile IP protocols, and also the most relevant to EPS. Therefore, Mobile IPv4 is only discussed briefly in this chapter, mainly pointing out the differences compared to MIPv6. The description also focuses on those aspects of MIPv6 that are most relevant for its use in EPS. Readers interested in a more complete description of all features and different options of Mobile IP in general and MIPv6 in particular should consult dedicated books on the topic or the relevant RFCs.

16.3.2 Host-Based and Network-Based Mobility Mechanisms

Before going into the details of how Mobile IP works, it is useful to take a high-level view of different mobility concepts. As was described in Section 6.4, IP-level mobility protocols can be roughly classified into two basic types: host-based mobility protocols and network-based mobility protocols. Mobile IP is a host-based mobility protocol where the UE has the functionality to detect movement and to exchange Mobile IP signaling with the network in order to maintain IP-level session continuity. The other type of mobility protocol, or mobility scheme, is the network-based mobility management scheme. In this case the network can provide mobility services for a UE that is not explicitly exchanging mobility signaling with the network. In this case it is a task of the network to keep track of the UE's movements and ensure that the appropriate mobility signaling is executed in the core network in order for the UE to maintain its session while moving. PMIPv6, described in Section 16.4, is an example of a network-based mobility protocol. GTP is another example of a network-based protocol that is used to support mobility.

16.3.3 Basic Principles of Mobile IP

Before going into the actual mechanisms of Mobile IP it is necessary to describe the terms and concepts used. The description in this section to a large degree covers Mobile IP concepts in general, but it is geared towards MIPv6 in specific matters; dual-stack MIPv6 is covered further below. Even though EPC supports both MIPv4 (in Foreign Agent mode) and DSMIPv6 for mobility between heterogeneous accesses, DSMIPv6 is the more general and future-proof protocol of the two. Mobile IPv4 in FA mode is supported primarily for interworking with legacy CDMA and WiMAX systems.

As mentioned above, Mobile IP allows a UE to be always reachable using the same IP address even when the UE moves between different IP subnetworks. This IP address is called the *Home Address* (HoA) and is an IP address assigned from the address space of the *home network* (also referred to as the *home link*).

Note that "home network" in Mobile IP terminology is not the same as the "home network" (or "Home PLMN") used when discussing roaming. A home network in the Mobile IP sense is the IP network where the UE's HoA has been allocated and is thus a term related to IP topology and IP routing. The "home network" in the case of roaming is, however, a term denoting the network of the home operator or business entity where the subscriber has its subscription. The Mobile IP "home network" may be located in the Home PLMN or the Visited PLMN depending on whether the PDN GW is allocated from the Home PLMN or Visited PLMN.

In Mobile IP terminology, the mobile UE is referred to as a *Mobile Node* (MN). However, in order to be consistent with the terminology used in the rest of this book, we will continue to use the term "UE" when referring to the Mobile Node also in this chapter.

When the UE is attached to its home network it can use the HoA in the usual way without any need for Mobile IP services. However, when the UE attaches to a different IP network where the HoA is not topologically located, this is no longer possible. In Mobile IP terminology, the UE in this case is attached to a "foreign link" (or "foreign network").

When the UE is attached to a foreign network it acquires a local IP address from that network. This IP address is, in Mobile IP terminology, called a *Care-of Address* (CoA). The CoA is topologically located in the network the UE is currently accessing.

When the UE is connected to this foreign network, IP packets addressed to the CoA will reach the UE, while packets addressed to the HoA will reach the home network instead and not the UE. To solve this problem, Mobile IP introduces a network entity that maintains an association between the CoA and the HoA. This entity is called a *Home Agent* (HA) and is a router that is located on the UE's home network. (For EPS, the HA functionality is located in the PDN GW.) The association between the two IP addresses is called *binding*. When the UE has attached to a foreign link, it informs the HA about its current point of attachment (i.e. its current local IP address, the CoA). The HA then intercepts packets that are routed to the home network addressed to the HoA, and forwards them in a tunnel to the UE's current location, i.e. its CoA.

This behavior, at least for the downlink, resembles the mail forwarding that can be used if a family moves from one city to another. The post office in the old city can be informed about the family's new address, and will "intercept" and forward all mail addressed to the old address by placing the mail in a new envelope addressed to the family's new address. However, as we will see below, this comparison does not really work when looking at uplink packets. In the Mobile IP case uplink packets are also typically sent via the HA in the home network, while in the example with the post office, the family can send letters from its new address without having to send them via the post office in its old home town. An exception to this principle is MIPv6 Route Optimization (RO), where traffic is not sent via the HA. However, since RO is not supported in EPS, this is only briefly discussed below.

Before describing in more detail how MIPv6 works, we should also introduce the third entity in the MIPv6 architecture, the *Correspondent Node* (CN). The CN is an IP node with which the UE is communicating. It could, for example, be a server of some kind or another UE with which the Mobile IP UE is communicating. The CN does not need any Mobile IP functionality.

Basic Mobile IP operation will be described below by going through an example use case where a binding is created and updated.

16.3.3.1 Bootstrapping

When the UE is powered on, it connects to a network and acquires an IP address from the local network. This IP address becomes the CoA. In order to utilize Mobile IP, the UE needs to have the IP address of the HA, a security association with the HA and a HoA. The process for establishing this information is called *bootstrapping*. Even though this information may be statically preconfigured in the UE and HA, it is in many cases beneficial to establish this information dynamically. In particular, in an EPS deployment with a large number of subscribers the option to preconfigure the UE does not scale very well and would be difficult for operators to manage. Therefore, dynamic bootstrapping mechanisms are used.

The MIPv6-capable UE also needs to determine whether it needs to invoke Mobile IP or not. The UE does this by performing *home link detection* to determine whether it is attached to its home link or to a foreign link.

Several different methods have been defined for how the UE discovers an IP address of a suitable HA. Also, EPS supports different procedures for how the HA IP address is provided to the UE. It may be discovered using DNS or be provided to the UE using other means depending on which access technology the UE is using. For more details, see Section 9.2.6.

Once the UE knows the HA IP address, it can contact the HA to set up a security association. MIPv6 uses IPsec to protect the Mobile IP signaling and IKEv2 to establish the IPsec SA. During the IKEv2 procedure, the UE and HA perform mutual authentication and the HA can also deliver the HoA to the UE. See Section 16.3.4 for further aspects relating to MIPv6 security.

When the UE has acquired its HoA it performs home link detection by checking whether the HoA is "on-link" or not, i.e. whether or not the HoA belongs to the local network where the UE is currently attached. If the UE is attached to its home network no Mobile IP services are needed. The UE can use its HoA in the usual way.

16.3.3.2 Registration

If the UE is attached to a foreign network, the UE needs to inform the HA about the current CoA. The UE does this by sending a Mobile IP Binding Update (BU) message to the HA. The BU message contains the HoA and the CoA, and is protected using the IPsec SA previously established. The HA maintains a *Binding Cache* containing the HoAs and CoAs for each UE that have registered with the HA. When receiving the BU for a new UE, the HA creates a new entry in the Binding Cache and replies to the UE with a Binding Acknowledgement (BA). The MIPv6 registration is illustrated in Figure 16.12. For a more detailed call for initial attachment using DSMIPv6, see Section 17.6.5 and 17.6.6.

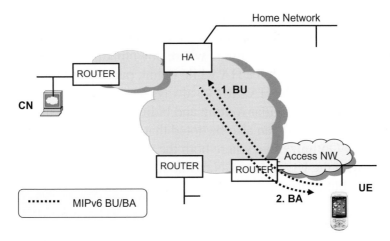

Figure 16.12: A Node Registers with the Home Agent by Sending a Binding Update.

16.3.3.3 Routing of Packets

When the UE is attached to a foreign network and a Binding Cache entry has been created in the HA, the HA intercepts all packets routed to the home network and destined for the UE's HoA. The HA then encapsulates the packets in a new IP header and forwards the packet to the UE's CoA. When receiving the packet, the UE decapsulates it and processes it in the normal way. When the UE sends packets, the UE tunnels the packets to the HA, which decapsulates the packets and forwards the packets to the final destination. This bidirectional tunneling of packets between the UE and the HA is illustrated in Figure 16.13.

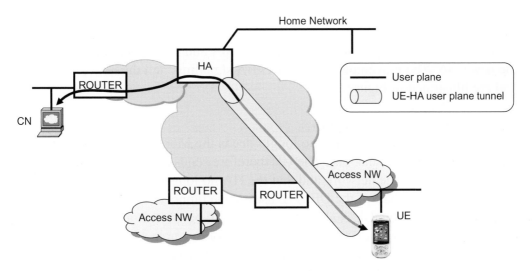

Figure 16.13: User Plane Tunneled Bidirectionally Between UE and Home Agent.

An alternative to bidirectional tunneling would have been for the UE to send the uplink packets directly to the destination, without tunneling them to the HA first. This would have created a "triangular routing" where downlink packets are routed to the home network and pass through the HA while uplink packets are routed directly to the final destination, bypassing the HA. MIPv6 always uses bidirectional tunneling, while Mobile IPv4 allows both triangular routing and bidirectional tunneling (called reverse tunneling for Mobile IPv4). It can also be noted that MIPv6 allows for a feature referred to as Route Optimization, where both the uplink and downlink user plane is sent directly between the UE and the CN, without passing the HA. Route optimization is not used in EPS, but is briefly described in Section 16.3.7.

16.3.3.4 Binding Lifetime Extension
A binding in the HA has a certain lifetime. Unless the binding is renewed before the lifetime expires, the HA will remove the binding. This is used, for example, to clean up bindings belonging to terminals that are no longer attached to the network and that did not cancel the binding properly when they were disconnected. In order for the terminal to refresh the binding, the terminal sends a new BU well before the expiry of the binding lifetime.

16.3.3.5 Movement and Update of the Binding
If the UE moves to a different point of attachment and receives a new local IP address, the UE again performs home-link detection to determine whether it is now connected to the home link. If the UE determines that it has moved to another network different from its home network, the UE needs to inform the HA about the new CoA acquired in the new network. If not, the HA would continue to forward the IP packets to the old foreign network. The UE thus sends a new BU to the HA containing its HoA and the new CoA. When receiving the BU, the HA updates the Binding Cache entry for the HoA with the new CoA and starts forwarding traffic to the new CoA. The movement, MIPv6 BU/BA signaling, and the new user-plane tunnel are illustrated in Figure 16.14.

16.3.3.6 Movement and De-Registration
If the UE moves to its home link it no longer needs the Mobile IP service since it can use the HoA in the usual way. The UE therefore sends a BU to inform the HA that it is now on its home network and that the HA no longer needs to intercept and forward packets on behalf of the UE. The user-plane tunnel between UE and HA is also removed. In EPS the UE is always considered to be on its home link when using a 3GPP access. Therefore, de-registration occurs, for example when the UE moves from a non-3GPP access where S2c is used to a 3GPP access (Figure 16.15).

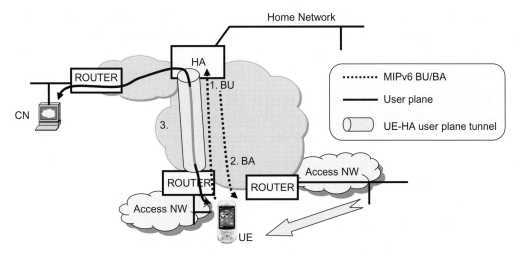

Figure 16.14: Movement, Update of Binding, and Switch of Tunnel Towards New Point of Attachment.

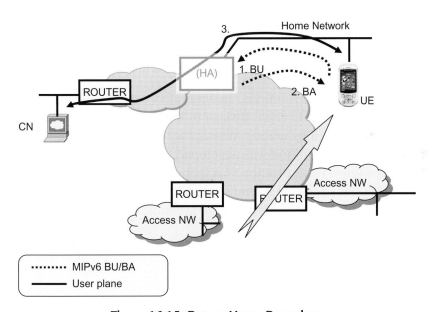

Figure 16.15: Return Home Procedure.

16.3.3.7 Binding Revocation

If the UE is located on a foreign link with a binding registration in the HA, the HA may in some cases want to terminate the Mobile IP session. This may, for example, happen if the user is no longer authorized to use Mobile IP. In this case the HA can send a Binding Revocation Indication (BRI) to the UE. The UE then replies with a Binding Revocation Acknowledgement (BRA) and the Mobile IP session is terminated. The BRI and BRA messages are defined in IETF RFC 5846.

16.3.4 Mobile IPv6 Security

The Mobile IP signaling extends between the UE and the HA. It is therefore important to ensure that this signaling is properly protected. Mobile IPv4 and MIPv6 use different security solutions, and in line with the rest of this chapter we will focus on MIPv6.

Even for MIPv6 there exist different security solutions. The MIPv6 RFC requires that IPsec is used to protect the BU and BA between the UE and the HA. Originally, MIPv6 security was based on the old IPsec architecture, either using manual configuration or using IKEv1 to establish the IPsec security association (see Section 16.9 for more information on IPsec). This is described in IETF RFC 3775 and IETF RFC 3776. Recently, the MIPv6 specification has been updated to also support the revised IPsec architecture and IKEv2. The usage of MIPv6 with the revised IPsec architecture is described in IETF RFC 4877.

The UE and the HA must support the use of ESP in transport mode to protect the BU and BA messages. Integrity protection is mandatory while ciphering is optional.

In addition to the IPsec-based solutions, an alternative security mechanism has been documented in IETF RFC 4285. Instead of using IPsec, this security solution provides integrity protection by adding message authentication mobility options to the MIPv6 signaling. This solution was developed for use in networks based on the legacy 3GPP2 standard. The motivation was that it would be more lightweight than the IPsec-based solutions and that it would provide sufficient security in the specific 3GPP2 deployments. However, for EPS-based systems, only the IKEv2-based security solution is supported.

For more details on Mobile IP security in EPS, see the description of the S2c interface in Section 15.5.1.

16.3.5 Packet Format

16.3.5.1 Mobile IP Signaling (Control Plane)

In order to understand the MIPv6 packet format, it is useful to recapture a few basic aspects about the IPv6 header. IPv6 defines a number of extension headers that can be used to carry the "options" of the IP packet. The extension headers, if they are present, follow after the "main" IPv6 header and before the upper layer header (e.g. TCP or UDP header). One of the extension headers, the hop-by-hop header, contains information intended for each router on the path. This header therefore has to be examined by each router on the path. In general, however, the extension headers contain information only intended for the final destination of the packet. This means that these extension headers do not need to be examined by every router on the path. Examples of extension headers containing information for the final destination of the packet are the ESP header (for IPsec) and the fragmentation header (if the packet is fragmented). The ESP and fragmentation headers are extension headers defined for explicit purposes. Another way to provide options to the final destination is to use the Destination Options extension header. This header can contain a variable number of options. Figure 16.16 illustrates an IPv6 packet containing a "main" header, extension headers, and payload.

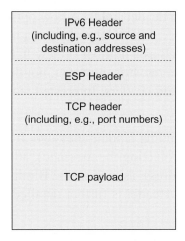

Figure 16.16: Example IPv6 Packet Containing Main IPv6 Header and an Extension Header as Well as an Upper Layer Header and Payload.

MIPv6 defines a new extension header, called the *Mobility Header* (MH), to carry the MIPv6 messages. All messages used in MIPv6, including the BU and BA, are defined as MH types. The MH format is shown in Figure 16.17.

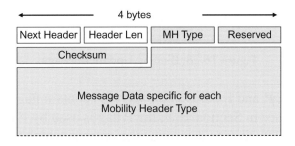

Figure 16.17: Mobility Header Format.

The Next Header and Header Length fields are not specific to the MH but are present in all extension headers. The Next Header field indicates what type of header (e.g. extension header or upper layer header) follows this header. The Header Length field contains the length of the header. The MH type field indicates what particular MIPv6 message this is, for example a BU, BA, BRI, or Binding Error. The Checksum field contains a checksum of the MH. The Message Data part contains information specific for each message (see below).

This means that MIPv6 messages are carried as part of the IPv6 header information and not as payload of the IPv6 packets. This is different from Mobile IPv4, where these are carried as UDP encapsulated payload in an IPv4 packet.

MIPv6 also defines other IPv6 header fields. A new option for the Destination Options header is used to carry the HoA. MIPv6 also defines a new routing header variant (Routing Header type 2) as well as a number of new ICMPv6 messages. Below we describe the BU and BA messages. It is, however, not the intention of this book to go through all MIPv6 messages and headers. Interested readers are referred to IETF RFC 3775 for more details.

Figure 16.18 illustrates the Binding Update message. It contains the main IPv6 header, the ESP header (for protecting the message), and a Destination Options header carrying the Home Address, as well as the Binding Update Mobility Header. The Binding Update Mobility Header is further detailed in Figure 16.19.

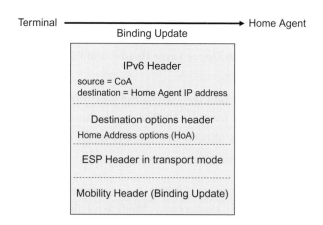

Figure 16.18: Binding Update Message.

The A, H, L, K, M, R, P, and F fields in the BU MH contain flags for different purposes. As we will see in Section 16.4, the P flag is used by PMIPv6. The Sequence number is used by the receiver to determine the order in which the BUs were sent by a UE. This is useful, for example, if the UE rapidly moves between different accesses

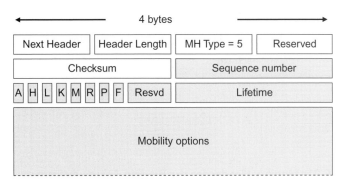

4 bytes

| Next Header | Header Length | MH Type = 5 | Reserved |

| Checksum | Sequence number |

| A | H | L | K | M | R | P | F | Resvd | Lifetime |

Mobility options

Figure 16.19: Mobility Header for the Binding Update Message.

and sends multiple BUs within a short interval. The Lifetime is the time that remains before the binding expires. The Mobility Options field may contain additional options. One example is the Alternate CoA mobility option. The CoA is used as the source address of the BU, but including it in the CoA mobility option as well allows it to be protected by ESP (ESP in transport mode does not protect the IP header). In response to a BU, the HA sends a BA: Figure 16.20 illustrates the BA message. It contains the main IPv6 header, the ESP header (for protecting the message), and a type 2 Routing Header carrying the HoA, as well as the BA MH.

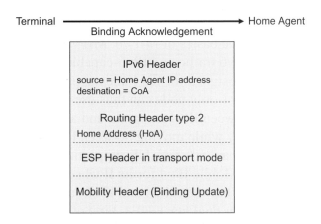

Terminal ⟶ Home Agent

Binding Acknowledgement

IPv6 Header

source = Home Agent IP address
destination = CoA

Routing Header type 2

Home Address (HoA)

ESP Header in transport mode

Mobility Header (Binding Update)

Figure 16.20: Binding Acknowledgement Message.

The MH for the BA message is illustrated in Figure 16.21. The Status field indicates the result of the BU.

The sequence number sent in the BA is the same as that received in the BU. This allows the Mobile IP client to match Updates with Acknowledgements. The Lifetime includes the time granted by the HA until the binding expires. To maintain the binding

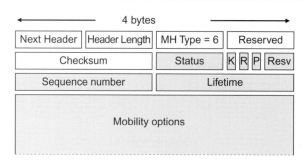

Figure 16.21: Mobility Header for the Binding Acknowledgement Message.

in the HA, the UE must refresh the binding before it expires by sending a new BU message to the HA.

16.3.5.2 User Plane

When the MIPv6 session has been established, all user-plane packets for the HoA are tunneled between the UE and the HA (except in the case where Route Optimization is used – see Section 16.3.7). This tunneling is performed using IPv6 encapsulation defined in IETF RFC 2473. Note, however, that additional encapsulation protocols are defined for the dual-stack version of MIPv6.

16.3.6 Dual-Stack Operation

The text above has described the basics of MIPv6. MIPv6 was designed for IPv6 only and thus supports only IPv6 traffic and IPv6-capable networks. Mobile IPv4, on the other hand, was designed for IPv4 and supports only IPv4 traffic and IPv4-capable networks. An IPv4-only node can thus use Mobile IPv4 to maintain connectivity while moving between IPv4 networks and an IPv6-only node can use MIPv6 to maintain connectivity while moving between IPv6 networks. However, this situation is not optimal for a dual-stack UE supporting both IPv4 and IPv6. Such a UE would need to use Mobile IPv4 for its IPv4 stack and MIPv6 for its IPv6 stack, so that it can move between IPv4 and IPv6 subnets. There are a few drawbacks with this solution for dual-stack UEs. First of all it requires that the dual-stack UE needs to support two sets of mobility management protocols, which increases the complexity of the UEs. Also, it needs to send two sets of Mobile IP signaling messages on every handover, to inform both the Mobile IPv4 HA and the MIPv6 HA about the move. Furthermore, since Mobile IPv4 requires an IPv4 CoA and MIPv6 requires an IPv6 CoA, all access networks need to be dual-stack in order to provide mobility for both the IPv4 and IPv6 sessions. For example, a dual-stack UE attempting to connect via an IPv4-only network would not be able to maintain connectivity of its IPv6 applications and vice versa. Also, to the operator this scenario has drawbacks since the operator needs to run and maintain two sets of mobility management systems on the same network.

The dual-stack extensions of MIPv6 avoid these drawbacks, by enhancing the protocol to also support the IPv4 access network (i.e. IPv4 CoA) and IPv4 user-plane traffic (i.e. using an IPv4 HoA). The dual-stack version of MIPv6 is usually referred to as DSMIPv6 and is specified in IETF RFC 5555. The solution defines extensions for carrying the mobile node's IPv4 CoA, IPv4 HoA, and IPv4 address of the HA in the MIPv6 signaling messages. It should be noted that DSMIPv6 requires that the terminal is always assigned an IPv6 HoA.

As indicated above, DSMIPv6 supports more network scenarios than basic MIPv6. Scenarios supported by DSMIPv6 are illustrated in Figure 16.22. A requirement to support both IPv4 and IPv6 traffic is that the HA supports both IPv4 and IPv6. Even though only single-stack foreign networks are shown in the figure, the foreign network may of course support both IPv4 and IPv6. In the latter case, the terminal should prefer using an IPv6 CoA.

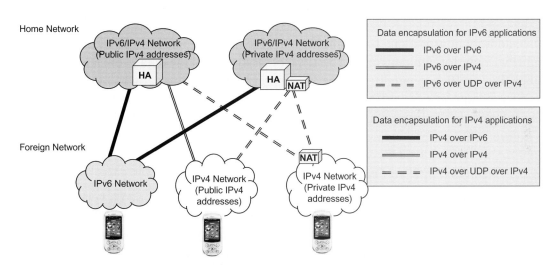

Figure 16.22: Network Scenarios for Dual-Stack MIPv6.

As explained in the previous section, all MIPv6 messages are defined as native IPv6 packets (using IPv6 extensions headers, etc.). In an IPv4-only foreign network the UE can, however, only acquire an IPv4 CoA and send IPv4 packets. In order to send an MIPv6 message to the HA, the MIPv6 packets must be encapsulated in IPv4 and sent to the IPv4 address of the HA. An example of a BU message for an IPv4-only foreign network is shown in Figure 16.23. In order to support private IPv4 addresses and NAT traversal, UDP encapsulation may be used.

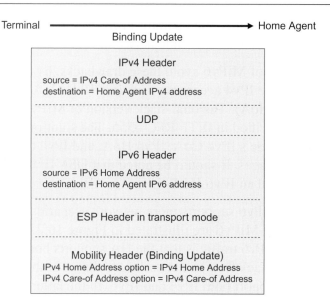

Figure 16.23: DSMIPv6 Binding Update Message for an IPv4 CoA.

Also, additional user-plane tunneling formats are needed to support IPv4 user data and IPv4-only foreign networks. The IPv4 or IPv6 user-plane data is encapsulated in either IPv4 or IPv6 depending on the IP version of the foreign network. Furthermore, in order to support private IPv4 addresses and NAT traversal, UDP encapsulation is also supported. The user data tunneling formats for the different scenarios are shown in Figure 16.22.

16.3.7 Additional MIPv6 Features – Route Optimization

MIPv6 is a quite extensive protocol and so far we have only provided a very-high-level description of a few main functions. One feature that was only briefly mentioned above is Route Optimization (RO). RO is supported for MIPv6 but not available in Mobile IPv4. It is an alternative to the bidirectional tunneling between UE and HA. With RO, the user-plane traffic is sent directly between the UE and the Correspondent Node (CN) without passing the HA.

RO is not supported by EPS and there are different reasons for this. In EPS the HA is located in the PDN GW and it is always assumed that the user-plane traffic goes through one PDN GW where charging, policy enforcement, and lawful intercept can take place. Furthermore, MIPv6 RO is limited to IPv6 traffic and IPv6 foreign networks. RO is thus not supported for IPv4 traffic even when DSMIPv6 is used. EPS provides other solutions that can enable efficient routing. In roaming situations it is, for example, possible to assign a PDN GW located in the visited PLMN, thus avoiding the transport of all user-plane traffic to the home PLMN. Also, PDN GW selection functions in EPS have an impact on the routing, for example by selecting a PDN GW that is geographically close to the UE.

MIPv6 RO allows a UE to inform a CN about its current CoA. The UE basically sends a BU to the CN and the CN in turn creates a binding in between the HoA and the CoA. When the CN sends a packet to a specific IP address, it checks its bindings for an entry (i.e. a HoA) that matches the IP address. If a match is found, the CN can communicate with the UE using the CoA. Traffic sent by the CN will thus be routed to the foreign network directly without passing the home network. MIPv6 defines special messages as well as security mechanisms to set up the route binding in the CN. Considering that RO is not used in EPS, and this is a book on EPS, we will not go into further details on this topic. The interested reader is instead referred to IETF RFC 3775.

16.4 Proxy Mobile IPv6

16.4.1 General

As explained in Section 16.3, mobility schemes can often be classified as being either host-based or network-based. MIPv6, described in the previous section, is a host-based mobility management solution where the UE has functionality to detect movement and to exchange IP mobility signaling with the network in order to maintain IP-level session continuity.

PMIPv6 (defined in IETF RFC 5213), on the other hand, is a network-based mobility management protocol that has a similar purpose to that of MIPv6, i.e. to facilitate IP-level session continuity. PMIPv6 reuses many of the concepts and packet formats that have been defined for MIPv6. A key difference between MIPv6 and PMIPv6 is, however, that with PMIPv6 the UE does not have Mobile IP software and does not participate in IP mobility signaling. A key intent of PMIPv6 is in fact to also enable IP-level mobility for those UEs that do not have Mobile IP client functionality. Instead, it is mobility agents in the network that track the UE's movement and perform IP mobility signaling on behalf of the UE. A mobility agent in the network acts as a proxy for the UE when it comes to IP mobility signaling, hence the name Proxy Mobile IPv6.

Since PMIPv6 reuses many parts of MIPv6 such as packet format, the description of PMIPv6 in this section will, to a large extent, build on the description of Mobile IP in Section 16.3.

PMIPv6 is used on the S2a and S2b interfaces and as a protocol alternative on the S5/S8 interface. Specific aspects relating to PMIPv6 use in EPS have been described in previous chapters, for example regarding EPS bearers (Section 6.3), mobility (Section 6.4), PCC (Section 8.2), and so on. For more details on PMIP-based interfaces, see Chapter 15. Below we describe the PMIPv6 protocol as such.

16.4.2 Basic Principles

PMIPv6 introduces two new network entities: the Mobile Access Gateway (MAG) and the Local Mobility Anchor (LMA). The MAG is the mobility agent that acts essentially as the Mobile IP client on behalf of the UE. The LMA is the mobility anchor point and its role is similar to that of the HA for MIPv6, i.e. to maintain a binding between the HoA of the UE and its current point of attachment. The MAG is located in the access network while the LMA is located in the network where the HoA is topologically located. The PMIPv6 architecture is illustrated in Figure 16.24.

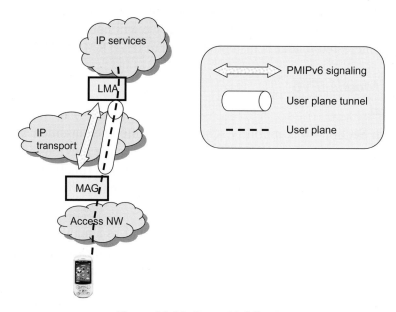

Figure 16.24: Proxy Mobile IP.

The responsibility of the MAG is to detect the movement of a UE and to initiate the appropriate IP mobility signaling. A key function of the MAG is also to emulate the UE's home network, i.e. to make sure that the UE does not detect any change in its layer-3 attachment even after the UE has changed its point of attachment. The UE should be allocated the same IP address and other IP configuration parameters after the move as it had before the move. Furthermore, the target MAG is also updated with other parameters such as IPv6 link-local address to be used by the MAG. This is to ensure that, in a handover, the target MAG uses the same link-local address as the source MAG. This gives the UE the impression that it is still on the same local network even after the handover. How this works will be seen in an example scenario below.

Below we provide an example of how PMIPv6 works and can be used in a network.

When the UE connects to an access network it typically performs access authentication and authorization for that access. During the access authentication, the UE also provides the user identity (the IMSI in case of EPS) and the security (e.g. encryption) may be set up. When the UE has successfully attached to the access network and provided its identity, it may, for example, request an IP address using DHCP. It should be noted that the details regarding the signaling between UE and the MAG depend very much on the type of access used. For example, access authentication and IP address allocation may be done in different ways in different access systems. See Chapter 7 and Section 6.3 for further details on access-specific aspects.

The MAG in the access network now initiates PMIPv6 signaling towards the LMA to inform the LMA of this user's current point of attachment. To do this the MAG first has to select a suitable LMA (this is similar to how the UE must discover a suitable HA in the case of Mobile IP). When PMIPv6 is used in EPS, the MAG performs LMA discover by resolving the APN using the DNS functions (for more details, see Chapter 9). The PMIPv6 IETF RFC 5213 also describes other means for the MAG to find a suitable LMA.

Once an LMA has been selected, the MAG sends a Proxy Binding Update (PBU) message to the LMA. The PBU contains the Proxy CoA, which is the IP address of the MAG. This allows the LMA to create an association (binding) between the Proxy CoA and the UE's HoA in a very similar way as the MIPv6 HA creates a binding between the CoA and the HoA. The difference here is that with PMIPv6 the UE is not aware of the Proxy CoA.

The LMA then replies with a Proxy Binding Acknowledgement (PBA). This message contains the IPv6 prefix allocated to the terminal. (With the dual-stack amendments described further below, the PBA may also contain an IPv4 address allocated to the terminal.) The PBA also carries other IP parameters associated with the home network, such as the MAG IPv6 link-local address. Once the MAG receives the PBA it can provide the allocated IP address/prefix to the UE (e.g. using DHCP).

When this is done, the MAG and LMA establish a tunnel where the user plane for the UE is forwarded. All user-plane data sent by the UE is intercepted by the MAG and forwarded to the LMA via the MAG–LMA tunnel. The LMA in turn decapsulates the packets and forwards them to their final destination. In the other direction, all traffic addressed to the HoA is intercepted by the LMA in the home network and forwarded to the MAG (via the tunnel), which in turn sends it to the UE (Figure 16.25).

To the UE it appears as though it was really connected to its home network since it is allocated the same IP address (the HoA) and other IP parameters associated with the

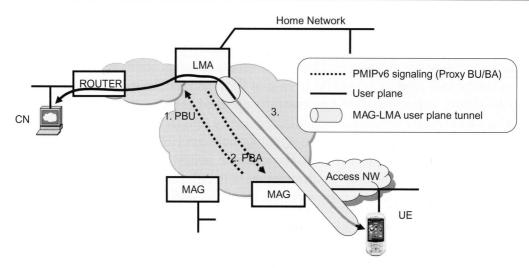

Figure 16.25: UE Connected to a Network Using PMIPv6.

home network. The MAG is emulating the home link towards the UE and the UE can continue its IP sessions as if it were still connected to the home network. See Section 17.2 for a more detailed call flow for an attachment procedure using PMIP.

If the UE now moves and connects to a different access network, using either the same access technology or a different access technology, the access attachment procedure will again be performed. The MAG in the new access network will detect that the UE has attached. In order to provide session continuity, the new MAG has to send a PBU to update the binding in the LMA. The new MAG therefore sends the PBU to the same LMA informing about the Proxy CoA of the new MAG. The LMA updates its binding and replies with a PBA containing, for example, the HoA and other parameters. The new MAG now assigns the same IP address (i.e. the HoA) as in the old access. The user-plane tunnel is moved to the new MAG. The UE again thinks it is connected to the home network and can thus continue to use the HoA as before the change in point of attachment (Figure 16.26).

For a more detailed call flow of a PMIP-based handover procedure in EPS, see Section 17.7.

If the LMA for some reason wants to disconnect the terminal, it sends a Binding Revocation Indication (BRI) to the MAG. The MAG in turn removes its mobility context, disconnects the terminal, and sends a Binding Revocation Acknowledgement (BRA) to the LMA. The format of the BRI and BRA messages are the same for both MIPv6 and PMIPv6 (IETF RFC 5846).

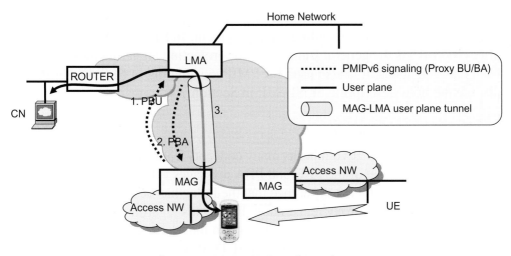

Figure 16.26: PMIP-Based Handover.

One important aspect to note in the example above is that the signaling between UE and MAG is access specific. There is no IP-level mobility signaling between the UE and the network as was the case for MIPv6. PMIPv6 is used inside the network in order to establish, to modify, or to remove the mobility session and to provide the MAG in the access network with the home network information (e.g. HoA). In this way the MAG can emulate the home network by assigning the same IP-layer parameters as they would have been assigned in the home network. The UE is thus unaware of the topology of the network. The user-plane tunnel between MAG and LMA enables the UE to use the HoA from any access link where PMIPv6 is used.

16.4.3 PMIPv6 Security

Since PMIPv6 is a network-based mobility protocol it has different security requirements than MIPv6, which is a host-based mobility protocol.

PMIPv6 performs mobility signaling on behalf of a UE that has attached to its network. PMIPv6 therefore requires that proper access authentication and authorization have been performed so that there is a trusted connection between the UE and the MAG before the MAG initiates PMIPv6 signaling. If this trusted connection is not required, a malicious UE might, for example, trigger an MAG to perform mobility signaling on another user's behalf. See Chapter 7 for further details on access security.

Also, the PMIPv6 signaling itself needs to be properly protected. The PMIPv6 RFC supports protection of the PMIPv6 signaling between the MAG and the LMA using

IPsec (ESP in transport mode with integrity protection). It is also possible to use other security mechanisms depending on deployment. In EPS, the Network Domain Security (NDS/IP) is the general framework for protecting signaling messages between network nodes. It is also used for PMIPv6 signaling in EPS. For further details, see the brief description of NDS/IP in Section 7.4.

16.4.4 PMIPv6 Packet Format

As already mentioned, PMIPv6 inherits many concepts from MIPv6, including the packet format. The format of the PBU and PBA messages is the same as for the MIPv6 BU and BA messages, with the difference that the P flag has been introduced to indicate that the BU/BA refers to a proxy registration. For details regarding the format of the BU and BA messages, see Section 16.3.5. PMIPv6 also introduces new Mobility Options for use with the PBU and PBA messages. See IETF RFC 5213 for further details.

The user plane is sent between MAG and LMA in a bidirectional tunnel. The tunnels may be unique per UE or shared by multiple UEs depending on deployment. The tunnels may be statically configured at the MAGs and LMAs in the network or be dynamically established and torn down. The MAG and LMA may use IPv6 encapsulation (IETF RFC 2473) for this tunnel but GRE tunneling is also supported in IETF RFC 5845. EPS requires that GRE tunneling is used where the GRE key field is used to uniquely identify a specific PDN connection (see Section 16.6).

In some environments there is a need to include specific information elements not defined as part of the main PMIPv6 specifications in the PMIPv6 messages. In this case it is possible to use vendor-specific Mobility Options that can be included with the Mobility Header. The general format of these vendor-specific options is defined in IETF RFC 5094. 3GPP EPS makes use of such vendor-specific options to transport, for example, the Protocol Configuration Option (PCO) fields, Charging ID, and 3GPP-specific error codes, as defined in TS 29.275.

16.4.5 Dual-Stack Operation

The dual-stack enhancements for PMIPv6 reuse the dual-stack features defined for DSMIPv6. This means that the PBU and PBA may contain IPv4 CoA options and IPv4 HoA options. This allows PMIPv6 to also support IPv4-only access networks as well as IPv4 HoAs. A key difference compared to DSMIPv6 is, however, that with PMIPv6 the IPv6 HoA is not mandatory. It is allowed to assign only an IPv4 HoA to the terminal. The dual-stack extensions for PMIPv6 are defined in RFC 5844. In a similar way as for DSMIPv6, when PMIPv6 is used over an IPv4 transport network, the PMIPv6 signaling messages are encapsulated in IPv4 packets. The user-plane tunnel MAG and LMA are encapsulated in IPv6, IPv4, UDP-over-IPv4, or

GRE-over-IPv4/IPv6, depending on the network scenario. As mentioned above, EPS uses GRE tunneling over IPv4 or IPv6 transport. Dual-stack scenarios for PMIPv6 are illustrated in Figure 16.27. Note that the access network may support IPv6 only (for simplicity this is not shown in the figure).

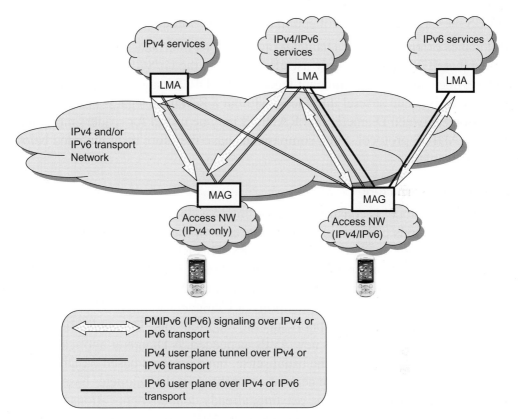

Figure 16.27: Example Scenarios for Dual-Stack PMIPv6.

16.5 Diameter

16.5.1 Background

Diameter is a protocol originally designed for Authentication, Authorization, and Accounting (AAA) purposes. It is an evolution of its predecessor, the RADIUS protocol. The Diameter protocol name is "derived" from RADIUS in the sense that the diameter is twice the radius. The RADIUS protocol has been commonly and successfully deployed to provide AAA services for fixed dial-up accesses as well as for cellular CDMA systems. It is also used in GPRS networks on the Gi interface.

The Diameter protocol was designed to overcome several shortcomings of RADIUS. For example, Diameter supports improved failure handling, more reliable message delivery, bigger size information elements, improved security, better possibilities for extensibility, more flexible discovery of other Diameter nodes, etc. Furthermore, in contrast to RADIUS, Diameter provides a full specification of intermediate entities such as proxies. At the same time Diameter was constructed to provide an easy migration and compatibility with RADIUS. For example, a Diameter message, like a RADIUS message, conveys a collection of Attribute Value Pairs (AVPs).

3GPP makes extensive use of Diameter on numerous interfaces. It should, however, be noted that 3GPP basically only uses RADIUS on the Gi/SGi interface and, therefore, does not have a significant legacy of RADIUS usage. Instead, Diameter is used exclusively on several interfaces, without a RADIUS legacy. Nevertheless, comparisons between Diameter and RADIUS may be useful for readers more familiar with the RADIUS protocol. Such comparisons have therefore been included below where applicable.

16.5.2 Protocol Structure

The Diameter protocol is constructed according to a single base standard and additional extensions called applications. The core of the Diameter protocol is defined in the Diameter base standard, IETF RFC 3588. This RFC specifies the minimum requirements for a Diameter implementation and includes a few general Diameter messages (called Commands in Diameter) as well as AVPs that can be carried by the commands. Extensions (called Applications in Diameter) are then created on top of the Diameter base protocol to support specific requirements. The applications may define new commands as well as new AVPs as needed. A Diameter application is thus not a program or application in the usual sense, but a protocol based on Diameter. The applications benefit from the general capabilities of the Diameter base protocol. Applications may also be based on existing, already defined, applications. In this case they inherit Diameter commands and mandatory AVPs from the application(s) they are based on, but they use new application identifiers, add new AVPs, and modify the protocol state machines according to their own procedures. Figure 16.28 illustrates the protocol structure of Diameter, including a few example Diameter applications.

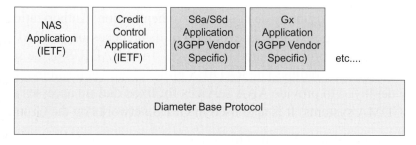

Figure 16.28: Structure of the Diameter Protocol, Consisting of a Base Protocol and Extensions Called Applications.

Several Diameter applications have been standardized by the IETF, but it is also possible to define vendor-specific applications. A "vendor" in this context is not necessarily a vendor making products; it is instead an entity (e.g. an organization or company) that has requested a Diameter application identifier from the Internet Assigned Numbers Authority (IANA). As will be seen in the interfaces section, 3GPP has defined several vendor-specific Diameter applications that are used over Diameter-based interfaces such as S6a, S6b, SWa, SWx, and so on. In many cases the 3GPP vendor-specific applications are based on existing Diameter applications defined by the IETF. Several of these applications were discussed together with the corresponding interfaces in Chapter 15.

16.5.3 Diameter Nodes

The network entities implementing Diameter act in a certain role in the network architecture. Diameter defines three types of Diameter nodes depending on the role the node plays: client, server, and agent. The role a certain Diameter node plays depends on the network architecture.

The client is typically the entity requesting a service from a Diameter server and thus originates the request to initiate a Diameter session with a server.

Diameter agents are Diameter nodes that can bring flexibility into the network architecture. They can be used to support a system where different parts of the network are operated by different administrations, such as in a roaming scenario. They are also used in the routing of Diameter messages to aggregate Diameter requests destined for a specific realm. The agents examine the received requests and determine the right target. This can provide load balancing features and simplified network configurations. Certain agents can also perform additional message processing.

There are four types of agents: relay, proxy, redirect, and translation agents. This can be compared to RADIUS, where basically only a single type of intermediary node, the RADIUS proxy, exists.

A relay agent is used to forward a message to the appropriate destination, depending on the information contained in the message. More information on Diameter message routing is provided below. The relay agent needs to understand the Diameter base protocol but need not understand the Diameter application used.

A proxy agent is similar to a relay agent, with the difference that it can perform additional processing of the Diameter messages, for example implementing certain policy rules. Since the proxy agent can modify messages it needs to understand the service being offered and thus also understand the Diameter application being used.

A redirect agent also provides a routing function, for example to perform realm-to-server resolution. The redirect agent, however, does not forward a received message to the destination, but, rather, replies with another message to the node that sent the request. This reply contains information that allows the node to send the request again, but now directly to the server. The redirect agent is thus not on the routing path of the Diameter messages.

A translation agent may perform translation between Diameter and other protocols. A typical example that can be used to support migration scenarios is a translation agent translating between Diameter and RADIUS.

16.5.4 Diameter Sessions, Connections, and Transport

Diameter uses either the Transmission Control Protocol (TCP) or Stream Control Transport Protocol (SCTP) to transport messages between two Diameter peers. Since TCP and SCTP are both connection-oriented protocols, a connection between the two peers has to be established before any Diameter command can be sent. Both TCP and SCTP provide reliable transport. This can be compared to RADIUS, which uses UDP as transport protocol, providing a connection-less and unreliable transport. For more details on SCTP and the differences compared to TCP, see Section 16.11.

The Diameter *connection* between two peers should be distinguished from the Diameter *session* being established between a client and a server. While a connection is a transport-level connection between two peers, the Diameter session is a logical concept describing the application-level association between a client and server (possibly spanning Diameter agents) identified by a session identifier. The Diameter peer connection and Diameter session are illustrated in Figure 16.29.

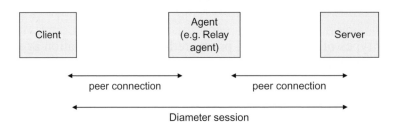

Figure 16.29: Diameter Connections and Diameter Sessions.

Diameter messages are protected using Transport Layer Security (TLS) or IPsec. The Diameter base specification mandates that all Diameter nodes must support IPsec while TLS is optional to support in the Diameter client. This protection is provided hop-by-hop between the Diameter peers.

In a 3GPP environment, however, the general framework of NDS/IP is used for all IP-based control signaling, including Diameter. There is thus no need to provide a specific security association between the Diameter nodes in EPS. For more details on NDS/IP, see Chapter 7.

In order to handle the flexibility of Diameter in terms of applications, security features, etc. in a dynamic fashion, two Diameter peers establishing a connection also perform a capability exchange. This exchange allows each peer to learn about the other peer's identity and its capabilities (protocol version number, supported Diameter applications, supported vendor-specific attributes, security mechanisms, etc.).

16.5.5 Diameter Request Routing

As mentioned above, Diameter agents may assist in the routing of a Diameter command towards its final destination, the Diameter server.

A Diameter agent forwarding a command typically performs the routing based on the destination realm as well as the application used. It may thus use a different destination based on the application identification. The Diameter node maintains a list of supported realms and known Diameter peers, as well as the peer's capabilities (e.g. supported applications).

An agent can perform realm-to-server resolution and can thus be used to aggregate requests from different sources destined for a specific destination realm. This allows the agent to act as a centralized routing entity.

This feature can also be utilized in EPS if a network deploys multiple Diameter servers, for example multiple HSS or PCRF nodes. For the HSS, the Diameter client may not know which HSS node handles the subscription record for a specific user. In this case a Diameter redirect agent or proxy agent can perform the resolution from a realm and user name into a server name for the HSS holding the subscription record. Diameter agents also have a specific use for PCRF selection, as is further described in Chapter 9.

16.5.6 Peer Discovery

Each Diameter entity must be able to find the next hop Diameter node. With RADIUS, each RADIUS client/proxy has to be statically configured with information about the RADIUS servers/proxies with which it may need to communicate. This can cause a high burden on the network management system to keep these configurations up to date. Diameter still supports the option to statically configure Diameter peers, but in addition it is possible to use more dynamic peer discovery mechanisms, for example by utilizing the DNS.

Diameter clients can then depend on the realm info together with the desired Diameter application and security level to look up suitable first-hop Diameter nodes to which they can forward Diameter messages. The discovered peer location as well as routing configuration will be stored locally and used when making routing decisions.

The Diameter base protocol also includes mechanisms to support transport failure handling between peers, for example using watchdog messages to detect transport failures as well as peer failover/fallback mechanisms.

16.5.7 Diameter Message Format

The Diameter messages are called commands. The content of the Diameter commands consists of a Diameter header followed by a number of AVPs. The Diameter header contains a unique command code that identifies the command and consequently the intention of the message. The actual data is carried by the AVPs contained in the message. The Diameter base protocol defines a set of commands and AVPs but a Diameter application can define new commands and/or new AVPs. The base protocol, for example, defines a set of base AVP formats that can be reused, essentially in an object-oriented fashion, when defining new AVPs. In an application it is thus possible not only to define new AVPs but also new commands, which makes Diameter very extensible and allows the construction of applications to suit the needs of 3GPP. The Diameter message format is shown in Figure 16.30.

The Application ID identifies which Diameter application the message is for. The hop-by-hop identifier is used for matching requests with responses. The end-to-end identifier is used to detect duplicate messages. Each AVP is identified using a unique AVP code. If the AVP is vendor specific, the Vendor ID is also used to uniquely identify the AVP. For more detailed information on the Diameter header and basic AVP formats, see IETF RFC 3588.

Each Diameter application may define its own command, or define its own variant of a command that was defined by another application. Since the number of applications is large, it is not possible to give a complete list of all available Diameter commands. In Table 16.4 we list the Diameter commands that are part of the Diameter base specification. Diameter applications typically support these Diameter commands, even though some of the commands, e.g. the accounting-related commands, may not be used by all applications. In Chapter 15 the EPC interfaces are described, including the Diameter applications defined for the different Diameter-based interfaces. In that chapter, similar tables of Diameter commands are available, in which are listed the Diameter commands specific to each Diameter application.

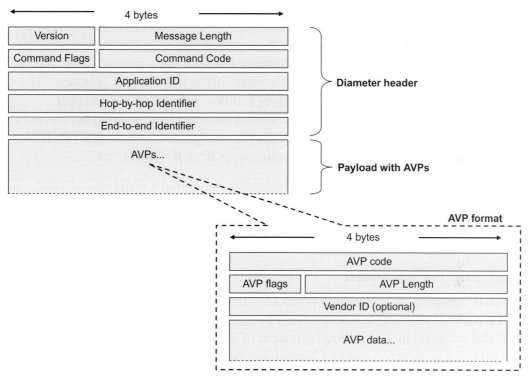

Figure 16.30: Diameter Message Format.

Table 16.4: Diameter Commands

Command Name	Abbreviation	Comment
Abort-Session-Request	ASR	Sent by a server to the client that is providing service, to request that the session is stopped
Abort-Session-Answer	ASA	Sent in response to the ASR message
Accounting-Request	ACR	Sent by a client, in order to exchange accounting information
Accounting-Answer	ACA	Sent in response to the ACR message
Capabilities-Exchange-Request	CER	Sent to exchange local capabilities, e.g. about supported Diameter applications
Capabilities-Exchange-Answer	CEA	Sent in response to the CER message
Device-Watchdog-Request	DWR	Sent to a peer to detect transport failures
Device-Watchdog-Answer	DWA	Sent in response to the DWR message
Disconnect-Peer-Request	DPR	Sent to a peer to inform its intentions to shut down the transport connection
Disconnect-Peer-Answer	DPA	Sent in response to the DPA message
Re-Auth-Request	RAR	Sent by a server to the client that is providing service, to request re-authentication and/or re-authorization
Re-Auth-Answer	RAA	Sent in response to the RAR message
Session-Termination-Request	STR	Sent by a client to inform the server that an authenticated and/or authorized session is being terminated
Session-Termination-Answer	STA	Sent in response to the STR message

16.6 Generic Routing Encapsulation

16.6.1 Background

The GRE is a protocol designed for performing tunneling of a network layer protocol over another network layer protocol. It is generic in the sense that it provides encapsulation of one arbitrary network layer protocol (e.g. IP or MPLS) over another arbitrary network layer protocol. This is different from many other tunneling mechanisms, where one or both of the protocols are specific, such as IPv4-in-IPv4 IETF RFC 2003 or Generic Packet Tunneling over IPv6 IETF RFC 2473.

GRE is also used for many different applications and in many different network deployments outside the telecommunications area. It is not the intention of this book to discuss aspects for all those scenarios. Instead, we focus on the properties of GRE that are most relevant to EPS.

16.6.2 Basic Protocol Aspects

The basic operation of a tunneling protocol is that one network protocol, which we call the payload protocol, is encapsulated in another delivery protocol. It should be noted that encapsulation is a key component of any protocol stack where an upper layer protocol is encapsulated in a lower layer protocol. This aspect of encapsulation, however, should not be considered as tunneling. When tunneling is used, it is often the case that a layer-3 protocol such as IP is encapsulated in a different layer-3 protocol or another instance of the same protocol. The resulting protocol stack may look like that shown in Figure 16.31.

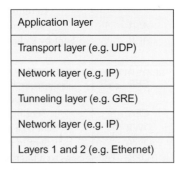

| Application layer |
| Transport layer (e.g. UDP) |
| Network layer (e.g. IP) |
| Tunneling layer (e.g. GRE) |
| Network layer (e.g. IP) |
| Layers 1 and 2 (e.g. Ethernet) |

Figure 16.31: Example of Protocol Stack when GRE Tunneling is Used.

We use the following terminology:

• Payload packet and payload protocol: The packet and protocol that needs to be encapsulated (the three topmost boxes in the protocol stack in Figure 16.31).

- Encapsulation (or tunnel) protocol: The protocol used to encapsulate the payload packet, i.e. GRE (the third box from the bottom in Figure 16.31).
- Delivery protocol: The protocol used to deliver the encapsulated packet to the tunnel endpoint (the second box from the bottom in Figure 16.31).

The basic operation of GRE is that a packet of protocol A (the payload protocol) that is to be tunneled to a destination is first encapsulated in a GRE packet (the tunneling protocol). The GRE packet is then encapsulated in another protocol B (the delivery protocol) and sent to the destination over a transport network of the delivery protocol. The receiver then decapsulates the packet and restores the original payload packet of protocol type A. Figure 16.32 shows an example of tunneling an IPv6 packet in a GRE tunnel over an IPv4 delivery protocol.

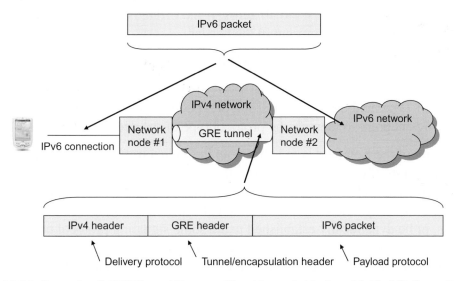

Figure 16.32: Example of GRE Tunnel Between Two Network Nodes with IPv4 Delivery Protocol and IPv6 Payload Protocol.

GRE is specified in IETF RFC 2784. There are also additional RFCs that describe how GRE is used in particular environments or with specific payload and/or delivery protocols. One extension to the basic GRE specification that is of particular importance for EPS is the GRE Key field extension specified by IETF RFC 2890. The Key field extension is further described as part of the packet format below.

16.6.3 GRE Packet Format

The GRE header format is illustrated in Figure 16.33.

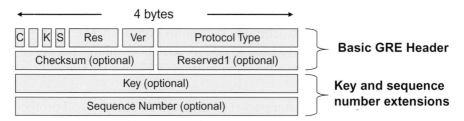

Figure 16.33: GRE Header Format Including the Basic Header as Well as the Key and Sequence Number Extensions.

The C flag indicates whether or not the Checksum and Reserved1 fields are present. If the C flag is set, the Checksum and Reserved1 fields are present. In this case the Checksum contains a checksum of the GRE header as well as the payload packet. The Reserved1 field, if present, is set to all zeros. If the C flag is not set, the Checksum and Reserved1 fields are not present in the header.

The K and S flags respectively indicate whether or not the Key and/or Sequence number is present.

The Protocol Type field contains the protocol type of the payload packet. This allows the receiving endpoint to identify the protocol type of the decapsulated packet.

The intention of the Key field is to identify an individual traffic flow within a GRE tunnel. GRE in itself does not specify how the two endpoints establish which Key field(s) to use. This is left to implementations or is specified by other standards making use of GRE. The Key field could, for example, be statically configured in the two endpoints, or be dynamically established using some signaling protocol between the endpoints. One specific use of the Key field in EPS occurs when GRE is used as tunnel format for the PMIP-based interfaces. The Key field is here dynamically established as part of the PMIP signaling and is used to identify a particular PDN connection between the MAG and LMA (see also description of PMIP in Section 16.4).

The Sequence number field is used to maintain the sequence of packets within the GRE tunnel. The node that performs the encapsulation inserts the Sequence number and the receiver uses it to determine the order in which the packets were sent.

16.7 S1-AP

The S1-AP protocol is designed for a single interface, namely the MME-to-eNodeB interface. The protocol name is derived from the interface name (S1) with the addition of AP (Application Part), which is the 3GPP term for a signaling protocol between two nodes.

S1-AP supports all mechanisms necessary for the procedures between MME and eNodeB, and it also supports transparent transport for procedures that are executed between the UE and the MME or other core network nodes.

S1-AP consists of Elementary Procedures. An Elementary Procedure is a unit of interaction between the eNodeBs and the MME. These Elementary Procedures are defined separately and are intended to be used to build up complete sequences in a flexible manner. The Elementary Procedures may be invoked independently of each other as standalone procedures, which can be active in parallel. In some cases the independence between some Elementary Procedures is restricted; in this case the particular restriction is specified in the S1-AP protocol specification 3GPP TS 36.413. An Elementary Procedure consists of an initiating message and possibly a response message.

The S1-AP protocol supports the following functions:

- Setup, modification, and release of E-RABs.
- Establishment of an initial S1 UE context in the eNodeB (to set up the default IP connectivity, to set up one or more E-RAB(s) if requested by the MME, and to transfer NAS signaling-related information to the eNodeB if needed).
- Provision of the UE capability info to the MME (when received from the UE).
- Mobility functions for UEs in order to enable intra- and inter-RAT HO.
- Paging: This functionality gives the EPC the capability to page the UE.
- S1 interface management functions, for example Reset, Error Indication, Overload Indication, Load Balancing, and S1 Setup functionality for initial S1 interface setup.
- NAS signaling transport function between the UE and the MME:
  - S1 UE context release
  - UE context modification function.
- Status transfer functionality (transfers PDCP SN status information from source eNodeB to target eNodeB in support of in-sequence delivery and duplication avoidance for intra-LTE handover).
- Trace of active UEs.
- Location reporting.
- S1 CDMA tunneling (to carry CDMA signaling between UE and CDMA RAT over the S1 interface).
- Warning message transmission.

There is no version negotiation in S1-AP. The forwards and backwards compatibility of the protocol is instead ensured by a mechanism where all current and future messages, and IEs or groups of related IEs, include ID and criticality fields that are coded in a standard format that will not be changed in the future. These parts can always be decoded regardless of the standard version.

S1-AP relies on a reliable transport mechanism and is designed to run on top of SCTP.

16.8 Non-Access Stratum (NAS)

NAS denotes the protocols between the UE and the MME, which implements the mobility management and session management procedures. The NAS protocols, EPS Mobility Management (EMM) and EPS Session Management (ESM), are tailor-made for E-UTRAN access and are defined by 3GPP in TS 24.301.

The EPS NAS protocols are defined as new protocols but they have many similarities with the NAS protocols used for 2G/3G. The EMM procedures are used to support UE mobility, security, and signaling connection management services for ESM. The ESM procedures are used to activate, deactivate, or modify EPS bearers.

16.8.1 EPS Mobility Management

The EMM procedures are used to keep track of the UE, to authenticate the UE, to provide security keys, and to control integrity protection and ciphering. The network can assign new temporary identities and also request the identity information from the UE. In addition, the EMM procedures provide the UE capability information to the network and the network may also inform the UE of information regarding specific services in the network. The EMM procedures are:

- Attachment
- Detachment
- Tracking area update
- GUTI reallocation
- Authentication
- Security mode control
- Identification
- MM information
- NAS message transport (used for SMS to CS fallback-enabled UE).

An optimization compared to 2G/3G is that the attachment procedure is always combined with an ESM procedure that activates a default bearer. This means that the UE will receive at least one bearer to a PDN by the completion of the combined procedure.

EMM procedures can only be performed if an NAS signaling connection has been established between the UE and the network. If there is no active signaling connection the EMM layer has to initiate the establishment of an NAS signaling connection. The NAS signaling connection is established by a service request procedure from the UE. For downlink NAS signaling the MME first initiates a paging procedure that triggers the UE to execute the service request procedure. The connection management procedures rely on services from the underlying S1-AP protocol on the S1-MME interface and RRC on the E-UTRAN-Uu interface to establish connectivity.

16.8.2 EPS Session Management

The ESM procedures are used to manage the bearers and PDN connections for a UE. This includes procedures for default and dedicated bearer establishment, bearer modification, and deactivation. As noted above, the default bearer establishment is always combined with the attachment procedure but it can be used as standalone procedures to establish additional default or dedicated bearers. The ESM procedures used for E-UTRAN are in principle network-initiated, but the UE may also request the network to modify the bearer resource or ask the network to execute the EPS bearer activation and deactivation procedures.

ESM procedures are:

- Default EPS bearer context activation procedure
- Dedicated EPS bearer context activation procedure
- EPS bearer context modification procedure
- EPS bearer context deactivation procedure
- UE requested PDN connectivity procedure
- UE requested PDN disconnect procedure
- UE requested bearer resource modification procedure.

The NAS protocol is implemented as standard 3GPP L3 messages according to 3GPP TS 24.007. Standard 3GPP L3 according to TS 24.007 and its predecessors have been used for NAS signaling messages in GSM and WCDMA/HSPA. The encoding rules have been developed to optimize the message size over the air interface and to allow extensibility and backwards compatibility without the need for version negotiation.

16.8.3 Message Structure

Each NAS message contains a Protocol Discriminator and a Message Identity. The Protocol Discriminator is a 4-bit value that indicates the protocol being used, i.e. for EPS NAS messages either EMM or ESM. The Message Identity indicates the specific message that is sent.

EMM messages also contain a security header that indicates if the message is integrity protected and/or ciphered. ESM messages, on the other hand, contain an EPS Bearer Identity and a Procedure Transaction Identity. The EPS Bearer Identity indicates the assigned bearer identity and the Protocol Transaction Identifier indicates a particular NAS message exchange between the UE and the MME.

The rest of the information elements in the EMM and ESM NAS messages are tailored for each specific NAS message.

The organization of a normal EMM NAS message is shown in Figure 16.34. Each row corresponds to one octet and bit 8 is the most significant bit in the octet.

8	7	6	5	4	3	2	1	
Security header				Protocol discriminator				octet 1
Message type								octet 2
Other information elements as required								octet 3
								octet n

Figure 16.34: General Message Organization Example for a Normal EMM NAS Message.

The EMM Service Request message is an exception that breaks the normal rules, since it has been tweaked to fit into a single initial RRC message and hence optimizes the performance of the system. The structure of an EMM Service Request message is shown in Figure 16.35.

8	7	6	5	4	3	2	1	
Security header				Protocol discriminator				octet 1
Security parameters and message authentication information								octet 2
								octet 3
								octet 4

Figure 16.35: General Message Organization Example for a Normal NAS Message.

The structure of ESM messages is shown in Figure 16.36.

8	7	6	5	4	3	2	1	
EPS bearer identity				Protocol discriminator				octet 1
Procedure transaction identity								octet 2
Message type								octet 3
Other information elements as required								octet 4
								octet n

Figure 16.36: General Message Organization Example for an ESM NAS Message.

16.8.4 Security-Protected NAS Messages

When an NAS message is security protected the normal EMM and ESM messages above can be ciphered and/or integrity protected and encapsulated as shown in Figure 16.37.

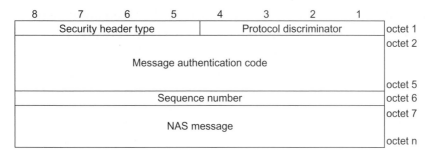

Figure 16.37: General Message Organization Example for a Security Protected NAS Message.

Note that the Service Request message is handled as a special case and is never sent as a security-protected NAS message.

Further details on the EPS NAS messages and the information elements are available in 3GPP TS 24.301 and 24.007.

16.8.5 Message Transport

The NAS messages are transported by S1-AP between MME and eNodeB, and by RRC between eNodeB and UE. The underlying protocols S1-AP (MME-eNodeB) and RRC (eNodeB-UE) provide reliable transport for NAS messages as long as the UE remains within a cell. The NAS protocol includes reliability mechanisms to cater for events like mobility and loss of coverage.

16.8.6 Future Extensions and Backward Compatibility

The UE and network are in principle specified to ignore information elements that they do not understand. It is hence possible for a later release of the system to add new information elements in the NAS signaling without affecting the UEs and network that implement earlier releases of the specifications.

16.9 IP Security

16.9.1 Introduction

IPsec is a very wide topic and many books have been written on this subject. It is not the intention and ambition of this chapter to provide a complete overview and tutorial on IPsec. Instead, we will give a high-level introduction to the basic concepts of IPsec focusing on the parts of IPsec that are used in EPS.

IPsec provides security services for both IPv4 and IPv6. It operates at the IP layer, offers protection of traffic running above the IP layer, and it can also be used to protect the IP header information on the IP layer. EPS uses IPsec to secure communication on several interfaces, in some cases between nodes in the core network and in other cases between the UE and the core network. For example, IPsec is used to protect traffic in the core network as part of the NDS/IP framework (see Section 7.4). IPsec is also used on the SWu interface to protect user-plane traffic between the UE and the ePDG, as well on the S2c interface to protect DSMIPv6 signaling between the UE and the PDN GW. For more details on S2c and SWu, see Sections 15.5.1 and 15.10.1 respectively.

In the next section we give an overview of basic IPsec concepts. We then discuss the IPsec protocols for protecting user data: the ESP and the AH. After that we discuss the Internet Key Exchange (IKE) protocol used for authentication and establishing IPsec Security Associations (SAs). Finally, we briefly discuss the IKEv2 Mobility and Multi-homing Protocol (MOBIKE).

16.9.1.1 IPsec Overview

The IPsec security architecture is defined in IETF RFC 4301. RFC 4301 is an update of the previous IPsec security architecture specification found in IETF RFC 2401. The set of security services provided by IPsec include:

- Access control
- Data origin authentication
- Connection-less integrity
- Detection and rejection of replays
- Confidentiality
- Limited traffic flow confidentiality.

By access control we mean the service to prevent unauthorized use of a resource such as a particular server or a particular network. The data origin authentication service allows the receiver of the data to verify the identity of the claimed sender of the data. Connection-less integrity is the service that ensures that a receiver can detect if the received data has been modified on the path from the sender. However, it does not detect if the packets have been duplicated (replayed) or reordered. Data origin authentication and connection-less integrity are typically used together. Detection and rejection of replays is a form of partial sequence integrity, where the receiver can detect if a packet has been duplicated. Confidentiality is the service that protects the traffic from being read by unauthorized parties. The mechanism to achieve confidentiality with IPsec is encryption, where the content of the IP packets is transformed using an encryption algorithm so that it becomes unintelligible. Limited traffic flow confidentiality is a service whereby IPsec can be used to protect some information about the characteristics of the traffic flow, e.g. source and destination addresses, message length, or frequency of packet lengths.

In order to use the IPsec services between two nodes, the nodes use certain security parameters that define the communication, such as keys, encryption algorithms, and so on. In order to manage these parameters, IPsec uses Security Associations (SAs). An SA is the relation between the two entities, defining how they are going to communicate using IPsec. An SA is unidirectional, so to provide IPsec protection of bidirectional traffic a pair of SAs is needed, one in each direction. Each IPsec SA is uniquely identified by a Security Parameter Index (SPI), together with the destination IP address and security protocol (AH or ESP; see below). The SPI can be seen as an index to a Security Associations database maintained by the IPsec nodes and containing all SAs. As will be seen below, the IKE protocol can be used to establish and maintain IPsec SAs.

IPsec also defines a nominal Security Policy Database (SPD), which contains the policy for what kind of IPsec service is provided to IP traffic entering and leaving the node. The SPD contains entries that define a subset of IP traffic, for example using packet filters, and points to an SA (if any) for that traffic.

16.9.2 *Encapsulated Security Payload and Authentication Header*

IPsec defines two protocols to protect data, the Encapsulated Security Payload (ESP) and the Authentication Header (AH). The ESP protocol is defined in IETF RFC 4303 and AH in IETF RFC 4302, both from 2005. Previous versions of ESP and AH are defined in IETF RFC 2406 and 2402 respectively.

ESP can provide integrity and confidentiality while AH only provides integrity. Another difference is that ESP only protects the content of the IP packet (including the ESP header and part of the ESP trailer), while AH protects the complete IP packet, including the IP header and AH header. See Figures 16.38 and 16.39 for illustrations of ESP- and AH-protected packets. The fields in the ESP and AH headers are briefly described below. ESP and AH are typically used separately but it is possible, although not common, to use them together. If used together, ESP is typically used for confidentiality and AH for integrity protection.

The SPI is present in both ESP and AH headers, and is a number that, together with the destination IP address and the security protocol type (ESP or AH), allows the receiver to identify the SA to which the incoming packet is bound. The Sequence number contains a counter that increases for each packet sent. It is used to assist in replay protection. The Integrity Check Value (ICV) in the AH header and ESP trailer contains the cryptographically computed integrity check value. The receiver computes the integrity check value for the received packet and compares it with the one received in the ESP or AH packet.

ESP and AH can be used in two modes: transport mode and tunnel mode. In transport mode ESP is used to protect the payload of an IP packet. The Data field as depicted

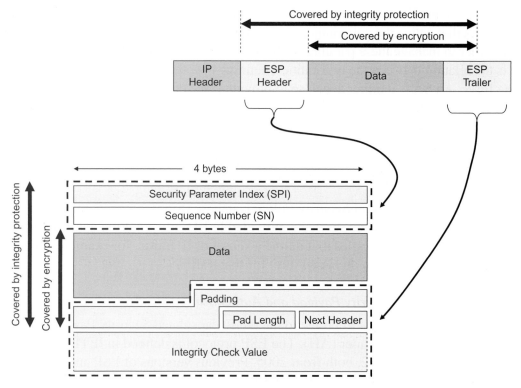

Figure 16.38: IP Packet (Data) Protected by ESP.

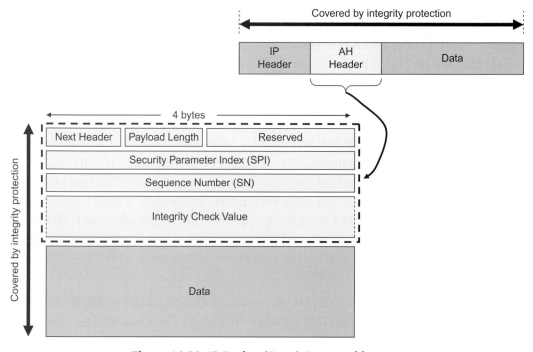

Figure 16.39: IP Packet (Data) Protected by AH.

in Figure 16.38 would then contain, for example, a UDP or TCP header as well as the application data carried by UDP or TCP. See Figure 16.40 for an illustration of a UDP packet that is protected using ESP in transport mode. In tunnel mode, on the other hand, ESP and AH are used to protect a complete IP packet. The Data part of the ESP packet in Figure 16.38 now corresponds to a complete IP packet, including the IP header. See Figure 16.41 for an illustration of a UDP packet that is protected using ESP in tunnel mode.

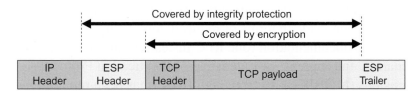

Figure 16.40: Example of IP Packet Protected Using ESP in Transport Mode.

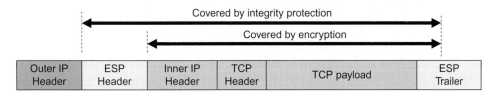

Figure 16.41: Example of IP Packet Protected Using ESP in Tunnel Mode.

Transport mode is often used between two endpoints to protect the traffic corresponding to a certain application. Tunnel mode is typically used to protect all IP traffic between security gateways or in VPN connections where a UE connects to a secure network via an unsecure access.

16.9.3 Internet Key Exchange

In order to communicate using IPsec, the two parties need to establish the required IPsec SAs. This can be done manually by simply configuring both parties with the required parameters. However, in many scenarios a dynamic mechanism for authentication, key generation, and IPsec SA generation is needed. This is where Internet Key Exchange (IKE) comes into the picture. IKE is used for authenticating the two parties and for dynamically negotiating, establishing, and maintaining SAs. (One could view IKE as the creator of SAs and IPsec as the user of SAs.) There are in fact two versions of IKE: IKE version 1 (IKEv1) and IKE version 2 (IKEv2).

IKEv1 is based on the Internet Security Association and Key Management Protocol (ISAKMP) framework. ISAKMP, IKEv1, and their use with IPsec are defined in IETF

RFC 2407, RFC 2408, and RFC 2409. ISAKMP is a framework for negotiating, establishing, and maintaining SAs. It defines the procedures and packet formats for authentication and SA management. ISAKMP is, however, distinct from the actual key exchange protocols in order to cleanly separate the details of security association management (and key management) from the details of key exchange. ISAKMP typically uses IKEv1 for key exchange, but could be used with other key exchange protocols. IKEv1 has subsequently been replaced by IKEv2, which is an evolution of IKEv1/ISAKMP. IKEv2 is defined in a single document, IETF RFC 4306, which thus replaces the three RFCs used for documenting IKEv1 and ISAKMP. Improvements have, for example, been made in terms of reduced complexity of the protocol, simplification of the documentation (one RFC instead of three), reduced latency in common scenarios, and support for Extensible Authentication Protocol (EAP) and mobility extensions (MOBIKE). Even though IKEv1 has been replaced by IKEv2, IKEv1 is still in operational use.

The establishment of an SA using IKEv1 or IKEv2 occurs in two phases. (On this high level, the procedure is similar for IKEv1 and IKEv2.) In phase 1 an IKE SA is generated that is used to protect the key exchange traffic. Also, mutual authentication of the two parties takes place during phase 1. When IKEv1 is used, authentication can be based on either shared secrets or certificates by using a public key infrastructure (PKI). IKEv2 also supports the use of the EAP and therefore allows a more wide range of credentials to be used, such as SIM cards (see Section 16.10 for more information on EAP). In phase 2, another SA is created that is called the IPsec SA in IKEv1 and child SA in IKEv2 (for simplicity we will use the term IPsec SA for both versions). This phase is protected by the IKE SA established in phase 1. The IPsec SAs are used for the IPsec protection of the data using ESP or AH. After phase 2 is completed, the two parties can start to exchange traffic using EPS or AH.

EPS makes use of both IKEv1 and IKEv2. The NDS/IP standard allows both IKEv1 and IKEv2 to be used (see Section 7.4). On other interfaces in EPS, however, it is primarily IKEv2 that is used. For example, on the SWu interface between UE and ePDG, and on the S2c interface between UE and PDN GW, IKEv2 is used.

16.9.4 IKEv2 Mobility and Multi-Homing

In the IKEv2 protocol, the IKE SAs and IPsec SAs are created between the IP addresses that are used when the IKE SA is established. In the base IKEv2 protocol, it is not possible to change these IP addresses after the IKE SA has been created. There are, however, scenarios where the IP addresses may change. One example is a multi-homing node with multiple interfaces and IP addresses. The node may want to use a different interface in case the currently used interface suddenly stops working. Another example is a scenario where a mobile UE changes its point of attachment to a network and is assigned a different IP address in the new access. In this case the UE would have to negotiate a new IKE SA and IPsec SA, which may take a long time and result in service interruption.

In EPS, this may occur if a user is using WLAN to connect to an ePDG. The user traffic between the UE and the ePDG (i.e. on the SWu interface) is protected using ESP in tunnel mode. The IPsec SA for ESP has been set up using IKEv2 (see Section 10.10 for more details). If the user now moves to a different network (e.g. to a different WLAN hotspot) and receives a new IP address from the new network, it would not be possible to continue using the old IPsec SA. A new IKEv2 authentication and IPsec SA establishment have to be performed.

The MOBIKE protocol extends IKEv2 with possibilities to dynamically update the IP address of the IKE SAs and IPsec SAs. MOBIKE is defined in IETF RFC 4555.

MOBIKE is used on the SWu interface to support scenarios where the UE moves between different untrusted non-3GPP accesses.

16.10 Extensible Authentication Protocol

16.10.1 Overview

The Extensible Authentication Protocol (EAP) is a protocol framework for performing authentication, typically between a UE and a network. It was first introduced for the Point-to-Point Protocol (PPP) in order to allow additional authentication methods to be used over PPP. Since then it has also been introduced in many other scenarios, for example as an authentication protocol for IKEv2, as well as for authentication in Wireless LANs using the IEEE 802.11i and 802.1x extensions.

EAP is extensible in the sense that it supports multiple authentication protocols and allows for new authentication protocols to be defined within the EAP framework. EAP is not an authentication method in itself, but rather a common authentication framework that can be used to implement specific authentication methods. These authentication methods are typically referred to as EAP methods.

The base EAP protocol is specified in IETF RFC 3748. It describes the EAP packet format as well as basic functions such as the negotiation of the desired authentication mechanism. It also specifies a few simple authentication methods, for example based on one-time passwords as well as a challenge-response authentication similar to CHAP. As well as the EAP methods defined in IETF RFC 3748, it is possible to define additional EAP methods. Such EAP methods may implement other authentication mechanisms and/or utilize other credentials such as public key certificates or (U)SIM cards. A few of the EAP methods standardized by IETF are briefly described below:

- EAP-TLS is based on TLS and defines an EAP method for authentication and key derivation based on public key certificates. EAP-TLS is specified in IETF RFC 5216.
- EAP-SIM is defined for authentication and key derivation using the GSM SIM card. EAP-SIM also enhances the basic GSM SIM authentication procedure by adding support for mutual authentication. EAP-SIM is specified in RFC 4186.
- EAP-AKA is defined for authentication and key derivation using the UMTS SIM card and is based on the UMTS AKA procedure. EAP-AKA is specified in IETF RFC 4187.
- EAP-AKA′ is a small revision of EAP-AKA that provides for improved key separation between keys generated for different access networks. EAP-AKA′ is defined in RFC 5448.

In addition to the standardized methods, there are also proprietary EAP methods that have been deployed in corporate WLAN networks.

EPS makes extensive use of EAP-AKA and EAP-AKA′ on various interfaces. EAP-AKA is supported for access authentication in untrusted non-3GPP accesses interworking with the EPC (SWa interface), for tunnel authentication towards the ePDG (SWu and SWm interfaces), as well as for establishing the IPsec SA to be used for DSMIPv6 (S2c and S6b interfaces). EAP-AKA′ is supported for access authentication in trusted and untrusted non-3GPP accesses interworking with the EPC (STa and SWa interfaces). When EAP-based access authentication over STa/SWa is performed it occurs prior to invoking the mobility protocol (PMIPv6, DSMIPv6, or MIPv4). For more details, see the corresponding interface descriptions as well as Chapter 7.

16.10.2 Protocol

The architecture for the EAP protocol distinguishes three different entities:

1. The EAP peer. This is the entity requesting access to the network, typically a UE. For EAP usage in WLAN (802.1x), this entity is also known as the supplicant.
2. The authenticator. This is the entity performing access control, such as a WLAN access point or an ePDG.
3. The EAP server. This is the back-end authentication server providing authentication service to the authenticator. In EPS, this is the 3GPP AAA server.

The EAP protocol architecture is illustrated in Figure 16.42.

EAP is often used for network access control and thus takes place before the UE is allowed access and before the UE is provided with IP connectivity. Between the UE

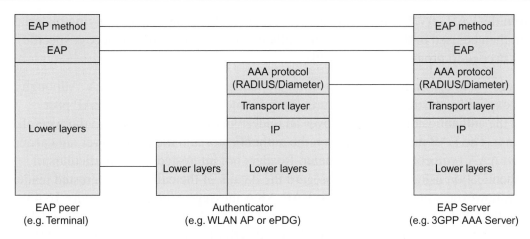

Figure 16.42: EAP Protocol Structure.

(EAP peer) and the authenticator, EAP messages are typically transported directly over data link layers such as PPP or WLAN (IEEE 802.11), without requiring IP transport. Instead, the EAP messages are encapsulated directly in the underlying link-layer protocol. There are different specifications for how this transport is provided. For example, IETF RFC 3748 describes EAP usage for PPP, while IEEE 802.1x describes EAP usage over IEEE 802 links such as WLAN. EAP can also be used for authentication with IKEv2, and in this case is transported over IKEv2 and IP (the IKEv2 and IPsec layers are not illustrated in Figure 16.42).

Between the authenticator and the EAP server, the EAP messages are typically carried in an AAA protocol such as RADIUS or Diameter.

The EAP communication between the peer and the server is basically transparent to the authenticator. The authenticator therefore need not support the specific EAP method used but needs only to forward the EAP messages between the peer and the server.

An EAP authentication typically begins by negotiating the EAP method to be used. After the EAP method has been chosen by the peers, there is an exchange of EAP messages between the UE and the EAP server where the actual authentication is performed. The number of round trips needed and the types of EAP messages exchanged depend on the particular EAP method used. When the authentication is complete, the EAP server sends an EAP message to the UE to indicate whether the authentication was successful or not. The authenticator is informed about the outcome of the authentication using the AAA protocol. Based on this information from the EAP server, the authenticator can provide the UE with access to the network, or continue blocking access.

Depending on the EAP method, EAP authentication is also used to derive keying material in the EAP peer and the EAP server. This keying material can be transported from the EAP server to the authenticator via the AAA protocol. The keying material can then be used in the UE and in the authenticator for deriving the access-specific key needed to protect the access link.

Figure 16.43 provides an example of an authentication using EAP-AKA. Although not explicitly illustrated in the figure, the EAP messages between the EAP peer and the authenticator are carried over an underlying protocol specific to the type of access. The EAP messages between the authenticator and the EAP server are carried in an AAA protocol such as Diameter. Readers not interested in the particulars of authentication using AKA can disregard the details of the call flow. Interested readers may, however, want to compare the EAP-AKA message exchange in Figure 16.43 with the EPS-AKA message exchange over E-UTRAN described in Section 7.3.1. The EPS-AKA over E-UTRAN and the EAP-AKA for accesses supporting EAP are two ways to perform AKA-based authentication.

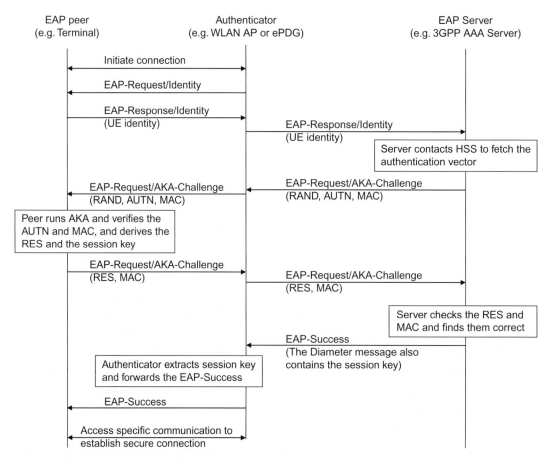

Figure 16.43: Example Message Exchange for Authentication Using EAP-AKA.

16.11 Stream Control Transmission Protocol

16.11.1 Background

The SCTP is a transport protocol, operating at an equivalent level in the stack as UDP (User Datagram Protocol) and TCP. Compared to TCP and UDP, SCTP is richer in functionality and also more tolerant against network failures. Even though both TCP and UDP are used as transport protocols in EPS, we will not describe them in any detail in this book since we assume that most readers have a basic understanding of these protocols. The SCTP, on the other hand, also used as transport protocol at several interfaces in EPS, is a less known transport protocol and therefore is presented briefly in this section.

SCTP is used on several interfaces in the EPC architecture – in particular, the mandated Diameter transport protocol on the S6a/S6d interfaces. SCTP is also used for the transport of S1-AP on the S1-MME interface. More details can be found, together with the description of the interfaces, in Chapter 10.

Compared to UDP (IETF RFC 768) from 1980 and TCP (IETF RFC 793) from 1981, SCTP is a rather new protocol originally specified in IETF RFC 2960 in 2000. The SCTP specification has since then been updated in IETF RFC 4960 (in 2007). The motivation for designing SCTP was to overcome a number of limitations and issues with TCP that are of particular relevance in telecommunication environments. These limitations, as well as similarities and differences between UDP/TCP and SCTP, are discussed below.

16.11.2 Basic Protocol Features

SCTP shares many basic features with UDP or TCP. SCTP provides (similarly to TCP and in contrast to UDP) reliable transport ensuring that data reaches the destination without error. Also, similarly to TCP and in contrast to UDP, SCTP is a connection-oriented protocol, meaning that all data between two SCTP endpoints are transferred as part of a session (or association, as it is called by SCTP).

The SCTP association must be established between the endpoints before any data transfer can take place. With TCP, the session is set up using a three-way message exchange between the two endpoints. One issue with TCP session setup is that it is vulnerable to so-called SYN flooding attacks that may cause the TCP server to overload. SCTP has solved this problem by using a four-way message exchange for the association setup, including the use of a special "cookie" that identifies the association. This makes the SCTP association setup somewhat more complex but brings additional robustness against these types of attacks. An SCTP association, as well as the position of SCTP in the protocol stack, is illustrated in Figure 16.44. As is also indicated in the figure, an SCTP association may utilize multiple IP addresses at each endpoint (this aspect is further elaborated below).

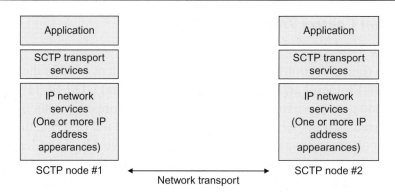

Figure 16.44: SCTP Association.

Similarly to TCP, SCTP is rate adaptive. This means that it will decrease or increase the data transfer rate dynamically, for example depending on the congestion conditions in the network. The mechanisms for rate adaptation of a SCTP session are designed to behave cooperatively with TCP sessions attempting to use the same bandwidth.

Similarly to UDP, SCTP is message-oriented, which means that SCTP maintains message boundaries and delivers complete messages (called chunks by SCTP). TCP, on the other hand, is byte-oriented in the sense that it provides the transport of a byte stream without any notion of separate messages within that byte stream. This is desirable to deliver, for example, a data file or a web page, but may not be optimal to transfer separate messages. If an application sends a message of X bytes and another message of Y bytes over a TCP session, the messages would be received as a single stream of X + Y bytes at the receiving end. Applications using TCP must therefore add their own record marking to separate their messages. Special handling is also needed to ensure that messages are "flushed out" from the send buffer to ensure that a complete message is transferred in a reasonable time. The reason is that TCP normally waits for the send buffer to exceed a certain size before sending any data. This can create considerable delays if the two sides are exchanging short messages and must wait for the response before continuing.

A comparison between SCTP, TCP, and UDP is provided in Table 16.5. More details on multi-streaming and multi-homing are provided below.

16.11.3 Multi-Streaming

TCP provides both reliable data transfer and strict order-of-transmission delivery of data, while UDP does not provide either reliable transport or strict order-of-transmission delivery. Some applications need reliable transfer but are satisfied with only partial ordering of the data and other applications would want reliable transfer but do not need

Table 16.5: Comparison Between SCTP, TCP, and UDP

	SCTP	TCP	UDP
Connection oriented	Yes	Yes	No
Reliable transport	Yes	Yes	No
Preserves message boundary	Yes	No	Yes
Sequential delivery	Yes	Yes	No
Unordered delivery	Yes	No	Yes
Data checksum	Yes (32-bit)	Yes (16-bit)	Yes (16-bit)
Flow and congestion control	Yes	Yes	No
Multiple streams within a session	Yes	No	No
Multi-homing support	Yes	No	No
Protection against SYN flooding attacks	Yes	No	N/A

any sequence maintenance. For example, in telephony signaling it is only necessary to maintain the ordering of messages that affect the same resource (e.g. the same call). Other messages are only loosely correlated and can be delivered without having to maintain a full sequence ordering for the whole session. In these cases, the so-called head-of-line blocking caused by TCP may result in unnecessary delay. Head-of-line blocking occurs, for example, when the first message or segment was lost for some reason. In this case the subsequent packets may have been successfully delivered at the destination but the TCP layer on the receiving side will not deliver the packets to the upper layers until the sequence order has been restored.

SCTP solves this by implementing a multi-streaming feature (the name Stream Control Transmission Protocol comes from this feature). This feature allows data to be divided into multiple streams that can be delivered with independent message sequence control. A message loss in one stream will then only affect the stream where the message loss occurred (at least initially), while all other streams can continue to flow. The streams are delivered within the same SCTP association and are thus subject to the same rate and congestion control. The overhead caused by SCTP control signaling is thus reduced.

Multi-streaming is implemented in SCTP by decoupling the reliable transfer of data from the strict order of transmission of the data (Figure 16.45). This is different from TCP, where the two concepts are coupled. In SCTP, two types of sequence numbers are used. The Transport Sequence Number is used to detect packet loss and to control the retransmissions. Within each stream, SCTP then allocates an additional sequence number, the Stream Sequence Number. Stream Sequence Numbers determine the sequence of data delivery within each independent stream and are used by the receiver to deliver the packets in sequence order for each stream.

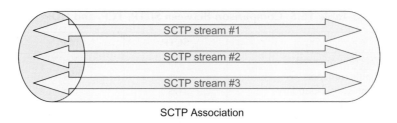

Figure 16.45: Multi-Streaming with SCTP.

SCTP also makes it possible to bypass the sequenced delivery service completely, so that messages are delivered to the user of SCTP in the same order they successfully arrive. This is useful for applications that require reliable transport but do not need sequential delivery, or have their own means to handle sequencing of received packets.

16.11.4 Multi-Homing

Another key aspect of SCTP that is an enhancement compared to TCP is the multi-homing features. In a telecommunications network it is very important to maintain reliable communications paths to avoid service outage and other problems due to core network transmission problems. Even though the IP routing protocols would be able to find alternative paths in the case of a network failure, the time delays until the routing protocol converge and the connectivity is recovered are typically unacceptable in a telecommunications network. Also, if a network node is single homed, i.e. it has only a single network connection, the failure of that particular connection would make the node unreachable. Redundant network paths and network connections are thus two components in widely available telecommunications systems.

A TCP session involves a single IP address at each endpoint and if one of those IP addresses becomes unreachable, the session fails. It is therefore complicated to use TCP to provide widely available data transfer capability using multi-homed hosts, i.e. where the endpoints are reachable over multiple IP addresses. SCTP, on the other hand, is designed to handle multi-homed hosts and each endpoint of an SCTP association can be represented by multiple IP addresses. These IP addresses may also lead to different communication paths between the SCTP endpoints. For example, the IP addresses may belong to different local networks or to different backbone carrier networks.

During the establishment of an SCTP association, the endpoints exchange lists of IP addresses. Each endpoint can be reached on any of the announced IP addresses. One of the IP addresses at each endpoint is established as the primary and the rest become secondary. If the primary should fail for whatever reason, the SCTP packets can be sent to the secondary IP address without the application knowing about it. When the primary IP address becomes available again, the communications can be transferred back. The primary and secondary interfaces are checked and monitored using a heartbeat process that tests the connectivity of the paths (Figure 16.46).

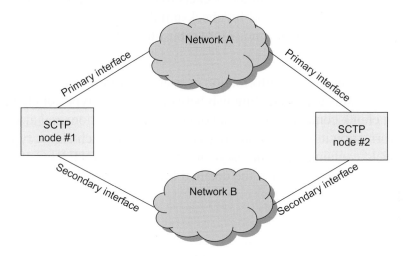

Figure 16.46: Multi-Homing with SCTP.

16.11.5 Packet Structure

The SCTP packet is composed of a Common Header and chunks. A chunk contains either user data or control information (Figure 16.47).

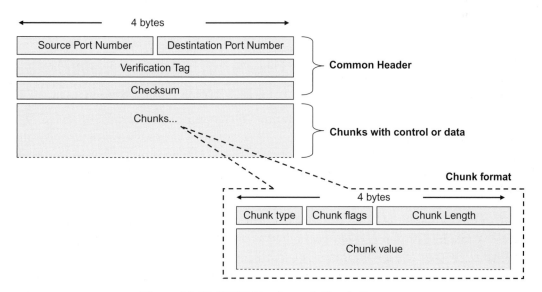

Figure 16.47: SCTP Header and Chunk Format.

The first 12 bytes make up the Common Header. This header contains the source and destination ports (SCTP uses the same port concept as for UDP and TCP). When an SCTP association is established, each endpoint assigns a Verification Tag. The Verification Tag is then used in the packets to identify the association. The last field of the Common Header is a 32-bit checksum that allows the receiver to detect transmission errors. This checksum is more robust than the 16-bit checksum used in TCP and UDP.

The chunks containing control information or user data follow the Common Header. The chunk type field is used to distinguish between different types of chunks, i.e. whether it is a chunk containing user data or control information, and also what type of control information it is. The chunk flags are specific to each chunk type. The chunk value field contains the actual payload of the chunk. IETF RFC 4960 defines 13 different chunk type values and the detailed format of each chunk type.

CHAPTER 17

Procedures

Procedures are a very important tool for describing and understanding how a telecommunication system works. Some procedures have already been described in previous chapters of this book, together with the presentation of the EPC key concepts. In this chapter, we provide a brief introduction to some additional procedures used in EPS. It should be noted, however, that it is not feasible within this book to include a complete description of all procedures that exist in EPS. Instead, we have chosen a few key procedures that should give a good overview of some of the most important use cases. Interested readers can consult the 3GPP technical specifications TS 23.060, 23.401, and 23.402 for complete descriptions.

In the first few sections (17.1–17.5), we describe procedures for 3GPP accesses, including attachment, detachment, and QoS procedures. We also describe handover procedures within a 3GPP access and between different 3GPP accesses. In the last few sections (17.6–17.8), procedures involving non-3GPP accesses such as WLAN and HRPD are described. Here we include, for example, attachment and detachment in non-3GPP accesses, as well as handover procedures between 3GPP and non-3GPP accesses.

17.1 Attachment and Detachment for E-UTRAN

17.1.1 Attachment Procedure for E-UTRAN

Attachment is the first procedure the UE executes after being switched on. The procedure is performed to make it possible to receive services from the network. An optimization in the SAE system is that the attachment procedure also includes the establishment of a default EPS bearer ensuring that always-on IP connectivity for UE/users of the EPS is enabled. An example of the attachment procedure is outlined in Figure 17.1.

EPC and 4G Packet Networks.
DOI: http://dx.doi.org/10.1016/B978-0-12-394595-2.00017-7

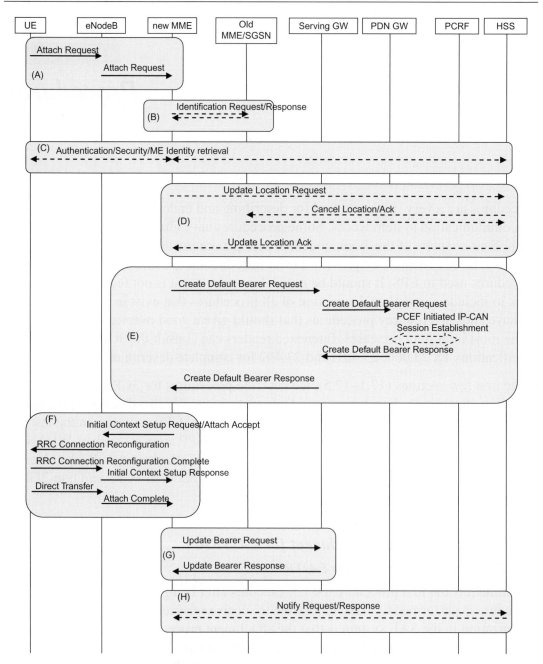

Figure 17.1: Attachment Procedure.

The procedure is briefly described in the following steps:

A. The UE sends an Attach Request message to the eNodeB. The eNodeB checks the MME ID transferred in the Radio Resource Control (RRC) layer. If the eNodeB has a link to the identified MME, it forwards the Attach Request to that MME. If not, the eNodeB selects a new MME and forwards the Attach Request.
B. The MME has changed and the MME uses the old MME ID in the GUTI to find the old MME and retrieves the UE context.
C. Authentication and security procedures are performed. The ME identity is also retrieved in conjunction with this step.
D. If the MME has changed, the MME informs the HSS that the UE has moved. The HSS stores the MME address and it instructs the old MME to cancel the UE context.
E. The default bearer is authorized by the PCRF and established between Serving GW and PDN GW.
F. The default bearer is established over the radio interface and the Attach Accept is sent to the UE.
G. MME informs the Serving GW of the eNodeB Tunnel Endpoint Identifier (TEID), which completes the setup of the default bearer as it can now be used in both uplink and downlink.
H. If the MME has selected a PDN GW that is not the same as the one in the received subscription information, it will send a notification of the new PDN GW identity to the HSS.

In addition, there are some additional steps that may be executed together with the attachment procedure. For example, if the UE's temporary ID (GUTI) is unknown in both the old MME and new MME (after steps A and B), the new MME will request the UE to send its permanent subscription identity (IMSI), as shown in Figure 17.2.

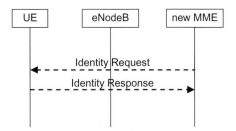

Figure 17.2: Identity Request.

The MME may check the ME identity with an Equipment Identity Register (EIR; after step C). The EIR can be used to blacklist, for example, stolen UEs. Depending on the response from the EIR, the MME may continue the attachment procedure or reject the UE (see Figure 17.3).

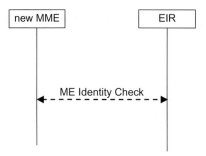

Figure 17.3: ME Identity Check.

If the UE would like to send an APN or PCO, it sets a flag in the initial Attach Request message. The MME will then request the information from the UE after the ciphering has started in step C. This way, there is no need to send the APN or PCO unencrypted over the radio interface. The ciphered options request procedure is used to transfer the APN and/or PCO to the MME (see Figure 17.4).

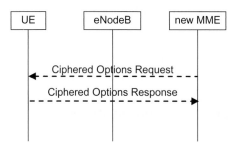

Figure 17.4: Ciphered Options Request.

17.1.2 Detachment Procedure for E-UTRAN

The detachment procedure is used to remove bearers and clear states in the network when, for example, the UE is turned off. The detachment procedure can also be used by the network to remove the bearers and states for a UE that has not performed a TA update because it is out of coverage. There may also be subscription or maintenance reasons to detach the UE, for example if an MME is taken out of service.

Note that, in the normal case, the MME address is not removed from the HSS and the MME can retain the UE context. This saves signaling with HSS, since it is rather likely that the UE will reattach in the same MME and then there is no need to inform the HSS and download the subscription data.

17.1.2.1 UE-Initiated Detachment Procedure

The procedure is shown in Figure 17.5 and is briefly described in the following steps:

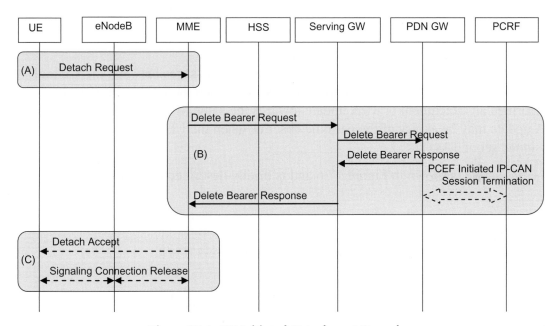

Figure 17.5: UE-Initiated Detachment Procedure.

A. The UE sends a Detach Request to the MME because it is turned off.
B. The MME instructs the Serving GW and PDN GW to delete any bearers for the UE, and the PCEF in the PDN GW informs the PCRF that the bearers are being removed.
C. The MME may confirm the detachment with a Detach Accept message and remove the signaling connection.

If the UE has not communicated with the network for a long time (longer than the TA update timer), the MME may initiate the detachment procedure. In that case, the MME may try to inform the UE with a Detach Request message and then it deletes the bearers as in the UE-initiated detach flow step B (Figure 17.5).

In some special cases, the HSS may also use the detachment procedure by sending a Cancel location with cause value subscription withdrawn. This will trigger the MME to remove the UE context, send a Detach Request to the UE, and delete the bearers as in step B (Figure 17.5).

17.2 Tracking Area Update for E-UTRAN

The TA update procedure is one of the mobility management procedures used to ensure that the MME knows in which TA, or set of TAs, the UE is currently located. TA update is performed when the UE moves to a TA outside the list of TAs it was assigned in the previous TA update or attachment procedure. TA update is also performed periodically even if the UE remains in the assigned TAs.

17.2.1 Tracking Area Update Procedure

In its simplest form, the TA update is just a couple of messages between the UE and the MME. The procedure in the simple form is used when the UE has previously performed an attachment or a TA update towards the same MME. The trigger for the TA update may be expiry of the periodic timer or when the UE has moved outside the assigned set of TAs.

The procedure is shown in Figure 17.6 and is briefly described in the following steps:

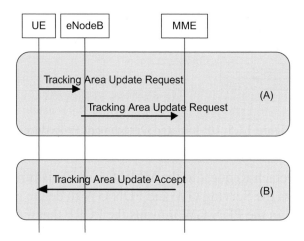

Figure 17.6: Tracking Area Update.

A. The UE decides to perform a TA update; in this case, the trigger can be either expiry of the periodic update timer or when the UE has moved outside its assigned set of TAs. When the eNodeB receives the TA update request, it routes the message to the correct MME based on the GUMMEI that the UE has placed in the RRC message that transports the TA Update message.

B. The MME recognizes the UE, resets the periodic update timer, and sends a TA Update Accept message to the UE. The TA Update Accept message may contain a new list of TAs for the UE. The UE stores the new list of TAs.

17.2.2 TA Update with MME Change

When moving to LTE from 2G/3G or when the MME is changed, the TA update procedure needs to cater for the change of nodes and update all related nodes accordingly.

The procedure is shown in Figure 17.7 and is briefly described in the following steps:

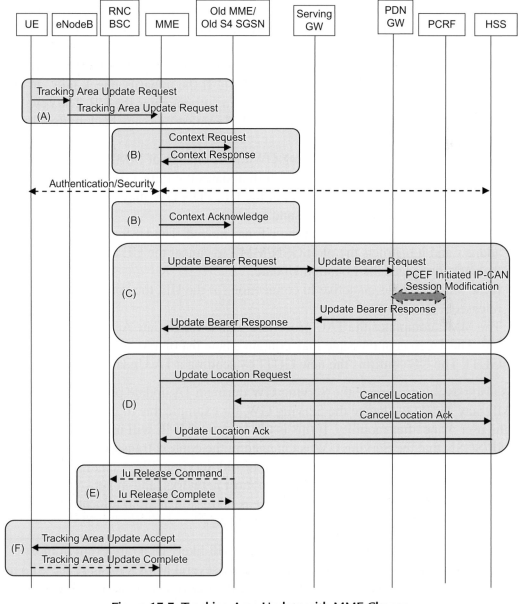

Figure 17.7: Tracking Area Update with MME Change.

A. The UE decides to perform a TA update; in this case, the trigger could be that the UE has moved outside its assigned set of TAs or that the UE has moved from 2G/3G into LTE. When the eNodeB receives the TA update request, it detects that the indicated MME is not associated with the eNodeB and hence a new MME needs to be selected. The eNodeB performs an MME selection and forwards the TA update to the selected MME.

B. The new MME uses the GUTI (the MME Identity inside the GUTI) received from the UE to determine where the UE context resides. The MME requests the UE context from the old SGSN or old MME. The request also includes the TA Update message, which allows the old MME/old SGSN to validate the integrity of the message. If the message passes the integrity check, the old MME or old SGSN sends the UE context to the new MME. If the integrity check fails, an error message is sent to the new MME, which then authenticates the UE. The new MME acknowledges the reception of the UE context message back to the old MME or old SGSN.

C. The new MME informs the Serving GW of the change of MME with an Update Bearer Request. The Serving GW in turn updates the PDN GW with the current RAT type and possibly the location of the UE. PDN GW may also inform the PCRF of the change in RAT type and location.

D. The MME informs the HSS that the UE has moved. The HSS stores the MME address and it instructs the old SGSN/MME to cancel the UE context. The HSS then acknowledges the Location Update.

E. If the old SGSN has an active Iu connection for the UE, the connection is released.

F. The MME completes the TAU procedure with a TA Update Accept message to the UE. The TA Accept message may contain a new list of TAs for the UE and a new GUTI. The UE confirms the new GUTI by sending a TA Update Complete.

It is also possible to change the Serving GW during a TA update procedure. The MME may decide to change the Serving GW, for example due to non-optimal routing of the user plane after mobility. If this is the case, the MME will inform the old MME/SGSN that the Serving GW is changed in the context transfer procedure. The old MME/SGSN will delete the bearers in the old Serving GW. A change of Serving GW is more than likely a very rare event in most networks, and in networks with collocated Serving GW and PDN GW, there is no reason at all to relocate the Serving GW (since the PDN GW will anyway remain fixed).

A change of MME due to mobility within LTE will likely be a rare event in most networks, since MME pools can be used to allow several MMEs to share the UEs in a larger area. As long as the UE remains in the pool coverage area, the UE can stay connected to the same MME.

17.2.2.1 Triggers for TA Update

TA update occurs when a UE experiences any of the following conditions:

- The UE detects it has entered a new TA that is not in the list of TAIs that the UE registered with the network
- The periodic TA update timer has expired
- The UE was in UTRAN PMM_Connected state (e.g. URA_PCH) when it reselected to E-UTRAN
- The UE was in GPRS READY state when it reselected to E-UTRAN
- The UE reselects to E-UTRAN when it has modified the bearer configuration modifications when on GERAN/UTRAN
- The radio connection was released with release cause "load rebalancing TAU required"
- The UE core network capability and/or UE-specific DRX parameters information is changed.

17.3 Service Request for E-UTRAN

17.3.1 UE Triggered Service Request

A Service Request is a procedure an UE executes when it is in Idle mode and needs to establish the bearers to send data or it needs to send signaling to the MME. The UE Triggered Service Request procedure is outlined in Figure 17.8.

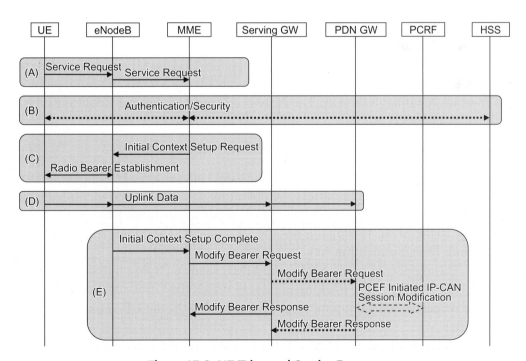

Figure 17.8: UE Triggered Service Request.

The procedure is briefly described in the following steps:

A. The UE sends an NAS message Service Request towards the MME encapsulated in an RRC message to the eNodeB. The eNodeB forwards the NAS message to the MME. The NAS message is encapsulated in an S1-AP: Initial UE Message.
B. Optionally the MME may trigger authentication and security procedures.
C. The MME sends an S1-AP Initial Context Setup Request to the eNodeB. This step activates the radio and S1 bearers for all the active EPS bearers. The eNodeB performs the radio bearer establishment procedure. When the user-plane radio bearers are set up, EPS bearer state synchronization is performed between the UE and the network.
D. Uplink data from the UE can now be forwarded by the eNodeB to the Serving GW. The eNodeB sends the uplink data to the Serving GW and the Serving GW forwards the uplink data to the PDN GW.
E. The eNodeB sends an S1-AP message Initial Context Setup Complete to the MME. The MME sends a Modify Bearer Request message for the accepted EPS bearers per PDN connection to the Serving GW. The Serving GW is now able to transmit downlink data to the UE. If dynamic PCC is deployed, the PDN GW may interact with the PCRF to get the PCC rule(s). If dynamic PCC is not deployed, the PDN GW may apply local QoS policy. The PDN GW sends the Modify Bearer Response to the Serving GW. The Serving GW returns a Modify Bearer Response.

17.3.2 Network Triggered Service Request

A Network Triggered Service Request is a procedure that the UE executes when the UE is in Idle mode and the network needs to establish the bearers to send data or it needs to send signaling to the UE. The Network Triggered Service Request procedure is outlined in Figure 17.9.

The procedure is briefly described in the following steps:

A. When the Serving GW receives a downlink data packet for a UE known not to be user plane connected, it buffers the downlink data packet and identifies which MME is serving that UE. The Serving GW sends a Downlink Data Notification message to the MME to which it has the UE context. The MME responds with a Downlink Data Notification Ack.
B. If the UE is registered in the MME, the MME sends a Paging message to each eNodeB belonging to the tracking area(s) in which the UE is registered. When eNodeBs receive the paging messages from the MME, the UE is paged by the eNodeBs.
C. Upon receipt of a paging indication, the UE initiates the UE triggered Service Request procedure as described in the previous section.

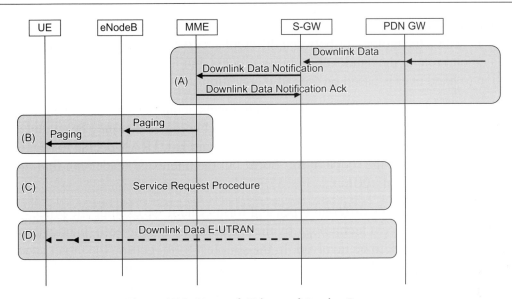

Figure 17.9: Network-Triggered Service Request.

D. The Serving GW transmits downlink data towards the UE once the S1-U bearers have been established.

17.4 Intra- and Inter-3GPP Access Handover

If we consider radio access and packet core network level handover without worrying about the service continuity and session continuity aspects, the following possible handover combinations can be found within and between 3GPP accesses:

a. Intra E-UTRAN
b. E-UTRAN to/from UTRAN (GTP or PMIP)
c. E-UTRAN to/from GERAN (GTP or PMIP)
d. Intra-GERAN, Intra-UTRAN, and GERAN to/from UTRAN (these handover cases are not covered in this book).

Note that all these scenarios can be Intra- or Inter-PLMN handover, as described below in more detail.

Within 3GPP (as well as in most other cellular technologies) a handover is defined as follows in very narrow terms (as per 3GPP TS 22.129):

Handover is the process in which the radio access network changes the radio transmitters or radio access mode or radio system used to provide the bearer services, while maintaining a defined bearer service QoS.

Handover is a key mobility mechanism for any cellular system, whether moving within an access technology or between different access technologies. Core networks play a crucial role in the handover mobility process, but in a majority of the cases the decision to hand over is based on the radio conditions.

The UE assists the network with radio measurements about the serving as well as neighboring cells in the same or different access technologies that may be candidates for handover. The details of how and when the UE and E-UTRAN decide to trigger a handover is far beyond the scope of this book, but the interested reader can find more details in Dahlman et al. (2011).

Handover can be of many different types and forms, and if we exclude the process of selecting the target access technology by the UE being handed over to, handover may cause the core network to be involved. In the simplest form, getting the target access network information the UE is connecting to and to a more complex form when one or more core network entities need to be relocated (e.g. MME or/and S-GW) to better serve the user. In addition to the process of actually changing the radio and/or core network entities, the handover process also needs to ensure service continuity, which entails maintaining the bearer characteristics for active services as much as possible. A system may use a handover/cell reselection mechanism to achieve service continuity for a UE actively involved in a session (transmitting and receiving data). Note that other mechanisms, such as SRVCC, also enable certain specific service continuity for specific types of service(s), which is addressed in Chapter 11.

So what are the different kinds of handovers possible from an overall network perspective? Let us consider the example diagram illustrated in Figure 17.10, depicting a simple scenario where Operator X and Operator Y have some of their radio networks connected to each other and their core network (here EPC) connected via a GRX/IPX interconnect. Operator X has two EPC networks connected to the RANs OPx1, OPx2, ..., OPxn, whereas Operator X has one EPC network connected to RANs OPy1 and OPy2. The first level of handover we can define may be whether the user is moving within Operator X networks, thus causing *Intra-PLMN handover*. If the user moves between RANs OPx1 and OPy2, then we have encountered *Inter-PLMN handover*. During Inter-PLMN handover, when the user also crosses radio access technology, such as E-UTRAN to/from UTRAN, the network has also performed *Inter-System Handover* by changing the radio access technology. Note that a UE will only be instructed to measure on neighboring cells to which handover is allowed.

In the case of Inter-PLMN handover, a number of aspects need to be understood and established before handover can be accomplished. First of all, in Inter-PLMN handover, a session may not only cross an operator's boundary, but it may also cross a national boundary. A call in the United States might, for example, arise in upstate

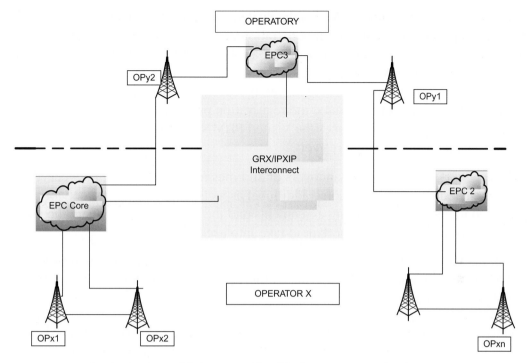

Figure 17.10: Multi-Operator Simplified 3GPP Network Diagram.

New York bordering Canada and continue inside Canada if Inter-PLMN handover is supported between the two operators involved. As such, Inter-PLMN handover very much depends on an individual operator's discretion whether it is supported or not. An operator may choose to drop the session and then require the UE to register in the new PLMN instead, and in this case the service continuity is not maintained. Before proceeding with Inter-PLMN handover, certain criteria need to be met, as specified in 3GPP TS 22.129:

- The ability to check with the home network whether the user is permitted to hand over from the visited network to a target network.
- Invocation of the handover procedure only occurring if the target network provides the radio channel type required for the respective call.
- The avoidance of "network hopping", i.e. successive handover procedures between neighboring networks for the same call.
- The possibility of user notification of Inter-PLMN HO (e.g. possible tariff change) when it occurs.

During handover procedures, a network may operate in one of three roles: Home PLMN, Serving PLMN, and Visited PLMN. Home PLMN is where the subscriber has his/her subscription. Visited PLMN is the visited network for a roaming user where the user has performed a successful registration process (i.e. the HSS is aware of where the user is located and has performed all the procedures necessary for updating the user's location). Serving PLMN/network is where the user may have been handed over to (e.g. the cell the user is being served by belongs to the Serving network operator) and has not yet performed the registration process. The most likely scenario is that the Serving PLMN becomes the Visited PLMN after registration unless the user moves out of the serving area/cell.

Shared networks also support handovers of all types mentioned above between shared and non-shared networks. But some additional aspects, such as the core network it is connected to and related roaming agreements and the home network where the user has a subscription, are also important to take into account.

Even though the most frequent cause of handover is the movement of the user (i.e. the UE), there are other additional triggers that can cause a handover to occur. Some examples of these are:

• Triggering due to a service requirement that may be met by a different RAT than the one the UE is being served by; the core network may trigger such a process.
• Various radio conditions such as change of radio access mode and the capacity of a cell to be able to provide the user's current services.

Even though, in principle, handover should not cause any significant loss/change of service or interruption of service, when multiple radio bearers are being handed over from one type of radio access to another, there may be a need to drop certain bearers and maintain others based on priority and relevant QoS information, data rate, delay constraints, error rate, etc. Also, sometimes, a certain QoS may be degraded to accommodate the handover of all PS bearers. Overall, instead of failing to hand over at all, it would be preferable to be able to hand over at least one bearer that is suitable for the target radio access. Usually, when moving from a higher bit rate to a lower one (e.g. UTRAN to GERAN), decisions have to be made that suit the serving network operator (HPLMN when not roaming) and an operator may choose to drop all active bearers or, based on certain predefined criteria, choose to hand over specific bearers.

So what has changed for EPS compared to what existing handover concepts did not already cover? For example, there is no longer a central entity controlling the RAN like an RNC for UTRAN and a BSC for GERAN. Instead, the eNodeBs in E-UTRAN connect directly to the EPC core network entities, MME for the control plane for signaling and Serving GW for user-plane traffic transferring data to/from a terminal for the user. In addition to E-UTRAN, the EPS must also support handover to/from non-3GPP accesses. These procedures are described later in this chapter.

Even though Inter-System handover is usually used in conjunction with 3GPP accesses, this term can easily be expanded to also cover Intra-PLMN handover between 3GPP and non-3GPP accesses. For 3GPP accesses, Inter-PLMN and Intra-PLMN handovers as well as Inter-System handovers are supported, with special emphasis on service continuity between UTRAN and E-UTRAN radio access. In addition, EPS also supports handover from E-UTRAN to pre-Release 8 3GPP networks, but note that handover in the opposite direction is not supported. In this case, the source network (i.e. EPS) has to adopt the target network requirements due to the fact that the target network cannot understand/interpret the EPS information because the pre-Release 8 networks are not upgraded.

Another special aspect of E-UTRAN radio access is that it is a packet-only system and thus there is no support for circuit-switched bearers and the CS Domain in the evolved system. So a handover from IMS voice on E-UTRAN EPS to CS voice on 2G/3G has been developed in 3GPP, known as Single Radio Voice Call Continuity (SRVCC). It is also possible to provide dual radio-based service continuity between 3GPP and other non-3GPP accesses with IMS, when the non-3GPP access is connected via EPS.

17.4.1 Phases of the Handover Procedures

PS handover procedures for GSM were developed with the basic principles of two main phases: handover preparation phase and handover execution phase. The same principles apply in the handover procedures for EPS as well.

We will briefly go over some of the principles of handover in the existing 2G/3G 3GPP Packet Core (GPRS PS Domain) and then go into the details of the EPS handovers. As in any handover case, there is a source cell in the RAN and a target cell in the RAN (within the same radio access in the case of Intra-RAT and in a different RAN in the case of Inter-RAT handover) where the terminal is planned to be moved to. For example, for a 2G system, Intra-BSS handover may be performed maintaining the same SGSN (known as Intra-SGSN handover) or also changing the SGSN (known as Inter-SGSN handover). Handover may also be performed between different radio access types, such as between BSS and UTRAN, which is known as Inter-RAT, where an Inter-RAT HO can also be Intra-SGSN or Inter-SGSN handover. In the case of 3G radio access with PS domain GPRS, there can be Inter-RNC HO (where RNC functions including the SRNS function are moved) with Intra- or Inter-SGSN HO, SRNS relocation procedures with Intra- or Inter-SGSN HO, and Inter-System HO with BSS to/from RNC handover is performed with Intra- or Inter-SGSN HO. In all these handover scenarios, the Packet Core network is involved and updated during the handover process. Note that the GGSN is not relocated or changed in any handover procedure.

The overall handover process may be described as the source access handling the procedures, such as UE and radio network measurements, to determine handover needs to be initiated, preparing the resources in the target radio and core network side, directing the UE to the new radio resources, and releasing the resources in the source radio and core network where applicable, and, additionally, handling gracefully any failures that may occur to get back to a stable status, and, additionally, ensuring that all control- and user-plane entities are properly connected in the new network. During this process, the uplink and downlink data may be buffered and then forwarded according to the most appropriate path determined by the specific process/handover type itself, thus minimizing any possible loss of user data.

We will now briefly analyze the high-level view of the two phases (Preparation and Execution) for handover. Figures 17.11 and 17.12 outline the preparation phase and the execution phase respectively.

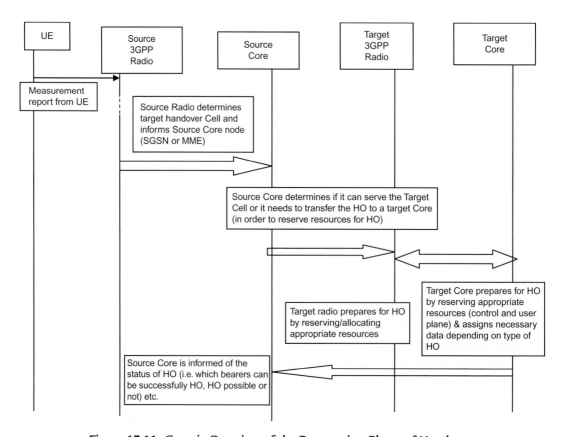

Figure 17.11: Generic Overview of the Preparation Phase of Handover.

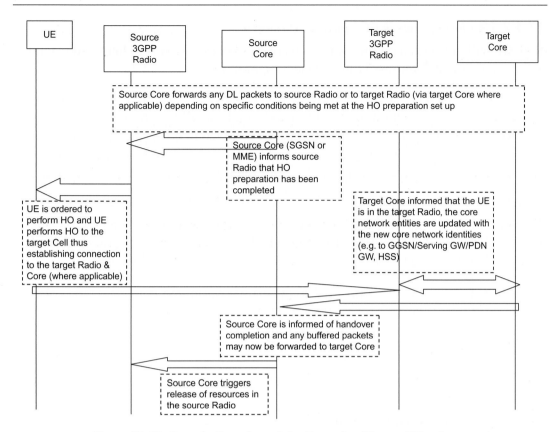

Figure 17.12: Generic Overview of the Execution Phase of Handover.

Handover may fail at any time during the handover procedure and handover may also be rejected by the target RAN; the source RAN may also cancel the handover due to conditions that may deem that the handover will not succeed or that the process has failed somewhere. If the handover is rejected or canceled by the target radio or source radio respectively, all acquired resources will be released and cleared of the handover process both in radio and in core networks. In the case of handover failure, depending on whether the failure occurred during the preparation or execution phase, the resources affected and actions needed would, of course, be different. If the UE fails to connect to the target cell during the execution phase, it returns to the source radio and sends an appropriate message of failure. If the UE has lost radio contact, then it is the source radio's responsibility to inform the source core and then the rest of the path in the target network is cleared/released. If failure is caused due to core network procedures, then an appropriate cause code would be sent to the UE and the subsequent actions determined (renegotiate bearers, etc.).

17.4.2 Handover Cases in EPS for 3GPP Accesses

Now we will get started on the main focus of this book, which is to help understand the process of handover involving E-UTRAN and EPS.

In the case of EPS and handover involving E-UTRAN, the MME and Serving GW are the entities involved for Intra-RAT (LTE) handover. If the PMIP protocol is used, the BBERF located in the Serving GW may also be involved in order to update the PCRF with the right BBERF information. Compared to 2G/3G handover using GPRS, the EPS provides change/relocation during handover for the following possible combinations:

1. Intra-E-UTRAN (between eNodeBs) only (Intra-MME handover can be accomplished via an X2-based or S1-based HO procedure).
2. Inter-E-UTRAN and MME change (Inter-MME).
3. Inter-E-UTRAN and MME and Serving GW change (Inter-MME and Inter-Serving GW).
4. Inter-E-UTRAN and Serving GW change (Inter-Serving GW).
5. Inter-RAT (E-UTRAN and GERAN/UTRAN) with core network entity combination of relocation (i.e. MME to/from SGSN relocation and then Serving GW may be relocated).

Handover between 3GPP and non-3GPP access is covered in the subsequent sections.

17.4.3 Handover within E-UTRAN Access

In certain extraneous conditions such as MME overload, an MME may trigger a relocation of users from the affected MME to a new MME. Note that, compared to 2G/3G, the EPC is clearly separated as control- and user-plane entities in relation to eNodeB: MME is the control-plane entity and Serving GW is the user-plane entity. Thus, depending on the type of handover performed, multiple entities may be relocated and updated with each other's information before the handover is completed. As a user's profile may restrict handover via roaming/area restrictions, MME provides this information to eNodeB via the S1 interface and then, during handover in active mode of the UE, eNodeB is responsible for verifying whether the handover is allowed or not.

As described for 2G/3G, LTE handover is also performed in two main phases, preparation and execution. Due to the nature of X2-based handover, the execution phase has been further divided into execution and handover completion phases. In the handover completion phase, core network entities like MME and Serving GW then become aware of the completed handover and complete the necessary update to the control- and user-plane paths throughout the bearer connection path. One of the simplest cases is the Intra-E-UTRAN handover case, described and illustrated in Figure 17.13.

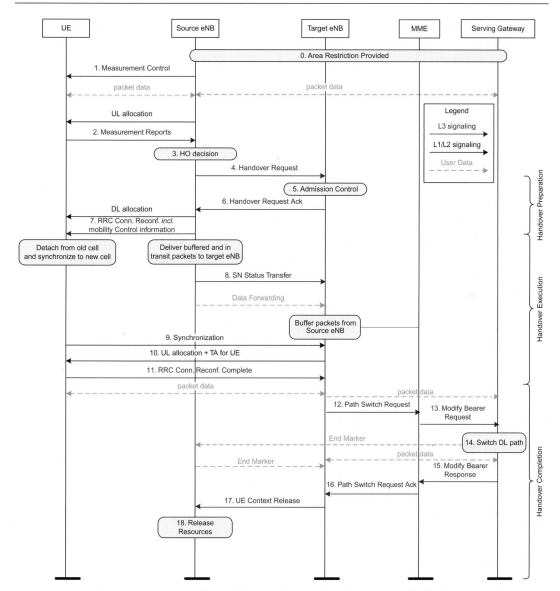

Figure 17.13: Simplified Intra-E-UTRAN HO Without Change of Core (Intra-MME/Serving GW Using X2 HO Procedure).

In this procedure, during the preparation phase, the core network entities are not involved. The source eNodeB makes the decision based on UE and radio-level information, as well as restriction data provided by the core network (i.e. MME) that handover needs to be performed, and selects the appropriate target eNodeB. eNodeBs are connected with each other via the X2 reference point over an IP infrastructure.

The source eNodeB is responsible for the decision which of the EPS bearers are subject to forwarding of packets from the source eNodeB to the target eNodeB; EPC maintains the decision taken, i.e. no changes are performed. Both source eNodeB and target eNodeB may need to buffer data and, during the handover phase, the source eNodeB establishes an uplink and downlink data forwarding path for user-plane traffic to the target eNodeB.

During the execution phase, the source eNodeB forwards any data received from the downlink path from the Serving GW towards the target eNodeB as long as the data is arriving and the source eNodeB is able to handle the data. Once the handover has been executed successfully (i.e. the UE is now connected to the target eNodeB), the target eNodeB informs the MME to switch paths and the MME then also informs the Serving GW to switch the user-plane path for the downlink data traffic (i.e. towards the UE) and informs the source eNodeB via an end marker about the end of data transfer. This is the simplest type of handover from the EPC perspective, as can be readily seen.

The procedure is more complex if MME determines that Serving GW relocation may be needed in this case for reasons such as appropriate user-plane connectivity. But in order to be able to perform Serving GW relocation, there must be full IP connectivity between source eNodeB and source Serving GW, between source Serving GW and target eNodeB, and between target Serving GW and target eNodeB.

On receipt of the Path Switch request from the target eNodeB, the MME requests the target Serving GW to create new bearers as required after handover completion and the target Serving GW updates the PDN GW with its address and updated user information, thus completing the new switched path between the UE and the target eNodeB and between the target Serving GW and the PDN GW. When the PDN GW gets the updated information, it starts sending downlink data via this updated path. The MME informs the Serving GW to release its resources for that UE and the target eNodeB releases the necessary resources in the source eNodeB, once path switch completion is indicated by the MME.

Even though it seems natural to have an X2-based handover, it is not reasonable to assume full IP connectivity between all eNodeBs directly from day 1 of E-UTRAN/EPS deployment and there may be other reasons why X2-based handover cannot be performed. In such cases, S1-based handover via the core network is the mechanism available for all listed handover cases above.

In the case of an S1-based handover, either the MME or both the MME and Serving GW can be relocated even though change of MME should not be done unless the target eNodeB belongs to a different MME pool area. If a source MME has selected a target MME for the handover, then it is the target MME's responsibility to decide if the source Serving GW needs to be relocated or not, otherwise it is the source MME's responsibility to make that decision. The sequence diagram below shows the necessary steps for the handover; note that it is the source eNodeB's decision whether direct data forwarding will be performed via the X2 interface or indirect data forwarding will be performed via the source or target Serving GW, depending on whether the Serving GW is relocated or not. The source MME releases source Serving GW resources only when it has confirmation that target Serving GW path switch has been successfully completed.

If the S1 handover is rejected for any reason, the UE then remains in the Source eNodeB/MME/Serving GW. Figure 17.14 illustrates detailed procedures that are specified in 3GPP TS 23.401; in this book we do not go into the detailed nodes or the procedural behavior of the flow. Readers interested in the protocol and sequence-level details of the various scenarios and flows should consult the 3GPP specifications that are indicated in Appendix A of this book. Note that points A and B are two occurrences where GTP-based procedures and PMIP-based procedures differ due to differences in how the GTP and on-path PCC and PMIP and off-path PCC work.

Note that we have focused on a GTP-based S5/S8 interface when describing the handover flows; if PMIP is used, then the interactions between the Serving GW and PDN GW would be slightly different and there is also an additional interaction due to the PCC using off-path signaling, causing additional interactions with PCRF. So when a change of Serving GW is required, the target Serving GW triggers the GW Session Control procedure to perform policy controlled functions, such as bearer binding, and also informs of an RAT change if applicable and then updates the PDN GW. When the source Serving GW is instructed to delete the bearers due to Serving GW relocation, it also ensures that PCRF association is removed between the BBERF in the source Serving GW and PCRF. If the Serving GW is not relocated, the Serving GW then updates PCRF with the updated QoS rules and session binding information, which then triggers PDN GW updates for PCC as well. In the case of GTP, since the policies are handled on-path, these interactions are not needed between Serving GW and PDN GW, though the PDN GW may trigger PCC procedures based on, for example, change in RAT type. Note that since the UE, the radio access networks, and the MME/SGSN do not differentiate whether GTP or PMIP is used over S5/S8 reference points, there is no effect on the main handover process due to the core network protocol selection process.

One aspect that is not supported during handover is dynamic protocol change (between GTP and PMIP) over S5/S8 reference points when handover may cause relocation of the Serving GW. The complexity of changing protocol "on the fly" seems unnecessarily complex and the necessity of such a dynamic change in the

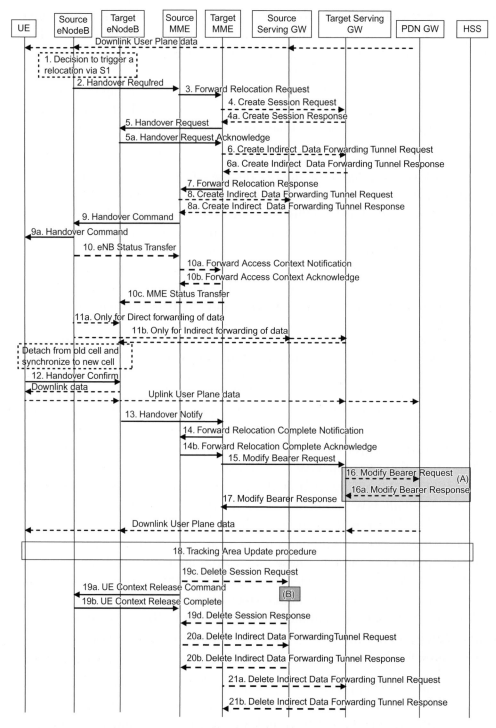

Figure 17.14: S1-Based Handover (Inter-E-UTRAN).

initial releases of LTE/EPS is not yet apparent. The situation can be made worse when a UE moves from one PLMN to another during handover and requires both the MME and Serving GW to be relocated; thus if a protocol change is also required at the same time, that affects the PDN GW as well as PCRF.

17.4.4 Handover between E-UTRAN and Other 3GPP Accesses (GERAN, UTRAN) with S4-SGSN

When one has to cope with such a wide deployment of GSM/UMTS and a huge subscriber base (which implies the number of terminals in the consumer's hands) as well as radio network equipment (GERAN, UTRAN), minimizing the impact on the system when performing handover (and in overall functions) is a key aspect to the overall success of EPS. Within 3GPP this type of inter-system handover was also developed during the specification of the 3G system. So the "know-how" and commitment were there to achieve the goal of an efficient inter-system handover. However, the changes compared to 2G to 3G and 2G/3G to LTE/EPS are much more substantial both from a radio and a core network perspective, especially in the core network. Since the differences for handover procedures compared to existing handovers are minimal, we will emphasize the differences and any additional aspects that would benefit readers' understanding of the process.

In the case of IRAT handover where E-UTRAN is the radio access technology and EPC is the core network, the same principles apply as before: perform the preparation directed by the source radio network and in the case of Inter-RAT handover, always perform the habdover via the core network. The source RAN adapts the content and information flow that suits the target access network (this is crucial in the case of handover from E-UTRAN to pre-Release 8 core networks), and as before the source decides to start the preparation of the handover. It is also source RAN that makes the final decision on the execution of the handover procedure.

Inter-RAT HO is considered to be a backwards handover where the radio resources are prepared in the target 3GPP radio access before the UE is commanded by the source 3GPP radio access to change to the target radio. The target access system is responsible for giving exact guidance to the UE on how to make the radio access there (radio resource configuration, target cell system information, etc.) during the preparation phase. Since the RAN connections are not yet established, the signaling and information transfer are done via the core network through the source radio access transparently to the terminal during the preparation phase of the handover.

In Inter-RAT HO involving E-UTRAN access, the core network entities involved during the preparation and transfer of data are the S4 SGSN and corresponding MME/Serving GW. Whether indirect forwarding is to be applied or not is configured in the MME and S4 SGSN as part of the operator-specific data: the configuration always indicates whether to perform indirect forwarding, whether to perform indirect forwarding only for Inter-PLMN Inter-RAT handover, or if it should not be performed.

The main aspect of this handover is that the MME to/from S4 SGSN relocation must always be performed in the case of E-UTRAN to 2G/3G handover.

Handover between E-UTRAN and UTRAN and handover between E-UTRAN and GERAN are very similar; we have chosen to illustrate the procedure by using handover from E-UTRAN to UTRAN as an example. The following steps outline how the generic procedures above apply to handover between E-UTRAN and UTRAN.

- Once the source eNodeB triggers the handover request to the MME, based on the information received, the MME determines that it is UTRAN Iu mode handover. It then selects the appropriate SGSN for the target RNC and initiates the appropriate resource reservation process in the target system. The MME is also responsible for mapping the EPS bearers that the source eNodeB has selected for handover to the PDP context bearers as applied to 2G/3G. This mapping is specified in 3GPP in order to provide a consistent outcome during handover.
- In the target core network, the SGSN prioritizes the PDP context bearers and decides whether the Serving GW needs to be relocated. If the Serving GW is to be relocated, the SGSN selects the appropriate target Serving GW and triggers the appropriate resources allocation.
- The SGSN also provides the target RNC with all relevant information provided by the source network in order to establish the radio resources required for handover. Once the target RNC has completed all necessary radio resource allocation towards the UE and user-plane resources have been established for the Iu, the target SGSN is informed of the completion of the process. At this point, the target RNC is prepared to receive user-plane data from either the SGSN or Serving GW, depending on whether direct tunneling is applied or not (in the case of no direct tunnel, SGSN remains in the user-plane path).
- The source MME and source eNodeB are then informed of successful handover resources reserved and the preparation phase is completed.
- In the case of indirect forwarding, the path is established by the SGSN in the target Serving GW and by the source MME in the source network.
- Once the source MME informs the source eNodeB about handover preparation completion, the source eNodeB starts forwarding data, commands the UE to perform handover to the target RNC, and provides the UE with all the necessary information provided by the target RNC. At this time, the UE no longer receives/ sends any data via the source eNodeB.

- The UE tunes into the target cell in which the radio bearers are established and the rest of the process continues in the core network, where the target SGSN informs the source MME of handover completion when the target RNC confirms it to the target SGSN. After that, the source MME and target SGSN need to release any forwarding resources. The source MME also needs to release source Serving GW resources in the case of Serving GW relocation and the Serving GW updates the PDN GW with appropriate information to establish the user-plane path between the RNC, SGSN, Serving GW, and PDN GW.

Note that any EPS bearers that were not successfully transferred are deactivated by the SGSN and Serving GW, and any downlink data is dropped in the Serving GW. Also note that the UE only re-establishes the bearers that were accepted by the target network for handover during the preparation phase.

The procedure flow in Figure 17.15 is included for illustration purposes only, detailing the preparation and execution phases of Inter-RAT handover from E-UTRAN to 3G UTRAN access.

Further information flows and a much more detailed description of the handover procedure between E-UTRAN and GERAN/UTRAN are given in 3GPP TS 23.401.

17.4.5 Handover for Gn/Gp-Based SGSN

Interoperation scenarios for operating E-UTRAN with a PLMN maintaining Gn/Gp SGSNs are supported for GTP-based S5/S8 interfaces. Thus, the PDN GW then acts as a GGSN supporting Gn/Gp interfaces towards the SGSN and the MME supports the Gn interface towards SGSN. Note that the HSS must also be able to work with the Gr interface or be supported via an interworking function to enable interworking between S6a and Gr functions. Figure 17.16 illustrates the architecture for an EPC network maintaining the Gn/Gp interface to SGSN.

The main principles for handover between E-UTRAN and UTRAN/GERAN connected to Gn/Gp core networks are as follows:

- The handover procedures within E-UTRAN involving MME and Serving GW are the same as in the case of using the EPC core for GERAN/UTRAN.
- The handover procedures within GERAN/UTRAN involving existing GPRS procedures are the same as those currently specified without EPC.
- A handover from E-UTRAN to UTRAN/GERAN would imply executing Inter-SGSN handover with the source SGSN represented by the MME for E-UTRAN.
- A handover from UTRAN/GERAN to E-UTRAN would imply executing Inter-SGSN handover with the source SGSN represented by the Gn/Gp SGSN for UTRAN/GERAN and the target SGSN represented by the MME for E-UTRAN.

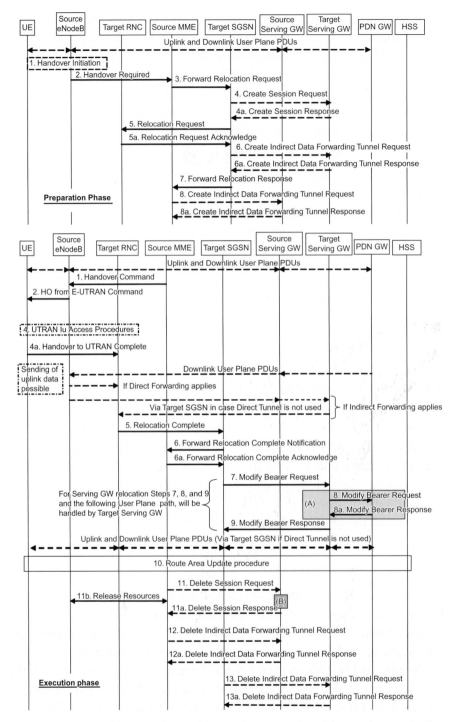

Figure 17.15: Combined Preparation and Execution Procedure for Inter-RAT Handover from E-UTRAN to UTRAN Access.

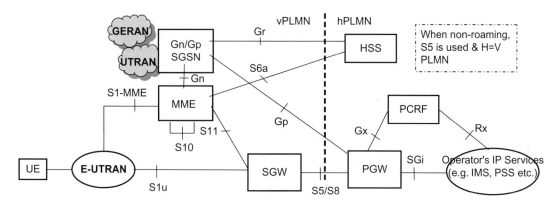

Figure 17.16: E-UTRAN and Gn/Gp SGSN Handover Interworking Architecture.

In addition, the MME has to select the Serving GW and the Serving GW must also update the PDN GW with the appropriate S5/S8-related information.

- The Serving GW establishes the appropriate setup towards eNodeB.
- If indirect forwarding is to be performed, the MME performs the same procedure as for Inter-E-UTRAN handover described above.
- The PDN GW represents the GGSN function and the GGSN selected for the UTRAN/GERAN must be a PDN GW acting as a GGSN.
- Mapping between PDP context and EPS bearers as well as other parameters is also handled in the appropriate entities (such as the MME taking care of handling the security and QoS parameters).

Figure 17.17 provides an illustration of an end-to-end procedure as described in 3GPP TS 23.401. We will not go into the details of these procedures as interested readers can consult the 3GPP specifications.

Note that in order to allow for early implementation of UEs and early deployment of networks there is an additional possibility to support mobility between E-UTRAN and UTRAN/GERAN. This mobility solution is based on RRC connection release with redirection information. RRC connection release with redirection information is implemented in the Source eNodeB and does not require any additional support in the network. The redirection information points the UE towards GERAN or UTRAN, where the UE will use the existing routing area update procedure to recover the connection. This procedure has worse performance than the other handover solutions and may create a significant break in the connection. It may, however, have acceptable performance for early deployments with primarily data-only users and a low number of devices.

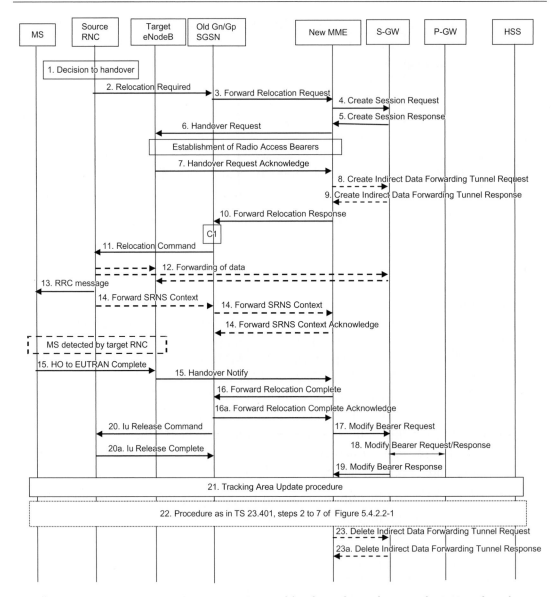

Figure 17.17: 3G Gn/Gp SGSN to MME Combined Hard Handover and SRNS Relocation Procedure.

17.4.6 Handover between GERAN and UTRAN Access Using S4 SGSN and GTP/PMIP Protocol

For GERAN and UTRAN access, handovers have been developed and deployed for a long time, using GPRS and GTP as the core protocol. During the SAE development process, there was no significant interest from any operators in developing support for handover between S4-SGSN with PMIP, or even GTP-based 2G and 3G and 2G and 3G (via GPRS) access. But during the course of the standardization work, interest grew in completing an EPC-only architecture for all 3GPP accesses, including the ability to handover between 2G and 3G radio networks with GPRS and 2G/3G access using EPC. The work did not require too much extra effort from a technical point of view, since it can be easily extended from the EPC support for 2G/3G. Specifications were developed such that S4 SGSN and Serving GW- and PDN GW-based EPC architecture supports handover between GERAN and UTRAN, as well as Intra-GERAN and Intra-UTRAN mobility with and without SGSN relocation and Serving GW relocation. The extra signaling required is to interact with the Serving GW at the appropriate time of handover for bearer establishment; also, in the case of PMIP, additional PCRF interactions are required. 3GPP TS 23.060 and 3GPP TS 23.401 Annex D describe the procedural differences in a nutshell, including the appropriate message name used between the SGSNs following EPC messages; the EPS bearer is used as parameter and the Serving GW and PDN GW are updated with appropriate bearer information as in the E-UTRAN to/from 2G/3G case.

In Figures 17.18–17.20, the procedure illustrates an example of how the interactions differ when handover occurs between Gn/Gp SGSN to a new S4-SGSN. Figure 17.18 shows the procedure with Gn/Gp SGSN, while Figures 17.19 and 17.20 show the steps in boxes (A) and (B), where the procedures differ when using the S4 interface instead of Gn/Gp.

Box (A) in the figure illustrates the changes in the basic Gn/Gp SGSN messages when S4-SGSN is the new SGSN (target). This also means that a new Serving GW first needs to be selected according to Serving GW selection function procedures (Chapter 9) and then, via a new S4-SGSN, establish the user-plane path between target RNC and Serving GW. The Protocol Type over S5/S8 is provided to this Serving GW regarding which protocol should be used over the S5/S8 interface.

Box (B) indicates update of the establishment of the user-plane path for all EPS bearer contexts (that are available after HO) between the UE, target RNC, Serving GW (for Serving GW relocation this will be the Target Serving GW and for Gn/Gp to S4-SGSN it will be the new Serving GW), and PDN GW. Steps (B2) and (B3) would be performed for each PDN GW the user has connectivity with and thus the EPS bearer contexts need to be established.

17.5 Bearer and QoS-Related Procedures

As has been described in Section 6.2, the 3GPP accesses use the concept of a "bearer" to manage the QoS between the UE and the PDN GW. For E-UTRAN, the EPS bearer extends between the UE and the PDN GW (for a GTP-based system), and the UE and

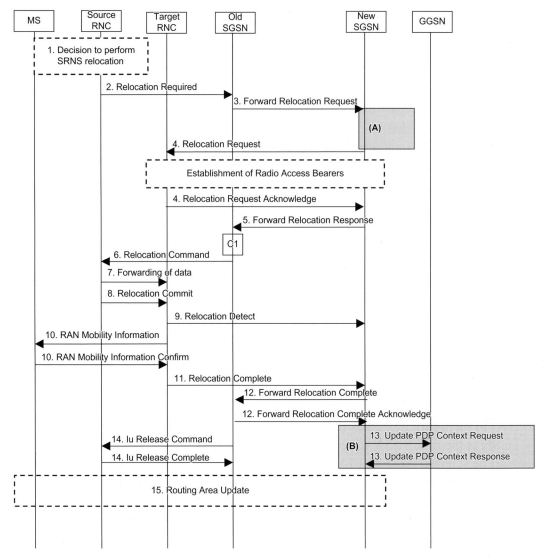

Figure 17.18: Inter-RNC Handover Showing Modifications Compared to GPRS When Using S4-SGSN/S-GW/P-GW.
This figure shows the procedure using Gn/Gp. Steps in boxes (A) and (B) differ for SGSN using S4 and are shown in the subsequent figures.

Serving GW (for a PMIP-based system). For GERAN and UTRAN, the bearers are implemented as PDP contexts between the UE and the SGSN. When GERAN and UTRAN are connected to the EPS via the S4 interface, the PDP contexts are mapped to EPS bearers between SGSN and the PDN GW or Serving GW. When Gn/Gp is used between the SGSN and the PDN GW, the PDP contexts extend all the way between the UE and the PDN GW.

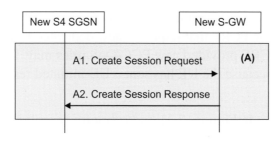

Figure 17.19: Box (A): S4-SGSN Causes Inclusion of Serving GW in the Path at HO.

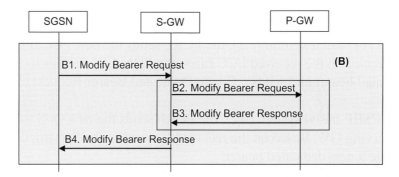

Figure 17.20: Box (B): Update of Relevant Parameters in the GW Using EPC-Based Procedure.

Each EPS bearer is associated with a well-defined QoS class described by the QCI, as well as packet filters that determine which IP flows are transported over the particular bearers. Certain bearers also have associated GBR and MBR parameters.

The bearers can be dynamically established, modified, or removed depending on the needs of the applications being used by the UE. In E-UTRAN, the dedicated bearer procedures are always initiated by the network. The UE may send a request for certain resources (QCI, bit rates, packet filters) to the network and this request may result in bearer operations being initiated by the network. In GERAN/UTRAN, the PDP context procedures are initiated either by the network or by the UE.

In this section, we will describe some of the dynamic procedures available in EPS to handle bearers for 3GPP accesses.

17.5.1 Bearer Procedures for E-UTRAN

17.5.1.1 Dedicated Bearer Activation
The dedicated bearer activation procedure is used when the network decides that a new dedicated bearer needs to be established. The procedure is initiated by the PDN GW (when using GTP-based S5/S8) or the Serving GW (when using PMIP-based

S5/S8). The trigger is typically that the PDN GW/Serving GW has received new PCC/QoS rules from the PCRF that require a new dedicated bearer to be established. The reason why the PCRF provided new PCC/QoS rules may, for example, be due to an Rx interaction, or because the UE has sent a UE-initiated resource request that was provided to the PCRF.

The procedure is shown in Figure 17.21 and is briefly described in the following steps:

A. The PCRF makes a policy decision because it received session information from an AF over the Rx interface or because it received a request for resources from the access network.
  - For a GTP-based system (A1), the PCRF sends the new PCC rules to the PDN GW. Based on the received PCC rules, the PDN GW decides to activate a new dedicated bearer and sends a Create Dedicated Bearer Request to the Serving GW.
  - For a PMIP-based system (A2), the PCRF sends the new QoS rules directly to the Serving GW. Based on the received QoS rules, the Serving GW decides to activate a new dedicated bearer.
B. The Serving GW sends a Create Dedicated Bearer Request to the MME.
C. The MME sends a command to the eNodeB to initiate the appropriate E-UTRAN procedures to establish an appropriate radio bearer. The appropriate reconfiguration of the RRC connection between UE and eNodeB is performed.
D. The MME acknowledges the bearer activation to the Serving GW by sending a Create Dedicated Bearer Response (EPS Bearer Identity, S1-TEID) message.
E. The Serving GW acknowledges the dedicated bearer setup.
  - For a GTP-based system (E1), the Serving GW sends a Create Dedicated Bearer Response to the PDN GW. The PDN GW sends an acknowledgement to the PCRF.
  - For a PMIP-based system (E2), the Serving GW sends an acknowledgement directly to the PCRF.

17.5.1.2 UE-Initiated Resource Request, Modification, and Release

As described in Section 8.1, there are two concepts for how QoS is allocated in the NW, either triggered by the network or triggered by an explicit request from the UE. The procedure described in this section supports the latter scenario where, for example, an application in the UE would like to have premium QoS and triggers the E-UTRAN interface in the UE to make a corresponding request to the network. The procedure can also be used when the UE wants to modify or release a previously granted resource. If accepted by the network, the request invokes either the dedicated bearer activation procedure, the dedicated bearer modification procedure, or a

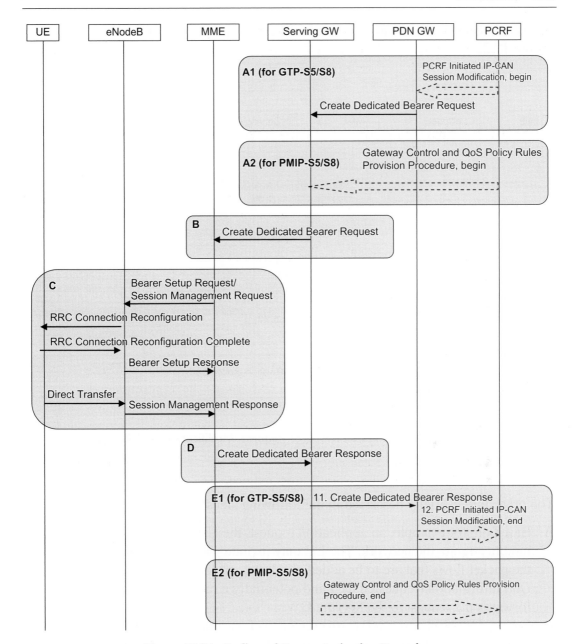

Figure 17.21: Dedicated Bearer Activation Procedure.

dedicated bearer is deactivated. The flow diagram for this procedure is illustrated in Figure 17.22.

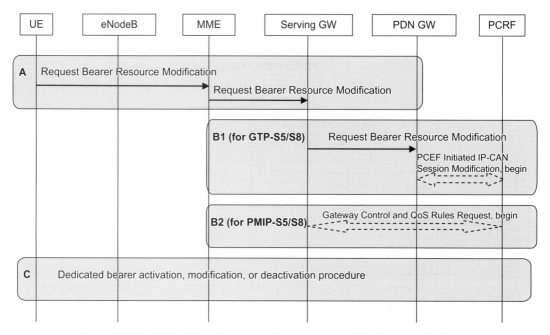

Figure 17.22: UE-Initiated Resource Request, Modification, and Release.

The procedure is briefly described in the following steps:

A. Based on, for example, an application request, the UE sends a request for resource modification to the network. The UE may include packet filter information for the packet flows that are to be added, modified, or deleted, as well as associated QoS information (requested QCI and potentially a requested GBR). The MME forwards the request to the Serving GW.

B. For a GTP-based system (B1), the request is forwarded to the PDN GW, which in turn informs the PCRF about the request. For a PMIP-based system, the Serving GW informs the PCRF about the request, without involving the PDN GW.

C. The PCRF makes a policy decision and provides the policy decision to the PDN GW and potentially the Serving GW. Depending on the new, modified, or deleted PCC/QoS rules, the PDN GW (for GTP-based S5/S8) or the Serving GW (for PMIP-based S5/S8) invokes the appropriate EPS bearer procedures.

17.5.2 Bearer Procedures for GERAN/UTRAN

17.5.2.1 UE-Initiated Secondary PDP Context Establishment

When the UE is camped on GERAN or UTRAN, it uses the secondary PDP context activation procedure in order to establish a new PDP context for the same PDN connection. The procedure is triggered by the UE for similar reasons as the UE-initiated bearer resource modification procedure over E-UTRAN (see above).

The S4-SGSN maps between the PDP context procedures used over GERAN/UTRAN and the EPS bearer procedures used towards the Serving GW and PDN GW. As can be seen below, the S4-SGSN maps from the secondary PDP context activation/modification procedure to a UE-initiated bearer resource modification procedure over S4.

The procedure is shown in Figure 17.23 and is briefly described in the following steps:

A. Based on, for example, an application request, the UE sends a secondary PDP context activation request. The S4-SGSN maps this into a UE-initiated bearer resource modification request and sends it to the Serving GW.
B. For a GTP-based system (B1), the request is forwarded to the PDN GW, which in turn informs the PCRF about the request. For a PMIP-based system, the Serving GW informs the PCRF about the request, without involving the PDN GW. The PCRF makes a policy decision and replies to the PDN GW or the Serving GW.
C. Depending on the new, modified, or deleted PCC/QoS rules, the PDN GW (for GTP-based S5/S8) or the Serving GW (for PMIP-based S5/S8) invokes the appropriate EPS bearer procedures. In this case, the UE has requested a new PDP context; this means that the PDN GW (or Serving GW in PMIP case) must initiate the establishment of a new dedicated EPS bearer in order to maintain the 1:1 mapping between PDP contexts and EPS bearers. When the S4-SGSN receives the Create Dedicated Bearer Request, it initiates the appropriate GERAN or UTRAN procedures. The S4-SGSN also completes the procedure towards the Serving GW.
D. The S4-SGSN responds to the UE with an Activate Secondary PDP Context Response.
E. The Serving GW completes the dedicated bearer procedure.

17.5.2.2 Network-Requested Secondary PDP Context Activation

The network-requested secondary PDP context activation procedures for GERAN/UTRAN correspond to the network-initiated dedicated bearer procedures for E-UTRAN. This means that it is the network that decides to establish a new PDP context. There is, however, one key difference between the GERAN/UTRAN and the E-UTRAN procedures. With E-UTRAN, the network-initiated bearer procedures are

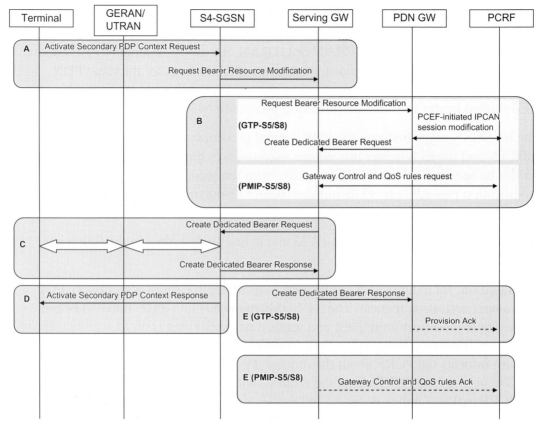

Figure 17.23: Secondary PDP Context Activation (UE Initiated).

the "native" bearer procedures. A UE-initiated bearer resource modification request over E-UTRAN is not a bearer procedure as such, but may trigger a network-initiated bearer procedure. With GERAN/UTRAN, it is the other way around. The secondary PDP context procedures are UE initiated, as illustrated in the previous section. As shown in this section, the network may trigger such a procedure by sending a request to the UE for activating a secondary PDP context.

Similarly to the UE-initiated procedures in the previous section, it is the SGSN that maps between PDP context procedures and EPS bearer procedures.

The procedure is shown in Figure 17.24 and is briefly described in the following steps:

A. Based on a trigger from the PCRF, the PDN GW (for GTP-S5/S8) or Serving GW (for PMIP-S5/S8) decides to establish a new dedicated bearer. In the case of

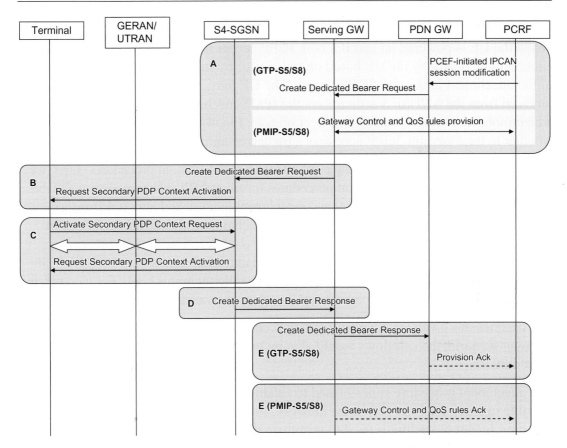

Figure 17.24: Network-Requested Secondary PDP Context Activation.

GTP-S5/S8, the PDN GW sends a Create Dedicated Bearer Request message to the Serving GW. In the case of PMIP-S5/S8, the PCRF interacts directly with the Serving GW.

B. The Serving GW sends the Create Dedicated Bearer Request message to the S4-SGSN. The S4-SGSN maps this into a request from the UE to activate a secondary PDP context.

C. The UE initiates the secondary PDP context activation procedure. The appropriate RAN procedures are performed and the S4-SGSN responds to the UE.

D. The S4-SGSN also responds to the Serving GW when the dedicated bearer (i.e. the secondary PDP context) has been activated.

E. For GTP-based S5/S8, the Serving GW forwards the response to the PDN GW, which in turn may inform the PCRF. For PMIP-based S5/S8, the Serving GW responds to the PCRF directly.

17.6 Attachment and Detachment for Non-3GPP Accesses

When the UE is powered on and if a non-3GPP access is available, it may be decided, either automatically by the UE based on policies or by manual choice, to make the initial attachment in the non-3GPP access. The UE may, for example, use policies received from the ANDSF. Furthermore, the IPMS mechanisms are used to decide which mobility protocol is used (see Section 6.4). Finally, it has to be determined whether to treat the non-3GPP access as a trusted or untrusted non-3GPP access.

In this section, we show three examples of attachment/detachment procedures. First, we describe attachment and detachment procedures for trusted WLAN access using GTPv2 on S2a. We then describe attachment and detachment in untrusted non-3GPP access using PMIPv6 on S2b. Finally, we describe the attachment and detachment procedure for a trusted non-3GPP access when using DSMIPv6 (S2c). Note that the other combinations, i.e. trusted access with PMIPv6 and untrusted access with DSMIPv6, are of course also possible, but they are not described explicitly in this book. For a complete description of the available attachment and detachment procedures, see 3GPP TS 23.402.

The descriptions given are for non-roaming scenarios. It may be noted that, in a roaming scenario, the call flows would also include a 3GPP AAA proxy as well as a visited PCRF.

The attachment procedure is performed to establish the IP connection and to make it possible to receive services from the network, for example when the UE is turned on. The detachment procedure is used to terminate the IP connections and to clear states in the network when, for example, the UE is turned off.

17.6.1 Attachment Procedure in Trusted WLAN Access Network (TWAN) Using GTPv2 on S2a

A simplified procedure illustrating how a UE connects via WLAN and the S2a connection to EPC is established is shown in Figure 17.25 and is described below. The details internal to the TWAN are not shown. The reason is that there are many options for how TWAN can be implemented and deployed, and 3GPP does not specify how it is done.

A. The UE finds a WLAN AP and connects using normal WLAN procedures.
B. The UE initiates 3GPP-based access authentication using IEEE 802.1X and EAP-AKA'. For more details on the authentication and security procedures in non-3GPP access in general and EAP-AKA' in particular, see Chapter 7.
C. When the authentication has been successfully completed, the UE asks for an IPv4 address using DHCPv4. The scenario with IPv6 is somewhat different, since the IPv6 stateless address autoconfiguration (SLAAC) is used rather than

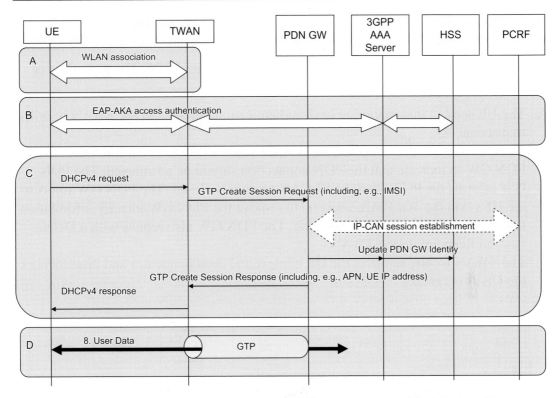

Figure 17.25: Attachment (for IPv4) in Trusted WLAN Access with GTPv2 on S2a.

DHCPv6. Triggered either by step B (authentication) or by the DHCPv4 request in this step C, the TWAN initiates tunnel establishment on S2a. In this example GTP is used and the TWAN sends a Create Session Request. The PDN GW creates a PDN connection, allocates an IPv4 address for the UE, and initiates IP-CAN session establishment with the PCRF. The PDN GW also sends the PDG GW Identity to the HSS (via the AAA server). The PDN GW replies to the TWAN with a Create Session Response. This completes the tunnel setup on S2a. The Create Session Response message includes the IPv4 address that was allocated to the UE. The TWAN then completes the DHCPv4 procedure with the UE and provides the IP address that was received from PDN GW.

D. All user data sent to/from the UE is forwarded inside the GTP tunnel on S2a.

17.6.2 Detachment Procedure in Trusted WLAN Access Network (TWAN) Using GTPv2 on S2a

The UE and the network may trigger a detachment and PDN disconnection. The HSS may, for example, trigger a termination of the PDN connection if the subscription has been canceled. The UE may also trigger a detachment if the UE wants to disconnect from the network. For a description of the available procedures, see 3GPP TS 23.402.

The UE-initiated detachment procedure is shown in Figure 17.26. The procedure is briefly described in the following steps:

A. The UE sends a disassociation or deauthentication message to the TWAN in order to disconnect.
B. Since the UE has disconnected, the TWAN sends a Delete Session Request to the PDN GW to indicate that the PDN connection should be terminated. The PDN GW informs the PCRF that the IP-CAN session is closed. The PDN GW informs the HSS (via the 3GPP AAA server) to remove the PDN GW identity information that is stored for this PDN connection. The PDN GW also replies with a Delete Session Response to the TWAN.
C. The TWAN locally removes the UE context and deauthenticates and disassociates the UE as necessary.

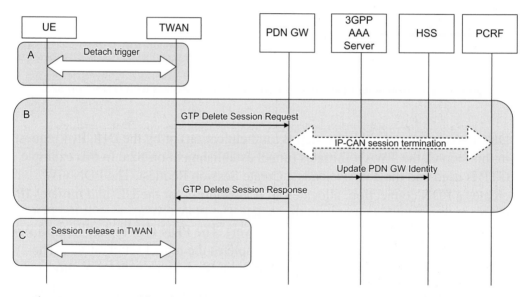

Figure 17.26: UE-Initiated Detachment in Trusted WLAN Access with GTPv2 on S2a.

17.6.3 Attachment Procedure in Untrusted Non-3GPP Access Using PMIPv6 (S2b)

The procedure is shown in Figure 17.27 and is briefly described in the following steps:

A. The UE establishes a connection to the untrusted non-3GPP access (e.g. WLAN) and receives a local IP address from the access network. There may be access

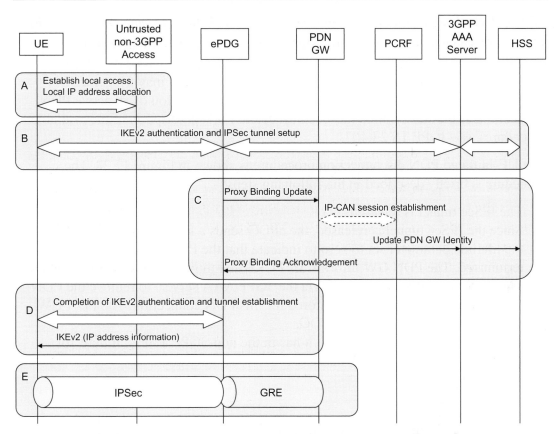

Figure 17.27: Attachment in Untrusted Non-3GPP Access with PMIPv6 (S2b).

authentication performed (e.g. based on EAP-AKA), but this is optional and is not shown in the call flow for simplicity.

B. The UE discovers the IP address of an ePDG using DNS and then initiates the IKEv2 procedure towards the ePDG. EAP-AKA is used for authentication, as described in Section 7.3.7. Diameter is used between ePDG and the AAA server, and between the AAA server and the HSS.

C. When the IKEv2 procedure has progressed, the ePDG sends a PBU to the PDN GW. The PDN GW notifies the PCRF about the new connection. The PDN GW also sends the PDG GW Identity to the HSS (via the AAA server). The PDN GW then responds with a PBA to the ePDG. The PBA contains the IP address for the PDN connection.

D. The final steps of the IKEv2 procedure are executed. The ePDG includes the IP address of the PDN connection in an IKEv2 message to the UE.

E. When the attachment procedure is complete, an IPSec tunnel has been established between UE and ePDG, and a GRE tunnel between ePDG and PDN GW.

17.6.4 Detachment Procedure in Untrusted Non-3GPP Access Using PMIPv6 (S2b)

There is no specific detachment procedure in untrusted non-3GPP access. Instead, each PDN connection is terminated separately. In order to detach a UE, all active PDN connections have to be closed. The UE and the network may trigger a PDN disconnection. The HSS may, for example, trigger a termination of all active PDN connections if the subscription has been canceled. For a description of the available procedures, see 3GPP TS 23.402.

The UE-initiated PDN disconnection procedure is shown in Figure 17.28. The procedure is briefly described in the following steps:

A. The IPSec tunnel is released.
B. Since the IPSec tunnel is released, the ePDG sends a PBU to the PDN GW with the lifetime parameter set to zero to indicate that the PDN connection should be terminated. The PDN GW informs the PCRF that the IP-CAN session is closed. The PDN GW informs the HSS (via the 3GPP AAA server) to remove the PDN GW identity information that is stored for this PDN connection. The PDN GW also replies with a PBA to the ePDG.
C. The UE may release any resource it has in the non-3GPP access. It may, for example, release the local IP address.

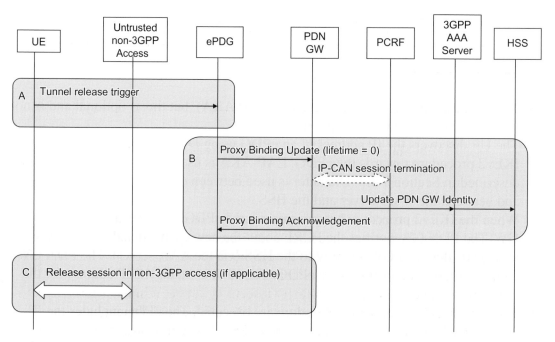

Figure 17.28: UE-Initiated Detachment in Untrusted Non-3GPP Access with PMIPv6 (S2b).

17.6.5 Attachment Procedure in Trusted Non-3GPP Access Using DSMIPv6 (S2c)

The procedure is shown in Figure 17.29 and is briefly described in the following steps:

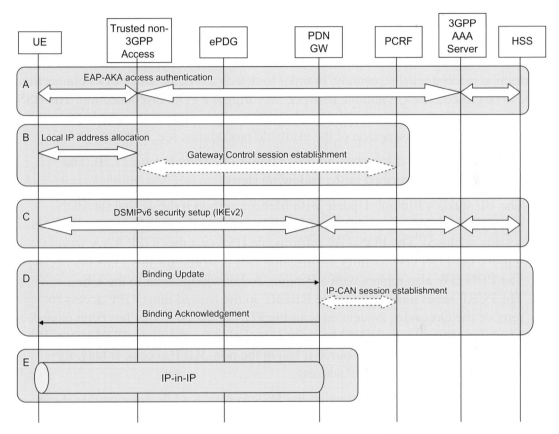

Figure 17.29: Attachment in Trusted Non-3GPP Access with DSMIPv6 (S2c).

A. The UE performs access authentication in the trusted non-3GPP access.
B. The local IP connection in trusted non-3GPP access is stabled and the UE receives a local IP address from the non-3GPP network. The trusted non-3GPP access also initiates gateway control session establishment with the PCRF.
C. If the UE has not bootstrapped DSMIPv6 security before, it initiates the IKEv2 procedure at this point. Authentication and IPSec SA establishment based on IKEv2 and EAP-AKA are performed. The IPSec SA will be used to protect the DSMIPv6 signaling.

D. The UE sends a Binding Update to the PDN GW. The PDN GW initiates IP-CAN session establishment towards the PCRF. The PDN GW replies with a Binding Acknowledgement to the UE.
E. When the attachment procedure is complete, there is an IP-in-IP tunnel between UE and the PDN GW. For more details on the operation of DSMIPv6, see Section 16.3.

17.6.6 Detachment Procedure in Trusted Non-3GPP Access Using DSMIPv6 (S2c)

There is no specific detachment procedure when DSMIPv6 is used. Instead, each PDN connection is terminated separately. In order to detach a UE, all active PDN connections have to be closed. The UE and the network may trigger a PDN disconnection. The HSS may, for example, trigger a termination of all PDN connections if the subscription has been canceled. For a description of the available procedures, see 3GPP TS 23.402.

The UE-initiated PDN disconnection procedure is shown in Figure 17.30. The procedure is briefly described in the following steps:

A. The UE sends a Binding Update with lifetime zero to indicate that the PDN connection should be closed. The PDN GW informs the PCRF that the IP-CAN session is closed. The PDN GW informs the HSS (via the 3GPP AAA server) to remove the PDN GW identity information that is stored for this PDN connection. The PDN GW also replies with a Binding Acknowledgement to the UE.
B. The PCRF sends a message to the BBERF in the trusted non-3GPP access to remove the QoS rules associated with the PDN connection that has been closed.
C. The UE terminates the IKEv2 security association for the given PDN connection.
D. The UE may release any resource it has in the non-3GPP access. It may, for example, release the local IP address.

There are also procedures for allowing the HSS to detach a UE, for example if the subscription has been canceled. For a description of the available procedures, see 3GPP TS 23.402.

In addition to E-UTRAN, the EPS must also support handover to/from non-3GPP accesses such as HRPD network, WiMAX, and Interworking WLAN (or I-WLAN as referred to in 3GPP). Thus, it is expected that EPS must be able to support handover between heterogeneous access systems, where the non-3GPP access networks are not developed in 3GPP. These handovers also need to support service continuity; thus, the ability to maintain IP connectivity within the EPC when moving between these heterogeneous access networks has taken a great deal of effort to accomplish.

If we consider radio access and packet core network-level handover without worrying about the service continuity and session continuity aspects, the handover combinations below can be found within and between 3GPP accesses.

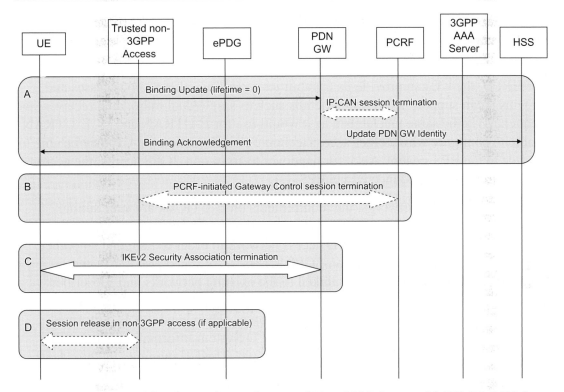

Figure 17.30: UE-Initiated Detachment in Trusted Non-3GPP Access with DSMIPv6 (S2c).

17.7 Intersystem Handover Between 3GPP and Non-3GPP Accesses

17.7.1 Overview

If we consider radio access and packet core network-level handover without worrying about the service continuity and session continuity aspects, the following possible handover combinations can be found within and between 3GPP accesses and non-3GPP accesses:

a. Optimized handover E-UTRAN to/from HRPD (for GTP and PMIP).
b. Basic non-optimized handover – trusted non-3GPP access (including HRPD) to/from GERAN/UTRAN/E-UTRAN (with GTP/PMIP on 3GPP access and PMIP/MIPv4FA/DSMIPv6 on non-3GPP access).
c. Basic non-optimized handover – untrusted non-3GPP access (including HRPD) to/from GERAN/UTRAN/E-UTRAN (with GTP/PMIP on 3GPP access and PMIP/DSMIPv6 on non-3GPP access).

Note that all these scenarios can be Intra- or Inter-PLMN handover, as described in more detail below. In the case of handover, we have chosen to focus on trusted non-3GPP access via the S2a interface, which implies network-based mobility using the PMIP protocol for non-3GPP accesses and either the GTP or PMIP protocol for 3GPP accesses.

17.7.1.1 HRPD and 3GPP Access

For HRPD, the UE can provide its capabilities regarding other radio accesses and systems it can support and relevant details such as single/dual radio, dual receiver, and frequency via the same mechanism as it would use for E-UTRAN access. E-UTRAN is able to configure which other access technology information the UE may be able to provide to the EPS system. Also, for handover to/from non-3GPP access, the access technologies must be connected via EPS.

In the case of HRPD networks, the requirements on E-UTRAN are very similar to those for the existing 2G/3G 3GPP access networks. The following need to be supported by E-UTRAN and HRPD access network in order to facilitate smoother performance compared to other non-3GPP accesses. As stated previously, the operator requirements are directly linked to the existing networks and subscriber base for CDMA systems. E-UTRAN controls the trigger for the UE to measure the HRPD information for preparation of handover, when handover is performed in the E-UTRAN to HRPD direction. The HRPD System Information Block (SIB) has to be sent on the E-UTRAN broadcast channel. The UE monitors the broadcast channel in order to retrieve the HRPD system information for the preparation of cell reselection or handover from the E-UTRAN to the HRPD system. HRPD system information may also be provided to the UE by means of dedicated signaling. HRPD system information contains HRPD neighboring cell information and CDMA timing information, as well as information controlling HRPD pre-registration. Note that pre-registration is used only when optimized handover is supported in the EPS (for more details see later sections below).

17.7.1.2 General Non-3GPP and 3GPP Access

In the case of general non-optimized handover, the UE needs to perform an access attachment procedure with target access. For handing over to a non-3GPP access, both network-based and host-based IP mobility management solutions are supported (see Section 6.4). For handovers from a non-3GPP access to a 3GPP access, only network-based mobility is supported in target non-3GPP access.

For handover using network-based mobility in target access, the terminal performs access attachment with an indication that the attachment is of "handover" type in order for the target network (i.e. radio access, MME, Serving GW) to establish the necessary resources for the handover and also, where possible, maintain the IP connectivity by maintaining the PDN GW and the terminal's IP address(es). For host-based mobility in target non-3GPP access, the terminal establishes a local IP connectivity in target access, and the handover and IP address preservation are then executed using the host-based IP mobility mechanism. Depending on the number of PDNs the UE was connected to before handover, and depending on what the target system can support, the PDN connectivity may be re-established in the target access by the network or by the UE itself, or some of them are dropped during handover. So the handover, in this case, has very few interactions with access networks and is performed by the UE and the core network. For more details on these handover procedures, see the sections below.

We therefore conclude that the basic handover requirements remain the same as we move towards E-UTRAN and EPS, but we also continue to enhance the system to accommodate non-3GPP accesses as well as IMS service continuity as we evolve the systems. In the sections below, we will show how EPS-level handovers are supported by the radio and evolved core networks.

17.7.2 Details of Handover in EPS with Non-3GPP Accesses

17.7.2.1 Optimized Handover for eHRPD Access

For HRPD networks, there is a significant subscriber base already out there with major North American and Asian operators having networks. Even though the two technologies (one developed in 3GPP and the other in 3GPP2) have been competing over the last 20 years, the two bodies have also cooperated in many areas in order to develop common standards that are strategically important to operators overall; examples of these are IMS and PCC development. For CDMA a number of companies cooperated extensively inside and outside the standards forum in order to develop special optimized handover procedures between E-UTRAN and HRPD access. The resulting handover procedure has efficient performance and reduced service interruption. This work was brought into mainstream 3GPP standards under the SAE umbrella, and further enhanced and aligned with the mainstream 3GPP work ongoing for SAE, and thus produced the so-called Optimized Handover between E-UTRAN and HRPD. HRPD networks then became known as evolved HRPD (eHRPD) to highlight the changes required for interoperability and connectivity with EPC and E-UTRAN, though there are no changes to the actual radio network and its functions.

The optimized handover has been defined to work in two modes of operation: idle mode where the UE is idle (i.e. does not have any active radio connectivity in the system, ECM-IDLE in E-UTRAN, and Dormant in HRPD) and active mode where the UE is active (i.e. active data transmission ongoing between the UE and the network). The actual handover is performed in two phases: the pre-registration phase, where the target access and specific core network entity for the specific access (MME for E-UTRAN and HRPD Serving GW or HSGW for eHRPD access) are prepared ahead of time anticipating a possible handover (but there is no time association of how long a UE may be pre-registered in the system); and the handover preparation and execution phase, where the actual access network change occurs.

Note that in the early deployment of E-UTRAN in a CDMA network, it was considered more prevalent and thus important to support E-UTRAN to HRPD handover than the reverse direction, since it was assumed that the HRPD networks would have sufficient coverage to keep a user within the HRPD system.

It should be noted that currently the HSGW only supports the PMIP protocol (S2a interface), whereas the E-UTRAN access may use either the GTP or PMIP protocol and, as such, the GRE keys must be provided to the HSGW even in the case of GTP-based EPC for E-UTRAN access.

Figure 17.31 outlines the steps for the E-UTRAN to HRPD access handover.

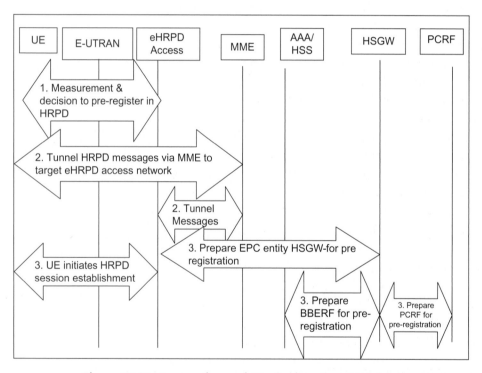

Figure 17.31: Preparation and Pre-Registration Overview.

1. Some of the basic requirements on the UE and the E-UTRAN when supporting handover from E-UTRAN to HRPD networks are that the E-UTRAN provides the HRPD information (e.g. neighboring cell information, CDMA timing, and HRPD pre-registration control information) on the broadcast channel for UE

measurement purposes. HRPD system information may also be provided to the UE by means of dedicated signaling. E-UTRAN is responsible for configuration and activation of HRPD measurements towards the UEs. The UE performs measurements in active mode when directed by the E-UTRAN network using information provided via dedicated radio signaling. Note that for idle mode, the UE performs cell reselection procedures in addition to being pre-registered in the target system prior to perform idle mode optimized mobility.

2. Once the UE decides to perform pre-registration to HRPD, it needs to tunnel the HRPD messages over E-UTRAN radio. The HRPD messages are encapsulated in the appropriate uplink messages for pre-registration or for handover signaling and downlink messages for other HRPD messages. The handover signaling is given higher priority, and the RAT type and other identifying information is also provided to the HRPD network for correctinterpretation of the messages. In order for the MME to select the correct target HRPD system, where the messages for that UE should be tunneled to and also assist the HRPD network with the collected appropriate radio-related information, each eNodeB cell is associated with an HRPD Sector ID (also known as reference cell ID). This Sector ID is provided to the MME during the message transfer over S1-MME. The MME then uses this information to find the appropriate target HRPD entity and tunnels the messages over S101 towards that entity.

3. Based on the trigger from the E-UTRAN radio network, the UE initiates establishment of a new session in the eHRPD network. This process causes the HSGW to be connected with the HRPD access network and, based on the information provided by the UE via the EPC, the HSGW also initiates the process of establishing connection with PCRF for a non-primary BBERF connection in order to provide functions like bearer binding and QoS rules provision as the BBERF function is located in the HSGW for this access. At this point, HSGW has the latest bearer information, APN, PDN GW addresses, etc. for that user. HSGW acquires the information about the already allocated PDN GWs from HSS/AAA. When the source E-UTRAN makes the decision to trigger the UT to hand over to target HRPD access, the UE starts sending the appropriate preparation messages to the HRPD access network via the tunnels over E-UTRAN and MME.

Figure 17.32 outlines the steps for the EUTRAN to HRPD access handover completion phase:

1. Based on the measurement information, the source E-UTRAN instructs the UE to perform handover to the HRPD access network; UE has already pre-registered to the HRPD network prior to any active mode optimized handover.
2. The UE initiates the procedures to establish traffic channel connection towards the HRPD access network over the E-UTRAN access, and E-UTRAN forwards the

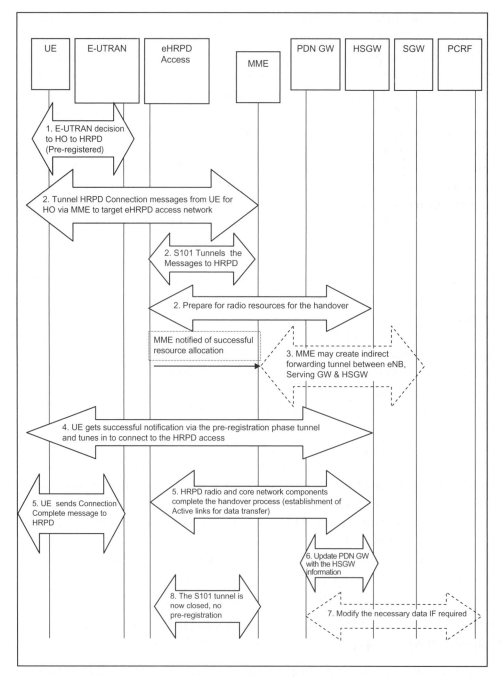

Figure 17.32: Active Mode Handover from E-UTRAN to HRPD in the EPC.

message to the MME, which then forwards the message to the HRPD node via the already established S101 tunnel, and also adds additional information such as uplink GRE keys, APNs, and PDN GW addresses associated with each APN.

3. The HRPD access network then establishes the necessary radio resources and requests HSGW for appropriate links and information in order to establish the connection between HRPD and HSGW. Once this process is completed, the MME is notified and then the MME may establish an indirect forwarding link between the Serving GW and HSGW by sending the Serving GW the necessary HSGW information. In this manner, there is a data forwarding path from eNodeB to Serving GW to HSGW.

4. The UE is informed of successful resource allocation completion and the UE tunes to the HRPD access network. The UE no longer communicates via E-UTRAN access.

5. The UE then sends a confirmation message directly to the HRPD network; at this time the UE has completely moved into the HRPD access network. The HRPD access network informs the HSGW of the UE's arrival and is prepared to receive/ send data.

6. The HSGW now establishes a link with the PDN GW where the PDN GW will now forward data towards the HSGW and not the Serving GW.

7. The PDN GW now interacts with the PCRF function to receive any modified data for the new access.

8. The S10 tunnel is terminated for that UE between HRPD and MME access.

In the case of multiple PDN connections existing for that UE, the HSGW must update the appropriate PDN GWs for each PDN connection individually.

The release of E-UTRAN resources is consistent with the E-UTRAN Inter-RAT handover scenario; note that the PDN GW can initiate resource release towards the Serving GW any time after it has successfully established the link with HSGW. Once the MME receives confirmation from the HRPD network about successful completion of handover, it initiates the release of the UE context to E-UTRAN.

The MME may have timers to trigger the Serving GW resources for the E-UTRAN bearers and the indirect forwarding tunnels to be released as in the Inter-RAT handover case. When the MME does trigger this release, it also indicates to the Serving GW not to release resources towards the PDN GW.

In the case of handover from HRPD to E-UTRAN access, the E-UTRAN access does not play any role in directing or assisting the UE. Due to the possible effects of the HRPD radio access part, it is unclear at the time of writing if the HRPD RAN will assist in any way in this handover or whether it will be up to the UE to decide when to perform the handover based on the measurements available to it. The steps are very similar to the E-UTRAN to HRPD handover, with some differences due to the access network type (Figure 17.33). The procedure is briefly described in the following steps:

1. The decision to hand over to E-UTRAN is made by the UE. The UE is in active mode in the HRPD network connected via EPC. Note that the UE may choose to leave HRPD and execute the E-UTRAN attachment procedure directly over E-UTRAN; in such a case the Attach type is not set to Handover.
2. The UE sends an NAS message of Attach to E-UTRAN with indication set to Handover, which is tunneled over HRPD access network and then forwarded to the MME via the pre-established S101 tunnel. The HRPD access node determines the MME and the TAI using the Sector ID to MME mapping function located in the HRPD network.
3. Based on the E-UTRAN attachment process, the MME needs to make the appropriate decision of whether to perform authentication or not. The MME also performs a location update procedure in order to update the HSS with user information, as well as retrieve the user's subscription data.
4. As in the normal E-UTRAN attachment procedure, the Serving GW is selected for that UE and the bearer creation process for a default bearer is initiated with an indication of handover. The Serving GW triggers the PDN GW to create the necessary information for default bearer establishment. The PDN GW performs this task and also triggers associated, dedicated bearer establishment procedures.
5. Once the MME is informed of the completion of the bearer establishment procedure in the GWs, the MME completes the attachment procedure towards the UE via the HRPD access network and associated established tunnels. The UE then completes the attachment procedure over the HRPD access.
6. Once the attachment procedure is completed, the UE moves to E-UTRAN access.
7. The UE then performs necessary radio connection establishment via the service request procedure, which triggers the establishment of radio links as well as the S1 user-plane setup, and then starts the UE-initiated bearer establishment procedures to complete the process. For any bearers that cannot be established during this process, the UE removes their resources.
8. Once the bearer establishment procedure is complete, the MME informs HRPD access via an S101 tunnel that handover is complete.
9. The HRPD access network initiates its resources release procedure.
10. The PDN GW triggers the resources to be released towards HSGW any time after it has established the Serving GW and PCRF relationship for E-UTRAN access.

If there are multiple PDN connections to be established, the UE must initiate them via a UE-initiated PDN connectivity process, as described in Section 6.2.

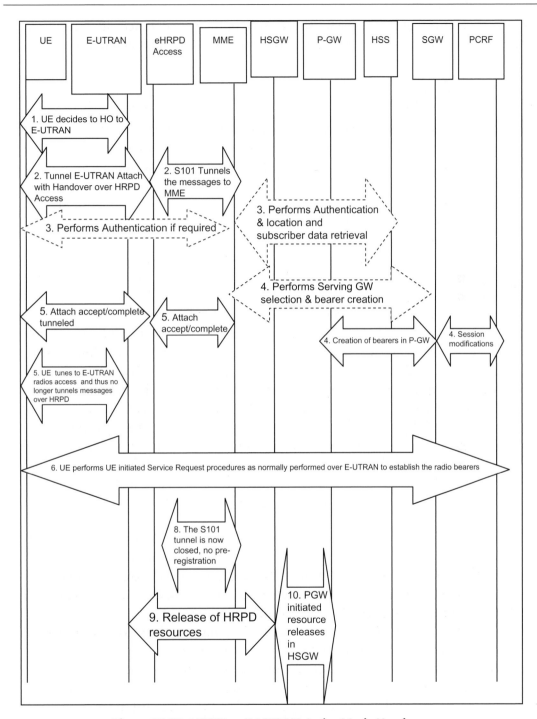

Figure 17.33: HRPD to E-UTRAN Active Mode Handover.

What is evident is that by pre-establishing some of the connections within the target access network any time prior to the actual handover itself, the execution of the actual handover process is reduced significantly during the E-UTRAN to HRPD scenario. Even though the benefits may not be as significant in the reverse direction, for consistency and simplicity of network behavior, the process has been kept consistent.

Even though this section focuses on active mode handover, it may benefit readers to briefly consider the idle mode handover process between E-UTRAN and HRPD.

In the E-UTRAN to HRPD direction, triggered by the cell reselection process, the idle UE, based on internal triggers or from E-UTRAN access, selects the HRPD access network where the UE is in a dormant state due to either pre-registration or previous attachment to that access. Then the UE follows the HRPD procedures to connect to the access and, in the core network, the HSGW establishes appropriate bearers towards the PDN GW and PCRF when applicable. The PDN GW then triggers the release of the resources in the E-UTRAN access.

In the HRPD to E-UTRAN direction, when the UE determines to perform an attachment procedure to E-UTRAN access, the process followed is the same as the active mode handover process until it tunes to the E-UTRAN access. The UE then performs a TA update procedure over the E-UTRAN access. The resource release process is the same as in the reverse direction, triggered by the PDN GW.

17.7.2.2 Basic Non-Optimized Handover with Non-3GPP Access

When considering non-optimized handover between and within non-3GPP accesses using EPC, the main requirement is to be able to preserve the IP address(es) and maintain the IP connectivity/service continuity when handover occurs. Since the handover decision is determined and executed from the UE, and there is no coordination between the access networks, this procedure is considered non-optimized compared to the optimized handover procedure between E-UTRAN and HRPD access.

Due to the number of protocol options, as well as the capability to support both host-based and network-based mobility, the handover procedure can be quite complex in terms of selecting the right combination to ensure that IP address preservation and session continuity is possible. As the reader may by now already be familiar with, the IP mobility management selection process as described in previous sections of this book plays a significant role in the handover process as well.

In the case of handover between 3GPP access and non-3GPP access and between non-3GPP accesses, the IPMS function makes the decision of how the IP connectivity will be performed (i.e. whether preservation is possible or not). In the case of network-based mobility (i.e. PMIP), the PMIP protocol supports these functions and UE capability is either indicated by the UE explicitly at handover or based on preconfigured information.

In the case of host-based mobility, the decision is taken in a slightly different manner. The decision may depend on whether the network is aware of the UE capability to support DSMIPv6 or MIPv4. This information may be acquired by the target non-3GPP access from the HSS/AAA (e.g. in the case of DSMIPv6, the UE performed S2c bootstrap before moving to the target non-3GPP access). If the IP mobility management protocol selected is DSMIPv6, the non-3GPP access network provides the UE with a new IP address, local to the access network. In an untrusted non-3GPP access, the terminal also has to set up an IPSec tunnel with the ePDG. In these cases, in order to get IP address preservation for session continuity, the UE has to use DSMIPv6 over the S2c reference point. The local IP address, allocated either by the trusted non-3GPP access network or the ePDG, is then used as a Care-of Address for DSMIPv6 within EPS. If the IP mobility management protocol selected is MIPv4, the address provided to the UE by the trusted non-3GPP access network is an FACoA and IP address preservation is performed over S2a using MIPv4 FACoA procedures. Note that MIPv4 is not supported when an ePDG is used. A basic handover flow when using DSMIPv6 is provided below. For readers interested in further details on handover flows using host-based mobility protocols, we recommend further detailed reading of 3GPP specifications TS 23.402 and TS 24.303. Further details on the basic operation of DSMIPv6 are also described in Section 16.3.

As can be understood from the IP Mobility Selection function, the protocol choice made by the UE and the protocol choice supported by the network need to be in sync in order to perform handover; the UE indicates in the attachment procedure that the Attach function is being performed due to handover and it may also indicate its preference of host- or network-based mobility and the protocol choice for the host-based mobility case. Note that in the case of network-based mobility in 3GPP, the UE is unaffected by the network choice of protocol (GTP or PMIP). If UE does not indicate any preference, then PMIPv6 is the selected protocol and, based on PMIP protocol principles, there are two ways of making the decision of preserving the IP and maintaining session continuity or not. An operator may configure its local policies at the PDN GW whether to preserve the IP address based on a timer, which allows the existing IP address to be maintained if and only if the source/old access system tears down its connection before the timer expires and then assigns a new IP prefix, or assigns a new IP prefix immediately and thus no IP address preservation is performed.

In addition, the UE may also use the information made available to it from ANDSF regarding its home operator's preference and other policies to assist it to select the preferred access network during handover.

Now let us focus on the scenario that has generated wider participation and generated operators' interest during the development and standardization of SAE (i.e. S2a interface using PMIPv6). This is also the default system behavior if operators choose not to configure any specific IPMS process in the UE and in the EPC network. Note that multiple PDN connectivity with different mobility protocols for the same UE is not supported, which means, for example, it is not possible for a UE that is connected to single or multiple PDNs over a 3GPP access to perform a handover to a non-3GPP access and then use different mobility protocols for the PDNs connected over non-3GPP access. We consider the flow in Figure 17.34 to cover the most likely scenario to be seen in the deployed network, where a handover is triggered from a trusted non-3GPP access network with S2a supporting PMIP to GTP-based (S5/S8 interface) EPC network.

The handover steps can be described as follows:

- *Steps 1–3.* The UE is connected to a non-3GPP access trusted by the operator and, in active mode, the UE detects the E-UTRAN access and determines to hand over (or rather transfer the sessions, as it is more a transfer of ongoing active sessions than "handover" in the true meaning of the word as used in existing mobile systems) to E-UTRAN from its current serving access system based on policies and other information available to the UE. Note that the UE is connected to an EPS network and has now moved to E-UTRAN access to perform an attachment. The UE sends an Attach Request to the MME over the E-UTRAN access with attach type indicating "Handover attach". This attachment procedure is handled as in the case of normal attachment in E-UTRAN and the UE should also include one of its APNs in this message.
- *Steps 4–6.* The MME may perform authentication of the UE as per E-UTRAN access via the HSS. Once the authentication is successful, the MME continues as for an E-UTRAN access, where it may perform a location update procedure and subscriber data retrieval from the HSS. The MME receives information on the PDNs the UE is connected to over the non-3GPP access in the subscriber data obtained from the HSS. The MME selects an APN, either the default APN or the APN provided by the UE. Since the attach type sent by UE is "Handover", the MME selects the PDN GW provided as part of the subscription data from the HSS. The MME then continues to select a Serving GW. The MME sends a Create Default Bearer Request to the selected Serving GW and includes the PDN GW address and handover indication.
- *Steps 7–10.* The Serving GW sends a Create Default Bearer Request with handover indication to the PDN GW, causing the PDN GW not to switch the tunnel from non-3GPP IP access to a 3GPP access system at this point. The PDN GW also interacts with PCRF to obtain the rules for the network IP-CAN

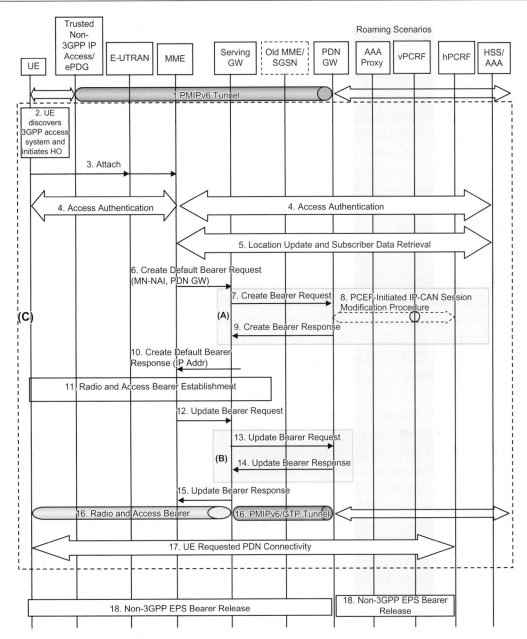

Figure 17.34: Handover from a Trusted Non-3GPP Access Network.

and PDN connection for all established bearers due to handover. Due to the handover, the PDN GW stores the new PCC rules for E-UTRAN access as well as maintaining the old PCC rules for trusted or untrusted non-3GPP IP access and still applies the old PCC rules for charging. PDN GW returns the UE's IP address/prefix assigned for the non-3GPP access it is handing over from. This information is then passed on to the MME indicating successful bearer establishment and setup of S5 tunnel establishment. Additional dedicated bearers may also be established during this process by PDN GW as in the case of normal attachment.

- *Step 11.* Radio and access bearers are established for E-UTRAN access as in the normal attachment case.
- *Steps 12–15.* The MME sends an Update Bearer Request (eNodeB address, eNodeB TEID, handover indication) message to the Serving GW. Based on the presence of the handover indication, the Serving GW sends an Update Bearer Request message to the PDN GW to prompt the PDN GW to tunnel packets from a non-3GPP IP access to a 3GPP access system and immediately start routing packets to the Serving GW for the default and any dedicated EPS bearers established. PDN GW can now route the packets to E-UTRAN access and stop data transfer to non-3GPP access.
- *Step 16.* The UE sends and receives data at this point via the E-UTRAN system.
- *Step 17.* For connectivity to any remaining PDNs from the old non-3GPP access, the UE establishes connectivity to each PDN, by executing the UE-requested PDN connectivity procedure.
- *Step 18.* The PDN GW initiates the resource allocation deactivation procedure in the non-3GPP IP access.

If the E-UTRAN access is using PMIP-based EPC, the difference in signaling sequences is illustrated by the steps (A) and (B), where dynamic PCC interaction based on off-path policy control-related signaling is executed as seen in general where the Serving GW interacts with PCRF. In the case of PMIP-based S5/S8, instead of a Create Bearer Request and Update Bearer Request, the PBU/PBA is sent from the Serving GW to the PDN GW.

If multiple PDN connections need to be established, the UE-initiated PDN connectivity procedures are executed either in sequence or in parallel in order to establish these additional PDN connections.

If the handover is performed towards a 2G/3G 3GPP access network, the procedures are very similar in the sense that the attachment and PDP context activation procedures are executed according to the specific 3GPP access itself, with the handover indication in the PDP context activation procedure, which allows the Serving GW and PDN GW to appropriately preserve the IP address/prefix and handle the sessions as in the E-UTRAN case.

Next, we will illustrate a handover scenario from a 3GPP access to a non-3GPP access, where S2c (i.e. DSMIPv6) is used in the target non-3GPP access (see Figure 17.35).

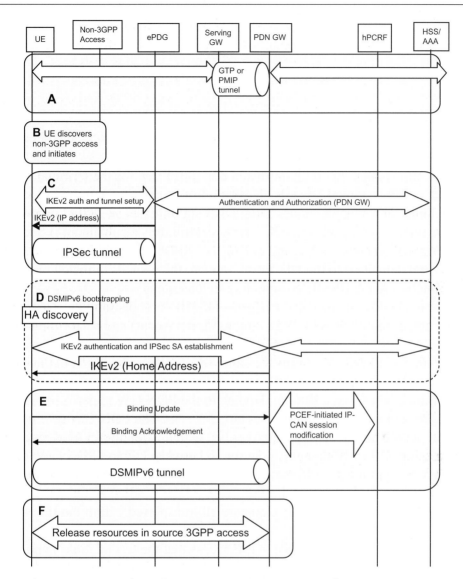

Figure 17.35: Handover from 3GPP Access to Untrusted Non-3GPP Access.

The session starts in a 3GPP access (e.g. E-UTRAN), where either GTP or PMIP is used on the S5/S8 reference point. The session is then handed over to a non-3GPP access. The IP mobility mode selection is made, resulting in DSMIPv6 being used. The terminal thus receives a local IP address in the target non-3GPP access. If the access is treated as untrusted, the terminal also sets up an IPSec tunnel towards the ePDG. In that case, the terminal also receives another local IP address from the ePDG. The terminal then invokes DSMIPv6 to the PDN GW in order to maintain the IP session.

Note that the same PDN GW has to be used for the target access and the source access in order to maintain IP session continuity.

The procedure is shown in Figure 17.35 and is briefly described in the following steps:

A. The terminal is attached over a 3GPP access, for example E-UTRAN.
B. The terminal discovers a non-3GPP access, for example WLAN, and decides to hand over the session.
C. If the decision is that this is an untrusted non-3GPP access, the terminal needs to establish an IPSec tunnel towards an ePDG. In this case, the terminal discovers an IP address of a suitable ePDG using DNS and initiates an IKEv2 procedure to authenticate and set up an IPSec SA. If successful, this results in an IPSec tunnel being established between UE and ePDG. The ePDG also allocates a local IP address and delivers it to the UE. From now on, the plane is tunneled in the IPSec tunnel, including the DSMIPv6 signaling in later steps.
D. If not done already, the terminal performs DSMIPv6 bootstrapping. This includes discovering a suitable PDN GW (acting as Home Agent) and performing the IKEv2 procedure with that PDN GW to set up an IPSec Security Association for DSMIPv6. The PDN GW returns the same IP address as was used in the source 3GPP access.
E. The terminal then sends a Binding Update to the PDN GW to perform the actual user-plane path switch from source to target access. The PDN GW informs the PCRF about the new access type and replies with a Binding Acknowledgement. A bidirectional IP-in-IP tunnel is now set up between UE and PDN GW. The terminal can continue its IP sessions using the same IP address as was used in the source 3GPP access. Note that the Binding Update, Binding Acknowledgement, and the DSMIPv6-tunneled user plane are all transported within the IPSec tunnel between terminal and ePDG.
F. The PDN GW informs the source 3GPP access that the terminal has handed over to a non-3GPP access. The resources in the 3GPP access can now be released.

The message flow above illustrates a handover from a 3GPP access to an untrusted non-3GPP access. In the case of handover from a 3GPP access to a trusted non-3GPP access using DSMIPv6, the procedure is similar with a few key differences:

• The tunnel setup with ePDG in step B is not performed. Instead, step B is replaced by an access authentication in the trusted non-3GPP access. The non-3GPP access also establishes a gateway control session with the PCRF via Gxx as part of step B.
• When the path switch has taken place in step E, the PCRF may provide the trusted non-3GPP access with updated QoS rules via the Gxx reference point.

17.8 QoS-Related Procedures in Non-3GPP Accesses

As has been described in Section 6.2, the 3GPP accesses use the concept of a "bearer" to manage the QoS between the UE and the PDN GW.

Each EPS bearer is associated with a well-defined QoS class described by the QCI, as well as packet filters that determine which IP flows are transported over that particular bearer. Certain bearers also have associated GBR and MBR parameters.

Other accesses not defined by 3GPP, such as HRPD or WiMAX, may have similar procedures for allocating resources in the access network. The terminology may be different but the purposes of the procedures are roughly the same – that is, to set up a (logical) transmission path between the UE and the network to transport traffic with certain QoS requirements. The mechanisms and procedures used between the UE and the access network are specific to each access and will not be described in detail in this book. Using the PCC architecture, EPS is able to interwork with such access-specific procedures.

The QoS procedures in the non-3GPP access network depend on the particular access technology being used. EPS defines generic procedures that can be used to interwork with different non-3GPP accesses. In this section, we illustrate a network-initiated QoS reservation procedure, similar to the network-initiated dedicated bearer and PDP context activation procedures described above. The interface between the UE and the access network is, however, not described since it varies between accesses (Figure 17.36).

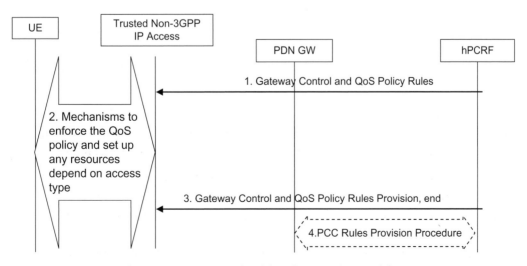

Figure 17.36: Network-Initiated Resource Provision.

PART V

Conclusion and Future of EPS

CHAPTER 18

Conclusions and Looking Ahead

The rapid uptake worldwide of 4G mobile services based on LTE/EPC technology proves that the work on SAE and the development of the EPC specifications was not only a major achievement carried out by 3GPP and its partners in the global mobile industry community; it was also very important in regards to the marketplace. The SAE work targeted a significantly broader scope than previous 3GPP releases, extending the functionality of the 3GPP packet core architecture to encompass interworking with access technologies standardized outside of 3GPP, as well as providing an evolved packet-only core for the next generation of mobile broadband access technology, LTE.

Inclusion of CDMA interworking capabilities and integration of eHRPD in the 3GPP architecture was naturally a major breakthrough, paving the way for network deployments of LTE and EPC that can be shared across an even wider operator community, as well as closer cooperation between 3GPP and 3GPP2. Global uptake of a single common technology means large volumes of handsets and network equipment, a highly competitive worldwide market leading to focus on cost-efficient solutions, and increased attention from service and application developers. A global technology also means excellent roaming possibilities in that users can access and utilize services in a large number of countries while using their own personal mobile devices ubiquitously.

Commercial deployment of LTE/EPC is well under way globally, with millions of users utilizing the technology. The unparalleled performance and characteristics offer exciting opportunities in providing services to end-users, be it private consumers looking for an excellent mobile broadband experience, corporate users looking to seamlessly connect their premises and their mobile workforce, high-speed broadband services being offered as an alternative to fixed-line installations, or machines connecting to machines without end-user interaction or participation. All predictions for the future include a rapid growth of the number of connected persons and devices on the planet, and LTE/EPC has a key role to play here.

EPC and 4G Packet Networks.
DOI: http://dx.doi.org/10.1016/B978-0-12-394595-2.00018-9

Going forward from the EPC solution and specifications as of 3GPP Release 11, there are obviously several areas that may be exploited and developed in the future. The authors of this book are convinced of the necessity that the decisions on the next steps to take should remain focused on strong commercial aspects, ensuring that 3GPP remains dedicated to features and functions of interest to the global community of network operators, consumers, enterprise customers, and application developers. With the specifications developed for EPC, 3GPP has provided an excellent platform for future core network evolution. The early years of commercial use have proved to be a success for the EPC technology in real-life deployments, and the prediction of rapid growth of numbers of connected users and devices, as well as rapid growth of global coverage, will be facilitated by the wide range of spectrum options defined for LTE.

Going forward, 3GPP is focusing on the development of new and enhanced LTE features and preparing for many billions of devices connected to access networks served by the EPC.

It will certainly continue to be a most interesting journey.

Appendix A: Standards Bodies Associated with EPS

SAE History and Background

SAE is a work item developed within 3GPP, but it also incorporated protocols from other standards bodies where necessary in order to prevent overlap, in particular protocols developed by the IETF and similar standards bodies. This may sound like a relatively easy task, but the process can become quite difficult due to both political as well as completion timing issues. Sometimes, the process of standardization within the IETF can take longer than expected, or different companies within 3GPP itself may have different priorities regarding the content and functions for the selection of protocols or the functionality required from the protocols. All of these aspects can affect the resulting standards developed within 3GPP and the SAE work item was no exception. This appendix provides an overview of the development of the SAE work item as background for interested readers.

Impact of Standardization Processes on SAE

Chapter 1 outlined the process of standardization, including the organization of the standards bodies and the approval processes involved within those organizations. As with most standardization efforts within 3GPP, the SAE work item builds upon existing technologies; in the case of SAE/EPC, the base was the existing 3GPP packet core system used for GSM/GPRS and WCDMA/HSPA. The progression of the LTE was a work item closely related to SAE. As the development of LTE progressed within the RAN groups of 3GPP, the opportunities to develop improvements to the Packet Core were identified by several different interested parties; the introduction of a work item to create an "all-IP" network naturally meant that there was a tremendous amount of interest in 3GPP's work in this area from new participants. In fact, the number of participants within 3GPP SA WG2 dealing with Architecture and Overall System aspects rose dramatically from around 100 people to about 180-225 participants during the peak period; approximately 75% of these participants were actively involved in attempting to shape the SAE work and EPS architecture.

One major driving force in the development of the work item was the mix of incumbent vendors and newcomers to the 3GPP system. The enormous interest in LTE/SAE work drew a large number of companies who had previously not participated in the 3GPP standardization process; many had indeed spent time working on other systems in various other standardization forums. These new entrants to 3GPP forged alliances and joined the SAE work. This created quite a contradictory vision for the future that was initially difficult to reconcile, since it is the vision between "continuity" and a completely "new beginning of sorts".

Some vendors and operators, with experienced of the existing 3GPP systems, considered it important that continuity should be maintained with the 3GPP packet core when evolving the system, whereas others initially wanted to explore new avenues based on technology used by other standardization forums rather than 3GPP. There were also a few existing 3GPP operators and vendors who focused solely on creating an architecture that was not based on the existing 3GPP architecture. All these different inflection angles and viewpoints were fed into the work on SAE, resulting in a very dynamic and diverse working environment. Initial investigations as documented in the 3GPP Technical Report 23.882 reflect various options with one common theme: that it makes sense to separate control- and user-plane entities. Some of the key architecture options that were discussed during the initial stages of the standardization process may be viewed as follows:

1. Evolution of the existing 3GPP packet core architecture, i.e. evolving the GTP protocol, but not necessarily reusing the architecture, with a single User-Plane Gateway (GW) for non-roaming cases, a local anchor and a GW for roaming, and using the network-based mobility protocol developed in 3GPP.
2. In principle, follow a very similar architecture as outlined in point 1 above, but where the two GW entities; thus, the roaming and non-roaming architecture is exactly the same, which is slightly different than 1. The protocols between these GWs would be developed in IETF.
3. An overlay model where a control-plane entity and a GW/Home Agent are used in the architecture with client-based mobility protocols.

A significant amount of time and effort were expended in discussing the functional division between core and radio networks, and the pros and cons of the different approaches. In the end, the 3GPP community made the decision to go with the architecture option without an RNC-like entity for the Radio Access Network and the community then focused on settling the functions belonging to Radio and Core aspects and moving forward with the investigations to develop the new architecture. Considerable effort was put into the work on arriving at the preferred functional split between the LTE RAN (now only consisting of base stations) and the packet core network as defined by EPC. 3GPP finally arrived at a decision where the functional split between RAN and CN for LTE are similar in nature as for WCDMA, with few exceptions.

Another difficult architecture decision was the selection of the Policy Control and Charging mechanism. In 3GPP systems up to this time, the GTP protocol carried not only mobility information, but also QoS and Charging and Policy Control information. This way of transferring information that supports the PCC infrastructure is also known as the "on-path model". Since the two options for handling mobility within EPC were decided to be supported using GTP and PMIP respectively, and the IETF PMIP protocol is unable to carry QoS and charging information itself, the on-path model used for GTP could not be supported for PMIP. The two options then considered in the case of PMIP PCC were a form of on-path model where a protocol, for example Diameter, may be used directly between the two involved entities (see the subsequent section on PCC for more details on the two models) to carry the necessary data, and another model, also known as the off-path model, where the data is then carried via the PCC infrastructure, thus taking some extra hops on its way. There were some additional subtleties between the two models where QoS enforcement and management are not handled by the same entities for GTP vs. the PMIP variant of the architecture. In the end, the community chose to use the off-path model for the PMIP variant, and thus the major hurdles for the architecture work were removed.

As work progressed, it became clear that the SAE work would not emerge with one set of protocols and design choices for the EPS. With somewhat divergent operator requirements and migration/evolution strategies, 3GPP needed to take a difficult decision: either follow one architecture alternative (which was virtually impossible to achieve), or allow for multiple alternatives. In the end, 3GPP emerged with multiple protocol variants used within one overall architecture framework to satisfy these requirements. An aspect worth noting is that not only did 3GPP select two protocol options with slightly different architectural variants for network-based mobility, including GTP evolution and PMIP track based on the IETF-developed PMIP protocol suite and 3GPP specific extensions to this mobility mechanism, 3GPP also continued to develop a terminal-based mobility option, though with rather limited interest from the operator community.

A final twist came in the conclusion phase, when it was decided that the existing packet core architecture should be maintained in parallel with EPC, while the original assumption and understanding of the work was that EPC would replace the existing packet core architecture. The consequence of this decision was the creation of two variants of how to interconnect to GERAN and WCDMA/HSPA access networks to packet data networks (e.g. GPRS and EPC).

One milestone accomplishment during the architecture work is that, regardless of the choice of network-based mobility model (GTP or PMIP), the user device acts in the same way. There is no dependency between the mobility protocol used by the network entities and how the terminal connects to the network. This transparency was intended to help the future development of EPS as a whole.

The following sections provide an overview of the different standards involved in EPC.

Third Generation Partnership Project (3GPP)

As mentioned in Chapter 1, GSM was originally developed as a European standard within ETSI. 3GPP was established in 1998 in order to unite a number of different regional standardization bodies for the creation of a global cellular standard. These regional bodies are referred to as Organizational Partners. As of 2009, the partners were ETSI (Europe), ARIB (Japan), ATIS (USA), CCSA (China), TTA (Korea), and TTC (Japan).

Originally, 3GPP was to produce specifications and reports for a 3G Mobile System based on evolved GSM core and radio networks. It rapidly evolved, however, to also have responsibility for the development of GSM technologies such as GPRS and EDGE.

More recently, 3GPP has led the work on a set of common IMS specifications and, in parallel with SAE, the Long-Term Evolution of the radio network, also known as LTE. The core network evolution, SAE, is naturally closely related to the LTE work item.

Structure of 3GPP

3GPP is organized into several different Technical Specification Groups (TSGs). These TSGs are responsible for the technical work and production of the specifications. There are four TSGs, split across the different areas that 3GPP works with:

- **TSG GERAN**: Responsible for the specification of the radio access part of GSM/EDGE.
- **TSG RAN**: Responsible for the definition of the functions, requirements, and interfaces of the UTRA/E-UTRA (WCDMA and LTE) radio networks covering both FDD and TDD variants.
- **TSG SA**: Responsible for the overall architecture and service capabilities of systems based on 3GPP specifications, and also for coordination between the different TSGs.
- **TSG CT**: Responsible for specifying terminal interfaces and capabilities. Also responsible for specifying the core network protocols of 3GPP systems.

Each of these TSGs has a number of Working Groups (WGs) associated with them; each WG is responsible for a certain number of tasks within the mandate of the TSG that they fall under. For example, WG SA1 is responsible for system requirements, while WG SA2 is responsible for the system architecture. WG CT4, meanwhile, handles the protocol definition for Basic Call Processing and protocols between nodes within the network. The relationship between the TSGs and the different WGs is explained below.

The overall management of 3GPP, for example organization and allocation of work, is handled by the highest decision-making body in 3GPP–the Project Coordination Group (PCG).

Each WG in 3GPP is responsible for producing Technical Specifications (TSs) and Technical Reports (TRs)–that is, the actual technical documents that can be downloaded from the 3GPP website (www.3gpp.org)–but they are not allowed to simply create whatever specification they like and publish it; each TS and TR must go through an approval process within the TSG responsible for their particular WG. Once this is completed, the specifications can then be taken to regional organizations (such as ETSI for Europe) to be approved as formal standards, or the ITU for use in their set of standards.

Each TS and TR is referred to by a set of digits, "xx.yyy", where xx refers to the so-called series number, while yyy refers to the particular specification within that series. As an example, 23.401 indicates that it is a system architecture document (23 series), while 401 refers to the actual specification, in this case GPRS enhancements for E-UTRAN access.

LTE specifications are handled in RAN WG1, RAN WG2, RAN WG3, RAN WG4, and RAN WG5. SAE specifications, meanwhile, are handled in SA WG1, SA WG2, SA WG3, CT WG1, CT WG3, and CT WG4. Each of these WGs is responsible for a different section of the SAE work and is covered in more detail in various chapters of this book.

Stages in 3GPP Standardization

When 3GPP develops a new standard or amends an existing standard, the work proceeds in three logical phases: stage 1, stage 2, and stage 3. It has been adopted by 3GPP and is also used by several other standardization organizations. It can be noted that work on the three stages often takes place simultaneously, or at least with significant temporal overlap, in order to make the standardization work time efficient.

During stage 1 the service requirements are specified, i.e. the functions that the system as a whole is intended to support. In the next step, stage 2, the architectural requirements are specified, taking the stage 1 requirements into account. This means that the different logical network entities and the reference points between the network entities are defined. The purpose and functions of each network entity and each reference point are also specified. The procedures–that is, the logical message flow between network entities–are defined, including what information is transferred across the reference points. In stage 3, the actual protocols are defined based on the architectural work done during stage 2. Each message is specified in detail and the message content, such as parameter formats, information element structure and so on, is defined. The stage 3 work also has the very important task of making sure that any errors will be handled appropriately by the system. This includes, for example, defining relevant result codes in response messages.

Tracking Down the Right Specification in 3GPP

All 3GPP specifications are freely available online for anyone to read. In order to find the right specification, it is generally easiest to search for it by its specification number, for example 23.401, or via the working group responsible for its development, for example SA2.

An important thing to remember when searching for specifications is to ensure that the release number is correct. 3GPP uses a parallel release mechanism in order to ensure that any developer has a stable platform to work upon. Once a release is frozen, it means that no more functional changes may be made to the specifications; they may, however, be changed if errors are found or for maintenance purposes.

Different WGs often work towards different releases at the same time. For example, the requirement specifications developed in SA1 are often frozen at a much earlier date than for the System Architecture documents created in SA2. So, SA1 may be working on Release 9 requirements, while SA2 is working on Release 8 architecture documents. This is quite natural; in order to ensure that SA2 has a stable set of requirements to work upon, they need to be frozen before work can commence.

A full list of the different specifications and the working groups responsible for them at the time of writing is available in Appendix A.

Internet Engineering Task Force (IETF)

The IETF, in comparison to 3GPP, is more loosely organized. The IETF is made up of individuals, rather than companies with participants from all different areas of the industry; it does not take membership fees and as a result anyone can participate. It takes care of IP-related protocols and has developed most of the protocols in use on the Internet. It handles only protocols, however, and does not define the network architectures that combine the different protocols together. Nor does it define the functionality of nodes on a network.

Structure of the IETF

The IETF is split into different areas: Applications Area, General Area, Internet Area, Operations and Management Area, Real-Time Applications, and Infrastructure Area, Routing Area, Security Area, and Transport Area. Each of these areas has several WGs under its directorship and each WG has a particular technical subject that it works on and produces a set of documents for. The WGs are therefore referred to as "Area Directorates". WGs are created for specific purposes and after their documents are complete, they are either disbanded or "rechartered" with a new set of deliverable documents. As a result, the active WGs in the IETF change; the latest list of WGs, at the time of writing, is available at: http://www.ietf.org/html.charters/wg-dir.html.

WGs are assigned a unique acronym that identifies the task that they are working on; for example, mip4 relates to Mobility for IPv4, or sipping refers to the group that handles the SIP, the protocol that forms a key component of the IMS.

In a similar fashion to 3GPP, the IETF has an overall group that forms the technical management team of the IETF; this is called the Internet Engineering Steering Group (IESG). Each Area Directorate has one or two directors who join the IETF chairman in the IESG. It is the IESG's responsibility to review all specifications that the WGs produce and also to decide on the overall technical direction that the IETF will take– that is, what areas the IETF should work on.

A high-level view of the IETF is illustrated below:

Open Mobile Alliance (OMA)

The OMA was created in 2002, comprising over 200 companies, including wireless vendors, IT companies, mobile operators, and application and content providers. The OMA is intended to be the focal point for the development of mobile service enabler specifications, supporting the creation of interoperable end-to-end mobile services.

OMA specifications are independent of the underlying network architectures. Within the 3GPP specifications for EPS, references to OMA are made for several different reasons, for example Device Management (DM).

More information about the OMA can be found on their website: http://www.openmobilealliance.org.

References

Beming, P., Frid, L., Hall, G., Malm, P., Noren, T., Olsson, M., et al. October, 2007. LTE-SAE architecture and performance. Ericsson Review 3. http://www.ericsson.com/ericsson/corpinfo/publications/review/2007_03/files/5_LTE_SAE.pdf

Blanchet, M., December, 2005. Migrating to IPv6: A Practical Guide to Implementing IPv6 in Mobile and fixed Networks. John Wiley & Sons, 418 pp. ISBN-10: 0471498920/ISBN-13: 9780471498926.

Brenner, M., Unmehopa, M., April 4, 2008. The Open Mobile Alliance: Delivering Service Enablers for Next-Generation Applications. John Wiley & Sons, 530 pp. ISBN-10: 0470519185/ISBN-13: 978-0470519189.

Camarillo G., Garcia-Martin, M.-A., November, 2008. 3G IP Multimedia Subsystem (IMS). ISBN-10: 0470516623.

Dahlman, E., Parkvall, S., Sköld, J., 2011. 4G – LTE/LTE-Advanced for Mobile Broadband. Academic Press/Elsevier. ISBN-10: 012385489X/ISBN-13: 978-0123854896.

Ericsson Interim Traffic Report, November 2011.

Ericsson Traffic Report, February 2012.

Ericsson White Paper on LTE Positioning, September 2011: http://www.ericsson.com/news/110909_positioning_with_lte_244188809_c

Hagen, S. May 2006. IPv6 Essentials, second edition, O'Reilly Media. ISBN-10: 0-596-10058-2/ISBN-13: 9780596100582, 436 pp.

HM Treasury, 2011. National Infrastructure Plan 2011, HM Treasury, UK. http://www.hm-treasury.gov.uk/national_infrastructure_plan2011.htm

Horn, G., 2010. 3GPP Femtocells: Architecture and Protocols. http://www.qualcomm.com/media/documents/3gpp-femtocells-architecture-and-protocols

Li, Q., Jinmei, T., Shima, K., January 2006. IPv6 Core Protocols Implementation. In: The Morgan Kaufmann Series in Networking. Morgan Kaufmann Publishers. ISBN-10: 0124477518/ISBN-13: 9780124477513.

Kaaranen, H., Ahtiainen, A., Laitinen, L., Naghian, S., Niemi, V., 2005. UMTS Networks. John Wiley & Sons, ISBN-10 0470011033(H/B)/ISBN-13 978-0470011034(H/B)

Noldus, R., Olsson, U., Mulligan, C., Fikouras, I., Ryde, A., Stille, M., September 2011. IMS Application Developer's Handbook: Creating and Deploying Innovative IMS Applications.

UNPD, 2009. World Urbanization Prospects. http://esa.un.org/unpd/wup/

3GPP Technical Specifications

3GPP TS 21.905 Vocabulary for 3GPP Specifications.

3GPP TS 22.011 Service accessibility.

3GPP TS 22.101 Service Aspects; Service principles.

3GPP TS 22.153 Multimedia priority services.

3GPP TS 22.168 Earthquake and Tsunami Warning System (ETWS) requirements.

3GPP TS 22.278 Service requirements for the Evolved Packet System (EPS).

3GPP TS 23.002 Network architecture.

3GPP TS 23.003 Numbering, addressing and identification.

3GPP TS 23.008 Organization of Subscriber Data.

3GPP TS 23.060 General Packet Radio Service (GPRS); Service description; Stage 2.

3GPP TS 23.122 Non-Access-Stratum (NAS) functions related to Mobile Station (MS) in idle mode.

3GPP TS 23.139 3GPP system – fixed broadband access network interworking; Stage 2.

3GPP TS 23.167 IP Multimedia Subsystem (IMS) emergency sessions.

3GPP TS 23.203 Policy and charging control architecture.

3GPP TS 23.216 Single Radio Voice Call Continuity (SRVCC); Stage 2.

3GPP TS 29.219 Policy and charging control: Spending limit reporting over Sy reference point.

3GPP TS 23.228 IP Multimedia Subsystem (IMS); Stage 2.

3GPP TS 23.237 IP Multimedia Subsystem (IMS) Service Continuity; Stage 2.

3GPP TS 23.246 Multimedia Broadcast/Multicast Service (MBMS); Architecture and functional description.

3GPP TS 23.272 Circuit Switched (CS) fallback in Evolved Packet System (EPS); Stage 2.

3GPP TS 23.292 IP Multimedia Subsystem (IMS) centralized services; Stage 2.

3GPP TS 23.401 General Packet Radio Service (GPRS) enhancements for Evolved Universal Terrestrial Radio Access Network (E-UTRAN) access.

3GPP TS 23.402 Architecture enhancements for non-3GPP accesses.

3GPP TR 23.882 3GPP system architecture evolution (SAE): Report on technical options and conclusions.

3GPP TS 24.007 Mobile radio interface signalling layer 3; General aspects.

3GPP TS 24.173 IMS Multimedia telephony communication service and supplementary services; Stage 3.

3GPP TS 24.285 Allowed Closed Subscriber Group (CSG) list; Management Object (MO).

3GPP TS 24.301 Non-Access-Stratum (NAS) protocol for Evolved Packet System (EPS); Stage 3.

3GPP TS 24.302 Access to the Evolved Packet Core (EPC) via non-3GPP access networks; Stage 3.

3GPP TS 24.303 Mobility management based on Dual-Stack Mobile IPv6; Stage 3.

3GPP TS 24.304 Mobility management based on Mobile IPv4; User Equipment (UE) – foreign agent interface; Stage 3.

3GPP TS 24.312 Access Network Discovery and Selection Function (ANDSF) Management Object (MO).

3GPP TS 24.604 Communication Diversion (CDIV) using IP Multimedia (IM) Core Network (CN) subsystem; Protocol specification.

3GPP TS 24.605 Conference (CONF) using IP Multimedia (IM) Core Network (CN) subsystem; Protocol specification.

3GPP TS 24.606 Message Waiting Indication (MWI) using IP Multimedia (IM) Core Network (CN) subsystem; Protocol specification.

3GPP TS 24.607 Originating Identification Presentation (OIP) and Originating Identification Restriction (OIR) using IP Multimedia (IM) Core Network (CN) subsystem; Protocol specification.

3GPP TS 24.608 Terminating Identification Presentation (TIP) and Terminating Identification Restriction (TIR) using IP Multimedia (IM) Core Network (CN) subsystem; Protocol specification.

3GPP TS 24.610 Communication HOLD (HOLD) using IP Multimedia (IM) Core Network (CN) subsystem; Protocol specification.

3GPP TS 24.611 Anonymous Communication Rejection (ACR) and Communication Barring (CB) using IP Multimedia (IM) Core Network (CN) subsystem; Protocol specification.

3GPP TS 24.615 Communication Waiting (CW) using IP Multimedia (IM) Core Network (CN) subsystem; Protocol Specification.

3GPP TS 25.331 Radio Resource Control (RRC); Protocol specification.

3GPP TS 25.913 Requirements for Evolved UTRA (E-UTRA) and Evolved UTRAN (E-UTRAN).

3GPP TS 26.346 Multimedia Broadcast/Multicast Service (MBMS); Protocols and codecs.

3GPP TS 29.060 General Packet Radio Service (GPRS); GPRS Tunnelling Protocol (GTP) across the Gn and Gp interface.

3GPP TS 29.061 Interworking between the Public Land Mobile Network (PLMN) supporting packet based services and Packet Data Networks (PDN).

3GPP TS 29.118 Mobility Management Entity (MME) – Visitor Location Register (VLR) SGs interface specification.

3GPP TS 29.168 Cell Broadcast Centre interfaces with the Evolved Packet Core; Stage 3.

3GPP TS 29.212 Policy and Charging Control (PCC) over Gx/Sd reference point.

3GPP TS 29.213 Policy and charging control signalling flows and Quality of Service (QoS) parameter mapping.

3GPP TS 29.214 Policy and charging control over Rx reference point.

3GPP TS 29.215 Policy and Charging Control (PCC) over S9 reference point.

3GPP TS 29.230 Diameter applications; 3GPP specific codes and identifiers.

3GPP TS 29.272 Evolved Packet System (EPS); Mobility Management Entity (MME) and Serving GPRS Support Node (SGSN) related interfaces based on Diameter protocol.

3GPP TS 29.273 Evolved Packet System (EPS); 3GPP EPS AAA interfaces.

3GPP TS 29.274 3GPP Evolved Packet System (EPS); Evolved General Packet Radio Service (GPRS) Tunnelling Protocol for Control plane (GTPv2-C); Stage 3.

3GPP TS 29.275 Proxy Mobile IPv6 (PMIPv6) based Mobility and Tunnelling protocols; Stage 3.

3GPP TS 29.276 Optimized Handover Procedures and Protocols between EUTRAN Access and cdma2000 HRPD Access.

3GPP TS 29.280 Evolved Packet System (EPS); 3GPP Sv interface (MME to MSC, and SGSN to MSC) for SRVCC.

3GPP TS 29.281 General Packet Radio System (GPRS) Tunnelling Protocol User Plane (GTPv1-U).

3GPP TS 29.303 Domain Name System Procedures.

3GPP TS 29.305 InterWorking Function (IWF) between MAP based and Diameter based interfaces.

3GPP TS 31.102 Characteristics of the Universal Subscriber Identity Module (USIM) application.

3GPP TS 33.106 Lawful Interception Requirements.

3GPP TS 33.210 3G security; Network Domain Security (NDS); IP network layer security.

3GPP TS 33.246 3G Security; Security of Multimedia Broadcast/Multicast Service (MBMS).

3GPP TS 33.320 Security of Home Node B (HNB)/Home evolved Node B (HeNB).

3GPP TS 33.401 3GPP System Architecture Evolution (SAE); Security architecture.

3GPP TS 33.402 3GPP System Architecture Evolution (SAE); Security aspects of non-3GPP accesses.

3GPP TS 36.101 Evolved Universal Terrestrial Radio Access (E-UTRA); User Equipment (UE) radio transmission and reception.

3GPP TS 36.300 Evolved Universal Terrestrial Radio Access (E-UTRA) and Evolved Universal Terrestrial Radio Access Network (E-UTRAN); Overall description; Stage 2.

3GPP TS 36.300 Evolved Universal Terrestrial Radio Access (E-UTRA) and Evolved Universal Terrestrial Radio Access Network (E-UTRAN); Overall description; Stage 2.

3GPP TS 36.304 Evolved Universal Terrestrial Radio Access (E-UTRA); User Equipment (UE) procedures in idle mode.

3GPP TS 36.305 Evolved Universal Terrestrial Radio Access Network (E-UTRAN); Stage 2 functional specification of User Equipment (UE) positioning in E-UTRAN.

3GPP TS 36.306 Evolved Universal Terrestrial Radio Access (E-UTRA); User Equipment (UE) radio access capabilities.

3GPP TS 36.321 Evolved Universal Terrestrial Radio Access (E-UTRA); Medium Access Control (MAC) protocol specification.

3GPP TS 36.322 Evolved Universal Terrestrial Radio Access (E-UTRA); Radio Link Control (RLC) protocol specification.

3GPP TS 36.323 Evolved Universal Terrestrial Radio Access (E-UTRA); Packet Data Convergence Protocol (PDCP) specification.

3GPP TS 36.331 Evolved Universal Terrestrial Radio Access (E-UTRA); Radio Resource Control (RRC); Protocol specification.

3GPP TS 36.355 Evolved Universal Terrestrial Radio Access (E-UTRA); LTE Positioning Protocol (LPP).

3GPP TS 36.401 Evolved Universal Terrestrial Radio Access Network (E-UTRAN); Architecture description.

3GPP TS 36.410 Evolved Universal Terrestrial Radio Access Network (E-UTRAN); S1 layer 1 general aspects and principles.

3GPP TS 36.411 Evolved Universal Terrestrial Radio Access Network (E-UTRAN); S1 layer 1.

3GPP TS 36.412 Evolved Universal Terrestrial Radio Access Network (E-UTRAN); S1 signalling transport.

3GPP TS 36.413 Evolved Universal Terrestrial Radio Access (E-UTRA); S1 Application Protocol (S1AP).

3GPP TS 36.414 Evolved Universal Terrestrial Radio Access Network (E-UTRAN); S1 data transport.

3GPP TS 36.420 Evolved Universal Terrestrial Radio Access Network (E-UTRAN); X2 general aspects and principles.

3GPP TS 36.421 Evolved Universal Terrestrial Radio Access Network (E-UTRAN); X2 layer 1.

3GPP TS 36.422 Evolved Universal Terrestrial Radio Access Network (E-UTRAN); X2 signalling transport.

3GPP TS 36.423 Evolved Universal Terrestrial Radio Access Network (E-UTRAN); X2 Application Protocol (X2AP).

3GPP TS 36.424 Evolved Universal Terrestrial Radio Access Network (E-UTRAN); X2 data transport.

3GPP TS 36.455 Evolved Universal Terrestrial Radio Access (E-UTRA); LTE Positioning Protocol A (LPPa).

3GPP TS 36.913 Requirements for further advancements for Evolved Universal Terrestrial Radio Access (E-UTRA) (LTE-Advanced).

3GPP2 Specifications

3GPP2 X.S0042-0 Voice Call Continuity between IMS and Circuit Switched System.

IETF RFCs

IETF RFC 768; User Datagram Protocol.

IETF RFC 793; Transmission Control Protocol.

IETF RFC 1035; Domain Names – Implementation and Specification.

IETF RFC 2003; IP Encapsulation within IP.

IETF RFC 2181; Clarifications to the DNS Specification.

IETF RFC 2401; Security Architecture for the Internet Protocol.

IETF RFC 2402; IP Authentication Header.

IETF RFC 2406; IP Encapsulating Security Payload (ESP).

IETF RFC 2407; The Internet IP Security Domain of Interpretation for ISAKMP.

IETF RFC 2408; Internet Security Association and Key Management Protocol (ISAKMP).

IETF RFC 2409; The Internet Key Exchange (IKE).

IETF RFC 2473; Generic Packet Tunnelling in IPv6 Specification.

IETF RFC 2606; Reserved Top Level DNS Names.

IETF RFC 2784; Generic Routing Encapsulation (GRE).

IETF RFC 2890; Key and Sequence Number Extensions to GRE.

IETF RFC 2960; Stream Control Transmission Protocol.

IETF RFC 3168; The Addition of Explicit Congestion Notification (ECN) to IP.

IETF RFC 3309; Stream Control Transmission Protocol (SCTP) Checksum Change.

IETF RFC 3344; IP Mobility Support for IPv4.

IETF RFC 3588; Diameter Base Protocol.

IETF RFC 3748; Extensible Authentication Protocol (EAP).

IETF RFC 3775; Mobility Support in IPv6.

IETF RFC 3776; Using IPsec to Protect Mobile IPv6 Signalling Between Mobile Nodes and Home Agents.

IETF RFC 3958; Domain-Based Application Service Location Using SRV RRs and the Dynamic Delegation Discovery Service (DDDS).

IETF RFC 4005; Diameter Network Access Server Application.

IETF RFC 4006; Diameter Credit-Control Application.

IETF RFC 4072; Diameter Extensible Authentication Protocol (EAP) Application.

IETF RFC 4186; Extensible Authentication Protocol Method for Global System for Mobile Communications (GSM) Subscriber Identity Modules (EAP-SIM).

IETF RFC 4187; Extensible Authentication Protocol Method for 3rd Generation Authentication and Key Agreement (EAP-AKA).
IETF RFC 4285; Authentication Protocol for Mobile IPv6.
IETF RFC 4301; Security Architecture for the Internet Protocol.
IETF RFC 4302; IP Authentication Header.
IETF RFC 4303; IP Encapsulating Security Payload (ESP).
IETF RFC 4306; Internet Key Exchange (IKEv2) Protocol.
IETF RFC 4555; IKEv2 Mobility and Multihoming Protocol (MOBIKE).
IETF RFC 4877; Mobile IPv6 Operation with IKEv2 and the Revised IPsec Architecture.
IETF RFC 4960; Stream Control Transmission Protocol.
IETF RFC 5094; Mobile IPv6 Vendor Specific Option.
IETF RFC 5213; Proxy Mobile IPv6.
IETF RFC 5216; The EAP-TLS Authentication Protocol.
IETF RFC 5447; Diameter Mobile IPv6: Support for Network Access Server to Diameter Server Interaction.
IETF RFC 5448; Improved Extensible Authentication Protocol Method for 3rd Generation Authentication and Key Agreement (EAP-AKA).
IETF RFC 5555; Mobile IPv6 Support for Dual Stack Hosts and Routers.
IETF RFC 5779; Diameter Proxy Mobile IPv6: Mobile Access Gateway and Local Mobility Anchor Interaction with Diameter Server.
IETF RFC 5844; IPv4 Support for Proxy Mobile IPv6.
IETF RFC 5845; Generic Routing Encapsulation (GRE) Key Option for Proxy Mobile IPv6.
IETF RFC 5846; Binding Revocation for IPv6 Mobility.

IETF Internet Drafts

IETF Internet Draft, Binding Revocation for IPv6 revocation, Mobility (draft-ietf-mext-binding-revocation).
IETF Internet Draft, GRE Key Option for Proxy option, Mobile IPv6 (draft-ietf-netlmm-grekey-option).
IETF Internet-Draft, IPv4 Support for Proxy ipv4-support, Mobile IPv6 (draft-ietf-netlmm-pmip6-ipv4-support).
IETF Internet Draft, Diameter Proxy Mobile IPv6: Mobile Access Gateway and Local Mobility Anchor Interaction with Diameter Server.
Note: At the time of preparing this book, several of the Internet Drafts listed above are close to becoming approved RFCs. The interested reader should consult the IETF web page (www.ietf.org) for the latest status.

ITU Recommendations

ITU-T Recommendation I.112; I.112 Integrated Services Digital Network (ISDN), general structure, vocabulary of terms for ISDNs.
ITU-R M.2134; Requirements related to technical performance for IMT-Advanced radio interface(s).
ITU-T recommendation H.325, Annex K Packet-based multimedia communications systems.

GSMA Specifications

IR.92 IMS Profile for Voice and SMS.
IR.92 Video Extensions to VoLTE IR.92.

OMA Specifications

Open Mobile Alliance, OMA AD SUPL: Secure User Plane Location Architecture (http://www.openmobilealliance.org).

IEEE Specifications

IEEE 802.11. IEEE Standard for Information technology–Telecommunications and information exchange between systems Local and metropolitan area networks–Specific requirements Part 11: Wireless LAN Medium Access Control (MAC) and Physical Layer (PHY) Specifications.

IEEE 802.1X. IEEE Standard for Local and metropolitan area networks – Port-Based Network Access Control.

Index